Sustainable Solutions

Sustainable Solutions

PROBLEM SOLVING FOR CURRENT AND FUTURE GENERATIONS

Richard Niesenbaum

Professor Muhlenberg College

New York Oxford
OXFORD UNIVERSITY PRESS

Oxford University Press is a department of the University of Oxford.
It furthers the University's objective of excellence in research, scholarship,
and education by publishing worldwide. Oxford is a registered trade mark of
Oxford University Press in the UK and certain other countries.

Published in the United States of America by Oxford University Press
198 Madison Avenue, New York, NY 10016, United States of America.

For titles covered by Section 112 of the US Higher Education
Opportunity Act, please visit www.oup.com/us/he for the latest
information about pricing and alternate formats.

CIP data is on file at the Library of Congress
978-0-19-939043-4

9 8 7 6 5 4 3 2 1

Printed by Sheridan Books, Inc., United States of America

This book is dedicated to my children Sophie and Jonah, and their children and their children and their children . . .

Contents

Preface

These are seemingly desperate times as we are bombarded with an endless supply of bad news billed as insurmountable problems that relate both to the declining state of our planet and to the human condition. This book is inspired by these times and the fact that despite their overwhelming nature, there exist realistic solutions to these problems. It serves to negotiate the tension between a growing sense of powerlessness and a palpable desire for effecting change that resides within us. It offers the optimistic possibility for workable solutions to the complex problems that confront us.

An Integrative, Problem-Solving Approach

Sustainability is an ideal framework for the development of solutions to many of these problems. Broadly defined as meeting the needs of the present without compromising the ability of future generations to meet their own needs, sustainability is a framework that integrates science with justice, economics, and policy. This book offers its readers opportunities to engage in this integration as part of a problem-solving approach. Policy, economics, and social elements are not isolated at the end of chapters, nor as separate chapters at the end of the book. Rather, each chapter provides the scientific basis to problems and their solutions; each does so within the context of economic development, policy, and social justice in order to reveal the complexity of the problems and sustainable solutions to them. This offers the interdisciplinary background that is required to understand many of these complex challenges, and it equips readers with the skills needed to develop solutions to them.

Readers will have opportunities to integrate across disciplines in order to better understand how things such as individuals and institutions, novel collaborations, innovation and entrepreneurship, new technologies, and policy decisions can be used to achieve sustainable solutions. The book offers readers opportunities for critical thinking, further action, and group work. Also, because of its intended interdisciplinary audience, the book is designed to be accessible to readers with varied backgrounds and interests in science and technology, public policy, economics, social justice, and public health. It provides substantial background for those interested in sustainability/social entrepreneurship, green business, and other careers that incorporate sustainability principles. It is also accessible to a broader audience of non-academic readers with an interest in sustainability.

Sustainable Solutions provides both the necessary background in science and a broad, interdisciplinary knowledge base that the reader can then connect to specific disciplines, such as environmental science, engineering, agriculture and food science, business, natural resource management, and public health. It prepares the reader to approach sustainable problem solving at a variety of scales from personal behavior to institutional- and community-level problem solving and up through regional, national, and global scales. Each chapter emphasizes skills such as how to identify and work with stakeholders on a particular issue, program assessment, and systems thinking. My hope is that this book will lay the groundwork for training of sustainability professionals and those in other careers that link to sustainability and that it will also empower everyday citizens to effectively act and engage with others in ways that further sustainability objectives.

Organization of the Book

Chapters are organized into three overarching sections. Part I, *The Problems We Must Solve,* introduces a variety of global concerns and issues as a context for sustainable problem solving. It then walks the reader through an exercise in developing a working definition of sustainability by first tracing its history and then extracting the core principles that emerge from that history. Once defined, a detailed approach to applying sustainability objectives to problem-solving skills and assessment of sustainable programs or projects is developed.

Part II, *The Earth as a Resource: Moving Toward a Sustainable Future,* focuses on the use of our planet for meeting the needs of humans. In each of the content areas (air, water, food, energy, non-fuel extractive resources, and waste), issues are identified and solutions are offered using the approach developed in Part I. Potential solutions are examined critically. Tools such as life cycle analysis and systems thinking are used to assess a variety of "sustainable technologies" to reveal both the positive and the negative aspects of things such as wind turbines, genetically modified organisms, and waste-to-energy. A goal of this book is not to get readers to blindly accept an approach as sustainable but, rather, to offer ways to critically examine potential solutions in order to assess their potential to contribute to sustainability and consider ways to improve upon them. Each chapter highlights the barriers to sustainability and ways to overcome them, and each ends with a discussion on assessment and of common elements for success.

Part III, *Integrating Sustainable Solutions at Different Levels of Organization,* focuses on sustainability, starting with the individual and then moving up hierarchically through institutions, communities, and cities to global sustainability and sustainable development. This section allows the reader to make connections among the prior chapters. By working at various levels of organization, the intersectionality of the themes presented in Part II becomes apparent. For example, a community is not simply faced with a single issue such as local food production. Rather, it faces multiple interconnected issues. Food, energy, water, forest and mineral resources, and waste production are all related. These in turn intersect with race, gender, culture, and poverty. It is through these connections that the systematic causes of problems are exposed, and proposals for larger scale reform such as new economic models, international policy, and approaches to development are explored.

Features

To help accomplish the goals of this book, a variety of features are employed. At the start of each chapter, the reader is guided with questions framed as **Planting a Seed: Thinking Before You Read**, and opportunities for critical thinking are offered at the end of every chapter under **Digging Deeper: Questions and Issues for Further Thought**. Each chapter concludes with specific ways for the reader to act in order to further sustainability objectives described as **Reaping What You Sow: Opportunities for Action**.

Also included are boxed elements that offer real examples of sustainable problem solving relevant to the focus of each chapter. Boxes on **Individual Action** challenge readers to think about what personal changes they can make to solve the problems by highlighting examples of successful individual actions, including behavioral change, effective dialog, and activism. This is done in a critical way, forcing readers to think about the complexity of their actions and the various implications of these actions. Boxes on **Stakeholders and Collaborators** focus on identifying and working with stakeholders, and they highlight examples in which individuals and groups that at first glance seem at odds with each other can come together on shared values to solve a current problem. Another feature, **Innovators and Entrepreneurs,** presents specific examples of how innovation and social entrepreneurship can be successful in achieving sustainability, offering a business perspective to problem solving. Each chapter also has a boxed element on **Policy Solutions** that emphasizes policy approaches and the role of government in working toward a sustainable future.

Acknowledgments

So many people have inspired, encouraged, helped, and supported me in writing this book. I have been fortunate to have spent the past 25 years at Muhlenberg College, a liberal arts college that is strong in the sciences and that values integrative teaching and learning across disparate disciplines. I have had the opportunity to collaborate with colleagues who are in academic programs in art, foreign languages, sociology, chemistry, media and communications, mathematics, political science, and entrepreneurial studies. This has made Muhlenberg my laboratory where I have been able to cultivate my expertise in the broad area of sustainability.

I have been surrounded and inspired by students dedicated to the cause of sustainability both as activists and as practitioners. Included among them are Hailey Goldberg, Hannah Bobker, Alison Barranca, Allie Heckerd, Felisa Wiley, and Chris Woods, who served as research assistants and made significant contributions to this book. The Muhlenberg College Provost's Office and Rita and Joseph Scheller and their family that endow my professorship through the RJ Foundation have provided me with both time and resources necessary to complete such an extensive project.

This project would not have been possible without the expert assistance of the many reviewers who offered constructive criticism and helped me get all the details as they should be. These scholars who have truly influenced the direction of this book include the following:

- Lisa Barlow, University of Colorado
- Shaunna Barnhart, Bucknell University
- Richard Bowden, Allegheny College
- John Cusick, University of Hawaii at Manoa
- Mark Davies, Hartwick College
- Chris Doran, Seaver College
- Michael Duncan, Washington and Lee University
- Jim Feldman, University of Wisconsin–Oshkosh
- Eric Fitch, Marietta College
- Robert Giegengack, University of Pennsylvania
- Greg Gordon, Gonzaga University
- Juliette Goulet, Brookdale Community College
- Patrick Hurley, Ursinus College
- Jon Jensen, Luther College
- Dwane Jones, University of the District of Columbia
- Nancy E. Landrum, Loyola University Chicago
- William McClain, Davis & Elkins College
- Sherie McClam, Manhattanville College
- Mark O'Gorman, Maryville College
- Jeffrey Paine, Indiana University–Purdue University Indianapolis
- Jason Patalinghug, Southern Connecticut State University
- Hamil Pearsall, Temple University
- Jay Roberts, Earlham College
- Suzanne Savanick Hansen, Macalester College
- Adrienne Schwarte, Maryville College
- Charles Scruggs, Genesee Community College
- Vincent Smith, South Oregon University
- Alicia Sprow, Alvernia University
- Ninian Stein, Tufts University
- Deborah M. Steketee, Aquinas College
- Jennie C. Stephens, Clark University
- Barbara Szubinska, Eastern Kentucky University
- Laine Vignona, Lake-Sumter State College
- Bill Vitek, Clarkson University
- Kenneth Worthy, University of California, Berkeley
- Jim Zaffiro, Central College

I also thank the many people at Oxford University Press who have been instrumental in making this project happen, particularly Executive Editor Dan Kaveney and Assistant Editor Megan Carlson. They served as my sounding board, my guides, my critics, and my cheerleaders. They have been great listeners and put a lot of faith in me as we boldly navigated into the new academic area of sustainability studies. In addition, I thank Rachel Schragis, the coordinator of the People's Climate March Arts Team, and the others who contributed to the image *Confronting Climate* for their permission to use part of it as the cover of this book. This work both celebrates the important role of art in social movements and embraces the complexity, breadth, intersectionality, and conflicting imperatives that are central to this book.

Brenda Casper, Daniel Janzen, Peter Petraitis, Andrew Friedland, Norris Muth, Emily Mooney, Graham Watkins, Jack Gambino, Tim Averill, Rita Chesterton, The Sea Education Association, The School for Field Studies, POWER Lehigh Valley, Beth Halpern, Mark Plotkin, David Orr, Kathrine Hayhoe, Kalyna Procyk, Tim Averill, Tammy Lewis, Carlos Fonseca, Elieth Montoya, Clark Gocker, Lora Taub-Pervizpour, and Kimberly Heiman have profoundly influenced my thinking and teaching about sustainability, the connections between people and environment, and social justice. I also owe my gratitude to Krista Bywater, S. Mohsin Hashim, Christopher Borick, Barry Rabe, Joseph Elliott, and Greta Bergstressor for making direct contributions of their own work in the form of boxed elements or images. Others, such as Barbara Niesenbaum, Elizabeth Dale, Vladimir and Jesse Salamun, Varina Doman, Mohammad Kahn, Nicole Karsch, Megan Leahy, Tracy Librick, Tim Averill, Tim Clarke, Jessie Gocker Nimeh, Jen Jarson, and Almut Hupbach provided other important forms of support.

I would not have been able to complete a work of this magnitude without the love and support of my close friends and family, many of whom have already been mentioned. In particular, I thank one of my closest friends, Jeffrey Frank, a farmer who stimulates and relentlessly challenges my thinking about sustainability, typically over a good meal. Finally, I thank my children Sophie and Jonah Niesenbaum, who taught me empathy, patience, and how to really listen. These abilities are not only essential for working in the area of sustainability but also serve to fuel my optimism and hope in these challenging times.

CHAPTER 1

THE PROBLEMS WE MUST SOLVE

We are living in complicated and often frightful times confronted by many serious environmental, economic, social, political, and geopolitical challenges that seem insurmountable. However, it is important to recognize that these problems we face are solvable. Changes in individual behaviors, advances in science and engineering, new technologies, innovative economic and business models, creative government policies, and novel collaborations among diverse members of society have great potential to generate workable solutions. Through innovation and action, we can make our planet a better place to live for all people, not just in the present but for future generations as well. The question, however, is will we make these changes and further develop and implement necessary solutions in order to do so?

PLANTING A SEED: THINKING BEFORE YOU READ

1. List what you believe to be the five most challenging problems that confront us.
2. Draw a diagram that shows the interrelatedness of these problems.
3. Make a list of the barriers that seem to stop us from solving these problems.
4. Propose some realistic ways to overcome these barriers and examples of possible solutions to these problems.

Optimism is a strategy for making a better future. Because unless you believe that the future can be better, you are unlikely to step up and take responsibility for making it so.
—Noam Chomsky

A recent study of US college and university students suggests that the answer to this question is no. It revealed that you, our current generation of young and emerging adults, the likely readers of this book, have disengaged from thinking about social and environmental problems. Your trust in others and interest in government have declined compared to prior generations.[1] Such disengagement and skepticism are apparently due to the fatigue and frustration from hearing about these problems as insurmountable. The majority of books, news, and lectures regarding the environment, the human condition, and sustainability discuss them within the context of impending calamity. They highlight our squandering of resources and the poisoning of our air, water, and children. They place us at a precipice teetering toward global disaster. They refer to conflicts between ecology and economics and to corporate control over government policy. They paint our global citizenry as greedy, ignorant, or both. Individuals are viewed as either powerless victims or selfish overconsumers who have no opportunity to make a difference. You, the generation with relatively fewer opportunities and higher levels of debt than previous generations, see the system as rigged. As a result, positive change often seems out of reach.

Today's students are left with a sense of powerlessness and disempowering guilt. They are justifiably taught that they, as "First World" capitalists, are the primary cause of this impending collapse. However, they are typically not taught that there are indeed solutions to these problems and that they—as the most tech-savvy, collaborative, innovative, connected, and resourceful generation—can contribute to them in significant ways.

Our educational system and even the popular media have typically been effective at making us aware of the problems that confront us, but they have done less to spawn and train new generations of committed problem solvers. Certainly, one must define problems in order to generate solutions. However, this dismal perspective about our future prospects, although very much justified, may be the exact thing that is inhibiting our potential to generate creative, practical solutions to global problems and, more important, to engage our citizenry in working toward those solutions.

Fortunately, there is much to be hopeful about. We have begun to understand that we as individuals, social groups, communities, institutions, and governments can play a major role in the creation of a just and sustainable future in terms of the health of our planet and the well-being of all people. We have come to recognize that our own education system fails us in that its graduates have been the ones most responsible for our mismanagement of planet Earth.[2] The seeds of transformation in the way we teach, do business, and act as individuals have been planted, and their growth is being amplified through the use of social media, the development of new technologies, and fundamental changes in values.[3]

This transformation is beginning to yield positive results. Here, I list just a few examples drawn from websites, reports, blogs, and news stories that appear across my screens on a daily basis:

- There are now more than 1,530 interdisciplinary, sustainability focused academic programs at 564 campuses in 67 states and provinces and 23 countries. These numbers will continue to rise.
- In 2015, more than 62.8 million US citizens volunteered more than 7.9 billion hours. This included one in five millennials.
- In 2014, China's top legislature approved major amendments to the country's 25-year-old Environmental Protection Law calling for much greater regulation of and stiffer penalties for polluters.
- In Asia, the cost of energy generated by solar photovoltaic cells has substantially decreased so that it is now competitive with fossil fuels without being subsidized.
- Between 1990 and 2010, more than 2 billion people gained access to improved drinking water sources, and the proportion of people using an improved water source increased from 76% to 89%.
- Since 1991, the United Nations has allocated $8.8 billion and provided $38.7 billion in co-financing for environmental projects in developing countries.
- One creative entrepreneur, Kavita Shukla, developed a simple, affordable product that can extend the life of fruits and vegetables by two to four times even for the billions of people who live without refrigeration.
- Through the collaborative efforts of diverse community members, Samsø Island in Denmark now generates more power from renewable sources than it consumes and has a net negative carbon footprint.

- From 2010 to 2020, the global market for environmental products and services is projected to double from $1.37 trillion to $2.74 trillion per year.

Of course, not all of the news is good, but this list does show tangible improvement in our environmental and human prospects and also that individuals, communities, organizations, and even governments can achieve positive results. The goal of this book is to inspire and empower you, the reader, to become an active participant in generating solutions to the problems we face. You are part of a diverse audience with a variety of backgrounds that might include business, science, engineering, education, international development, policy and law, communications, agriculture, technology, the humanities and arts, or any of the trade professions. This book will allow you to place the presented problems within the context of your own expertise or interest and to see how that expertise connects to these problems and their solutions.

This book will teach you how to develop solutions to environmental and social problems by first providing you with the interdisciplinary background that is required to understand the many complex problems that we face. Then we will provide you with a framework for defining problems and offer specific competencies to approach and solve them. Through globally diverse examples at many different levels of organization we will offer you recipes for success and concrete ways to overcome barriers to change. By critically examining these examples you will begin to see what works and what constrains success, and also how to assess or measure that success.

In this chapter, we first establish a broad context for examining global problems. Then in Chapter 2, we use a historical approach to develop a definition for sustainability that will serve as our lens for examining problems and developing solutions. The next set of chapters are thematically organized by the ways in which we depend on our planet and the impacts we have on it as we do so. These include air, water, food, energy, non-energy extractive resources, and waste. In the final section, we re-organize our thinking around different levels of organization to which we will apply what we have learned in the previous chapters. These levels include the roles of the individual and institutions and how problems can be solved by communities, cities, regions, and on a global scale.

Natural Resources and Environmental Issues

In this section, we introduce some of the basic background required to understand the environmental issues we face. First, we examine them in relation to the natural resources on which we depend. Then we explore global environmental issues such as climate change and the impacts of invasive species. We next consider the services that our ecosystems perform and the value of biodiversity and the urgent need to protect them and to incorporate their value into our economic system. As we introduce these environmental problems and issues, we begin to suggest the broader possibilities for solutions that will be elaborated on in subsequent chapters.

Natural Resources

To state the obvious, our lives depend on what the Earth provides. Natural resources are materials that can be found within the environment that are either necessary or useful to humans. Resources are often referred to as natural assets that can serve as raw materials to be used for economic production or consumption. **Ubiquitous resources** are those that can be found everywhere, such as sunlight, water, and air. Others, such as mineral resources, occur more locally. Resources can be classified as **biotic** or **abiotic**. Biotic resources come from live organisms and organic matter. Examples include forest and animal products and also fossil fuels, which come from decayed organic matter. Abiotic resources are non-living, inorganic materials such as minerals, metals, air, and water.

Matter can be neither created nor destroyed, and all resources cycle through components of the biosphere or environment occupied by living organisms. For example, nitrogen cycles from the atmosphere to soil and aquatic environments and into living organisms, in which it is often converted into different forms through biological processes (Figure 1.1). The temporal scale of this cycling is highly variable among different resources. The temporal scale of cycling determines the rate of recovery or regeneration. Resources with short periods of regeneration are considered **renewable resources**, and those with slow rates of replenishment are **non-renewable resources**. From a human use perspective, when the rate of extraction outpaces the rate of renewal for a particular resource, that resource is considered non-renewable (Figure 1.2).

Because most of our natural resources come in fixed amounts and their regeneration is a slow process, there is concern that we will deplete them and therefore will be unable to meet the needs of future generations. Resources on which humans depend are threatened either from overextraction and depletion or from destruction through pollution or other types of human disturbance. Some estimates reveal that our use of resources well exceeds the biosphere's capacity to regenerate them and that within the next 20 years, two planets will not be sufficient to meet human resource needs on our single globe.

Some question this finite view of resource availability. The case is often made that technological advances will allow us to do more with less or develop new alternative ways to extract resources. This notion was promoted by Julian Simon, who in the 1980s authored a controversial book titled *The Ultimate Resource*.[4] Simon argues that there is no resource crisis. His reasoning is that as a resource becomes scarce, its price will rise, creating an incentive for people to discover more of that resource, ration and recycle it, and eventually, stimulate innovators to find a replacement for that resource. The "ultimate resource" is not any particular physical object but, rather, the capacity for humans to invent and adapt.

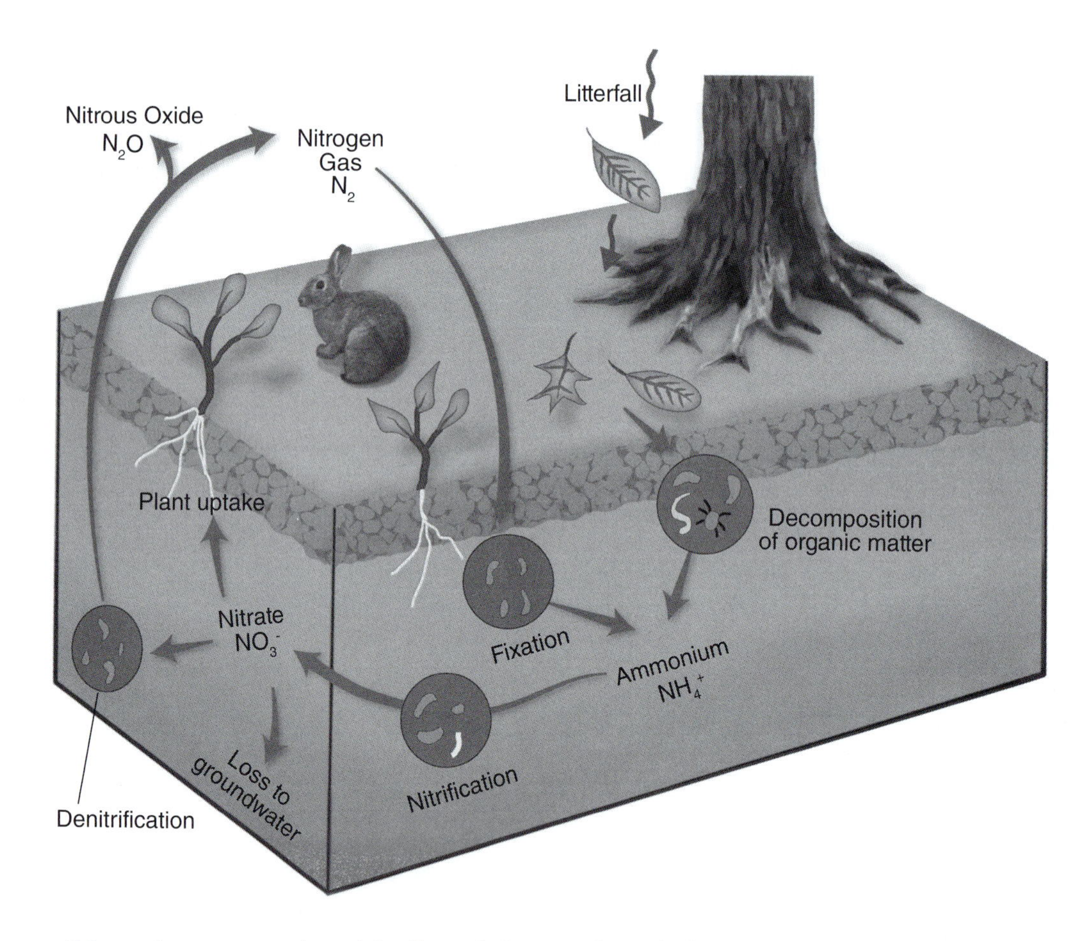

Figure 1.1 Schematic representation of the flow of nitrogen through the environment, including atmospheric, soil, and biotic components. The bulk of nitrogen is in the atmosphere and can only be used by microbes that convert it into useable nitrogen for plants and other organisms through the process of nitrogen fixation. Useable nitrogen is also added to the soil through the decomposition of organic material. Denitrifying bacteria convert nitrate back into atmospheric N_2 to derive enrgy.

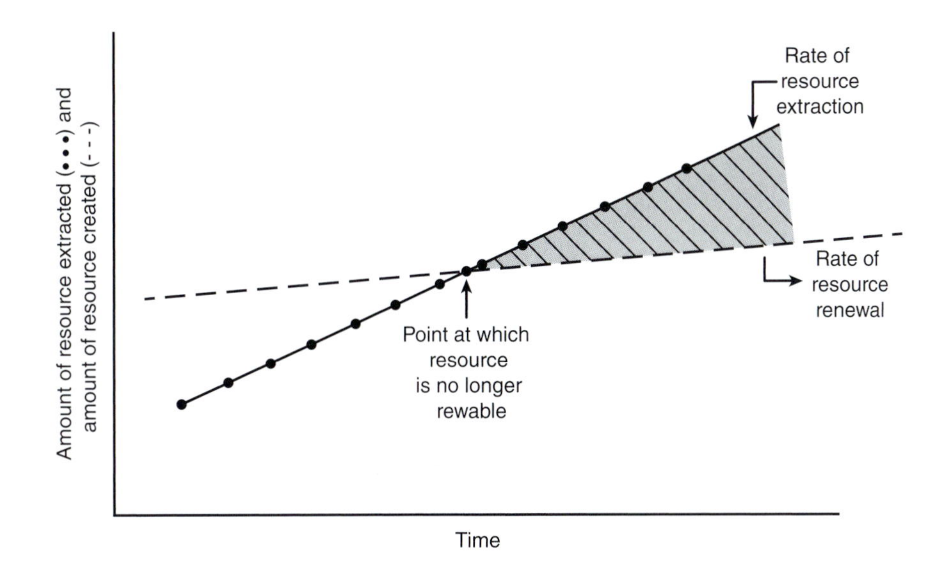

Figure 1.2 The amount of resource extracted (• • •) and the amount of resource regenerated (- - -) over time. The slope of each line represents the rate of use or renewal. In this example, the rate of resource extraction exceeds that of renewal. The point at which the lines meet is when the resource can no longer be considered renewable (shaded area). For many of our natural resources, that rate of extraction far exceeds the rate of renewal. *Courtesy R. Niesenbaum.*

These arguments are somewhat flawed in that they do not take into account impacts beyond the specific resource. For example, recycling technology, such as our ability to recycle aluminum, has allowed resources with slow rates of regeneration to become renewable, but not without the expense of energy. **Hydrofracturing** (fracking) technology has allowed for continued exploitation of fossil fuels beyond limits based on prior extraction technologies, but not without serious impacts and threats to other environmental components such as drinking water. It is useful to explore how technologies can both improve efficiency of extraction and use of natural resources and also how innovation can help us do this and develop alternative resources. However, this must be done within the broader context of natural resource management, energy use, and environmental impact. Next, some examples of natural resources are presented with explanations of how they cycle through the environment and the degree to which they may be renewable.

AIR

Air is an abiotic, renewable resource. The manner in which it cycles is more complicated because air consists of a variety of components that enter into different, somewhat complex cycles. By volume, dry air contains approximately 78% nitrogen, 21% oxygen, and smaller amounts of carbon dioxide, argon, water vapor, and other gases. Oxygen, carbon dioxide, and nitrogen cycle through biotic systems and are vital to living systems (Figures 1.1 and 1.3). Oxygen and carbon dioxide are cycled through the processes of photosynthesis, respiration, and decomposition. They are vital for living organisms. Carbon is stored in a variety of reservoirs, including soil, the ocean, biomass, rock, and as fossil fuels. Carbon dioxide is released during the burning of stored sources including fossil fuels and rock sources through volcanic activity. These have led to a steady increase in atmospheric carbon dioxide, which is a significant **greenhouse gas** contributing to **global climate change**, as will be addressed later in this chapter.

Nitrogen is the most abundant element in our atmosphere. It is also a vital element because compounds essential to living systems, such as protein and DNA, contain nitrogen. Ironically, most life forms, including all plants and animals, are unable to utilize nitrogen in its most abundant form, dinitrogen gas (N_2). Fortunately, specialized groups of microorganisms are capable of nitrogen fixation. **Nitrogen fixation** is the reduction of N_2 gas to ammonia NH_4^+ using a class of enzymes called nitrogenases to catalyze the reduction. This is an energy-intensive process, so these microbes are typically either photosynthetic or in symbiotic relationships with plants. Photosynthesis provides the needed energy. There are also denitrifying bacteria that oxidize useable nitrogen to generate energy and return N_2 back into the atmosphere (see Figure 1.1).

Although air and its constituents cycle over relatively short timescales, making it comparatively renewable, human activity has drastically influenced these cycles and negatively impacted air quality. Air pollutants in the form of carbon, nitrogen, and sulfur oxides; suspended particulates; ozone; and volatile organic and other types of compounds have been a serious threat to human and ecological health. Given the vital nature of air for living organisms, it is important that we develop solutions to the problem of air pollution. These must include new policies, changes in individual behavior, and technological innovation that are elaborated on in Chapter 3.

WATER

Water, like air, is an abiotic, renewable resource that is vital for all of life. It exists in three forms: liquid, solid ice, and gaseous water vapor. There are well over 300 million trillion gallons of water on Earth, but less than 1% of that is in the form of potentially drinkable fresh water (Figure 1.4). The oceans store approximately 97.5% of the planet's water as salt water, and approximately 2% is frozen in glaciers. Both physical and biological processes are responsible for cycling water. The physical processes include evaporation, freezing and thawing, and condensation and precipitation. Biological activity such as **transpiration** (evaporation of water through plants) and metabolism also contribute to the cycling of water.

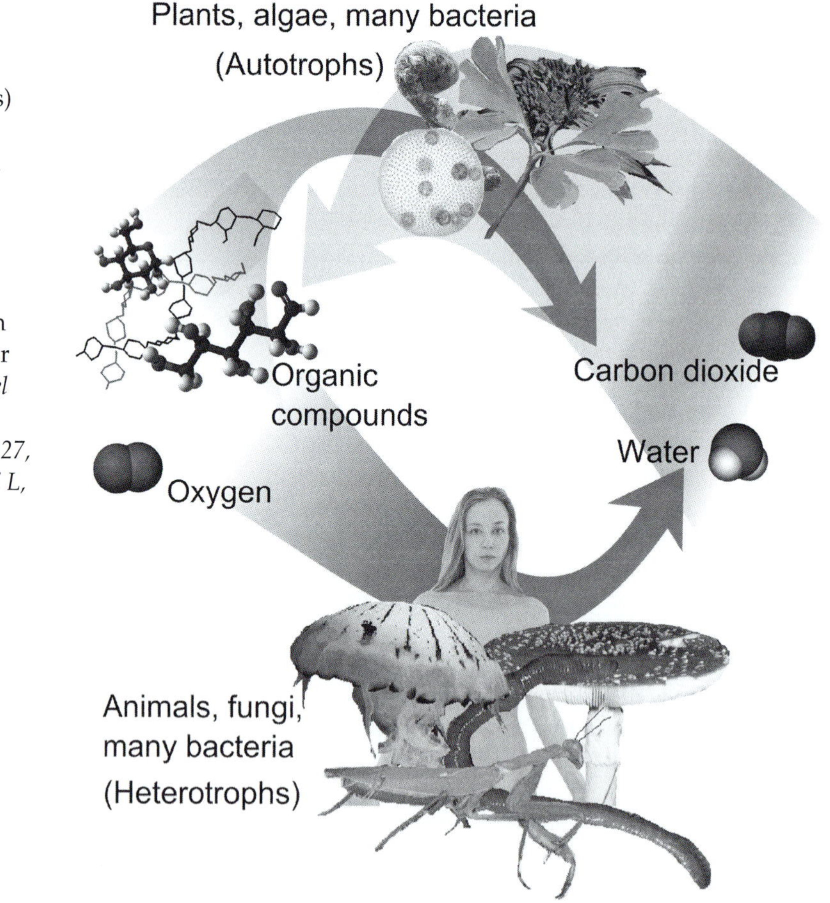

Figure 1.3 Carbon and oxygen cycle between autotrophs and heterotrophs. Plants, protists, and many bacteria (autotrophs) use carbon dioxide (CO_2) and water to form oxygen (O_2) and complex organic compounds through photosynthesis. Autotrophs and heterotrophs such as animals and fungi use such compounds to again form CO_2 and water through cellular respiration. *Derivative by Mikael Häggström, using originals by Laghi l, BorgQueen, Benjah-bmm27, Rkitko, Bobisbob, Jacek FH, Laghi L, and Jynto.*

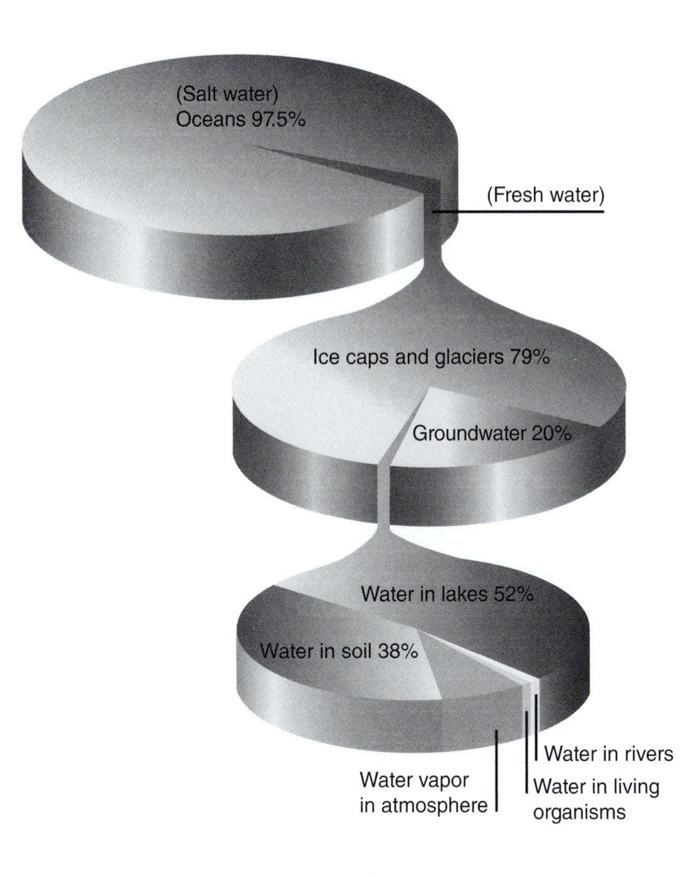

Figure 1.4 Most of the 300 million trillion gallons on Earth is salt water in our oceans (97.5%) and less than 1% is in the form of potentially drinkable freshwater. *Courtesy World Bank Group, https://goo.gl/images/Roa1rU*

Given the natural processes and rates of cycling of water, it is considered a renewable resource. However, the availability of water for animal and plant consumption, including that by humans, and for agriculture can be very limiting. The availability of potable water can be limited through overexploitation at local and regional levels. It can also be limited through contamination with either chemical pollutants or biological pathogens introduced through poor or absent sanitation infrastructure. Low access to clean drinking water availability is a crisis especially in developing countries. Globally, an estimated 663 million people lack ready access to improved sources of drinking water, and 2.4 billion people lack access to proper sanitation and hygiene. As a result, 340,000 children younger than age 5 years die annually from related diarrheal diseases, which equates to nearly 1,000 children per day.[5] In addition, aquatic ecosystems including rivers, lakes, estuaries, wetlands, and marine systems are sensitive to and readily impacted by human activity and habitat destruction.

There are ways to solve these water-related problems, including policies and technologies that reduce water pollution, improved water delivery and sanitation infrastructure, and new innovative approaches to generating and conserving clean water, such as desalination and purification. Enforced policies that protect water and aquatic habitats, promote water conservation, and allow for regional water sharing and distribution are also vital. These solutions are explored in Chapter 4.

ENERGY

Energy resources can be either non-renewable or renewable. Examples of non-renewable resources include **fossil fuels**, specifically oil, coal, and natural gas. As the name implies, these fuels are buried combustible geologic deposits of decayed or fossilized organic materials from microbes, plants, and animals. Over hundreds of millions of years, these organic materials have been converted to crude oil, coal, natural gas, or heavy oils by exposure to heat and pressure in the Earth's crust. Also referred to as **hydrocarbons** because they consist of compounds of bound hydrogen and carbon, fossil fuels represent a significant storage reservoir or sink for global carbon, which is released to the atmosphere as carbon dioxide when it is extracted and burned.

Uranium, used in nuclear power plants, is also considered non-renewable because its occurrence is finite and it cannot be replaced through natural processes within the time frame of human extraction. Nuclear power plants produce electricity using neutrons to bombard compounds such as uranium. This process of nuclear fission results in the release of energy in the form of heat that can boil water to create steam that can generate electricity by driving turbines. Nuclear energy allows for energy independence in regions that lack fossil fuel deposits, and it has reduced emissions compared to fossil fuels. However, there have been legitimate concerns about safety, what to do with radioactive wastes, and potential depletion of uranium resources.

Examples of renewable energy resources include solar, geothermal, hydropower, wood, and biofuels. They are considered renewable in that many of them, like solar energy, are not depleted when used or can be produced at rates equivalent to their rate of use. Currently, renewable energy resources represent approximately 20% of the global energy production, and their rate of expansion is exceeding predictions.

The use of non-renewable energies, although central to our lifestyles and economies, has a number of limitations. Our fossil fuels are being depleted. They are not geographically well distributed, resulting in unequal access, and they negatively impact the environment and public health. The world now consumes 85 million barrels of oil per day, or 40,000 gallons per second, and demand is rapidly growing. Oil and other fossil fuel reserves will begin to decline by 2020, but the more easily extracted, less expensive reserves are already nearly depleted. New techniques for extracting less accessible fossil fuels are being developed and expanded, including hydrofracturing, extraction from tar sands, and deep oceanic drilling. However, these have increased costs and extreme environmental and health risks. Because fossil fuels are concentrated in specific regions, there have been geopolitical consequences, including military conflict over access to those resources. In addition to the environmental consequences of fossil fuel extraction, their combustion also contributes to significant environmental change, including air pollution, and climate change, as described throughout this book.

The continued use of fossil fuels has implications for national security, energy independence, and the environment. One response to this has been the increase in development and use of renewable forms of energy. As the price of fossil fuels rises and that of renewables declines, the use of renewable energy will increase and could be part of potential solutions to our complex energy problems. In addition to the high cost of renewable infrastructure, renewables have other limitations that must be considered. For example, the use of land to grow biofuels may affect food production. Not all geographical locations have sufficient exposure to sunlight to exploit solar energy. Other renewables negatively impact aquatic systems or require non-renewable rare-earth metals in their production. As we develop renewable energy, all of these issues must be considered. Despite these limitations, the transition to renewable energy is essential for the future of our planet and people. In Chapter 6, we explore new advances in renewable energy and examine the barriers to this transition and ways to overcome them.

TERRESTRIAL AND MARINE EXTRACTIVE NON-FUEL RESOURCES

We rely on a number of terrestrial and marine resources for things other than fuels. These resources include soil, forest products, fisheries, and mined mineral and rock materials. Soil, particularly the upper fertile layer referred to as topsoil, is an essential natural resource. It provides a medium for plant growth and biological activity. Soil regulates the flow and storage of water and nutrients. Soil can also filter and buffer pollutants. Although one might think of it as an abiotic resource, it is actually a complex living system. One gram of soil contains as much as 10,000 different microbial species. There are 1.5 times as many organisms in a teaspoon of soil as people on Earth. Although soil is renewable, the time span of renewal well supersedes that of human use because the natural processes required to create 1 cm of topsoil take 100–400 years whereas its depletion through human activity occurs much more rapidly.

Certain practices, such as those used in organic agriculture, can both conserve and expedite the restoration of soil resources. Intensive soil use and human impact result in various forms of degradation, including erosion, organic matter loss, salinization, nutrient depletion, compaction, reduced biological activity, and chemical toxicity. Given our reliance on soils for the production of food and other plant products, protecting soil quality, like protecting air and water, should be a central goal of national and international environmental policy. Unfortunately, this has not typically been the case, but in Chapter 5 we will see that a variety of techniques in agriculture focus on soil health and new policies and technologies have been developed to protect soil.

Timber and non-timber forest products are biotic, renewable resources. Timber products are primarily lumber and pulp used in paper production. There are many other useable products from timber. These products are made from the cellulose and lignin in the wood and when refined are used in the manufacture of asphalt, rayon, flavorings, paint, detergents, and plastics. Non-timber forest products include fuel wood, charcoal, resins, fibers, decorative foliage, and edible nuts and fruits. Also, numerous cultures and communities rely almost entirely on the medicinal properties of forest plants for their health care. Furthermore, more than 25% of prescription medications in modern medicine are based on or derived from plant compounds, and there is much potential for future cures to come from plant sources.

In spite of or because of our dependence on forest systems, they are disappearing at an astonishing rate. The average annual net loss of forest is approximately 5.2 million hectares per year. Deforestation has been directly linked to population growth and the increased resource demands associated with it (Figure 1.5). The majority of forested land is cleared for agriculture and livestock grazing. The overextraction of forest resources, forest fires, illegal logging, and insecure tenure of forest land are also resulting in forest decline. The cutting of forest for firewood is also a significant driver of deforestation. In rural areas of developing countries, 95% of total energy consumption can be from firewood. This not only reduces forest cover but also deteriorates air quality, which negatively affects the health of those who breathe the air. Social and governance-related phenomena, such as increasing settlements, weak law enforcement, rising poverty, and civil conflict, also contribute to the rising rates of deforestation.

Urbanization and sprawl models of development also contribute to loss of forest habitat. There are a variety of approaches to managing forest resources that can enhance the degree to which they can be renewable and that can also decrease environmental damage while providing essential resources and economic potential. These are addressed in Chapter 7. Also, managed development and agriculture can further serve to protect forests (see Chapters 5 and 11).

Another terrestrial resource that often comes from forested ecosystems is hunted animal meat and other products. This is a biotic resource that should be renewable given that under ideal conditions, game or wildlife populations should either be growing or maintained at **equilibrium**, where birth rates equal death rates. Unfortunately, in many areas of the world, such conditions are not met. Most game populations are threatened, and many have been driven to the brink of extinction by habitat loss and overhunting. Even when protected in wildlife preserves, large animals are often subject to poaching or illegal hunting. To effectively protect these populations, there must be alternative ways to meet the nutritional and economic needs of those who resort to poaching in conjunction with enforcement and protection of the animals.

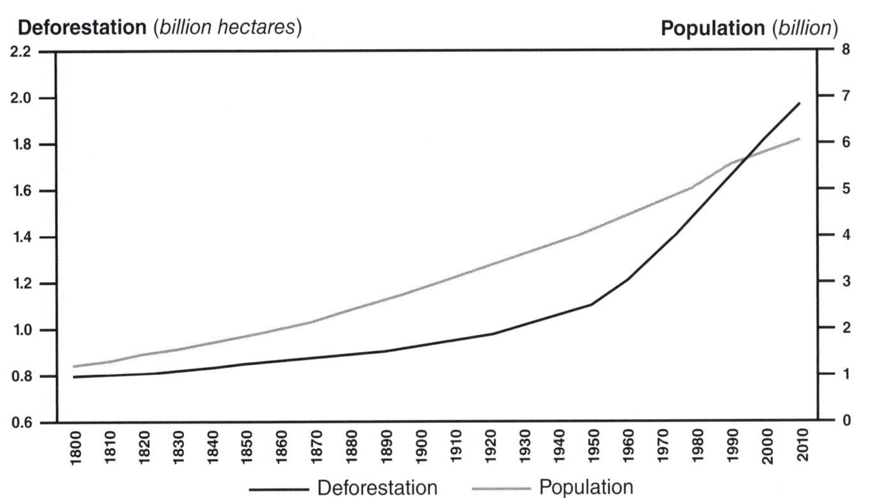

Figure 1.5 World population and cumulative deforestation, 1800 to 2010. These same trends have continued through 2016. *Source: UN FAO State of the World's Forests 2012, http://www.fao.org/docrep/016/i3010e/i3010e02.pdf - page 3, Figure 1.*

In contrast to those being driven toward extinction, there are some game populations that are growing out of control. An excellent example of this is deer in the eastern United States. Because their natural predators have been eliminated, their populations grow unchecked even in the face of hunting. These predators are referred to as **keystone species** because of their major effect on their environment relative to their abundance and also their critical role in maintaining the structure of their ecological communities. The loss of large predators such as wolves in eastern US forests has allowed the deer populations to expand, which in turn assert extreme browsing pressure on the vegetation of forested ecosystems, gravely impacting their ecology. There is evidence that overbrowsing by deer negatively impacts other species, such as **neotropical migrant birds** that nest in these habitats. Effective management programs can protect both the resource and the habitats and other species they are impacting.

Non-fuel mineral resources are materials that are concentrated in or on the Earth's crust that offer economic potential. These are typically considered separately from fossil fuel deposits. They are abiotic, non-renewable resources that are extracted from the Earth using a variety of mining techniques that involve either surface or subsurface excavation. Mined resources include precious metals, metallic ores, and other rock materials. Many industrial processes and products, including those produced by the technology industry, utilize various mineral resources. Historically, mining has had very negative environmental, public health, and social consequences, and it continues to do so. New approaches, policies, and the development of alternative materials that can reduce these consequences are discussed in Chapter 7.

Marine and aquatic resources also have great economic importance and potential. They can include biotic resources such as animal and plant life, including fish and seafood. In this regard, overfishing has been extremely problematic, and as a result we are experiencing a global collapse of natural fisheries, with more than 75–90% of stocks being depleted. In addition, because of marine pollution, many naturally caught fish contain toxins, including polychlorinated biphenyls and mercury. In addition to providing food, the marine environment can provide abiotic materials that are of significant value. Many minerals can be mined from the deep sea, such as gold, nickel, cobalt, copper, manganese, and zinc. With limited reserves on land, deep sea mining is becoming more attractive despite the high costs of extraction. Marine aggregates such as sand and gravel are extracted and used in the construction industry for the manufacture of building materials such as concrete and manufactured stone. Extraction of such resources may have negative consequences for the marine and aquatic environments and their biodiversity. In addition, both the development and the destruction of coastal habitats and wetlands has had negative impacts. Other types of impacts on marine and aquatic systems include the dumping of wastes, especially plastics; chemical pollution; agricultural and industrial run-off; and ocean acidification and coral reef bleaching associated with rising CO_2 and climate change. Reducing waste, regulating ocean dumping, and reversing climate change are solutions to these problems.

WASTE

Although not typically viewed as a natural resource, humans' production of waste and garbage must be considered when contemplating the future of our planet. Globally, we generate more than 3.5 million tons of solid waste per day, and this is expected to triple by 2100. This has drastic economic and environmental costs. Reductions in waste production will require us to rethink our current industrial and economic models; shifts in individual behaviors; and other innovative solutions, including the use of waste as a resource. These solutions are the focus of Chapter 8.

The Changing Climate

Climate change is directly linked to our resource use and most environmental issues. Thus, any approach to environmental problem solving must be done so within the context of global climate change. Climate change, also referred to as global warming, is either influenced by and/or impacts every other environmental issue. Climate change is real, it is scientifically documented, and its negative effects are being realized today. In this section, we examine the causes and consequences of climate change. We also examine general solutions to the problem of climate change, with the understanding that it will be dealt with more specifically in practically every other chapter in this book.

UNDERSTANDING CLIMATE CHANGE

One of the first steps in understanding climate change is the differentiation of weather from climate. **Weather** consists of short-term changes in atmospheric variables such as wind, temperature, cloudiness, moisture, and atmospheric pressure. **Climate**, on the other hand, is a measure of the average pattern of the previously mentioned atmospheric variables over a long period of time. Climate is different from weather in that weather only describes the short-term conditions of these variables within a narrow geographical region. Weather is highly variable, whereas climate changes more gradually, is more predictable, and is often considered over wider geographical ranges or even globally. The occurrence of an odd weather event, such as a very warm day in winter in a particular region, does not mean the climate is changing. Changes in global climate are long-term changes in average atmospheric variables over time and across the globe.

Visible light readily passes through the atmosphere and heats the surface of Earth. The Earth reflects or reradiates this energy as heat in the form of infrared thermal radiation (Figure 1.6). This **infrared thermal radiation** is at a longer wavelength than the visible light spectrum and does not readily pass through the atmosphere back into space. Rather, the gases that comprise the upper atmosphere absorb some of this heat and re-radiate it back to the lower atmosphere and the surface of the Earth. The gases that prevent the heat from leaving the atmosphere and re-radiated it back to Earth are referred to as **greenhouse gases** (GHGs).

GHGs include carbon dioxide (CO_2), water vapor, methane, nitrous oxide, and ozone. They are called greenhouse gases because like the glass of a greenhouse, shorter wave radiation from the sun readily passes through them and, like the glass, GHGs are responsible for retaining the heat and warming the atmosphere below, but the way heat is retained in a greenhouse is very different. The glass of a greenhouse reduces airflow and prevents loss of heat through convection, whereas the GHGs absorb and re-radiate the heat. Regardless of this difference, this analogy has stuck, and the process of shortwave radiation passing through the atmosphere and reflecting off of the Earth's surface as longer wave infrared radiation that is then captured by the GHGs and re-radiated back to the Earth is still referred to as the **greenhouse effect** (see Figure 1.6). As GHGs

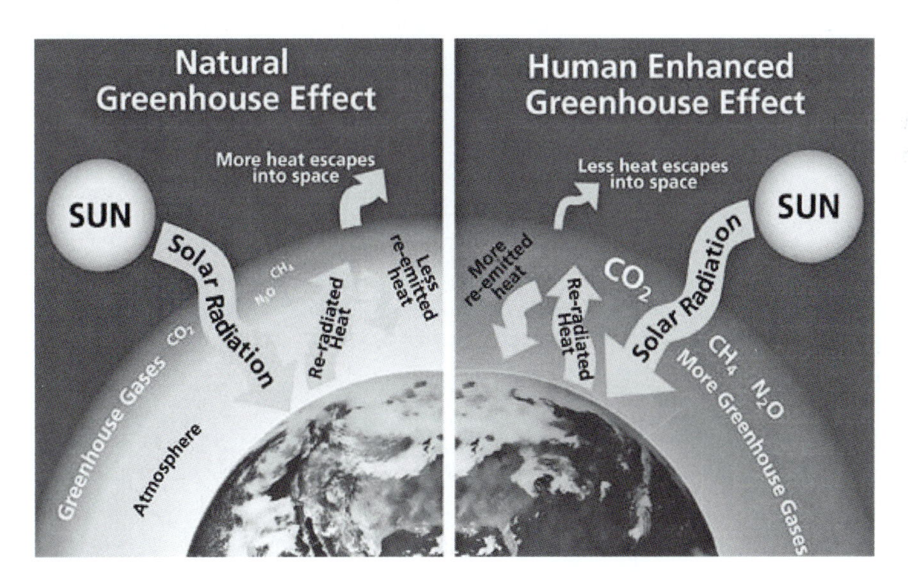

Figure 1.6 The greenhouse effect explained. Increased GHGs capture and re-radiate longer wave radiation or heat, warming the upper atmosphere and earth surface. Increasing GHG concentrations through human and natural process increase the amount of heat captured and re-radiated, thereby increasing global temperatures. *Image courtesy of the Field Museum, https://climatechicago.fieldmuseum.org/learn*

increase, the extent to which the atmosphere absorbs and re-radiates infrared radiation or heat back to the lower atmosphere and the Earth's surface increases, resulting in global warming.

There is great variation in the extent to which each GHG contributes to global warming. This depends on both the amount of each GHG in the upper atmosphere and the relative measure of heat that each gas traps or its **global warming potential (GWP)**. GWP for a specific gas is based on how long it remains in the atmosphere and how strongly it absorbs heat. It compares the amount of heat trapped by a certain mass of the GHG in question to the amount of heat trapped by the same mass of CO_2. We further address GWP as we more closely consider specific GHGs in Chapter 3.

The abundance of GHGs in the upper atmosphere—specifically carbon dioxide, methane, and nitrous oxide—has sharply risen during the past two centuries. This is made evident by examining ice cores from the Greenland and Antarctic ice sheets. These cores have in them bubbles of ancient air that can be dated, and their composition can be determined. These data, coupled with contemporary atmospheric measurements, show that GHG concentrations have dramatically increased since the industrial revolution.

Some have argued that the rise in atmospheric CO_2 is the result of a natural fluctuation rather than due to human causes. However, several pieces of evidence have convinced the scientific community that this is not the case. First, increases in anthropogenic CO_2 emissions starting in 1850 and rising rapidly after 1950 have been directly linked to increases in atmospheric CO_2 (Figure 1.7). Also, the increasing rate of fossil fuel consumption is directly related to the rise in CO_2 emissions. Plant material and things derived from plant materials, such as fossil fuels, have distinctive stable **carbon isotopic ratios** or proportions of different forms of carbon (C_{13}:C_{12}). Such analyses have revealed that the recent increase in atmospheric CO_2 is due to plant-derived sources such as fossil fuels rather than non-plant-derived carbon sources such as releases from volcanoes and the oceans. Contributions of other GHGs such as methane and nitrous oxides have been directly linked to human activities, including industrial agriculture and waste production. This represents solid scientific evidence that not only are GHGs steadily rising but also this rise is being caused directly by our actions.

Accumulated GHGs have the capacity to retain and reflect heat back to the Earth. Thus, as atmospheric GHGs rise in concentration, so do mean global temperatures. The Intergovernmental Panel on Climate Change (IPCC)[6] reported that during the 20th century, global temperatures rose approximately 0.1°C per decade. However, in the past 20 years, this warming rate has doubled. At current rates of fossil fuel combustion, average global temperatures will warm by 3°C by the middle of this century. The greatest levels of warming will be in the northern polar region, where average temperatures

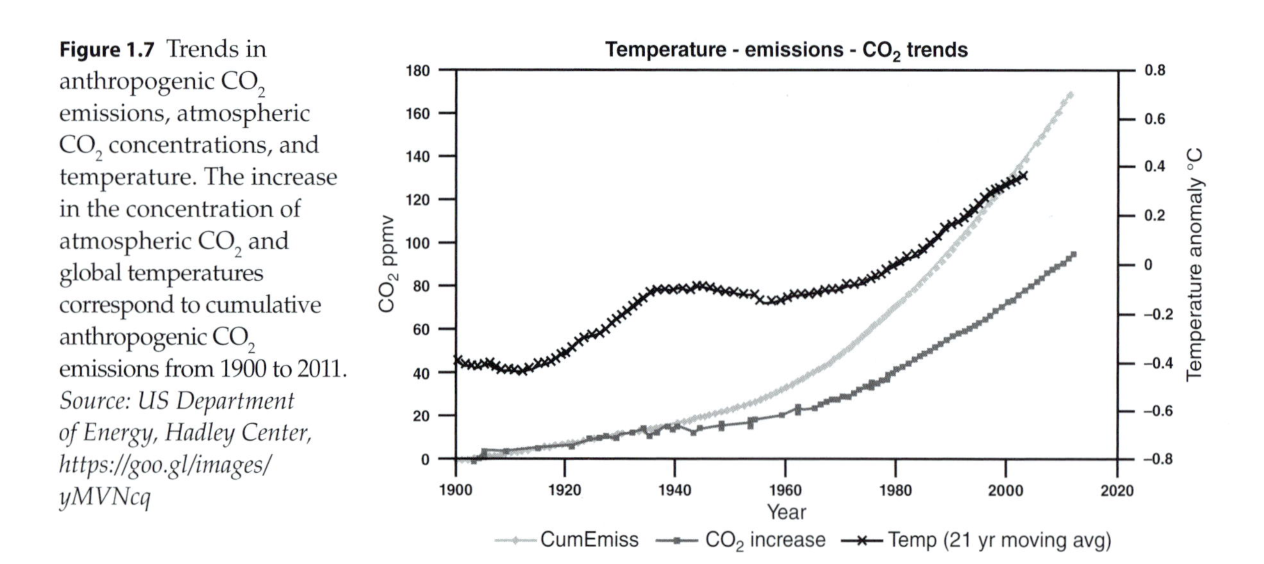

Figure 1.7 Trends in anthropogenic CO_2 emissions, atmospheric CO_2 concentrations, and temperature. The increase in the concentration of atmospheric CO_2 and global temperatures correspond to cumulative anthropogenic CO_2 emissions from 1900 to 2011. *Source: US Department of Energy, Hadley Center, https://goo.gl/images/yMVNcq*

are predicted to rise by nearly 7°C by next century. This is due to the feedback effect of the melting polar ice caps.

The consequences of this pattern of global warming are serious. We already experience some of these effects, including drought, crop loss, increased wildfire risk, extreme weather events, and flooding and storm damage. These types of impacts will increase in frequency and severity in the next 50 years. In addition, with the continued rise in average global temperatures, sea ice at the poles has been melting. This in turn is causing a rise in sea level that will be devastating to coastal regions and inhabited islands throughout the world, resulting in millions of climate refugees as people migrate from flooded areas (Figure 1.8).

Other impacts of rising CO_2 and climate change include ocean acidification, which is resulting in the bleaching and death of diverse and important ecosystems. It is estimated that 10% of coral reefs are already dead and that 60% of the world's reefs will be dead by 2050.[6] In addition to coral reefs, there will be high rates of extinction in other habitats. Amphibian species are among the most threatened by climate change.

Climate change will also have negative implications for global public health. This will include increased mortality from extreme heat and also deaths due to increased flooding and extreme high sea level events. More frequent droughts will result in food and water shortages, hunger and malnutrition, and increases in waterborne disease. Warming may favor the range expansion of insect vectors of diseases such as dengue fever and malaria, and it may increase the spread of allergens. There will also be significant economic losses of up to 5% of gross domestic product for most nations.[7] These economic impacts are likely to disproportionately affect the poorest people in the world because they live in regions that will be more heavily impacted and also because they lack the means to adapt or respond to this change.

Consensus among scientists is often difficult to achieve. This is not the case with global climate change and its cause and consequences, as described previously. Scientists are in agreement that climate change is occurring as the result of human activities, and many of the effects can already be seen. Many corporations are accounting for current and future impacts of global climate change in their long-range business plans. This is particularly true for the insurance industry, which is in the business of assessing risk. A recent report by the IPCC shows that GHGs have risen to unprecedented levels since 2000, and these levels will rise more quickly than previously predicted.[8] This is occurring even in the face of a growing number of policies designed to reduce climate change.

Despite this overwhelming scientific evidence and the early emergence of the devastating effects of climate change, there are groups with their own agendas that deny that climate change is occurring or that it is caused by human activity. This is particularly the case in the United States, where there is a small but adamant denial movement that is primarily politically motivated. However, this movement has begun to decline as both the scientific evidence and actual current impacts of climate change are convincing the broader public and even politicians of its clear and present danger.

What are the solutions to climate change? They can be divided into two types: mitigation and adaptation. **Mitigation** is the attempt to reduce GHG concentrations. This can occur by reducing or stopping emissions and through sequestration or active removal of GHGs from the atmosphere. Because plants actively take up CO_2, their removal through deforestation will result in increasing its atmospheric

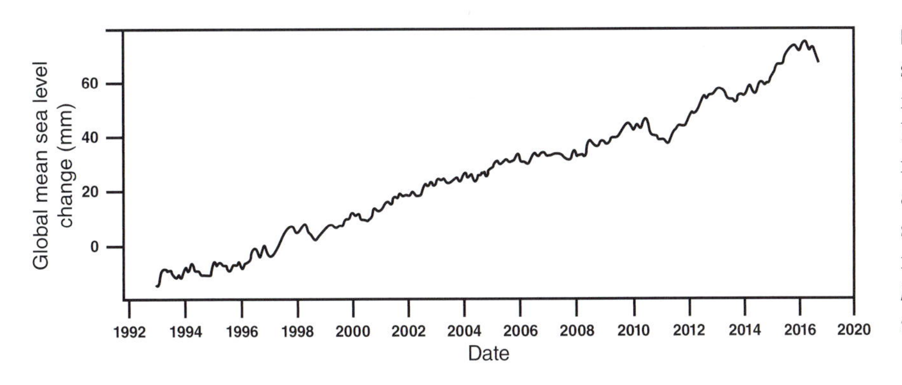

Figure 1.8 Global average sea level change (mm) from 1993 to 2017. Data are from satellite measurements. Values are 60-day averages with seasonal fluctuations removed. *Source: NASA, https://climate.nasa.gov/vital-signs/sea-level/*

concentration, but adding plants through reforestation can increase CO_2 uptake from the atmosphere and reverse the trend. **Adaptation** refers to how human communities must adjust their lifestyles in response to changing climate and its impacts. Within the framework of mitigation and adaptation, there are solutions to the problems of global climate change. Our capacity to both mitigate and adapt to climate change will determine our **resilience** or ability of our ecological, social, and economic systems to absorb or cope with the change. To mitigate climate change, we must find new ways to **decouple** greenhouse gas emissions from economic development, a phenomenon that is already occurring in some countries (Figure 1.9). Because of the overarching nature of climate change, it and potential solutions for mitigation and adaptation are addressed in nearly every chapter of this book. Solutions must run the gamut from policy implementation to innovation and entrepreneurship and individual action, and they must bring together diverse stakeholders. One of the major goals of this book is to develop solutions that meet these criteria. Another is to move beyond mitigation and adaptation and instead consider a **transformation** or radical change or alteration of the way we live our lives on this planet in order to protect it and ourselves from the devastating effects of climate change.

Invasive Species

Another global environmental problem is that of **invasive species**. Invasive species are those that have ecological characteristics such as high reproductive output and colonizing ability that allow them to invade new habitats once they are introduced. They are often non-native organisms that are introduced to a natural habitat either on purpose or by accident. In their new habitat, natural predators or, in the case of plants, herbivores that normally control them are missing. Left unchecked, these species proliferate and can quickly come to dominate an area. Disturbances

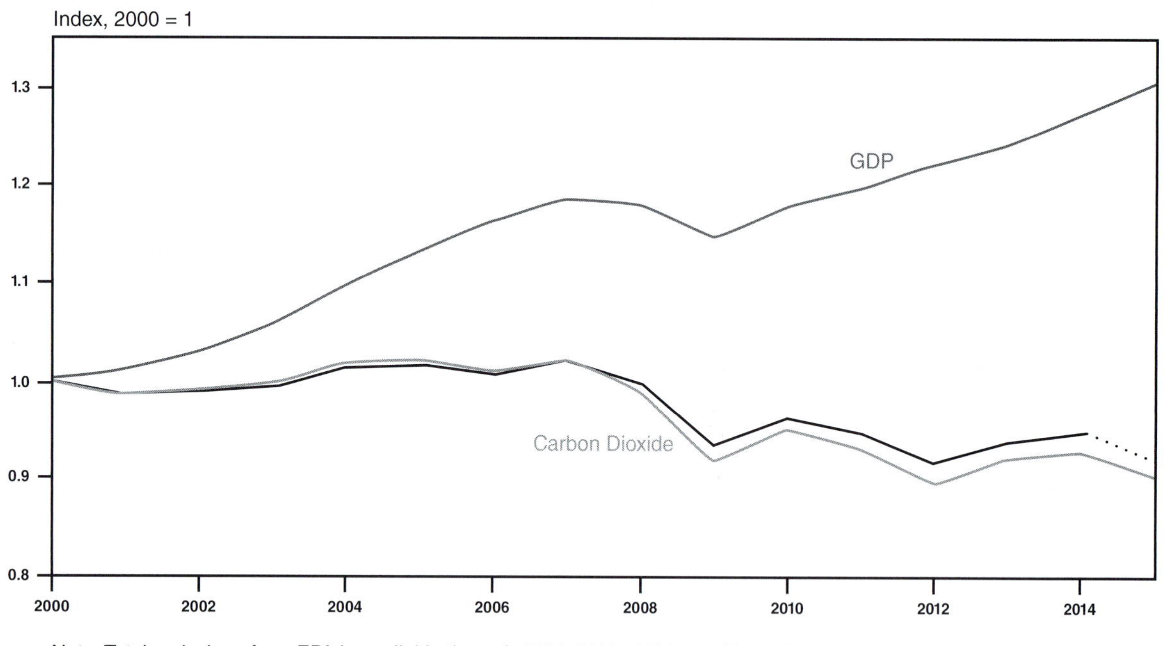

Figure 1.9 In the United States, GHG emissions are declining while the economy as measured by GDP grows. *Source: Environmental Protection Agency, Energy Information Administration, and Bureau of Economic Analysis, https://wattsupwiththat.com/2017/01/12/obama-whitehouse-gdp-has-been-decoupled-from-co2-emissions/*

including grazing by livestock and fragmentation of ecosystems further encourage the establishment and proliferation of invasive species.

Unintentional introductions can occur through global commerce as non-native species are transported as "hitchhikers" by air, water, railways, or roads and in wood and wood products. Ballast water taken up at sea and released in port by transoceanic vessels is a major vector for non-native aquatic species through the global transport of their propagules or larvae. Through this process, it is estimated that more than 3,000 different species of aquatic life may be transported on any given day.[9] An example of this is the zebra mussel *Dreissena polymorpha*. The larvae of this small freshwater mussel were transported in ballast from the Black, Caspian, and Azov seas to the Great Lakes and the St. Lawrence Seaway, where they have invaded hard aquatic substrates. In addition to North America, they have invaded much of costal Europe, where they disrupt the ecosystems through monoculture colonization and damage harbors, ships, and water treatment and power plants.

In addition to accidental occurrence, deliberate introductions have commonly occurred through the introduction of new horticultural varieties, crops, and plants for wildlife management. Failed attempts at biological control in which a potential predator is introduced to control a pest have also resulted in the introduction of invasive species. An example of this is the introduction of the cane toad *Rhinella marina* to Australia in an effort to control the grubs of beetles that were devastating Australia's sugar cane crops. The toad was generally unsuccessful in reducing the targeted pest, and the prolific nature of the toad allowed it to spread and become firmly established throughout much of the continent.

So why are invasive species problematic? Doesn't the introduction of new species increase diversity? The answer is generally no. Invasive organisms tend to drive down native diversity. In the United States, invasive species have contributed to the decline of 42% of all endangered and threatened animals and plants. One example is paperbark, *Melaleuca quinquenervia*, which was introduced to the Florida Everglades in the late 1800s in an effort to assist in drying out swampy land but established monocultures where it was planted. Its rapid proliferation reduced diversity in areas of the Everglades from 80 species of plants to 4 species.[10]

There are numerous other examples of the detrimental effects of invasive species. Those most commonly known that have not already been mentioned include kudzu or "the vine that ate the South"; the starling, which was originally introduced as four pairs in Central Park in New York City; the Asian long-horned beetle, which has caused approximately $10 billion worth of tree damage in the United States; and the Asian carp, which has devastated the ecology of Lake Victoria in East Africa.

Large economic costs are associated with invasive species. These include costs associated with reduced production in agriculture and forestry, negative impacts on ecological tourism, and the control and management of invasive species. It is estimated that damage and control of invasive species in the United States cost more than $120 billion per year and that invasive species cause $7.4 billion in annual productivity losses of 64 major crops.[11] Regarding the example of *Melaleuca* described previously, control efforts within the state of Florida cost more than $2.2 million annually. These costs do not take into account losses in biodiversity and losses of ecosystem services. As we explore and assess solutions in this book, it will be important to consider the potential for introduction and encouragement of invasive species through those actions.

Ecosystem Services and the Value of Biodiversity

People who value the natural world and all of its biodiversity often do so for intrinsic reasons. Many people are committed to protecting nature for spiritual, ethical, and even aesthetic reasons. However, when considering environmental problem solving, it is often useful to move beyond the esoteric value of nature and to take into account the services that ecosystems perform and the value of biodiversity. The ways that humans benefit from natural processes are referred to as **ecosystem services**. There are four broad types of ecosystem services. **Provisioning services** refer to the useable materials or energy that people obtain from ecosystems. These include food, water, genetic resources, and the other natural resources described in this chapter. **Supporting services** are those that play a role in the maintenance and function of all other types of ecosystem services. These include pollination and seed dispersal, nutrient cycling, and the provision of habitat for species that perform other services. **Regulating services** are those that control or maintain environmental conditions. Examples include carbon sequestration and climate regulation, purification of air and water, and flood control. Finally, **cultural services** provide spiritual, aesthetic, or historical value and also venues for science, education, and recreation.

When assigning value to biodiversity, we consider two types: direct and indirect values. **Direct values** include consumptive use values such as foods, medicines, and fuels that are used and consumed locally and not marketed nationally or internationally. **Productive use values** are direct use products that are commercially harvested or produced and sold. **Indirect values** of biodiversity are those that provide indirect benefits to humans that are more difficult to quantify. These include the social and cultural, ethical, and aesthetic value of biodiversity as well as the ecosystems services described previously. **Option values** include currently undiscovered potential benefits of biodiversity. For example, a yet to be discovered plant may offer a cure for cancer or diabetes.

Unfortunately, most economic systems do not take into account the value of ecosystems services and biodiversity. When a for-profit entity impacts the environment, it calculates its bottom line without including contributions from or costs of impacting natural systems that actually serve its interests or those of others. In other words, it reaps the benefits of these services, such as natural water treatment or pollution abatement, without paying for them. When the entity destroys or depletes these resources, it is often not held responsible for the economic losses incurred by this impact. For example, entities involved in industrial agriculture are able to increase their profit margin by producing "cheap food." One reason why this food is cheap and their profits are high is that large producers are not responsible for these sorts of external or indirect costs in the production of this food. These costs might include natural pollination on the use side and pollution of water resources or health or societal problems caused by their actions on the impact side.

These types of costs are termed **externalities**. These are ecosystem benefits or consequences of an industrial or commercial activity that affect other parties and are costs that are not accounted for in the production of goods or services. Externalities are not fully reflected in the price of a good or service or in the determination of profit. As we consider the impacts of such commercial activity and solutions to associated problems, we must do so within the context of these sorts of externalities.

The Human Condition

This book is not just about the environment but also about the link between humans and the environment. It is also important to consider our own condition and recognize that our quality of life is both dependent on and impacts the environment. In this section, we consider the problems related to population growth, rates of consumption, economic issues, and education and public health. Again, we begin to broadly establish solutions to these problems that will be specifically addressed in the chapters that follow.

The Growth of Human Populations and Consumption

Exponential growth is when the amount being added to a system is proportional to the amount already present. As an exponentially growing system increases in size, so does the rate of increase. Such populations remain small at first, but the rate of increase escalates with population size. Theoretically, when left unchecked, a population should grow exponentially, and this is essentially what has occurred historically with the global human population (Figure 1.10). In the first 200,000 years of human existence, the population growth rate slowly increased as 1 billion people were added during that time. It took only another 130 years to add the next billion, 14 years to add the next, and this rate of increase continues to rise as the current population size exceeds 7 billion people and rapidly approaches 10 billion.

Exponential growth should eventually level off as resources become limited. The **carrying capacity** of a population is defined as the population size at which the growth rate of a population becomes zero. As a population approaches its carrying capacity, growth rate begins to decline due to resource constraints, eventually leading to the point at which the birth rate equals the death rate. Theoretically and in nature, some populations actually crash and reduce in number well below the carrying capacity under limiting conditions.

Many have argued that the human population has reached its carrying capacity, but so far resource constraints have not seemed to limit population growth,

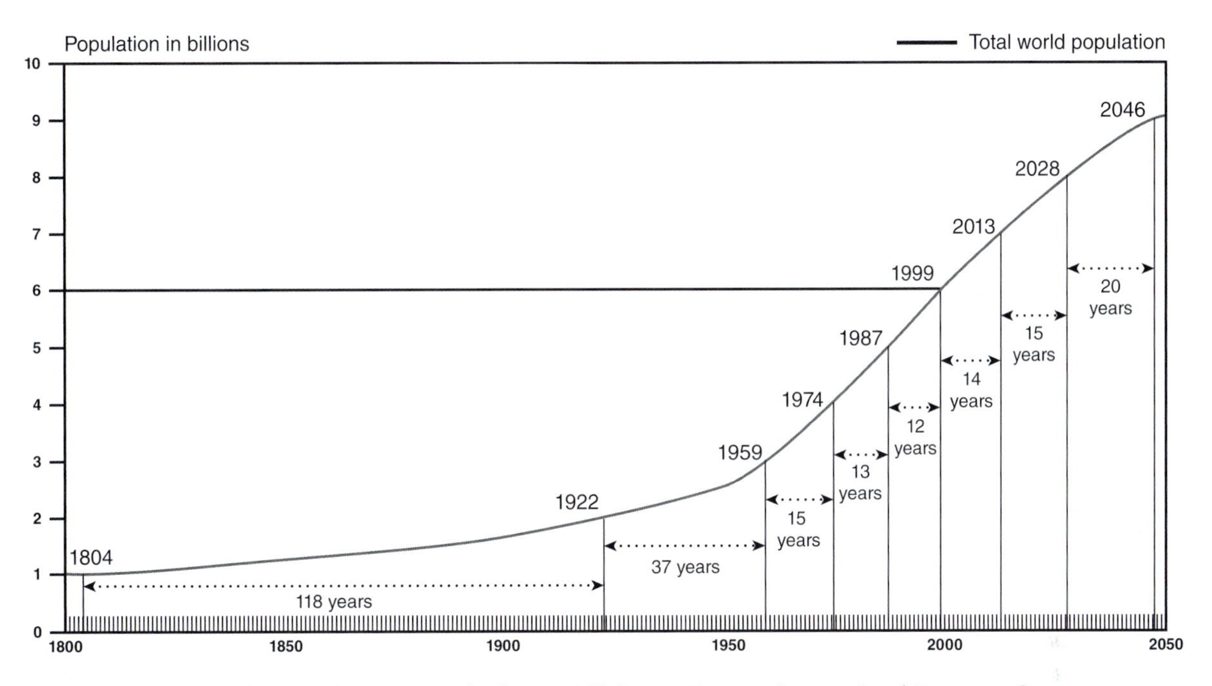

Figure 1.10 Total world population growth. *Source: US Census Bureau, International Programs Center, International Database.*

as the most constrained regions often have the highest growth rates. This lack of decline in growth rates suggests that the concept of carrying capacity is being misapplied to human populations. Moreover, applying the concept of carrying capacity to human populations is problematic for a number of reasons. The first is based on the argument that humans can defy limits through the discovery of new resources and technologies, and history has reflected this. Second, the point at which populations begin to decline probably should not be our first indicator of crisis. The quality of life at the Earth's carrying capacity may not be something we want to experience because there are likely to be harsh health, social, and environmental implications as death rates approach and perhaps even surpass birth rates.

Currently, approximately 70 million people are added to the planet per year. That is equivalent to adding more than 200,000 people every day, or approximately 140 people every minute. However, this is not occurring uniformly across the planet. In fact, in some areas, populations are declining. In developed, wealthier countries, populations are either holding stable or declining. However, in the least developed areas of the world, populations are rapidly expanding such that 98% of the world's population growth is occurring in middle- to low-income countries, where the potential for economic opportunity is the least.

There are a number of factors driving the differences in population growth between the industrialized, developed areas of the world and poorer, developing countries. The first is **fertility rate**, or the average number of children each woman has in her lifetime. A fertility rate of two is essentially a replacement rate that stabilizes population growth. In many developed countries, fertility rates are either at two or below, resulting in stable or declining populations. In developing countries, the fertility rate ranges from three to seven, contributing to their population expansion. A second contributing factor to the disparity is average age of first reproduction, which tends to be younger in less developed countries. A younger age at first reproduction not only increases fertility rate by lengthening the time of childbearing age but also decreases the generation time or the average time between two consecutive generations. A population in which the average age of first reproduction is 20 years will yield five generations in 100 years, whereas an average age of first reproduction of 25 years will yield four generations. Thus, the average age of first reproduction is one of the most important determinants of population growth rate. A third factor is the age structure of a population. If a population has a large number of young people just entering their reproductive years, the rate of growth of that population will rise even with low fertility rates; again, this is a common pattern in less developed, poorer nations.

Another important influence of population growth in relation to development is the rate of mortality. Developed countries tend to have reduced birth and death rates, which stabilize populations. Death rates are rapidly declining in many poorer countries as they begin to experience the benefits of development, such as improved medical care and public health; however, reductions in birth rate tend to lag behind. Today, this sharp decline in death rates followed by slowly declining birth rates is characteristic of most of the less developed regions of the world, resulting in rapid population growth. It is expected that if these countries continue to develop, birth rates will eventually decline and populations will begin to stabilize.

The changes in birth and death rates and hence population size with development are referred to as the **demographic transition** (Figure 1.11). This transition can be divided into essentially four phases. Phase 1 is referred to as "primitive stability," with equally high birth and death rates. Phase 2 exhibits declining mortality with increased medical care and public health but precedes any decline in birth rate. Population growth rapidly increases during this phase. In Phase 3, birth rate begins to decline and approach death rates, reducing population growth. In Phase 4, birth rates and death rates are both low, resulting in a stabilized population (see Figure 1.11).

Currently, most developing countries are in Phases 2 and 3, in which population growth rates are the highest. Few are transitioning into low growth phases, and approximately half of them have been "trapped" in this phase of rapid population growth for more than 50 years. This has evoked questions about the potential of this model to necessarily predict future population growth, particularly in the less economically developed countries. Demographic transition theory also does not account for recent phenomena such as HIV, which has increased mortality in a number of sub-Saharan African nations, nor does it necessarily take into account government interventions such as population control programs. The model also assumes that population changes are caused by increased wealth associated with industrial changes, and to a lesser extent it considers the role of social change in determining birth rates. But as a model, it serves to elucidate mechanisms by which patterns of birth, mortality, and population growth change with development. The following are two examples of developing countries that have made the transition to reduced birth and death rates, resulting in decreased population growth; these cases are used to flesh out the mechanisms by which population growth is stabilized:

Sri Lanka: Prior to World War II, advances in public health had been largely limited to affluent, industrialized countries. But since then, improvements in public health have been made in many of the poorer countries of the world, resulting in sharp declines in

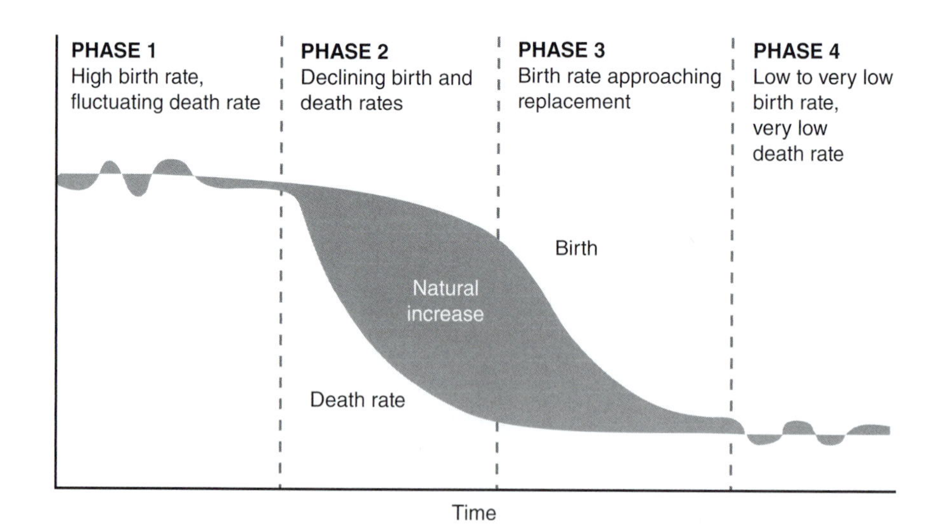

Figure 1.11 The four phases of demographic transition. *Courtesy of The Encyclopedia of Earth, https://editors.eol.org/eoearth/wiki/File:Figure_4_classic_stages_of_demographic_transition_438x0_scale.gif.jpeg*

death rate. An example of this is the South Asian country of Sri Lanka, which has a population of approximately 20 million. In 1945, the death rate in Sri Lanka (then called Ceylon) was high. In 1946, a large-scale program of mosquito control using DDT to eradicate malaria was started. This was quite successful and cut the death rate by more than 50%. From the 1940s until the mid-1970s, death rates were relatively low, while birth rates had not substantially declined. As a result, the population was increasing at an annual rate of 1.2%, with a doubling time of 57.5 years. But by the 1980s, fertility rates began to decline and life expectancy continued to rise, resulting in an aging population. Today, fertility rates average approximately two children per woman, and the population growth rate has fallen below 1% per year. This was achieved through government health and education reforms. This proved to be a wise investment because the demographic shift resulted in a growing working-age population and a decrease in the number of dependents per family. This ultimately allowed for continued economic development and improvements in the general standard of living.[12]

Bangladesh: Bangladesh is a small country with a population of 150 million people. It is ranked seventh in the world for population size, with more people than Russia. However, it is smaller in area than the US state of Florida so that it is ranked number one in population density, with 950 people per square kilometer. This density is equivalent to half of the people in the entire United States living only in the state of Florida. In the 1960s, fertility rates were as high as 7 or 8 children per family, and it was believed that culture and religion would make it impossible to reduce this number and slow population growth. However, after independence in 1972, the child mortality declined from 24% and fertility rates declined to approximately 2.5 children per family. This dramatic change is often referred to as the Bangladesh Miracle.[13]

These two case studies reveal several potential ways to stabilize population growth in developing countries. In Sri Lanka, health care and education reform was a clear priority. In Bangladesh, the completion of the demographic transition coincided with independence from Pakistan in 1971. This was followed by the development of an open culture of research-based innovation in community-based health care that brought key health interventions to every household that directly led to drastic reductions in population growth.

In addition to improvements in health care, a number of other common changes contributed to the demographic transition in these two countries. These included improvements in nutrition and education, including the introduction of family planning. Economic growth and poverty reduction have also played a major role. But probably the most important factor has been the empowerment of women. As women have obtained educational opportunities, improved literacy, and entry into the formal labor market, marriage and pregnancy have been delayed and fertility rates have decreased.

THE POPULATION BOMB

The Population Bomb, a bestselling book written by Stanford University Professor Paul R. Ehrlich in 1968, identified overpopulation as the single most important issue confronting humans.[14] It warned that the "population explosion" would result in mass starvation and major societal upheaval. The book advocated for immediate action to limit population growth as a means to save our species, and it argued that the only way to save the planet was to stabilize global population growth rates. The book served to bring the issue of overpopulation to a wider audience, but it has been criticized for its alarmist tone and its inaccurate predictions.

One additional problem is that the emphasis on population growth and disparities in rates between developed and developing countries places the focus of these global problems on the developing world. But population growth rate is not the only disparity between developing and developed nations. Rates of consumption vary and are often inversely related to population growth rates. Higher per capita rates of consumption by individuals in a population may do more to impact our planet and the depletion of its resources than the number of individuals in a population.

CONSUMPTION DISPARITIES

The most developed countries of the world consume approximately 32 times more total resources per capita compared to the least developed countries. The United States is home to approximately 4.5% of the world's population but consumes more than 25% of the world's resources. Per person per year, the most developed nations of the world, such as the United States, consume fossil fuels at 3.6 times the rate of the world average and 15.1 times the rate of Africa. It has been too easy to blame the developing world and its population pressures for its impact on the planet, but when we consider rates of resource consumption and depletion, true patterns of impact become clear.

One way to integrate population size with consumption to determine impacts is through the **IPAT model**.[15] This model takes into account per capita negative impact (I) on the environment, expressed as the product of population size (P), affluence (A), and technology (T). Affluence is assumed to be directly related to the average consumption of each person in the population. Technology represents how resource intensive the production of affluence is. The greater these two variables are, the greater the environmental impact.

There have been a number of criticisms of the IPAT model, including its simplicity, the interdependency of the variables, and that technology can also decrease impact through increased efficiencies. Regardless, this model illustrates that in industrialized countries, in which levels of affluence and technology are high, there is potential for large environmental impact even at lower population sizes. Solutions that reduce environmental impact must move beyond population control in the developing world and include reductions in consumption in industrialized countries. These are considered throughout the book, but an in-depth discussion of them is provided in Chapter 8.

Fiscal Crises and Economic Opportunity

According to the World Economic Forum,[16] 3 of the top 10 global risks of highest concern are severe economic disparity, structurally high unemployment or underemployment, and the potential for fiscal crises in key economies. If these global risks are not effectively addressed, there could be far-reaching social, economic, and political fallout. These issues are interconnected with resource use, climate change, and population growth and/or decline, as previously described in this chapter. The Forum suggests that these risks can only be addressed through long-term thinking and collaboration among businesses, governments, and civil societies.

The Forum reports that the wealth of the world is divided between two groups: the richest 1% and the remaining 99%. There is extreme economic disparity between these groups. This gap between the rich and the poor has been steadily increasing and is at its highest level in 50 years. The average income of the richest 10% of the population is approximately nine times that of the poorest 10%. This is up from seven times that of the poorest 10% 25 years ago. The poorest half of the world's population has a shared wealth equal to that of the richest 85 individuals. The **gross domestic product (GDP)**, a measure of the monetary value of all the finished goods and services produced within a country's borders in a specific time period, is an estimate of that country's wealth. The GDP of the 41 most indebted, poor countries with a total population of 567 million people is less than the wealth of the world's 7 richest people combined.

Almost half of the people in the world, more than 3 billion people, live in extreme poverty, living on less than $2.50 per day. Some countries, such as Brazil and a number of developing countries in Asia and Africa, have effectively reduced poverty. However, large inequities in wealth remain. Economic inequality disproportionately affects women and children. In the poorest areas of the world, the earning potential and educational opportunities for women are further reduced. Globally, 1 billion children live in poverty (1 in 2 children in the world), 640 million live without adequate shelter, 400 million have limited access to safe water, 270 million have no access to health services, and 10.6 million died in 2003 before they reached the age of 5 years (or roughly 29,000 children per day).

Economic inequity is caused by a variety of factors, including a regressive pattern of taxation that reduces the rate for the highest income groups. Since the 1970s, these rates have significantly declined in many countries throughout the world, but particularly in the United States and United Kingdom. Other factors contributing to inequity are deregulation of industry, international trade policy, exploitation and the disempowerment of labor, and unequal access to technology. There is still strong political pressure to continue all these trends and to maintain this inequity.

Another global risk is chronically high unemployment and underemployment. In contrast to cyclical patterns that have occurred in the past, today's globally high rates of unemployment appear to be structural, which means that employment opportunities will not increase without significant change or reform. One major cause of this is that education and training programs are either non-existent or have not adapted to the changing job market, resulting in a mismatch between jobs offered by employers and the skill set of potential workers. Other causes include reduction in the labor force due to automation, movement of industry to areas where labor is cheaper, and large-scale bankruptcy in locations where a single industry has been a dominant employer.

The World Economic Forum places the threat of major fiscal crisis at the top of its list of global risks. The cause of these crises will be increasing deficits and debt characteristic of the majority of both developed and developing nations of the world. A fiscal crisis occurs when investors begin to doubt a government's ability to repay its debt. This results in escalating interest rates on further borrowing. This in turn adds more debt and creates more fear and potential that a country will default on its debt. Debt development such as this leads to the instability of financial systems and eventually forces governments to cut expenditures and raise taxes, triggering recession and massive unemployment.

Advanced economies are currently in danger of such fiscal crisis, as indicated by the large amount of debt in the United States (>100% of GDP) and Japan (well over 200% of GDP). The collapse of key economies such as these will result in a cascading effect of global economic failure. Again, this failure will disproportionately impact the global poor. Such crises result in increased unemployment and reduced wages. At the same time, financially strapped governments and aid agencies are forced to cut funding for vital social and public health services, and environmental protection efforts, including options for adapting to climate change, are weakened. A number of realistic economic reforms that include collaboration between the public and private sector will help ensure that fiscal frameworks become more resilient.

Education and Public Health

In addition to global disparities in wealth and economic opportunity between developed and developing countries, there are also grave differences in education and public health. Nearly 1 billion people entered the 21st century unable to read a book or sign their names. More than 25% of all children in 18 countries have never been to primary school, and in 23 countries, more than 75% of adolescents have not completed lower secondary school. These numbers are considerably greater for lack of secondary and post-secondary education. Often, girls and women are the ones who lack these opportunities. Access to education has been shown to be linked to fertility rates and population growth, public health, economic opportunity and equity, and effective natural resource management. Accordingly, universal primary education that ensures access for all children everywhere has become an important objective of the United Nations and the world's leading development organizations. We focus on ways to improve global education in Chapter 12, and ways in which education connects to specific issues are considered throughout the book.

During the past several decades, people living in developing countries have experienced considerable improvements in health. For example, during the past 40 years, annual mortality among children younger than age 5 years has been reduced by 50%. Unfortunately, since 2000, these gains have slowed or even reversed in some areas of the world. The countries experiencing these setbacks are primarily in sub-Saharan Africa and some areas of South Asia, where patterns of disease and mortality in less developed countries have changed markedly. This includes the HIV/AIDS epidemic, the re-emergence of tuberculosis, increases in the incidences of insect-borne diseases, the emergence of new infectious diseases, and increases in violence associated with political unrest. Also, a number of chronic diseases that were thought to be of concern primarily in developed countries are exhibiting increased incidence in developing countries. These include the increased occurrence of coronary artery disease, cancers, diabetes, and cerebrovascular damage. Many of these trends can be reversed by increasing access to health care, including medicines and vaccines, and effective education programs specifically with regard to HIV/AIDS and mosquito control.

A number of public health problems are directly linked to environmental issues, including the limits of the environment to provide nutritious food and water to meet the needs of all people. These limits are primarily due to mismanagement. Globally, it is estimated that approximately 870 million people are undernourished, and more than 100 million children younger than age 5 years are undernourished and underweight. A sufficient supply of water that is of high enough quality for drinking is essential for the well-being of humans. Naturally occurring water systems are being depleted and contaminated at alarming rates. As noted previously, millions of people die each year from water, sanitation, and hygiene-related causes.

Pollution and climate change will impact the health and lives of people throughout the world in complex and very serious ways. For example, the increased use of toxic chemicals without sufficient environmental and occupational regulation has resulted in increases in a variety of acute and chronic diseases. The connection between the environment and human health is emphasized throughout this book.

Developing Sustainable Solutions

When many think of the environment or the natural world, they think of its awesome power and magnificent beauty; they think of it as something pristine; and they want to protect and save it for its intrinsic value. They are reminded of all the incredible species and their amazing adaptations that allow them to survive and reproduce in the natural world. But they often forget about one species. It is a wide-ranging species that occurs in tropical and temperate forests, on mountaintops, and in desert and polar regions. Like all species, its survival is dependent on the natural world. That species is *Homo sapiens*. Yes, our species is inextricably linked with the environment. Nature and the environment are worth saving not just for their intrinsic value but also because our lives depend on it. There is also a clear link between environmental factors and diseases such as cancer.[17]

Our planet has been in existence for 4.6 billion years. The evolution of early humans occurred less than 5 million years ago; thus, humans have occupied Earth for 0.003% of its existence. This is proportional to just a few seconds in an entire day. Despite our short time here on Earth, human activities have had a significant global impact on its ecosystems. Because of this, many scientists now argue that we have entered into a new geological epoch referred to as the **Anthropocene**, the portion of the Earth's history during which human activity has begun to cause significant global change by approaching or exceeding **planetary boundaries**. These boundaries are biophysical thresholds that define the safe operating space for planetary life support systems essential for human survival. According to Johan Rockström and his collaborators, three of nine of these interlinked planetary boundaries have already been overstepped (Figure 1.12). It is also argued that these planetary boundaries should be the basis for defining preconditions for human development.[18]

Recognizing that our actions threaten our own life support system is an important first step in beginning to protect and restore it. But we must also acknowledge the complicated connection among natural resources, biodiversity, climate change, human population growth and consumption, economic issues, education, and public health, and we must realize that they need to be considered in concert to solve the problems that confront us and our planet. History has taught us that preserving and protecting the environment without considering the needs of the people who depend on it will not work.

The notion that we need to protect our planet because we use it to meet our diverse needs is central to what we call sustainability. In this book, we examine the question of how we can continue to do this without ruining our ability to do so in the future by wrecking the planet on which we depend. Although defined in many ways, sustainability almost always refers to managing the environment for current and future generations so that members of those generations have opportunities to live full and healthy lives.

In Chapter 2, we examine the history of the sustainability movement and use the context from this

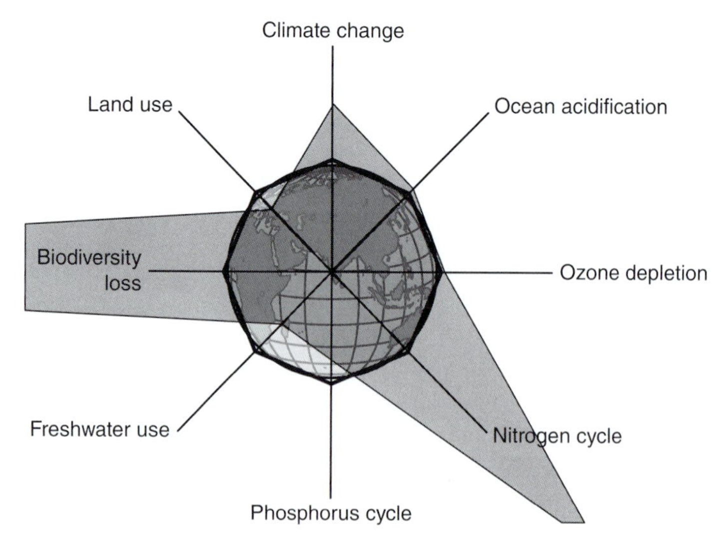

Figure 1.12 The nine planetary boundaries for Earth systems on which human survival depends. The shaded areas indicate the boundaries in the three systems that have already been exceeded. (Adapted from the Stockholm Resilience Center, https://commons.wikimedia.org/wiki/File:Planetary_boundaries.svg)

chapter along with that history to develop a useful definition of sustainability. We apply this definition to the variety of problems we face on planet Earth. Sustainability is presented as a way to analyze problems and develop solutions to them. We explore specific solutions, programs, and projects, and we examine how they are implemented and assessed. We do this through a critical lens and with an open mind to allow us to see the potential benefits and limitations of each approach. Our goal is to develop interdisciplinary solutions to complex problems in an effort to sustain our planet in order to meet the needs of the current and future generations of *our* species.

Chapter Summary

- The goal of this book is to offer the interdisciplinary background that is required to understand the many complex problems that we face and to do so in a way that will both inspire and equip its diverse audience to develop solutions to these problems.
- Natural resources are materials in the environment that are either necessary or useful to humans. They cycle naturally in the environment. When the rate of extraction of a particular resource outpaces its renewal in natural cycles, that resource is considered non-renewable.
- The fundamental resources of air and water are abiotic and theoretically renewable, but human pressure has negatively impacted their quality and accessibility.
- Humans rely on terrestrial and marine resources such as soil, forest and animal products, and mined materials. Current practice results in depletion of those resources and negative impacts on environment and public health.
- Terrestrial and marine environments also provide energy on which humans rely heavily. Non-renewable fossil fuels are rapidly being depleted and contribute to global climate change. Their continued use has implications for national security, energy independence, and the environment. Examples of renewable energy resources include solar, geothermal, wood, and biofuels. Currently, renewable energy resources represent less than 20% of global energy production and use.
- Climate change or global warming is linked to human activity through our contribution to atmospheric greenhouse gases and the removal of natural carbon sinks. The consequences of global climate change are serious and include drought, crop loss, extreme weather events, damage due to increased fires and flooding, coral reef destruction, and rising sea levels.
- Another global environmental problem is that of invasive species, which are non-native organisms that when introduced to an environment dominate and drive down native diversity.
- Natural ecosystems provide a diversity of services that benefit humans, and biodiversity has potential value as new foods and medicines. Most economic systems and business models do not take this value into account. This value should be accounted for when considering the impact of commercial activity and the cost of products and services.
- Developing countries tend to have higher population growth rates, whereas developed countries have disproportionately higher rates of consumption. These need to be considered in tandem as ways to reduce environmental impact and resource limitations and depletion.
- Other threats to the human condition include the potential for fiscal crisis, severe economic disparity, incomplete or ineffectual education, and public health issues. Environmental, economic, and social issues need to be considered in concert to solve the problems that confront our species and our planet.

Digging Deeper: Questions and Issues for Further Thought

1. Julian Simon argued that there is no resource crisis—that as resources become scarcer, they increase in cost and in turn there is economic incentive to discover more, ration, recycle, and develop alternatives to them. Can you think of examples in which such human ingenuity generated solutions that eliminated a resource crisis? Were there any other consequences associated with that solution? Are there areas in which solutions have not yet been generated? What are the limitations of this way of thinking?
2. By the end of this chapter, connections between the problems we confront and social justice have begun to emerge. Pick one problem (e.g., climate

change, population growth, and fiscal crises) and explore the linkage of the problem to social justice.
3. A number of conservation programs that protect specific areas of nature by strictly banning human economic activity have failed. Why? What might be a more effective approach to conservation?
4. What are the criteria for designating a resource as renewable? Give examples of renewable and non-renewable resources? Are there aspects or components of some renewable resources that are not completely renewable? What might be a better term for such resources?

Reaping What You Sow: Opportunities for Action

1. We have seen that there are large disparities in per capita consumption of resources between developed and developing countries. Either for yourself or for your community, develop a plan to reduce individual consumption of one or more resources. How would you measure or quantify that reduction?
2. For one or more of the issues presented in this chapter, identify some organizations that directly address them. What is their approach? How are they funded? Choose one that appeals the most to you, and explore ways to either directly get involved with it or support its efforts.

References and Further Reading

1. Twenge, Jean M., W. Keith, and Elise C. Freeman. 2012. "Generational Differences in Young Adults' Life Goals, Concern for Others, and Civic Orientation, 1966–2009." *Journal of Personality and Social Psychology* 102 (5): 1045–62.
2. Orr, David W. 1994. *Earth in Mind: On Education, Environment, and the Human Prospect*. Washington, DC: Island Press.
3. Safronova, Valeriya. 2014. "Millennials and the Age of Tumblr Activism." *New York Times*, December 19, 2014. http://www.nytimes.com/2014/12/21/style/millennials-and-the-age-of-tumblr-activism.html
4. Simon, J. 1983. *The Ultimate Resource*. Princeton, NJ: Princeton University Press.
5. WHO/UNICEF Joint Monitoring Programme. n.d. Accessed January 31, 2017. https://www.wssinfo.org
6. Intergovernmental Panel on Climate Change. 2013. *Climate Change 2013: The Physical Science Basis. Contribution of Working Group I to the Fifth Assessment Report of the Intergovernmental Panel on Climate Change*. Cambridge, UK: Cambridge University Press.
7. Kreft, S., D. Eckstein, L. Dorsch, and L. Fisher. 2015. *Global Climate Risk Index 2016*. Berlin: Germanwatch. https://germanwatch.org
8. Intergovernmental Panel on Climate Change. 2014. *Climate Change 2014: Mitigation of Climate Change. Contribution of Working Group III to the Fifth Assessment Report of the Intergovernmental Panel on Climate Change*. Cambridge, UK: Cambridge University Press.
9. National Research Council. 1995. *Understanding Marine Biodiversity: A Research Agenda for the Nation*. Washington, DC: National Academy Press.
10. DiStefano, J. F., and R. F. Fisher. 1983. "Invasion Potential of *Melaleuca quinquenervia* in Southern Florida, U.S.A." *Forest Ecology and Management* 7: 133–41.
11. Pimentel, David, Rodolfo Zuniga, and Doug Morrison. 2005. "Update on the Environmental and Economic Costs Associated with Alien-Invasive Species in the United States." *Ecological Economics* 52 (3): 273–88.
12. World Bank. 2012. "Sri Lanka—Demographic Transition: Facing the Challenges of an Aging Population with Few Resources." Accessed June 10, 2014. http://www.eldis.org/go/home&id=65540&type=Document#.U5cUuXbvjYh
13. Rosling, Hans. 2007. "The Bangladesh Miracle." https://www.gapminder.org/videos/gapmindervideos/gapcast-5-bangladesh-miracle
14. Ehrlich, Paul R. 1968. *The Population Bomb*. New York: Ballantine.
15. Ehrlich, Paul R., and John P. Holdren. 1971. "Impact of Population Growth." *Science* 171 (3977): 1212–17.
16. World Economic Forum. 2014. "Global Risks 2014, Ninth Edition." Geneva: World Economic Forum.
17. Steingraber, Sandra. 2010. *Living Downstream: An Ecologist's Personal Investigation of Cancer and the Environment*. Cambridge, MA: Da Capo Press.
18. Rockström, J., Steffen, W., Noone, K., Persson, Å., III, F. S. C., Lambin, E. F., . . . Foley, J. A. 2009. A safe operating space for humanity. *Nature* 461: 472–475.

CHAPTER 2

DEVELOPING SUSTAINABLE SOLUTIONS

In Chapter 1, we saw that the state of the environment and human actions and quality of life are all complexly interconnected. Thus, solving problems in either realm requires simultaneous consideration of the other. We must consider the environment as something that is required to sustain human life while recognizing that our ability to do this is contingent on how we live our lives. The principle of sustainability embraces the notion that lasting solutions to environmental problems are difficult to achieve if they do not also improve the quality of life of those who live in and depend on that environment. However, sustainability has had numerous loosely applied definitions. Also, because it has been co-opted and deceptively used to market products or services as environmentally friendly, it has at times been maligned or discredited.

Problem-solving frameworks require precise definitions and approaches so that strategies can be implemented and ultimately assessed. Given that our framework is that of sustainability, we must precisely define it in a practical way. This definition must be broad enough so that it can be applied to a wide range of issues while also providing specific objectives that can be used to develop measurable outcomes. This chapter establishes a robust definition of sustainability that will serve as a framework for identifying and solving problems. We begin by placing sustainability within a historical context. Then we examine the terminology and definitions that emerged from that history and extract those elements that will be most useful in solving problems at the interface of the environment and human needs. We then use these to develop a working definition of sustainability that will allow us to identify problems related to it, most effectively develop creative solutions that can solve those problems, and determine how best to assess the effectiveness of these solutions.

PLANTING A SEED: THINKING BEFORE YOU READ

1. List five key terms that come to mind when you think of the definition of sustainability.
2. Do an internet search and locate three rather different definitions of sustainability. Explain how they differ and what common ideas they share.
3. What are the "must have" elements of a good, working definition of sustainability?

The great challenge of the 21st century is to raise people everywhere to a decent standard of living while preserving as much of the rest of life as possible.

—Edward O. Wilson

From Environmentalism to Sustainability

The modern concept of sustainability has its origins from the 19th-century transcendentalist movement of the eastern United States. **Transcendentalism** is a religious and philosophical movement that held as its core belief that individuals have knowledge about themselves and the world around them that "transcends" or goes beyond the tangible and logical thinking. One obtains this knowledge through intuition and spiritual connection. Many of the early transcendentalist writers, such as Ralph Waldo Emerson, Henry David Thoreau, Walt Whitman, Nathaniel Hawthorne, Margaret Fuller, and Herman Melville, believed and wrote about a spiritual connection with nature as a path to self-knowledge. They saw opportunities for self-awareness, reflection, and liberation by connecting to their environment. Nature was viewed as a teacher. For Emerson, a spiritual connection to nature was a direct link to God.[1] Influenced by Emerson, Thoreau viewed nature as an awe-inspiring force that should be a way of life in which one can express his or her individuality and self-reliance, and he wrote about this in his most famous book *Walden*.[2] Transcendentalists also believed in the elimination of slavery, and they promoted women's rights, creative and participatory education for children, and labor and social reform through social protest. However, this was not overtly connected to their environmental ethic.

From this transcendentalist view of nature as a source of spiritual foundation and the elucidator of self-knowledge emerged an ethic that acknowledges and values the deep connection between humans and environment. This movement and its writers made strong spiritual and ethical arguments for the preservation of nature. Nature was to be protected, not to meet our pragmatic need for resources but, rather, for its fundamental intrinsic value. Nature was viewed as sacred and humans as intruders. This **preservationist** perspective continued to be reflected well into the 1900s by authors such as John Muir, who wrote about and fought to protect Yosemite and created the Sierra Club,[3] which served to further the preservationist environmental objectives (Figure 2.1).

Figure 2.1 A stereograph of Muir and Roosevelt in Yosemite taken from Glacier Point during their 1903 camping trip. *Courtesy of the Library of Congress Prints and Photographs Division Washington, D.C. 20540 USA.*

The Conservationist Movement

The **conservationist** movement that began in Europe as early as the 1600s primarily to develop and promote forest timber management spread to the United States in the early 20th century. Although inspired by the transcendentalists who exalted the inherent value of nature, the conservationists moved beyond the spiritual connection to the natural world and preservationist objectives to recognize other ways in which humans depend on their environment. In the early 1900s, President Theodore Roosevelt first made conservation part of the US national agenda by establishing the US Forest Service. Influenced by Muir, Roosevelt was responsible for expanding the number of national forests, parks, and monuments (Figure 2.1).

American forester and politician Gifford Pinchot furthered the conservation agenda by promoting scientific forestry with the aim of conserving forest resources through planned use and renewal. So with the conservationist movement a new land ethic was developed that valued nature, but not as something sacred that should be set apart from humans. Rather, nature was viewed as something that should be valued for the variety of ways that it could meet the needs of humans and should be managed accordingly. This later evolved into the "multiple use" approach to conservation that encouraged environmental management to maximize potential for outdoor recreation, including hunting and fishing; watershed protection; and the production of timber.

The contemporary environmental movement that began to take hold in the 1950s and 1960s seemed to walk the line between preservation and conservation in a growing response to observable declines in the environment. Two influential authors, Aldo Leopold and Rachel Carson, helped bring environmental issues to the forefront. In his book *Sand County Almanac*,[4]

Leopold further advocated a land ethic that consisted of a responsible relationship between people and the land they inhabit. Carson, in her book *Silent Spring*,[5] brought environmental concerns to the American public through her documentation of detrimental effects of chemical pesticides, particularly DDT, on the environment.

As awareness of environmental issues grew in the 1960s and 1970s, so did environmental activism. Through the work of environmentalists and authors such as Wendell Berry, David Brower, Paul Ehrlich, James Lovelock, and Edward Abbey, environmental issues came to the forefront, including the celebration of the first Earth Day in 1970. During this period, environmental awareness became mainstream. This led to the creation of the Environmental Protection Agency (EPA) and a proliferation of important environmental legislation, such as the Clean Air Act, the National Environmental Policy Act, and the Wilderness Act. With this came the expansion of the global environmental agenda through organizations such as the United Nations (UN) and international development agencies and also the birth of hundreds of **non-governmental organizations (NGOs)** focused on the environment.

The Emergence of Sustainability

In 1972, in order to develop common principles in the preservation and enhancement of the human environment throughout the world, the UN Conference on the Human Environment met in Stockholm, Sweden. Here, we saw for the first time a real emphasis on the environment as it relates to the human condition. It was where we first began to discuss the integration of environmental protection development and social well-being. The Stockholm conference declaration recognized the difference between renewable and non-renewable resources and also the need to manage them.[6] It encouraged policy that reduced pollution and promoted environmental research, education, and planning to reconcile the conflict between the need for development and environmental quality. The declaration also recognized that many environmental issues are global in nature and can only be addressed through multilateral cooperation among nations. Much of the language and thinking that comprises the notion of sustainability first emerged from this conference in Stockholm.

The Brundtland Definition

The further integration of development with environment as the root of sustainability emerged in 1983 when the UN formed the World Commission on Environment and Development. Gro Harlem Brundtland, then Prime Minister of Norway, was appointed to chair the commission, which became widely referred to as the Brundtland Commission. The publication that was generated by the Commission, *Our Common Future,* or the Brundtland Report as it is commonly known, addressed the conflict between the unfettered promotion of globalized economic growth and the accelerating ecological degradation that was occurring on a global scale. It made the case for a more sustainable development, which it defined as "the kind of development which meets the needs of the present without compromising the ability of future generations to meet their own needs."[7]

The Brundtland Report prioritized meeting the basic and essential needs of the world's poorest people, such as access to clean air and water, nutritious food, education, equitable economic opportunity, and others presented in Chapter 1. The report recognized that those needs are best met through economic growth that offers opportunity for all people. It also conveyed an understanding of the limits on the environment's ability to meet both present and future needs, and it recommended that if development is to be sustainable, these limitations must be addressed in order to preserve intergenerational equity. The report proposed long-term environmental strategies for achieving this.

The Brundtland Report was the first to articulate what is commonly referred to as the three pillar model of sustainability. Also referred to as the three E's, the three main pillars of sustainability and sustainable development include *e*conomic growth, *e*nvironmental protection, and social *e*quality. Typically, this is graphically represented as a Venn diagram with three intersecting circles representing environmental, economic, and social components (Figure 2.2). It is argued that development is only sustainable at the intersection of those circles. Historically, economic growth has been the priority for development, and the other two pillars have been neglected, especially environmental protection. There was considerable agreement on this conceptual framework that was put forth by the Brundtland Commission; however, translating this view of sustainability into effective initiatives and actions with measureable outcomes has been and continues to be a challenge.

The Earth Summit

Twenty years after the first global environment conference held in Stockholm, the UN sought to help governments rethink economic development and find

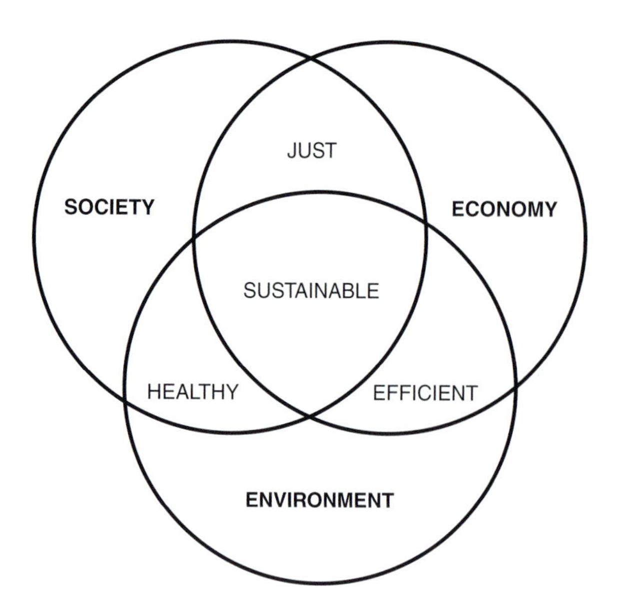

Figure 2.2 The three pillar model of sustainability from the Brundtland Commission, 1987.

ways to halt the depletion and destruction of natural resources and pollution of the planet. To address this, the UN Conference on Environment and Development (UNCED), or Earth Summit, was held in Rio de Janeiro, Brazil, in June 1992. It brought together leaders from 172 countries, 2,400 representatives from international NGOs, and more than 17,000 broader participants. The summit agreements included (1) Agenda 21,[8] a comprehensive program for global action in all areas of sustainable development; (2) the "Rio Declaration on Environment and Development,"[9] a series of principles defining the rights and responsibilities of states; and (3) "The Statement of Forest Principles,"[10] a set of standards to guide the sustainable management of forests worldwide. It was also the first convention to specifically address global climate change and the protection of biological diversity through legal conventions.

The message of the Earth Summit was that in order to achieve sustainable development, a transformation of our attitudes and behavior must occur. This message reflected the complexity of the problems of poverty and the impact of excessive consumption by affluent populations. It urged the development of international and national plans and policies to ensure that all economic decisions fully took into account any environmental impact. Priorities included (1) making eco-efficiency a guiding principle for business and governments; (2) reducing the production of toxic components, such as lead in gasoline, or poisonous waste; (3) the development of alternative sources of energy to replace the use of fossil fuels, which are linked to global climate change; (4) the development of public transit systems to reduce vehicle emissions and the health problems caused by polluted air and smog; and (5) directly addressing the growing scarcity of water.

At the Earth Summit, the case was made that if development is to be sustainable, the decision-making processes regarding that development must be open,-accountable, and participatory. Agenda 21 was the most comprehensive plan of action ever sanctioned by the international community. The Earth Summit influenced all subsequent UN conferences that then focused on the relationship between human rights, population, social and economic development, and environment—or what is now recognized as **sustainable development**.

Another outcome from the emergence of sustainable development has been the concept of **ecological footprint**. An ecological footprint is a quantitative measure of human demand on nature that is now in wide use by scientists, businesses, governments, agencies, institutions, and individuals who want to monitor ecological resource use and environmental impact to adjust activities, inform decisions, and guide sustainable development. It can be applied to specific resources, such as in the case of water or energy footprints, and at a variety of scales. We examine this and other indices of development, resource use, and environmental and social impact in later chapters.

The UN Millennium Project

After the Earth Summit in Rio, much of the focus by the UN on sustainable development was within the context of the Millennium Development Project, which got its start at the 2000 Millennium Summit. The declaration produced at this summit stated that every effort must be made to counter the irreparable damage caused by human activities that threaten our planet and our people. The participants of the summit resolved to adopt a new ethic of conservation and stewardship.

The Millennium Project was commissioned by the UN Secretary-General in 2002 to develop a concrete action plan for the world to achieve specific development goals. The Millennium Development Goals

(MDGs) were to (1) eradicate extreme poverty and hunger; (2) achieve universal primary education; (3) promote gender equality and empower women; (4) reduce child mortality; (5) improve maternal health; (6) combat HIV/AIDS, malaria, and other diseases; (7) ensure environmental sustainability; and (8) develop a global partnership for development.

In 2005, the independent MDG advisory body headed by economist Jeffrey Sachs presented its final recommendations to the Secretary-General in a synthesis report, "Investing in Development: A Practical Plan to Achieve the Millennium Development Goals."[11] This report outlined specific objectives for each goal to be achieved by 2015, and it detailed country-level processes and ways that the international community could support them to achieve the goals. By 2015, there was much success toward reaching these goals in what the UN declared as the most successful anti-poverty movement in history. We examine both the challenges and successes in achieving the MDGs in each chapter that relates to specific goals. In Chapter 12, we more closely examine the MDGs within the broader context of sustainable development.

The Sustainable Development Goals

Each of the MDGs reflects what we have already determined to be important elements of sustainability and sustainable development. Recognizing this in developing the post-2015 successor of the MDGs, the UN removed sustainability as a single goal focused on the environment and made it the overarching theme. Dubbed the **Sustainable Development Goals (SDGs)**, this newer program is more formally known as "Transforming Our World: The 2030 Agenda for Sustainable Development." Its 17 global goals include ending poverty and hunger; improving health and education, cities and communities, and infrastructure; protecting forests and oceans; combatting climate change; and fostering peace, justice, and strong institutions. Again, we should incorporate the values expressed in this agenda in our definition of sustainability. The SDGs and how to achieve them are central to our discussion of sustainable development in Chapter 12.

The Emergence of Sustainable Business

Another positive consequence of the Brundtland Report was the emergence of the concept of sustainable business. The report recognized that sustainable development could not be achieved through government regulations and policies alone, and it argued that industry had a significant role to play. Although it was clear that corporations drove economic development, the Brundtland Report emphasized that they needed to be more proactive in balancing their emphasis on profit with social equity and environmental protection.

The sustainable business community adapted the three pillar model of sustainability in what is referred to as the triple bottom line of planet, people, and profit. In his 1997 book *Cannibals with Forks: The Triple Bottom Line of 21st Century Business*,[12] John Elkington made the case that businesses can and must link the goals of profit with fair and beneficial business practices toward labor and the community while minimizing environmental impact. Corporate sustainability also has come to include the participatory model of development argued for in Rio by more broadly identifying corporate stakeholders not just as investors and shareholders but to include employees, customers, suppliers, and members of the communities in which they operate.

Sustainability has often been considered a burden on business profits. However, a report from *Harvard Business Review*[13] suggests that considering the other two "bottom lines" can lower costs, open new markets, and stimulate innovation. Some have criticized the triple bottom line as an ineffective way to enhance social conditions and preserve the environment, and that profit in this model does not consider full cost accounting or externalities such as impacts on the environment and public health. Another issue is that there are plenty of examples of businesses that have abused the label of sustainability as a way to falsely market their product as environmentally friendly or as promoting social justice. We more closely examine the idea of sustainable businesses in Chapter 10, and we consider the role of innovation and entrepreneurship in achieving sustainable solutions throughout the book while being mindful of their potential for abuse or detracting from sustainability principles.

Redefining Sustainability Within a Scientific Framework

Independent of the international development and business arenas, the scientific community has developed its own definition of sustainability. Scientists base their definition on the rate of resource use in relation to the rate of resource renewal. They have thus defined sustainability as a state in which humankind

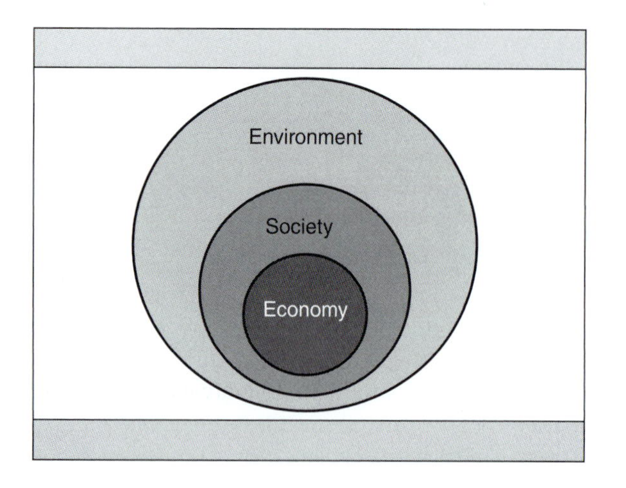

Figure 2.3 The concentric or three-nested-dependencies model reflects human society's complete dependence on the environment and states that without food, clean water, fresh air, fertile soil, and other natural resources, society—and therefore economic opportunity—will not exist. (Adapted from Doppelt, *The Power of Sustainable Thinking*.)

extracts natural resources at rates that do not exceed our capacity to discover replacement or substitute resources. Implicit in this is that we reuse those resources as much as possible and dispose of wastes that do not exceed the capacity of natural systems to assimilate and neutralize those wastes. Ways of achieving this often focus on a global strategy of maintaining human populations that will allow for the previously mentioned conditions.

A formal approach to further develop the scientific definition of sustainability was led by Karl-Henrik Robèrt[14] and his organization Natural Step, a non-profit group founded in Sweden in 1989. In a diverse participatory approach, Robèrt developed a framework that lays out the system conditions for sustainability that are based primarily on the laws of thermodynamics. His group developed four conditions for sustainability. He refers to them as first-order scientific principles that are required to achieve a sustainable society. The first three principles state that the following cannot be systematically increased over time: (1) the amount of substances extracted from the Earth's crust, (2) the amount of waste produced as a by-product of society, and (3) the rate of degradation of the planet by physical means. A fourth principle is that in society, people should not be subject to conditions that systematically undermine their capacity to meet their needs. Rather than three intersecting circles of economy, environment, and society Natural Step represents sustainability with inclusive concentric circles employing a nested dependency model (Figure 2.3).

This framework bases its approach to strategic planning and problem solving on the concept that is referred to as *backcasting* from principles. **Backcasting** begins planning with the end in mind rather than projecting forward current conditions (Figure 2.4). It is the act of envisioning ourselves in the future as having met the principles of sustainability and asking the question, "What do we need to do today to reach that successful outcome?"

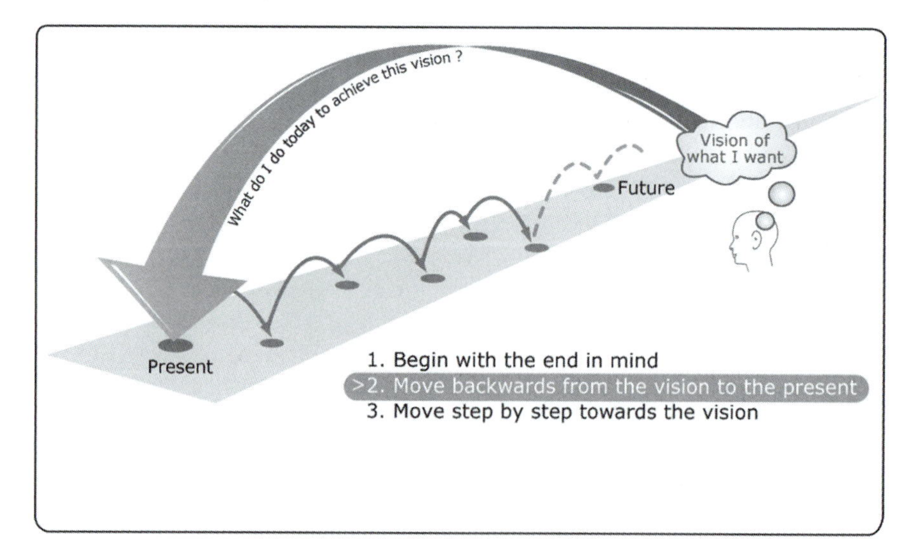

Figure 2.4 Backcasting, or planning with the end in mind. (Adapted from The Natural Step http://www.naturalstep.ca/backcasting)

An example of using backcasting principles is the development of greenhouse gas (GHG) emissions reduction goals based on a desired future outcome determined by scientific estimates on what that reduction should be to stop or reverse climate change. This is opposed to the current **forecasting** policy approach. Forecasting sets future goals by extrapolating current practice into the future and making adjustments based on that. For example, we typically extend per capita GHG production to projected future population sizes and then make adjustments to reduce future emissions without a specific end goal in mind other than to correct for increasing populations. It is believed that the backcasting approach will lead to more innovative, transformative results, whereas the goals based on forecasting will at best achieve only incremental change. One of the challenges with backcasting is achieving consensus on a desired future scenario. This can best be achieved through a participatory, stakeholder approach that is well informed by scientific data.

The Natural Step approach to strategic sustainable development is to rely on backcasting from scientifically based sustainability principles that represent something that a diverse group of stakeholders endorses. If these principles are violated, our global society is viewed as unsustainable. These principles have been adapted in Sweden and the United States to develop what are referred to as **eco-municipalities**. These are local communities and governments that have adopted ecological and social justice values in their charters.

LINKING SOCIAL AND PLANETARY BOUNDARIES

In 2012, economist Kate Raworth with Oxfam International developed the doughnut model of sustainability, which links scientific thinking with social condition.[15] This model integrates the nine planetary boundaries presented in Chapter 1 with the top social priorities identified by the world's governments leading up to the Rio Earth Summit. Resource use that exceeds the planetary boundaries threatens Earth's life support systems, whereas resource use that causes us to fall below social thresholds such as poverty, quality education, health, and gender equality results in deplorable human conditions. The model defines sustainability as the space between planetary and social boundaries—the doughnut—where environmental and social conditions are safe and just for humans. Raworth argues that this can be best achieved through a regenerative and distributive economy (Figure 2.5.)[16]

SYSTEMS THINKING

There is a predominant view that sustainable problem solving is best achieved through **systems thinking**. This is a holistic approach that focuses on the linkages and interactions between elements that compose an entire system, including social and environmental conditions as described previously. This is in contrast to more traditional, reductionist ways of studying processes as independent, causal chains of events (Figure 2.6). For example, rain can be viewed as a linear chain of events starting with the evaporation of water from the land resulting in more moisture in the air. This in turn results in cloud formation and eventually rain, suggesting that the ultimate cause of rain is evaporation with cloud formation as an intermediate. This event-oriented thinking is more simplistic, and it may cause us to miss how different parts of a complex system interact and influence each other. This can be remedied through systems thinking.

In systems thinking, we take into account the overall system as well as its parts. This elucidates how individual parts or stages in a system will act when isolated from each other or from the system's environment. It also reveals how each individual part can influence others either directly or indirectly through positive and/or negative feedback loops. Systems thinking allows us to see the whole as more than just the sum of its parts. It shows us that as the parts interact; novel properties emerge that give the system unique characteristics. These **emergent properties** are common to complex systems.

A systems approach to our rain example would view the causes of rain as a cyclical series of feedback loops that influence each other (see Figure 2.6). For example, rising temperatures may increase evaporation through a positive feedback loop that depletes the amount of surface available to evaporate. Increased evaporation and the movement of clouds through broader weather patterns could result in an overall loss of water from the system. This in turn could change the environment, resulting in further increases in temperature and even more evaporation, like what occurs in the process of desertification. Human actions, such as clear-cutting a forest or the introduction of irrigation-intensive agriculture, can also be added to the model to predict their influence on surface water, evaporation, and rain.

The holistic, interactive understanding of systems thinking can also provide insight about the ways in which we might change systems to better meet

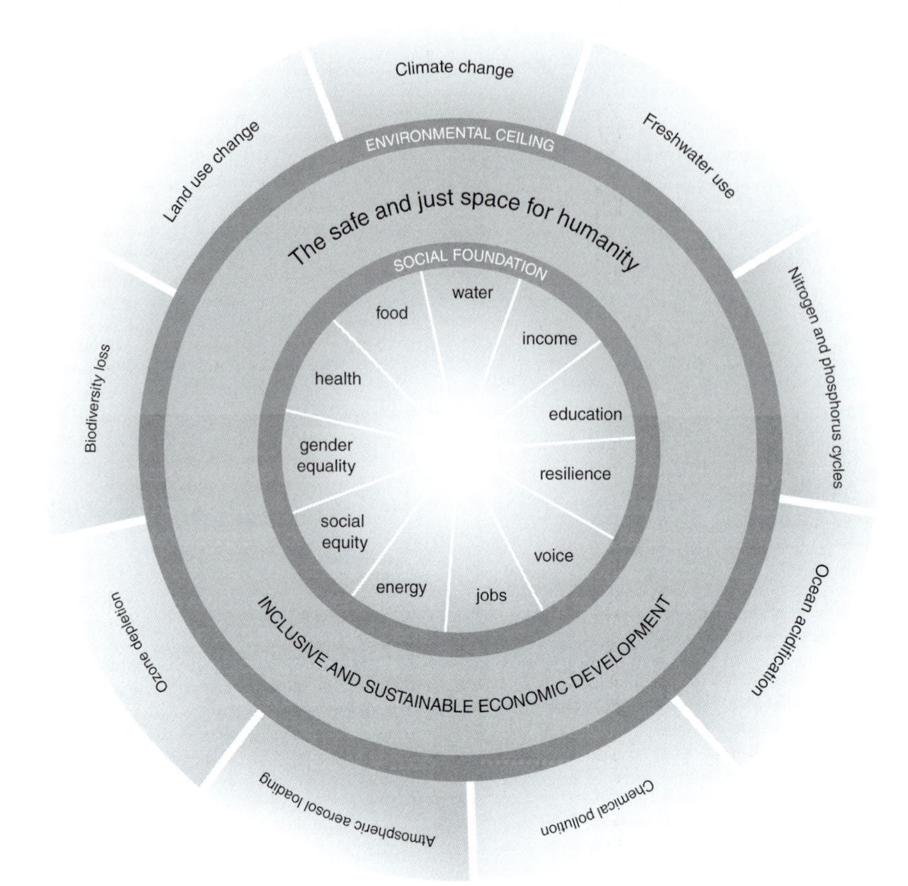

Figure 2.5 Integrating the 9 planetary boundaries with 11 top social priorities. Resource use that exceeds the planetary boundaries threatens the Earth's life support systems. Resource use that moves us below the social thresholds results in unacceptable human deprivation such as hunger, disease, and poverty. In the space between the planetary and social boundaries (the doughnut) lies an environmentally safe and socially just space in which humanity can thrive and sustainability is achieved. (Adapted from Raworth, K. 2012. A Safe and Just Space for Humanity, Oxfam International, p. 4, https://www.oxfam.org/sites/www.oxfam.org/files/file_attachments/dp-a-safe-and-just-space-for-humanity-130212-en_5.pdf)

sustainability objectives. In addition, it can help us anticipate unintended consequences from our actions. For example, a study in the tropical savanna region of Australia used systems thinking to predict how complex decision-making about land use might influence tree density.[17] By sharing and integrating different forms of knowledge from a diversity of stakeholders, scientists were better able to predict outcomes from various management actions under a variety of conditions. These interacting influences are best visualized by mapping all of the root causes and how they are linked (Figure 2.7). The links or causal relationships between different influences are indicated in Figure 2.7 by arrows. Quantitative data for each factor—for example, cattle pasture consumption—and knowledge about how factors interact are then used to generate predictions in these complex systems.

Life cycle analysis (LCA), also known as life cycle assessment, is often integrated with systems thinking in the process of sustainable design. LCA examines the impacts throughout the life cycle of a product, from resource extraction and construction to use and, ultimately, disposal. When systems thinking and LCA are used together, opportunities for innovation can emerge that reduce negative environmental and social impacts of what is being designed. This process defines the problem by first examining the whole system. By then examining life cycle impacts, the designer can

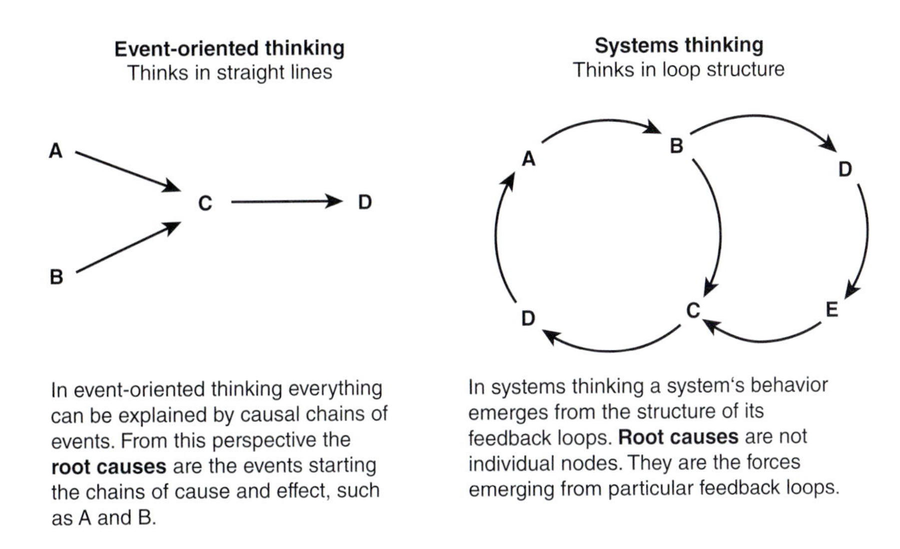

Figure 2.6 Event-oriented or linear thinking explains phenomena using causal chains of events. In systems thinking, root causes are the emergent properties or forces from specific interactions or feedback loops. (Adapted from http://www.thwink.org/sustain/glossary/EventOrientedThinking.htm)

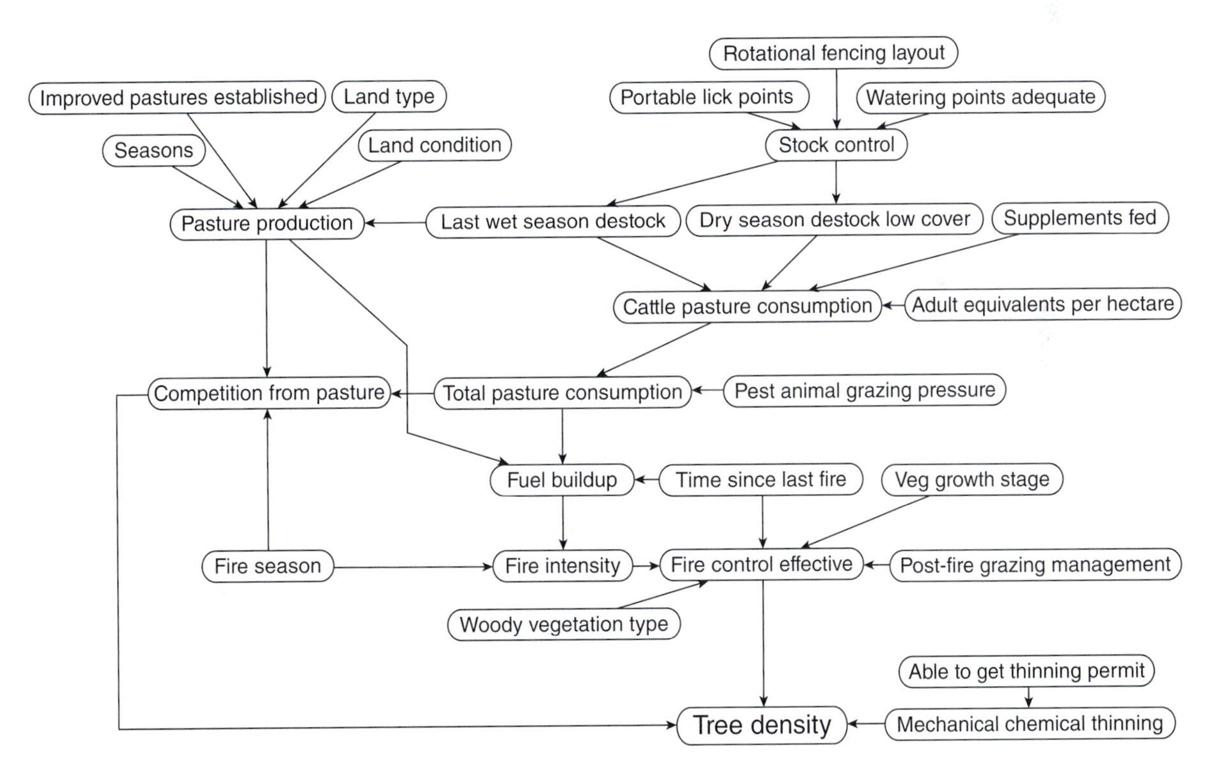

Figure 2.7 A systems thinking approach to predicting how multiple factors interact to influence tree density on an Australian Savannah. Arrows indicate linkages among factors. Probability data for each factor can then be incorporated into the systems model to predict tree density. (Adapted from Bosch, O. J. H., C. A. King, J. L. Herbohn, I. W. Russell, and C. S. Smith. 2007. "Getting the Big Picture in Natural Resource Management—Systems Thinking as 'Method' for Scientists, Policy Makers and Other Stakeholders." *Systems Research and Behavioral Science* 24 (2): 217–32. doi:10.1002/sres.818)

prioritize objectives to develop solutions that optimize conflicting factors within the entire system.

As an example of integrating systems thinking with LCA, we could apply it in the design of a more energy-efficient way to dry clothes. Initially, one might think that the best solution would be to make the clothes dryer itself more efficient. But by considering the product within the context of the whole system—that is, clothes get dirty, are washed, and are dried—and all of the life cycle consequences of each stage (Figure 2.8),

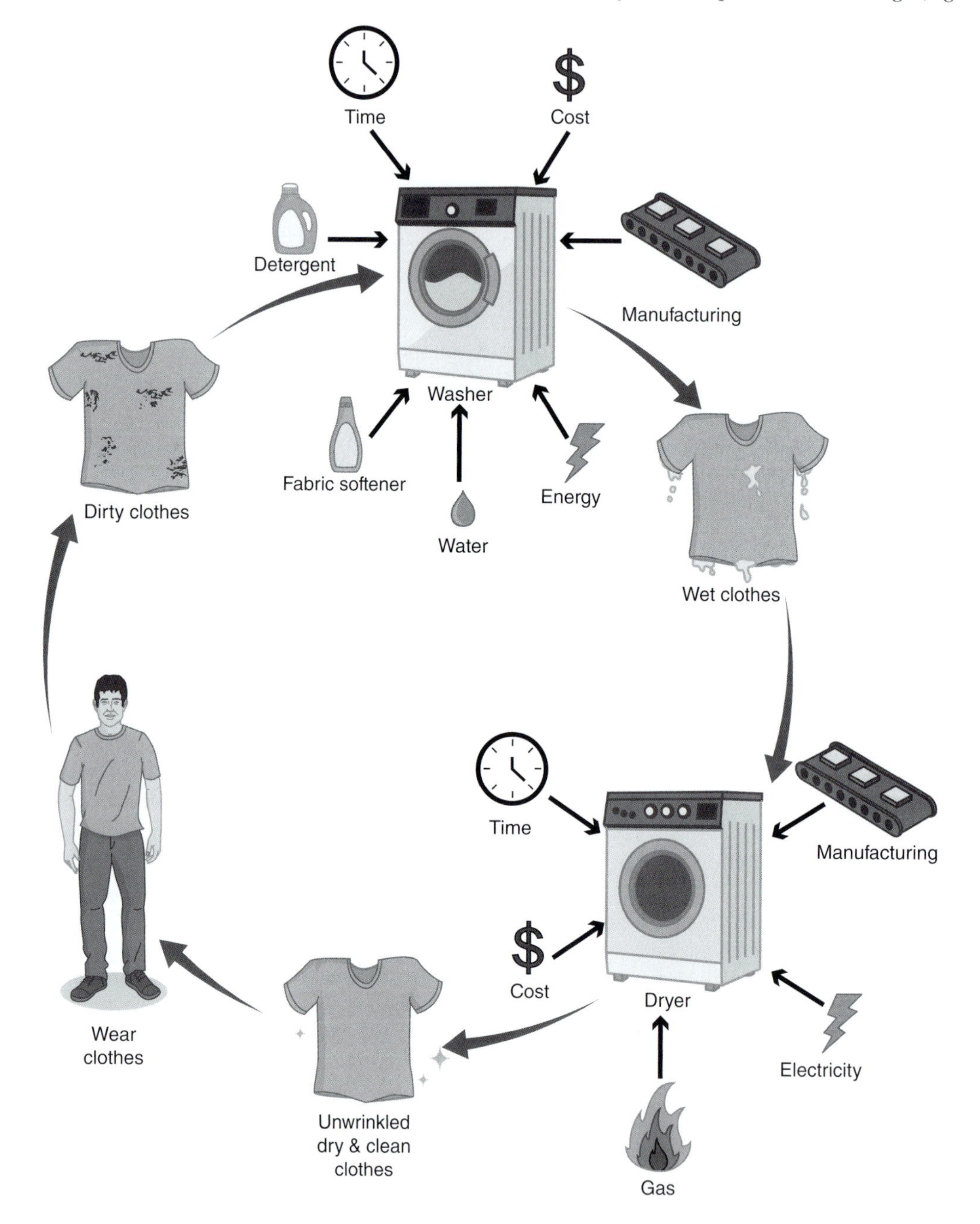

Figure 2.8 Linking systems thinking with LCA allows for a more complete understanding of the clothes washing system and provides greater insight into how to define problems and develop solutions to increase the efficiency of the clothes drying process. (Adapted from Autodesk® Sustainability Workshop, https://sustainabilityworkshop.autodesk.com and http://slideplayer.com/slide/5802549/)

a different solution emerges. With this systems LCA approach, it becomes evident that using a slightly more energy-intensive washing machine that provides clothes that have less moisture to the drier would save more energy (Figure 2.9). Other possible solutions to this problem also emerge, such as replacing the dryer with more conveniently used clotheslines or drying racks. Systems thinking and LCA allow us to move beyond making small improvements to one part of a process to generate a more effective solution for improving an entire process.

NATURE-BASED SOLUTIONS

A more recently emphasized approach to sustainable problem solving is the development and implementation of **nature-based solutions (NbS)**. This approach emerged from the recognition of the potential of nature to provide or inspire the development of solutions to many of the previously discussed societal challenges. These solutions provide simultaneous environmental, social, and economic benefits consistent with our definition of sustainability, and they are often more affordable than other types of solutions.[18] They rely on natural ecosystem, landscape, and ecological functions to achieve sustainability objectives. These solutions are achieved through ecosystem protection, management, and restoration; the "re-naturing" of cities and developed regions; and the development of natural infrastructure and encouragement of natural processes.

An example of an NbS is the protection of coastal mangrove habitats in order to limit flooding and damage to coastal communities from extreme weather and rising sea levels resulting from climate change. An alternative solution might be to construct a large sea wall. But in this case, the NbS performs the function of the wall while also supporting local fisheries, sequestering atmospheric carbon dioxide (CO_2), and preserving biodiversity. Another example of an NbS is the encouragement of natural soil processes in sustainable agriculture. An example of natural infrastructure is a planted green roof that insulates and cools buildings, manages rainwater, reduces air pollution, sequesters CO_2, increase biodiversity, and provides increased opportunities for people to connect to nature.

An NbS takes advantage of **ecosystem services** or the many ways in which ecosystems benefit humans. Examples include supplying clean water, providing pollinators for crop pollinations, and climate regulation and flood control. NbS are also often based on the principle of **biomimicry**, an approach to innovation that seeks sustainable solutions through design that emulates nature and natural processes. Examples of biomimicry include the design of climate-controlled buildings based on the structure of self-cooling termite mounds or designing sustainable agricultural systems based on the function of natural prairies.

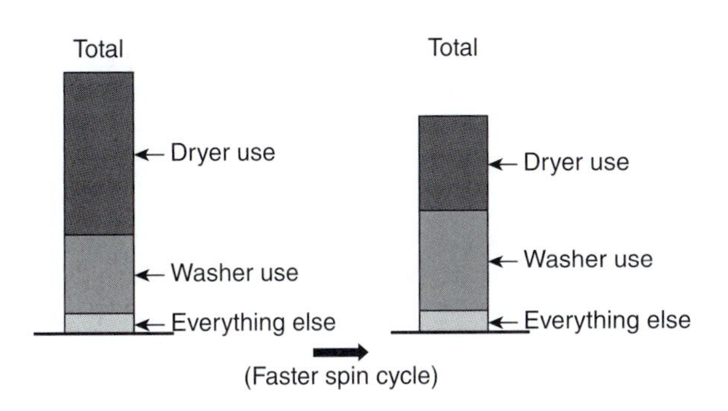

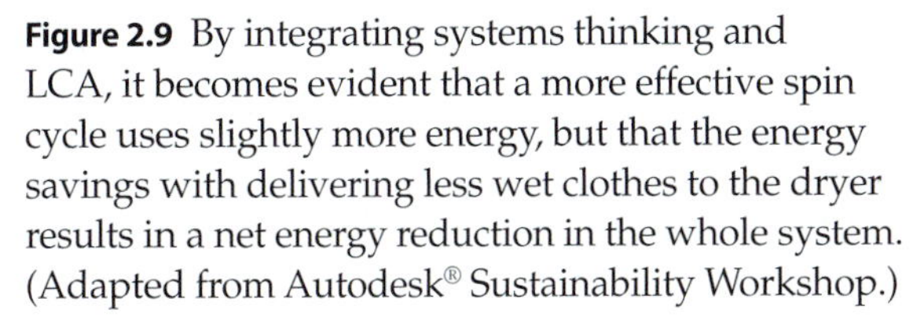

Figure 2.9 By integrating systems thinking and LCA, it becomes evident that a more effective spin cycle uses slightly more energy, but that the energy savings with delivering less wet clothes to the dryer results in a net energy reduction in the whole system. (Adapted from Autodesk® Sustainability Workshop.)

Sustainable Communities

Consistent with the concept of eco-municipalities, some have argued that the unit or level at which sustainability can best be achieved is the community level. There are several definitions of a "sustainable community." A sustainable community seeks to maintain and improve the economic, environmental, and social characteristics of an area so its members can continue to lead healthy, productive, enjoyable lives.[19] In a sustainable community, resource consumption is balanced by resources assimilated by the ecosystem. The sustainability of a community is largely determined by the availability of resources for things such as food, fiber, water, and energy needs and by the ability of natural systems to process its wastes. A community is unsustainable if it consumes resources faster than they can be renewed, produces more waste than natural systems can process, or relies on distant sources for its basic needs.[20] Sustainability at community and regional scales is considered in Chapter 11.

Indigenous Views on Sustainability

The evolution and conceptualization of sustainability presented here, although global, has been largely Euro-centric. It is important to recognize that many indigenous societies have long embraced the idea of sustainability well before the conception of the ideas thus far presented. For example, the concept of meeting the needs of the present without compromising the ability of future generations to meet their own needs is consistent with the Great Law of the Iroquois that mandates that the decisions they make today should benefit their children seven generations into the future.

It should be recognized that some critique models of sustainability that have evolved from more traditional approaches and have been modified to fit current consumptive ways of living. For example, Robin Kimmerer speaks to how the Potawatomi Nation, of which she is a member, views the modern concept of sustainability as a mere modification of the way we consume or take from the planet. She argues that the focus of sustainability should be on what we can give to the Earth and the practice of gratitude for what the planet provides in what she refers to as "the Honorable Harvest." As we develop our definition of sustainability and work with diverse groups, we should be aware of these critiques and sensitive to the different cultural views that people hold of their relationship with the environment.

Religion and Sustainability

Worldwide, more than 80% of people identify with a religious group. The notion of sustainability is rooted in values that are consistent with those of many religions. Thus, religious institutions have a large potential role in furthering sustainability and sustainable development. At the Earth Summit, religious scholars argued for a global ethic grounded in a kinship with other humans and non-human entities through religion. Most, if not all, of the world's religions have the central underpinnings of sustainability that we have been discussing written into their doctrines. Historically, they have been at the forefront of social justice, including supporting and reforming labor movements, immigration law, civil rights, protection of refugees, and justice for the poor and oppressed, among other important issues.

Christians believe that caring for the world is a display of respect for its creator. Muslims believe that care of animals, other people, and the environment is central to their obedience and gratitude toward Allah. The Torah, Kabbalah, and other important Jewish texts make a strong case for an environmental ethic and social activism, including the concepts of *bal tschchit* and *Tikkun olan*, which emphasize acts of kindness in order to perfect or repair the world. Eastern religions emphasize moderation in consumption. Buddhism, Hinduism, the B'hai, Confucianism, and Taoism all teach that moderation and restraint are the path to happiness and success. Accordingly, early American environmentalist John Muir was called "the Taoist of the West." Sustainability has been central to the Quaker religion because simplistic, just, and peaceful living are at its core. Other religions not mentioned here also embrace many of these same values.

Despite these shared values, religious institutions have not always acted in ways that furthered the values of sustainability that we have been discussing. In fact, their efforts have often contradicted these principles. Religious texts can be and have been interpreted in a variety of ways, some of which favor exploitation of the Earth over sustainable stewardship. Intolerance and exclusive claims to truth have contributed to tensions between peoples, including wars or forced conversion. Although missionary movements have had the good intention to serve elements of sustainable development, this has sometimes been imposed on people with the intent of conversion and forced cultural change. Emphasis on afterlife, objections to grappling with population growth, and the avoidance of other political issues have also hindered the role of religion in dealing with environmental and some social issues. The practices of many fundamentalists sects have contradicted the principles that we have been developing. Finally, there is often a disconnect between doctrine and the action of those who participate in religion. Religious institutions can and should have an important leadership role in translating their teachings and traditions into actions in order to improve the human and environmental condition.

Environmental Justice

At the same time that the inclusion of social justice with sustainability had been occurring, the environmental justice movement emerged. In the 1970s, sociologist Robert Bullard began to recognize that in the United States, landfills, incinerators, and other activities with negative environmental and health consequences had been disproportionately placed in communities of color and poverty. Stemming from Bullard's work, the environmental justice movement embraces the principle that all people and communities are entitled to equal protection under our environmental laws. The movement also integrates environmental justice with basic civil and human rights.

The social justice movement fathered by Bullard in the United States led to the full integration of social justice in global sustainability in what Julian Agyeman referred to as **just sustainabilities**. In a book co-edited by Bullard, *Just Sustainabilities: Development in an Unequal World*,[21] just sustainabilities are defined as the need to ensure a better quality of life for all, now and into the future, in a just and equitable manner while living within the limits of supporting ecosystems. This served to codify the issue of equitably improving the human condition across all peoples as central to sustainability and sustainable development. Agyeman promoted the concept of spatial justice, which links together social justice and space in a way that has caused us to think about making communities more sustainable by integrating environment, justice, cultural diversity, and democracy. We explore this further as we consider sustainable communities in Chapter 11.

A Working Definition of Sustainability

In this section, we work through an exercise to develop a definition of sustainability that we will then use to apply to a variety of problems throughout the book. It is based on the values espoused as the concept emerged in the history described in the previous sections. We essentially deconstruct the ways in which the word "sustainability" has been used throughout the years, and we extract the common principles. We then incorporate those principles into a definition that can be applied to establish and achieve desired results or to analyze and critique current practice. This is a useful exercise that can be done by any organization or class with sustainability objectives. The list of values can come from existing definitions, history, or an organization's vision or objectives. Thus, the definition that is derived can be tailored to meet more specific needs as long as it represents the common values and principles that emerged with the sustainability and environmental movements. Our goal here is to develop a more general definition that incorporates these values which can then be used to create a specific plan or set of objectives that can be assessed when implemented. The definition can also be applied to assess an existing entity or program.

In order to develop a practical definition of sustainability, we must first determine to what it is that we are applying the definition—that is, what we are saying meets the conditions for sustainability or what is sustainable. For simplicity, let's call that "X." So our definition could start with the statement,

> *X is sustainable if it meets the following conditions . . .*

We can easily identify a number of things that are intended to be sustainable that we could substitute for *X* as we apply our definition. These could include processes or activities such as development, resource use, food production, or ecosystem management. Another category includes implemented programs or projects, for example, in poverty reduction, renewable energy projects, or in health care or education reform. A third category includes organizational entities such as institutions, businesses, communities and regions, and even households or individuals. *X* could also be processes or objects designed and made by humans, such as products, buildings, technology, or innovations. We might also think of *X* as potential solutions to any problems presented in this book.

The next step is to determine the set of conditions or principles that will make the consequences of a particular entity sustainable. To do this, we can return to history to examine what has been valued or determined to be the conditions for sustainability. Table 2.1 lists the various criteria or conditions for sustainability that have been established as the concept evolved over time. Each is listed with its origin(s) and overarching category.

In our definition, we will try to represent each of these values. We also make central the Brundtland definition because of its importance in the sustainability movement, but here we add to it to make it more robust and inclusive of the other principles that have emerged. We also reword it to extend "meeting the needs of current generations" to include increasing opportunity for them to meet their own needs. It is also important to emphasize the approaches by which the conditions are determined. For example, they should be based on science, systems thinking, or consensus among stakeholders. Also, the definition is only useable if the conditions lead to measurable outcomes that

TABLE 2.1 Conditions for Sustainability and from Where They Were Drawn

CATEGORY	VALUE OR CONDITION FOR SUSTAINABILITY	WHERE ARTICULATED
Approach	Deal with global problems through multilateral cooperation	STOC, BRUN, ES, MDG, SDG
Approach	Participatory/inclusive stakeholder approach	ES, SCI, CS, SB, SDG
Approach	Use backcasting principles	SCI
Approach	Use systems thinking	SCI
Economic	Economic development	BRUN, ES, SB, CS
Economic	Equitable economic opportunity	BRUN, SB, SCI, CS, SDG
Economic	Profit	SB
Economic	Do not undermine people's capacity to meet their own needs	SCI, CS, MDG, SDG
Environment	Address global climate change	ES, SDG
Environment	Conserving resources	TR, SCI, CS, SDG
Environment	Develop public transportation systems to reduce vehicle emissions	ES
Environment	Eco-efficiency	ES, SB, SCI, CS
Environment	Effective scientifically based resource management	CON, STOC, BRUN, ES, SCI, MDG, B, SC, IND
Environment	Improve environmental research and education	STOC
Environment	Increase renewal and renewable resource use	CON, STOC, SCI, CS, SDG
Environment	Effective environmental management	CON, BRUN,ES, MDG, SB, SCI, CS
Environment	Environmental preservation/protection	TR, BRUN, ES, REL, MDG, SDG
Environment	Maximize ecosystem productivity	CON
Environment	Protection of biological diversity	TR, ES
Environment	Reduce disposal of waste	SCI, CS
Environment	Reduce the production and use of toxic chemicals, and pollution	STOC, ES, MDG, SCI
Environment	Replace fossil fuels with alternative energy to reduce GHG emissions	ES, SDG
Environment	Watershed protection	CON
Environment	Do not degrade the planet	CON, ES, SCI, MDG
Human	Address water quality and scarcity	BRUN, ES, MDG, SDG
Human	Combat disease	MDG, SDG, SC
Human	Increase access to and improve education	BRUN, TR, CS, MDG, SDG
Human	Reduce poverty and improve quality of life	TR, BRUN, ES, REL, MDG, CS, SDG
Human	Human rights, elimination of slavery	TS, ES, REL, MDG, SDG
Human	Improve maternal health	MDG
Human	Improved nutrition/less hunger	CON, BRUN, MDG, SDG
Human	Labor reform	TR, SDG
Human	Manage population growth	ES, SCI
Human	Multiple use	CON
Human	Promote women's rights, gender equality, and empower women	TR, MDG, SDG

CATEGORY	VALUE OR CONDITION FOR SUSTAINABILITY	WHERE ARTICULATED
Human	Increase recreation and spiritual opportunities in nature	TR, CON, IND
Human	Reduce child mortality	MDG
Human	Reduce consumption	ES, SCI, REL, SDG
Human	Social development	BRUN, ES, REL, MDG, SB
Human	Social justice	TR, BRUN, ES, MDG, SDG, SB
Human	Spiritual connection	TR, REL, IND
Overarching	Integration of environmental protection, development, and social well-being	STOC, BRUN, ES, SB, SC, SDG
Overarching	Intergenerational equity	CON, STOC, BRUN, ES, MDG, SDG, IND
Overarching	Meet the needs of humans	CON, STOC, BRUN, ES, MDG, SDG, CS, SC, IND

BRUN, Brundtland definition; CON, conservationist movement; ES, Earth Summit; IND, traditional, indigenous, or principles; MDG, UN Millennium Development Goals; REL, religious doctrine; SB, sustainable business; SCI, scientific definitions; SDG, UN Sustainable Development Goals; STOC, Stockholm; TR, transcendentalist/preservationist movements.

can be assessed. When we include all of these ideas, a resultant definition is as follows:

> *X is or will be sustainable if it meets the needs of current generations or increases their opportunity to meet their own needs without compromising the ability of future generations to do so. This is best achieved using a participatory, inclusive process based on scientific, system thinking, and backcasting principles that ensures:*
>
> *1. Extraction and use of resources that maximizes renewal, encourages reuse, minimizes waste, and protects the environment, including conservation and preservation of natural systems, reducing pollution, and slowing or reversing global climate change;*
>
> *2. Economic development and equitable economic opportunity including profit that does not undermine the capacity of people to meet their own needs; and*
>
> *3. An elevated standard of human well-being for all people, including but not limited to improved health, improved nutrition and access to clean air and water, increased access to education, and other basic human rights.*
>
> *The sustainability of X is best assessed through the development of measureable indicators that show improvement in each of these criteria.*

Some may argue that what we just went through is an arbitrary process. But it is not. It is rooted in context of the existing global problems described in Chapter 1, and it is based on the values of the environmental and sustainability movements that emerged in response to those problems. However, it is not a perfect definition, and readers are urged to modify it and develop a definition that might better suit their needs without compromising the fundamental principles of sustainability. When applying the definition, it is also important to differentiate between "needs" and "wants" within the context of consumption.

One of the problems with defining sustainability is that it might be easier to label something as *not* sustainable. Also, there is no quantifiable threshold of sustainability, and it cannot be represented with a single number. However, a set of values can be obtained and compared to baseline data, standards, and stated objectives, and one can look for a positive trend. In this way, sustainability is not an endpoint; rather, it is a trajectory for which we can assess whether there is a positive trend. As demonstrated later through examples, the previous definition lends itself to developing specific objectives for which indicators can be developed. These indicators can then be used to assess the degree to which the entity or process being assessed is moving positively in the direction of sustainability by meeting specific criteria. It can be used to examine specific approaches and determine which aspects of them meet the conditions of sustainability and which can be improved.

Implementing and Assessing Sustainable Solutions

Let's test the utility of our definition by applying it to two mini-case studies. The first is the state of Massachusetts wind energy initiative, and the second is an ecotourism project in the Nam Ha National Protected Area in the Lao People's Democratic Republic (PDR). We apply our working definition to each case to test its usefulness in assessing to what extent the project goals reflect the principles of sustainability. As we work though this exercise, we will introduce some important terms in this process, and we will offer their definitions in the following section.

Case 1: Massachusetts Wind Energy Initiative

The Commonwealth of Massachusetts has established a goal to install 2,000 megawatts (MW) of wind energy by 2020. Wind turbines would be installed both on- and offshore with the objectives of reducing fossil fuel dependence for electricity generation, reducing GHG emissions, and creating economic development and green jobs. So let's adapt our definition to this project and see if it helps us assess the sustainability of this project, determine its strengths, and perhaps identify areas that could strengthen its sustainability objectives or identify additional information that could be helpful. In this case, *X* is the Massachusetts wind energy initiative, allowing us to apply the definition as follows:

> *The Massachusetts wind energy initiative will be sustainable if it meets the needs of current generations or increases their opportunity to meet their own needs without compromising the ability of future generations to do so. This is best achieved using a participatory, inclusive process based on scientific, systems thinking, and backcasting principles that ensures:*
>
> *1. Extraction and use of resources that maximizes renewal, encourages reuse, minimizes waste, and protects the environment, including conservation and preservation of natural systems, reducing pollution, and slowing or reversing global climate change;*
>
> *2. Economic development and equitable economic opportunity including profit that does not undermine the capacity of people to meet their own needs; and*
>
> *3. An elevated standard of human well-being for all people, including but not limited to improved health, improved nutrition and access to clean air and water, increased access to education, and other basic human rights.*
>
> *The sustainability of this project will be best assessed through the development of measureable indicators that show improvement in each of the these criteria.*

So let's break down some objectives of the program that can be seen on the official website of the Commonwealth of Massachusetts (http://www.mass.gov) and determine if they are consistent with the principles in the definition. These objectives are as follows:

- Meet the Commonwealth's goals for a clean environment and a robust economy.
- Focus energy production on wind energy that is fueled by an infinitely renewable resource: moving air.
- Reduce dependency on fossil fuels that when extracted cause drastic alteration of landscapes; and when combusted release toxic pollutants and climate-changing GHGs into the atmosphere.
- Contribute to the growing clean energy industry in Massachusetts, an industry that is creating new jobs, reducing environmental impact, and decreasing dependence on imported sources of energy.
- Offer incentives for commercial, community, and micro-wind projects that have the potential to serve diverse groups.
- Minimize the negative health impacts of energy production and use. Though there may be potential health impacts of wind energy through sound and visual flickers generated by the spinning blades, these are minimal compared to the health effects of continued or expanding fossil fuel use.
- Create a wind energy working group composed of a diverse group of stakeholders representing developers; environmental groups; federal, state, and local officials; academic institutions; utility representatives; lawyers; advocacy groups; turbine manufacturers; and farming and rural interests with open membership to discuss the implementation of this program.

On the surface, it seems quite apparent that if this program is successful there will be many environmental gains. Once the 2,000 MW objective is achieved, annual CO_2 reductions are estimated to be 5.2 million tons, with water savings of nearly 3,000 million gallons per year. Overall air quality will improve, and there will be less global impact through extraction and combustion of fossil fuels.

When assessing sustainability, other potential environmental issues associated with wind turbines need to be addressed. First, wind turbines can potentially have adverse impacts on wildlife. Birds and bats may fly into the moving blades of a turbine, resulting in mortalities. So far, evidence suggests that this threat is minimal, especially compared to threats due to changes in habitat brought about by climate that will occur if the same amount of energy is produced using fossil fuels. In addition, turbines are being designed to further reduce the potential of impact, and placement with respect to migratory routes or nesting grounds is often a consideration.

Employing LCA, one must consider the resources and energy that go into making these turbines and the fate of turbines as they become obsolete. For example, the construction of wind turbines depends on rare-earth minerals mined primarily from China, where extraction has negatively impacted both the environment and public health. Finally, much of the complaints about wind energy have to do with the aesthetic impact on nature. This could have implications regarding placement of turbines that we will address later.

Massachusetts makes strong claims about the economic benefits of its wind energy initiative, including significant payments to landowners, local property tax revenue, and job creation from both construction and operation phases. The total economic benefit is expected to be $2.8 billion, with more than 900 long-term local jobs and 3,000 during the course of construction. Most jobs will require specialized training, so whether access to this economic opportunity will be equitable could be questioned. However, increased economic activity and levels of employment from this project should increase overall economic activity in service and retail areas. The local benefits could be enhanced by localizing the supply chain—that is, by increasing the use of locally produced materials and equipment required for the project.

In terms of human well-being, there are less obvious benefits. Claims for improved public health through reduced emission have been made. There is education value to the project, as a number of turbines have been installed at schools and associated curricula have been developed. One social justice issue may have to do with the placement of turbines. Many of the wealthier Massachusetts communities have fought their placement for aesthetic reasons primarily through zoning regulation both on land and at offshore sites. This could result in the placement of turbines primarily in poorer communities.

The approach taken in this program appears to be consistent with our definition of sustainability. The project and its predicted outcomes are based on science. The objective of developing a 2,000 MW system by 2020 was generated through a backcasting process in which an endpoint was agreed upon, and then a plan to achieve that goal was established. The Wind Energy Group is represented by a diversity of stakeholders, and membership is open to the public. However, the extent to which this group has been involved in the decision-making process is unclear.

So there are many aspects that make this project sustainable and others that are less apparent or may actually contradict sustainability principles. The next step is to rigorously assess this. Assessment starts with the development of indicators based on the stated goals of the project to assess whether or not these goals have been achieved and to track potential conflicts with the principles of sustainability. The latter may shed light on how the project could be more sustainable. Examples of each type of indicator are listed as follows:

Indicators based on stated goals:

- By 2020, Massachusetts will have accumulated 2,000 MW of new wind power generation.
- New wind energy development will create 900 long-term jobs by 2020.
- As wind energy replaces fossil fuel burning, there will be a 10% reduction in breathing-related illnesses.

Indicators based on potential weaknesses:

- Wind turbines will be equitably placed in low- and high-income areas in proportion to the occurrence of those areas.
- Job creation will occur at all income levels in proportion to the number of people at those levels, and salaries at each level will be consistent with national standards.

These are examples and the list is in no way complete, but they show how indicators based on our definition can be used to both track objectives and assess potential pitfalls. Adjustments can then be made based on ongoing monitoring and assessment. For example, if job creation is not occurring at lower income levels, a job training program might be created as part of this project to address this issue. It is important that we not only track our successes but also develop indicators that can be used to make adjustments to better meet our overall goals of sustainability. This was in no way meant to be a complete critique of this very highly regarded initiative, which is already reaching many of its interim objectives. We also know that much more data are available that we did not access in this particular analysis. We used this merely as an example of how our definition of sustainability can be used to examine a particular project, raise questions, and examine both strengths and potential weaknesses with regard to the principles in our definition. This is the approach we take throughout this book. Next, a very different example is presented in Case 2, which deals with an international ecotourism development project.

Case 2: Nam Ha Ecotourism Development Project in Laos

Established in 1999, the Nam Ha Project is a community-based ecotourism initiative implemented by the Lao National Tourism Administration. The Nam Ha National Protected Area is in the remote northern province of Luang Namtha, Lao PDR, or Laos as it is often called (Figure 2.10). The goal of the project has been to implement a community-based ecotourism development approach that places local communities at the center of tourism development and the management process. It has been funded by the Government of New Zealand and the United Nations Educational, Scientific and Cultural Organization (UNESCO).

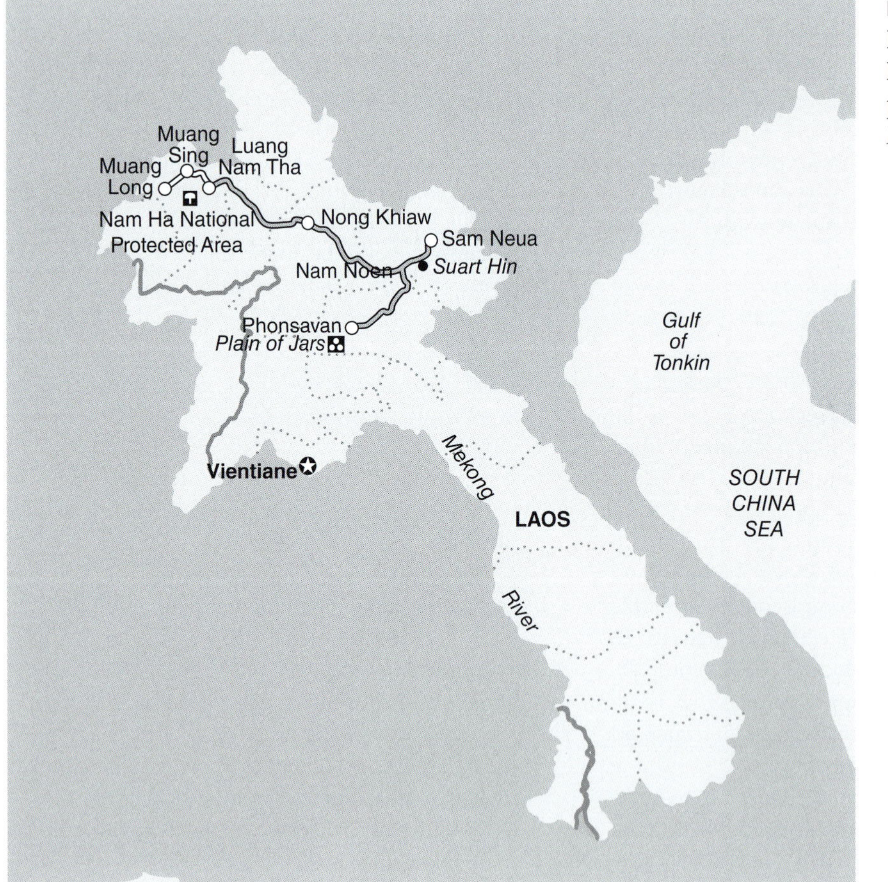

Figure 2.10 Map of Lao PDR (Laos) showing Nam Ha Ecotourism project. *Courtesy of Nam Ha Ecotourism project.*

They also provide technical assistance and monitoring of the project. Another main focus of the project has been to ensure that tourism makes measurable contributions to the conservation and protection of the region's outstanding natural and cultural heritage in an economically viable framework.[22]

The 222,400 hectare (ha) protected area includes a diversity of habitats with some of the region's most significant wilderness area supporting a rich biodiversity. Included among this biodiversity are mammals of significance such as clouded leopards, tigers, Asian elephants, and Muntjac deer. There are an estimated 288 bird species and very diverse plant communities. The region also contains 21,000 people in 57 villages who have depended on the resources of this region for their survival. Prior to the formation of the reserve, threats to the environment included slash-and-burn agriculture, overextraction of timber and non-timber forest products, and hunting by both the residents and outsiders.[22] In establishing the reserve, local communities worked closely with management authorities to design the project. Cooperative agreements that established stakeholder responsibilities set up a framework in which local communities play a primary role in managing and protecting natural and cultural resources. Community members also play a major role in ecotourism ventures and are employed as eco-guides and conservation and wildlife managers. Other new sources of local income generation include an increased service industry and the sale of handicraft and souvenirs. A transparent revenue-sharing agreement was established for visitor fees that contribute to village development and conservation efforts. The ultimate goal was to add value to natural resources through tourism revenue, thereby preserving a biologically diverse region while raising the standard of living of the people who live in that region. For example, a community member who used to harvest wild orchids for pennies per day can now lead a group of tourists to see those orchids for $5 per day.[22]

This project seems almost quintessentially sustainable, but applying our definition to this project will allow us to better determine the degree to which it is consistent with the principles in the definition and may shed light on areas for improvement. So once again, let's apply our definition as follows:

> *The Nam Ha Ecotourism Development Project will be sustainable if it meets the needs of current generations or increases their opportunity to meet their own needs without compromising the ability of future generations to do so. This is best achieved using a participatory, inclusive process based on scientific, system thinking, and backcasting principles that ensures:*
>
> *1. Extraction and use of resources that maximizes renewal, encourages reuse, minimizes waste, and protects the environment, including conservation and preservation of natural systems, reducing pollution, and slowing or reversing global climate change;*
>
> *2. Economic development and equitable economic opportunity including profit that does not undermine the capacity of people to meet their own needs; and*
>
> *3. An elevated standard of human well-being for all people, including but not limited to improved health, improved nutrition and access to clean air and water, increased access to education, and other basic human rights.*
>
> *The sustainability of this project will be best assessed through the development of measureable indicators that show improvement in each of these criteria.*

Let's turn to the United Nations Development Programme (UNDP) report on this project to determine how it is doing with regard to the previously mentioned principles. The report emphasizes the following project impacts:

- Through the development of alternative opportunities for individual income generation, this project has reduced pressure on natural resources within the reserve.
- A reserve of more than 200,000 ha of diverse habitats with significant wilderness has been established. Village-based sanctuaries protect forest and wildlife species. In many ways, these preservation efforts have been successful, but threats to the protected areas remain, including deforestation for rubber cultivation.
- Ecotourism has generated funds for protected area management and conservation work, and a wide network of local stakeholders are involved in conservation and wildlife protection rather than higher level agencies and personnel from outside of the communities, making it more effective.

- Ecotourism generates revenue for the local economy, increasing per capita income in one of the poorer areas of the country. This additional income has supplemented traditional village livelihoods, allowing for job diversification, time for child care, savings, and the ability to pay for education.
- Women and ethnic minorities have a high rate of participation in all tourism ventures in the region.
- Part of the project included the installation of clean water systems, providing healthy drinking water, and the creation bathing of sites that protect natural springs, which are the sources of water for drinking and shrimp production, an important source of protein.[22]

In terms of process, this project has embraced the principles in our definition. The participatory/stakeholder approach is central. Conservation planning and management are scientifically based, and the establishment of the project used a creative backcasting approach. This project has had extensive quantitative monitoring and evaluation. The previous summary is presented as part of the exercise to test the definition and practice developing indicators. Let's develop some indicators as follows based on stated goals:

- By 1990, the 222,400 ha Nam Ha National Protected Area will be established.
- The establishment of this protected area will significantly raise the standard of living of the residents in the 57 villages that are within its boundaries.
- Ecotourism will generate more income and have significantly less impact than traditional forest resource use.
- Women and ethnic minorities will realize economic benefits from ecotourism activities that are equivalent to those of other groups.
- The quality of drinking water will be maintained or improved by project efforts.

The project report shows that all of these indicators have been met to some degree. These indicators clearly reflect the economic, environmental, and social values that are central to our definition of sustainability. What would strengthen this assessment would be quantitative objectives—for example, percentage increases in income over time. These quantitative indicators can then be used to determine the extent to which those objectives have been met.

What else would help inform us about the sustainability of this project? The report indicates that deforestation is still occurring. It would be helpful to see trends in rates of forest lost. It also indicates that income from the tourism fees is being used to promote conservation and halt deforestation. It would be helpful to have indicators that link funding to reduced forest loss. Also, because one of our principles is the reduction of global climate change, and forests play a role in this as CO_2 sinks, it would be valuable to estimate how forest conservation has helped sequester a specific amount of CO_2. Finally, this project originally received external funding from New Zealand and other international parities. To be truly sustainable, the project should be able to maintain itself without continuing to depend on outside funding, so this type of information would be useful.

The Nam Ha Ecotourism Development Project in Lao PDR is a model for local sustainable development that strengthens people and communities, and it preserves nature by elevating its economic value in a preserved rather than exploited manner. As with the previous example, this was in no way a critique of the program but, rather, an exercise in articulating objectives, identifying indicators, and determining the kinds of information that would be useful in assessing sustainability. This project received the UNDP Equator Prize for best practices in community-based conservation and sustainable livelihoods. It continues to bring together local guides and an international community of ecologically minded tourists in a way that serves both groups and the environment well (Figure 2.11).

Applying Our Definition

In the chapters that follow, we will examine a number of specific problems and either actual or proposed solutions that increase our ability to meet the criteria for sustainability that are central to our definition. We will not, as we did in this chapter, place each solution in the definition to see how it fits. Rather, we will examine specific approaches, issues, and solutions within the context of our definition. Our goal will be to determine if the principles and conditions of sustainability are increasingly being met and how each solution might be improved to further us along the sustainability trajectory.

Figure 2.11 Locally guided forest treks at the international award-winning Nam Ha Ecotourism project. *Courtesy of Nam Ha Ecotourism project.*

For a given problem, there is a tendency to argue for one or a few sustainable solutions rather than examining a larger suite of possible solutions. For example, to deal with climate change, the primary and in some cases only solution offered is to eliminate the use of fossil fuels for energy and transportation. However, animal agriculture, which is responsible for more GHG emissions than those from all transportation sources, is often ignored because asking people to change their diet is viewed as less popular than attacking the fossil fuel industry. As in this case, a more holistic approach that examines how a variety of solutions can contribute to a specific objective would be taken. We will also view each solution with a critical eye and balance that with an open mind as we use scientific, social, and economic perspectives to assess them. Rather than dismissing a particular approach outright, we will consider how that approach may or may not be used to further sustainability objectives. For example, many activists in the sustainability arena completely reject the use of genetically modified organisms (GMOs). Although there are many serious problems associated with GMOs, the approach of this book is to acknowledge these problems but also to explore ways in which they could be used to further sustainability. Some may see this as controversial, but really it is the scientific, stakeholder approach to analysis that we value and will apply equally to everything we analyze.

Developing Competency in Sustainability Problem Solving

Although many examples of innovative solutions to sustainability related problems are presented in the chapters that follow, this book is not meant as a technical guide for specific projects, nor is it necessarily designed to prepare the reader for what are commonly referred to as "green jobs." Rather, the objective of this book is to build competencies in sustainable problem solving. These competencies were developed at a number of conferences and workshops, including the American Association for the Advancement of Science 2010 Forum for Sustainability Science Programs.[23] They have been promoted as key components to sustainability curricula by the Association for the Advancement of Sustainability in Higher Education and adopted as such by a number of programs, such as Arizona State University's School of Sustainability.

As you read this book, you will start to develop the aptitude required for solving real-world sustainability challenges that exist and interact at a variety of scales. Using an interdisciplinary, systems thinking approach, you will learn how to analyze problems from a holistic perspective (systems thinking competence). You will learn how to ethically assess a problem within the context of environmental integrity, social justice, and equitable economic opportunity while recognizing that this context may vary geographically and across cultures (normative competence). You will also develop the capacity to anticipate how specific sustainability-related interventions or the lack of intervention may impact the future and the time frame in which this may occur (anticipatory competence).

Through examples and exercises, you will learn how to create solutions that will further sustainability objectives (strategic competence), and you will come to recognize that these solutions are only obtainable through effective communication and collaboration among diverse stakeholders and experts from different disciplines (interpersonal competence). Within the context of these original core competencies and others that have since been proposed, the goal of this book is to create sustainability thinkers and doers who are effective at understanding and achieving real-world changes—transformational changes that move beyond isolated "Band-Aid"-type solutions to more systematic approaches that will lead us to a sustainable future.

A Lexicon for Sustainable Problem Solving

In addition to defining sustainability, there are a number of fundamental terms that we have and will be applying regularly. It will be important to consistently apply these definitions, so they are listed here. This will serve as our lexicon, or common language, for acquiring the competencies for effective sustainable problem solving. Throughout the book, many other important terms will be presented and defined, but those listed here are of particular importance.

DEVELOPMENT

Generally, development refers to growth and progress. It refers to transformation that is often based on scientific, technical, economic, or social processes. Most frequently, the term refers to economic progress, but it can apply to political, social, and technological progress as well. These sectors tend to be interconnected, so is difficult to separate them. We often define the state of development in particular countries as developed, developing, or undeveloped. Accordingly, we often use the terms the developed, developing, or undeveloped world. Previously, these have been referred to as the First, Second, and Third World, respectively, although this terminology is less frequently used today. Countries are also often referred to as industrialized or non-industrialized, again reflecting the degree of economic development. Developed countries are sometimes grouped as the Global North and less developed as the Global South, reflecting general geographical tendencies. The use of all these terms can be tricky because many countries are transitioning between states, and certain aspects of their economy could be highly developed while social progress lags behind.

GROSS DOMESTIC PRODUCT

Development is often measured using indicators such as gross domestic product (GDP). GDP is the monetary value of all final goods and services produced within a nation in a given year. It is the sum of the value of all consumer spending or consumption, all government spending, all investment within a nation, and the value of a nation's net exports. Its increase is one of many indicators of economic development, but there are many arguments against its uses as a measure of sustainable development that will be addressed in subsequent chapters.

COMMUNITY

When developing sustainable solutions, it is important to consider the scale or the level at which one might implement and assess sustainability. This can occur at the level of the individual or household, institution, community, or region on up through a global scale. In Part III of this book, we move through these different levels of organizations. Although each level is important, much of the work in the sustainability arena has been at the community level. Here, we define a community as a group of people who can have diverse characteristics yet are linked by social ties and are often located in the same geographical location or setting. A community is faced by common concerns and often shares common values, although a community can be divided over certain issues.

STAKEHOLDER

As discussed with regard to our definition of sustainability, the development, implementation, and assessment of sustainable solutions must be an open, participatory process that involves a diversity of stakeholders. A stakeholder can be any person, group, or organization that has interest or concern for the issue and any proposed remedy to it. Stakeholders may be part of the same community but can include those who are geographically removed from an issue or project but who still have an interest either because of the potential regional or global impacts of an activity or because they value what might be potentially impacted by a particular plan or action. The practice of involving stakeholders in sustainability related decision-making is emphasized throughout this book. However, there are a number of challenges to this approach. Identifying stakeholders and engaging them in the issue can be difficult. The representation of stakeholders, their role in the decision-making process, and whether all participants should have equal say are key issues that will need to be addressed.

THE COMMONS

The commons refers to shared resources such as those discussed in Chapter 1. Examples include the atmosphere, oceans, fishing stocks, terrestrial resources, and non-renewable fuels. Those working on sustainability issues often refer to the "tragedy of the commons," which Garret Hardin explored in his 1968 article in *Science*.[24] The "tragedy" is the tendency of individuals to act according to their own self-interests at the expense of the larger group's long-term best interests, resulting in the depletion of the common resource. This type of behavior represents a significant challenge to sustainability and will be discussed as we explore particular resources.

EXCLUDABILITY

When considering the commons, resources, and consumption, it is important to distinguish among excludable and non-excludable goods and services. When they are excludable, it is possible to prevent consumers who have not paid for them from having access to them—for example, consumable products such as food and clothing. By comparison, non-excludable goods and services can be accessed by non-paying consumers. Shared resources such as fish, air, and timber may be non-excludable and are what we consider to be the commons. Both excludable and non-excludable goods can be either rivalrous or non-rivalrous. A rivalrous good is one for which the use of that good by one person presents a significant barrier to others who desire to use that good at the same time. Durable as opposed to immediately consumable rivalrous goods do have the potential to be shared. For example, a car or bicycle could be shared over time, whereas a piece of fruit cannot be shared once it is consumed. Non-rival goods may be consumed by one consumer without preventing simultaneous consumption by others. Examples of non-rivalrous, excludable modes of consumption include entertainment such as movies or concerts, Wi-Fi, or cable TV. Something that is non-excludable and non-rivalrous is a public good or service such as a scenic view, street lights, or public safety.

TRUE COST ACCOUNTING AND EXTERNALITIES

True cost accounting is the practice that accounts for all external costs or externalities, including environmental, social, and economic, that are generated by the creation of a product. Typically, these are not accounted for when a corporation produces a product. For example, the manufacturing of a product may have been done in a sweatshop in which workers suffer social and health consequences. It may also have polluted local water sources, impacted the public health of a community, hastened global climate change by releasing GHGs, contributed to the depletion of non-excludable items without paying for them, and taken advantage of a variety of ecosystem services. These external costs must then be paid for by some other component of society. Because the corporation is not held responsible for these external costs, the product can be sold relatively cheaply while maximizing profit. If we are interested in sustainability, it will be important to consider these external costs in addition to the price of a product. They should also be considered as new economic and environmental policies are constructed.

LIFE CYCLE ANALYSIS

LCA is a complete inventory of relevant energy, material inputs, social justice issues, and environmental releases of a particular product during its complete life cycle. This includes the extraction of resources to make component parts and the product's construction, through its use, and finally its disposal. It is also referred to as life cycle analysis, eco-balance, and cradle-to-grave analysis. A simple comparison of an electric car to one that runs on fossil fuels could easily reveal that the electric is more sustainable. But when a full LCA is conducted, it may be revealed that many more toxic components go into the construction of the electric vehicle. Perhaps resources required for the construction of specific parts impact the environment and poorer communities. LCA will be an important tool as we examine many of our sustainable solutions and products.

SYSTEMS THINKING

A system is an organized collection of highly integrated parts that interact in complicated ways. When considering sustainability, we deal with systems composed of social and environmental parts and processes at a variety of scales that have the potential to influence one another. To fully understand how our actions might impact an ecosystem, we must employ systems thinking. Systems thinking allows us to consider human impact and sustainability from a broad perspective that includes seeing overall structures, patterns, and cycles as opposed to just specific events. This approach is useful in identifying cause and effect within and among systems so that they can be more effectively managed.

NATURE-BASED SOLUTIONS

These are solutions to societal challenges that are inspired by or supported by nature and natural processes. They tend to be relatively cost-effective and simultaneously provide environmental, social, and economic benefits that contribute to resilience.

ECOSYSTEM SERVICES

The ways that humans benefit from natural processes are referred to as ecosystem services. There are four broad types of ecosystem services. Provisioning services refer to useable material or energy outputs from ecosystems. Supporting or habitat services are those that have a role in the maintenance and function of all other types of ecosystem services. Regulating services are those that control or maintain environmental conditions. Cultural services provide spiritual, aesthetic, recreational, and other less tangible values.

BIOMIMICRY

Based on the idea that nature has already solved many of the problems that confront us, biomimicry is the design and production of materials, structures, and systems that are modeled on biological entities and processes. It is a form of nature-inspired innovation in the development of sustainable solutions.

THE PRECAUTIONARY PRINCIPLE

When an activity could potentially harm human health or the environment, precautionary measures should be taken even if some cause–effect relationships are not fully established scientifically. This is known as the precautionary principle. It justifies the use of cost-effective measures to prevent environmental degradation and protect human health even in the absence of full scientific certainty. The precautionary principle should be applied in an open, informed, and democratic manner. Its application must consider all potentially affected parties and an examination of the full range of alternatives, including no action.

MITIGATION

Generally, mitigation refers to the action of slowing or halting an ongoing process that has negative implications that will worsen over time. It is essentially the act of making a condition or consequence less severe. Mitigation is what we do when we try to reduce the impacts of an activity on the environment and human well-being. For example, climate change could be mitigated by reducing GHG emissions or by actively sequestering them.

ADAPTATION

Adaptation relates to the decision-making processes and actions that we undertake to survive changing conditions. In terms of sustainability, adaptation is the change that must occur if we are going survive environmental change. In the absence of mitigation, change will be inevitable, and the inhabitants of the planet will have to adapt to those changes in order to survive. For example, some argue that given the inevitability of climate change and the fact that its impacts are already far-reaching, we are going to have to adapt in order to sustain ourselves. Examples include the development of drought-resistant crops, the movement of communities to higher ground, and other innovations that will make life on a warming planet possible.

TRANSFORMATION

Whereas adaptation represents incremental change in response to changing conditions, transformation is a more radical and pervasive reorganization of a social–ecological system. It is a potentially fundamental change in anticipation of major stresses that creates new order and changes the way parts of systems interact. Many sustainability advocates argue that mitigation and adaptation will not be sufficient and that behavioral, economic, and political transformation will ultimately be required to achieve global sustainability.

RESISTANCE, RESILIENCE, AND STABILITY

Resistance is the property of ecosystems, communities, populations, or institutions to remain essentially unchanged when subject to disturbance or outside forces of change. The opposite of resistance is sensitivity. Resilience, on the other hand, is the capacity of a system to recover, adapt, and grow in the face of change. Resilience is a measure of the magnitude of disruption that is required to move a system out of equilibrium and the cost or effort required to restore a system to equilibrium after a disruption has occurred. Both resistance and resilience are measures of stability.

Given the inevitability of both environmental and social change, sustainability will require the development of innovative ways to resist or stop change or to respond to change through resilience.

Understanding resilience is important for ensuring the sustainability of economic, social, and environmental systems as major disruptions occur due to events such as climate change and globalization. The emerging field of resilience analysis will provide an assessment tool for how resilient a particular system or entity will be to change, which can then inform policy or action to minimize vulnerability to change.

ASSESSMENT

Assessment is a review process that involves the gathering and analysis of information from multiple and diverse sources in order to determine if specific goals or objectives have been met. It is typically based on developed indicators that can be either qualitative or quantitative. It is a way to both measure success and identify areas of improvement. It should be a continuous process that is revised and improved upon each time it is done.

INDICATOR

An indicator is a unit of information measured over time that documents changes in a specific condition in relation to a specific goal or objective. Indicators can be compared to a benchmark or point of reference that allows for assessment of progress or performance with others, such as comparisons of the rate of deployment of solar panels among countries or the energy efficiency of similar buildings. A baseline is the state of an indicator at the start of a process that is being assessed. Comparing change in baseline indicators over time reveals historical performance. Good indicators should be measurable, precise, and consistent. They should be developed collaboratively and in a transparent manner.

GOALS AND OBJECTIVES

To develop indicators, project goals and objectives need to be clearly stated. A goal is a general summary of a desired outcome from any solution we propose. Goals should be general, brief, and measurable. Objectives are more specific and reflect desired outcomes. They should be impact-oriented, measurable or quantitative, and measured over a specific length of time.

FORECASTING AND BACKCASTING

Forecasting sets future goals by extrapolating current practice into the future and making adjustments based on that. Backcasting is a planning method that starts with defining a desirable future outcome and then works backwards to identify policies and programs that will connect the future to the present. It is the act of envisioning ourselves in the future as having met a specific set of conditions (see Figure 2.4). Backcasting is the preferred way to set goals and objectives within the sustainability framework, but a combination of the two approaches could be beneficial and instructive.

Chapter Summary

- Because of the interconnection of environmental problems and the growing needs of humans, these two must be considered together; this is the essence of sustainability.
- To develop a working definition of sustainability, one can trace the origin and use of related terms throughout the history and evolution of the sustainability movement.
- The principles from each movement and step in the construction of the concept of sustainability can be extracted, organized, and placed into an overarching definition that reflects these principles.
- The definition that we developed first identifies the kinds of things that we might label as sustainable. It emphasizes intergenerational equity and participatory processes rooted in science. It is centered at the intersection of environmental, economic, and social priorities.

- Sustainability is best assessed through measurable indicators that show improvement in the criteria represented in our definition.
- Our working definition will be used to examine and consider actual or proposed solutions that detract from our ability to meet our criteria for sustainability.
- The goal of this book is to build competencies in sustainable problem solving that include systems thinking, normative, anticipatory, strategic, and interpersonal competencies.

Digging Deeper: Questions and Issues for Further Thought

1. What are the basic principles of sustainability?
2. What are the connections between poverty, affluence, and sustainability?
3. Interpersonal competencies such as collaboration, transparency, good governance, accountability, and participatory approaches are viewed as essential elements in successfully achieving sustainability objectives. What do these terms mean? For an organization that you most closely associate with, such as a club, your school, or your place of work, discuss how these approaches have or have not been used in a decision-making process and what the effects of that were.
4. How does your faith or religious background view sustainability principles, and how are members of that faith community working to achieve them?

Reaping What You Sow: Opportunities for Action

1. Take a walk in a nearby park, woods, or natural area. Be sure to use all of your senses as you take in your surroundings. Write down any observations you make during your walk. Upon return, write a description of the place or area where you walked and what you observed, and describe what you experienced and how you felt in the environment during your walk. Consider how what you experienced relates to our definition of sustainability. Assume that the site in which you walked is slated for development, and based on your experiences and your new understanding of sustainability, make a strong case for its protection.
2. Choose a park or natural area in your own community. Identify all of the stakeholders associated with it and the roles that they play. Identify a particular issue associated with that area as it might relate to sustainability, and try to predict how each stakeholder might perceive that issue.

References and Further Reading

1. Emerson, Ralph Waldo. 1836. *Nature.* Corvallis: Oregon State University.
2. Thoreau, H. D. 1854. *Walden or Life in the Woods.* Boston: Ticknor & Fields.
3. Muir, John. 1920. *The Yosemite.* New York: Century.
4. Leopold, A. 1949. *A Sand County Almanac: And Sketches Here and There.* New York: Oxford University Press
5. Carson, R. 1962. *Silent Spring.* Boston: Houghton Mifflin.
6. United Nations Environment Programme. 1972. "Declaration of the United Nations Conference on the Human Environment." Presented at the United Nations Conference on the Human Environment, Stockholm, Sweden.
7. The World Commission on Humans and Environment Development. 1987. *Our Common Future.* New York: Oxford University Press.
8. United Nations. 1992. "Agenda 21." United Nations Conference on Environment and Development, The Earth Summit, Rio de Janeiro, Brazil.
9. United Nations. 1992. "The Rio Declaration on Environment and Development." United Nations Conference on Environment and Development, The Earth Summit, Rio de Janeiro, Brazil.
10. United Nations. 1992. "The Statement of Forest Principles." United Nations Conference on Environment and Development, The Earth Summit, Rio de Janeiro, Brazil.
11. United Nations Millennium Development Project. 2005. *Investing in Development: A Practical Plan to Achieve the Millennium Development Goals.* New York: United Nations Development Programme.
12. Elkington, J. 1997. *Cannibals with Forks: The Triple Bottom Line of 21st Century Business.* North Mankato, MN: Capstone.

13. Nidumolu, R., C. K. Prahalad, and M. R. Rangaswami. 2009. "Why Sustainability Is Now the Key Drive of Innovation." *The Harvard Business Review*, September.
14. Robèrt, K.-H. 2008. *The Natural Step Story: Seeding a Quiet Revolution*. Gabriola Island, British Columbia, Canada: New Catalyst Books.
15. Raworth, K. 2012. *A Safe and Just Space for Humanity*. Oxford, UK: Oxfam International.
16. Raworth, K., 2017. *Doughnut Economics: Seven Ways to Think like a 21st Century Economist*. Chelsea Green Publishing.
17. Bosch, O. J. H., C. A. King, J. L. Herbohn, I. W. Russell, and C. S. Smith. 2007. "Getting the Big Picture in Natural Resource Management—Systems Thinking as 'Method' for Scientists, Policy Makers and Other Stakeholders." *Systems Research and Behavioral Science* 24 (2): 217–32. doi:10.1002/sres.818
18. Nesshöver, C., T. Assmuth, K. N. Irvine, et al. 2017 "The Science, Policy and Practice of Nature-Based Solutions: An Interdisciplinary Perspective." *Science of the Total Environment* 579: 1215–27.
19. Hart, M., 1999. Guide to sustainable community indicators. North Andover, Ma.: Sustainable Measures Inc.
20. Sustainable Community Roundtable. 1995. "State of the Community—A Sustainable Community Roundtable Report on Progress Toward a Sustainable Society in the South Puget Sound Region." Olympia, WA: Sustainable Community Roundtable.
21. Agyeman, J., R. D. Bullard, and B. Evans, eds. 2003. *Just Sustainabilities: Development in an Unequal World*. Cambridge, MA: MIT Press.
22. United Nations Development Programme (UNDP). 2012. "Nam Ha Ecotourism Project, Lao People's Democratic Republic, Equator Initiative Case Studies: Local Sustainable Development Solutions for People, Nature, and Resilient Communities." New York: UNDP.
23. Wiek, A., L. Withycombe, and C. L. Redman. 2011. "Key Competencies in Sustainability: A Reference Framework for Academic Program Development." *Sustainability Science* 6 (2): 203–18. https://doi.org/10.1007/s11625-011-0132-6
24. Hardin, Garret. 1968. "The Tragedy of the Commons." *Science* 162 (3859): 1243–48.

CHAPTER 3

THE AIR WE BREATHE

We interact with our environment with every breath we take. Breathing is the mechanism that enables air to be brought from the environment into our bodies, where oxygen is transferred into our bloodstream and carbon dioxide (CO_2) is transferred out. An average healthy adult at rest does this approximately 12 times per minute. Children and infants do this at a much greater rate, ranging from 20 to 60 breaths per minute. Without breathing air even only for a few minutes, we begin to starve our brains of oxygen, causing damage and a fairly rapid demise.

The layer of the atmosphere at the surface of the Earth, the **troposphere**, contains the air that we breathe. In addition to consisting of oxygen (20.95%) and CO_2 (0.04%), the most abundant component in air is dinitrogen gas (N_2), which comprises 78.08% of air by volume. In Chapter 1, we learned how oxygen and CO_2 are cycled through the processes of respiration, photosynthesis, and decomposition; we also discussed the more complex biological cycling of nitrogen. In addition to these major components, air also contains water vapor and a variety of other gases at much lower concentrations, including argon, neon, helium, methane, nitrogen and sulfur oxides, and ozone. Because of their normally low concentrations, these are referred to as trace gases. Air also has suspended in it very small liquid or solid particles with diameters often considerably less than 10 μm. These are referred to as aerosols or particulate matter, and they include substances such as mineral dust, sea salt, pollen, and microorganisms, as well as larger sulfur, nitrogen, and carbon compounds.

PLANTING A SEED: THINKING BEFORE YOU READ

1. Do you think the air is cleaner or more polluted than it was 50 years ago? Why? What have been the major sources of air pollution?
2. Has US environmental policy ever been effective in reducing pollution? How and when? What are the current barriers to passing effective environmental legislation in the United States and in other countries?
3. Give examples of how air pollution can be a local, regional, international, and global problem.
4. What is the precautionary principle? Should it be applied to clean air policy? Why?

Polluting is a choice; breathing is not.
—Clean Air Fairbanks

Air Pollution

The composition of our air is changing. Although in some areas of the world air quality has improved, in others it continues to decline through both natural and **anthropogenic**, or human, sources of harmful materials or pollutants. This poor air quality has health implications for humans. According to the World Health Organization (WHO), each year as many as 7 million people die prematurely because of poor air quality, and even more have a reduced quality of life due to the increased incidence of asthma and other ailments caused by exposure to these pollutants.[1] Poor air quality also impacts other organisms and threatens both natural and built environments.

The introduction of these harmful pollutants into the Earth's atmosphere is referred to as **air pollution**. Natural sources of air pollution include forest and grassland fires, dust storms, sea salt spray, and volcanic activity. For example, volcanoes release particulate ash, acid mists, and a variety of other toxic chemicals. These natural sources of air pollution can be significant. However, human-caused or anthropogenic sources of air pollution are the main cause of poor air quality, especially in cities and other populated sites.

Anthropogenic air pollutants are classified as either primary or secondary. A **primary air pollutant** is a harmful substance that is released into the air directly from a source. An example of a primary air pollutant is sulfur dioxide (SO_2) produced by coal-burning power plants. **Secondary pollutants** are not directly emitted in a hazardous form but are compounds that, once released, are converted into harmful pollutants through chemical reactions in the atmosphere. These reactions are often driven by light energy. An example of a secondary pollutant is **ozone**, a major component of smog that is formed through complex chemical reactions in the atmosphere involving volatile organic compounds (VOCs), oxides of nitrogen (NO_x), and ultraviolet (UV) radiation from the sun.

Anthropogenic air pollutants can come from point or mobile sources. **Point sources** are identifiable, stationary locations such as factories, power plants, agricultural sites, and landfills (Figure 3.1). **Mobile sources** are those that move and are typically related to transportation. Another common source of air pollution is unintentional leaks of gases from breaks or small cracks in seals, tubing, valves, or pipelines and other unintended or irregular releases, mostly from industrial or energy extraction activities. This type of unintended release is referred to as **fugitive air pollution**.

Figure 3.1 Factory at sunset in Cleveland, Ohio. Industry is one of the top contributors of air pollution. *Courtesy of Joseph Sohm/Shutterstock.*

The US Environmental Protection Agency (EPA) refers to the six principal air pollutants that are the most common and are known to be harmful to human health as **criteria pollutants**. These are carbon monoxide, nitrogen oxides, sulfur dioxide, particulate matter, ozone, and lead. Other common air pollutants include VOCs, air toxics, and toxic metals, many of which have been determined to cause cancer or other serious health effects, such as reproductive problems or birth defects, and adverse environmental effects.

Relatively recently, attention has focused on **greenhouse gas** (GHG) emissions as a form of pollution that is driving global climate change. Although a number of countries have dealt with the regulation and control of other kinds of air pollutants, limiting GHG emissions has only recently been seriously addressed and will require much more attention. The primary GHGs are water vapor, carbon dioxide, methane, nitrous oxide, and ozone. It was noted in Chapter 1 that GHGs play a role in warming the Earth as they absorb and re-radiate heat that is generated when light reaches the surface of the Earth, and also that increases in GHG emissions through human activity have contributed to climate warming. Unlike other types of air pollution, the concern about GHGs is not focused on direct impacts on human health but, rather, their role in climate change and the indirect negative impact it will have on humans and the environment. In the following sections, we examine the sources and impacts of the different types of pollutants described previously. We then examine a variety of solutions that can decrease emissions, improve air quality, and reduce the impact of air pollution on human and environmental health.

Criteria Pollutants

The six most common air pollutants are designated by the EPA as criteria pollutants. They are called "criteria" air pollutants because the EPA regulates them by developing human health and environmentally based criteria for setting permissible levels that are determined through scientific research. These limits based on human health are called **primary standards**. Limits developed to prevent environmental and property damage are called **secondary standards**. The criteria pollutants are carbon monoxide, nitrogen oxides, sulfur dioxide, particulate matter, ozone, and lead. Although these designations and limits have been made by the EPA, they represent common pollutants globally, especially where urbanization is occurring. The United Nations Environmental Programme (UNEP) and WHO have also designated these same pollutants as priorities for emission reductions throughout the world.

CARBON MONOXIDE

Carbon monoxide (CO) is a colorless, odorless, toxic gas emitted from the combustion of fuels or other carbon-based materials. The majority of CO emissions to ambient air come from mobile sources such as automobiles. CO also represents a significant and very dangerous indoor air pollutant through the use of poorly maintained or unvented heating equipment, improperly vented natural gas appliances such as stoves or water heaters, and improperly ventilated garages when automobiles are left idling. In less developed countries, indoor cooking and heating with wood or gas and improper ventilation can lead to high levels of indoor CO exposure. Because CO preferentially binds to the blood's oxygen-carrying molecule, hemoglobin, high amounts of CO in the air can replace oxygen in the blood. This reduces oxygen delivery to the body's organs such as the heart, brain, and other tissues. The most immediate symptoms of CO poisoning are headache, dizziness, weakness, nausea, vomiting, chest pain, and confusion. At higher levels of exposure, CO can cause loss of consciousness and death. Exposure to moderate and high levels of CO over long periods of time has also been linked to increased risk of heart disease. Less is known about the health effects of long-term exposure to low levels of CO.

NITROGEN OXIDES

Nitrogen oxides are a group of highly reactive compounds that include nitric oxide (NO), nitrogen dioxide (NO_2), nitrous oxide (N_2O), and nitric acid (HNO_3). These are commonly interconverted, so the general term used is nitrogen oxides (NOx). These nitrogen compounds are emitted through both natural processes and human activity. Natural sources include the production of NO during thunderstorms as lightning causes the splitting of N_2 gas in the atmosphere and through soil processes such as nitrogen fixation (for a review of the nitrogen cycle, see Figure 1.1). Human activity has drastically altered the balance of nitrogen on the planet such that we have far exceeded the particular planetary boundary (see Figure 1.12).

Anthropogenic sources of nitrogen oxides are primarily emissions from cars, trucks, and buses; power plants; and equipment powered by fossil fuels. NO_2 contributes to photochemical smog and gives it its reddish-brown color. It also contributes to the formation of ground-level ozone and fine particulate pollution, as described later. NO_2 is linked with a number of adverse effects on the respiratory system, including airway inflammation in healthy people and increased respiratory symptoms in people with asthma. Another nitrogen oxide, N_2O_2, is a GHG that is produced in the fertilization of agricultural systems, through the combustion of fossil fuels, and through industrial processes.

The addition of nitrogen into ecosystems by way of air pollution can also contribute to **eutrophication**, or nutrient enrichment of aquatic systems. The rest comes from terrestrial run-off. For example, 30–40% of the nitrogen in Chesapeake Bay comes from atmospheric deposition. Because nutrients such as nitrogen, which under normal conditions limit plant growth, are added to aquatic systems through eutrophication, this results in uncontrolled growth of algae and other aquatic plants. When these plants die and decompose, the bacteria consume a considerable amount of oxygen from the water. This **biological oxygen demand** (BOD) then results in the further death of aquatic animals. Through increased BOD, eutrophication can cause a chain reaction of events resulting in entire aquatic systems with reduced or no oxygen (**hypoxia** and **anoxia**, respectively).

SULFUR DIOXIDE

Sulfur dioxide and other oxides of sulfur are highly reactive gases that are primarily emitted during fossil fuel combustion at power plants, industrial facilities, and from mobile sources in transportation. SO_2 is also emitted when metals are industrially extracted from ore and during paper production. Natural volcanic and oceanic processes also contribute to atmospheric SO_2. Similar to NO_2, SO_2 has been linked to a number of adverse effects on the respiratory system. SO_2 also has a noxious odor.

PARTICULATE MATTER

Particulate matter (PM) consists of solid particles and liquid droplets suspended in the air. They are also referred to as aerosols. Some particles, such as dust, dirt, soot, and smoke, are large or dark enough to be seen with the naked eye, whereas others are microscopic. In addition to numerous natural sources including volcanic activity, primary anthropogenic sources of PM include construction sites, unpaved roads, agricultural fields, smokestacks, and fires, including open burning. PM is also produced as a secondary pollutant through complex atmospheric reactions of chemicals such as sulfur dioxides and nitrogen oxides that are emitted from power plants, industries, and automobiles. Indoor fires for cooking and heat are a significant source of PM as an indoor air pollutant. The size of particles is directly linked to their potential for causing health problems, with smaller particles (<10 μm in diameter) posing the greatest problems. This is because smaller particles are more readily drawn deep into the lungs, where they cause respiratory and cardiac problems, including asthma, heart attacks, and decreased lung function. PM also limits visibility and can cause environmental and aesthetic damage, including the staining of stone buildings and culturally valuable objects.

OZONE

Ozone (O_3) occurs as a secondary pollutant in the troposphere or ground-level atmosphere. It is also formed naturally in the upper atmosphere or stratosphere, where it serves to absorb harmful UV light. **Tropospheric O_3** is created when NO*x* and VOCs react in the presence of sunlight (Figure 3.2). Emissions from industrial facilities and electric utilities, motor vehicle exhaust, gasoline vapors, and chemical solvents are some of the major sources of NO*x* and VOCs and thus contribute to O_3 production. Breathing ozone can cause a variety of health problems, including chest pain, coughing, throat irritation, and congestion. It can worsen bronchitis, emphysema, and asthma. Repeated exposure can permanently scar lung tissue.

Stratospheric O_3 is naturally produced when solar UV radiation breaks the bonds of an O_2 molecule and then the single O atoms bond with another, unbroken O_2 molecule, forming O_3. In the upper atmosphere, ozone absorbs 97–99% of the high-energy UV radiation from the sun before it reaches the surface, where it can damage plants, cause skin cancer in humans, and induce genetic mutation in other animals, as documented in amphibians.[2] Thus, this layer of ozone extending upward from 6 to 30 miles above the Earth's surface serves as a protective shield for living organisms. However, this "good" ozone is gradually being depleted by chemicals referred to as ozone-depleting substances (ODSs). Examples of ODSs include refrigerants such as chlorofluorocarbons (CFCs) and hydrochlorofluorocarbons (HCFCs) and other industrial chemicals, such as halons, methyl bromide, carbon tetrachloride, and methyl chloroform.

The depletion of stratospheric ozone has had serious health and ecological implications through increased exposure to UV radiation, including increased rates of skin cancer, cataracts, and other health problems in humans; abnormal development in animals; and impaired photosynthesis and increased susceptibility to disease in plants.

LEAD

Lead is a metal found naturally and as an anthropogenic pollutant in the environment as well as in manufactured products. In the past, lead was a major component of gasoline, and motor vehicles were the major contributor of lead emissions to the air. Because of the recognition of the health hazards of atmospheric lead, there have been regulatory efforts to reduce or eliminate it in gasoline. As a result, lead emissions from automobiles have greatly declined during the past 30 years. However, a number of industrial processes, including waste incinerators, power plants, lead smelters, and lead-acid battery manufacturers, still emit significant amounts of lead into the air. The use of lead in paint, ceramics, pipes and plumbing materials, solders, batteries, and other products represents additional avenues for exposure. Lead can enter drinking water in a variety of ways, such as seen

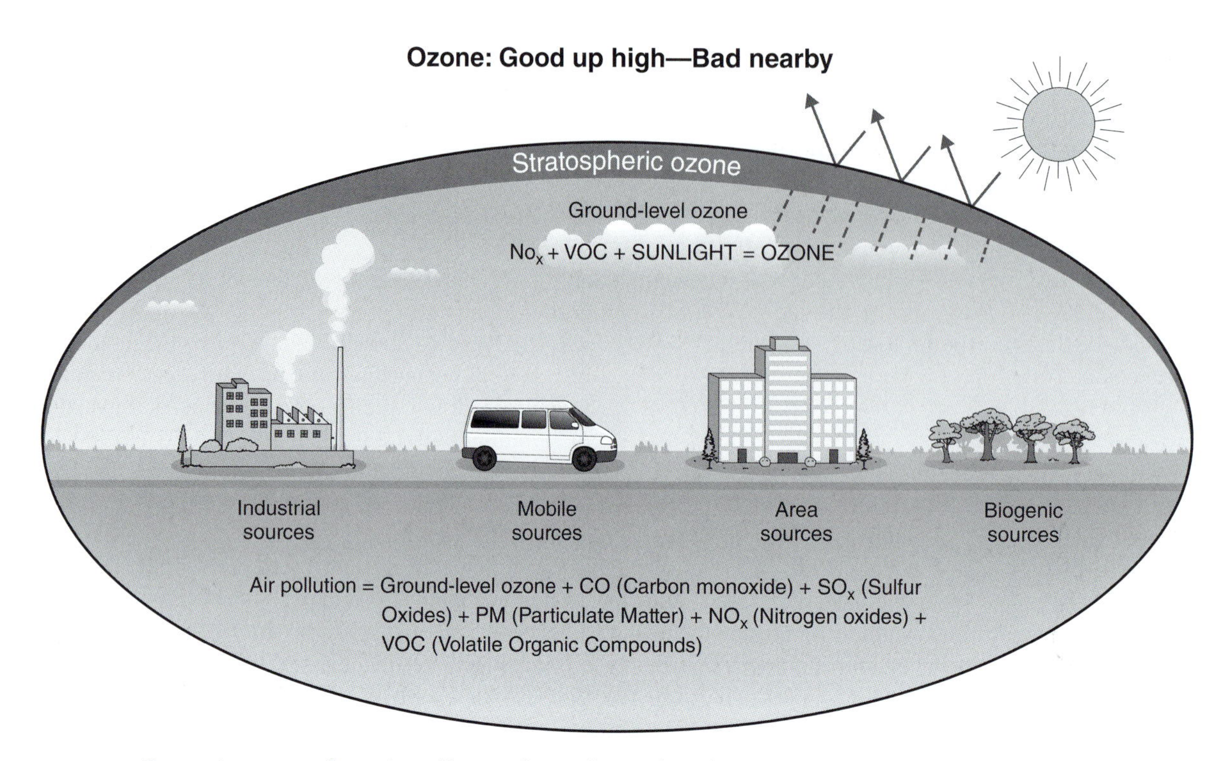

Figure 3.2 Ozone is a secondary air pollutant that is formed in the troposphere or ground-level atmosphere through complex chemical reactions involving volatile organic compounds, oxides of nitrogen, and ultraviolet light. This ground-level ozone is a harmful pollutant. In the stratosphere or upper atmosphere, ozone is formed naturally where it has the positive effect of filtering and decreasing the amount of dangerous UV radiation reaching the Earth's surface. However, ozone-depleting substances released from human activity break down the ozone in the stratosphere, resulting in increased exposure to this dangerous form of radiation. *Image courtesy of the Field Museum, climatechicago.fieldmuseum.org*

in 2016 in Flint, Michigan, where approximately 12,000 children were exposed to drinking water with high levels of lead, resulting in a range of serious health problems. The issue of lead in drinking water is further explored in Chapter 4.

Once taken into the body, lead enters the circulatory system and is accumulated in body tissues. Lead exposure and accumulation can adversely affect the nervous system, kidney function, the immune system, the reproductive and developmental systems, and the cardiovascular system. Lead exposure also affects the oxygen-carrying capacity of the blood. Lead causes neurological effects in children and cardiovascular disease in adults. Infants and young children are especially sensitive even to low levels of lead, which may contribute to behavioral problems, learning deficits, and lowered IQ. Lead is persistent in the environment and accumulates in soils and sediments through deposition from air sources. Ecosystems near anthropogenic sources of lead have experienced losses in biodiversity, decreased growth and reproductive rates in plants and animals, and neurological effects in animals.

Volatile Organic Compounds

VOCs are organic chemical compounds with low boiling points that evaporate under normal atmospheric conditions of temperature and pressure. They exist as gases in the air. Most scents or odors are VOCs. One natural source of VOCs is plants, which emit the bulk of these compounds. VOCs play an ecological role in plants' interaction with other organisms in defense, reproduction, or interplant "communication." In addition to natural sources, there are large numbers of synthetic VOCs that are emitted into the air through industrial processes and other human activities. Examples of VOCs are formaldehyde, toluene, benzene, and trichloroethylene, as well as a variety of pesticides used in agriculture.

In addition to playing a role in the production of photochemical oxidants such as ozone, VOCs are commonly used in household products and materials such as paints, carpets, building materials, and cleaning supplies, which represent a significant source of indoor air pollution. Studies by the EPA and other researchers have found that VOCs are common in indoor environments and that their levels may be up to 1,000 times higher than those outdoors.[3] There may be anywhere from 50 to hundreds of individual VOCs in the indoor air at any one time. Some may produce objectionable odors even at low concentrations, but many have no noticeable smell. Many VOCs are irritants and can cause headaches and dizziness, along with eye, nose, and throat irritation. Long-term exposure to certain VOCs may lead to chronic diseases including cancer. At high concentrations, some VOCs are toxic.

Air Toxics and Toxic Metals

Air toxics, also known as hazardous air pollutants (HAPs), are those that are known or suspected to cause cancer, birth defects, or other adverse health and environmental effects. Examples include benzene found in gasoline, perchloroethylene emitted from dry cleaning facilities, and other industrial compounds, such as methylene chloride, dioxin, asbestos, and toluene. Most air toxics originate from human-made sources. These include mobile sources in transportation and stationary sources such as factories, refineries, incinerators, and power plants. There are also a number of indoor sources of air toxics, including some building materials and cleaning solvents. Some air toxics are also released from natural sources such as volcanic eruptions and forest fires.

Another type of toxic air pollution is the emission of heavy metals. In addition to lead, which is a criteria pollutant, other heavy metals such as cadmium, arsenic, and mercury are common air pollutants emitted mainly as a result of industrial activity. Cadmium is also naturally released through volcanic activity and through human activity in the production of metals and cement, waste incineration, and fossil fuel combustion. Cadmium exposure has been linked to kidney and bone damage, and it has been identified as a potential carcinogen. Arsenic is released as arsine into the air by volcanoes, the weathering of arsenic-containing minerals and ores, and commercial or industrial processes including mining and smelters. The health effects of arsenic inhalation include headaches, vomiting, abdominal pain, and kidney failure.

Mercury is released into the air by coal-fired power plants, cement kilns, trash incinerators, as well as during gold mining. Once airborne, mercury can be deposited in soils and aquatic systems, in which it is converted to the organic form, methylmercury—the form of mercury that is most harmful to biological organisms. It can accumulate in the food chain and lead to human exposure. Humans risk ingesting dangerous levels of mercury when they eat contaminated foods that have accumulated it. This is particularly true for fish. Once in the human body, mercury acts as a neurotoxin, interfering with the brain and nervous system. Exposure to mercury can be particularly hazardous for pregnant women and small children. As with mercury, a major issue with most other heavy metal air pollution is their deposition and accumulation in soil and water environments. Once accumulated, they threaten both ecology and human health.

Greenhouse Gases and Climate Change

The primary GHGs are water vapor, carbon dioxide, methane, nitrous oxide, and ozone. Others include human-produced synthetic halocarbons such as CFCs. GHGs absorb and emit infrared radiation or heat as it rises from the surface of the Earth, resulting in overall warming. As explained in Chapter 1, the increasing concentrations of GHGs in the upper atmosphere increase its absorptive and reflective capacity. This results in the retention of heat within the upper atmosphere, which in turn results in further warming and climate change.

The relative contribution of each GHG to global warming is determined by both the amount emitted and the **global warming potential (GWP)** of each GHG (Table 3.1). GWP is a relative measure of how much heat a certain mass of each GHG traps in the atmosphere in relation to the same amount of CO_2. This is determined by the atmospheric lifetime of the gas and its ability to absorb and re-radiate infrared radiation (heat). The potential contribution of each GHG to global warming can be better understood by considering both rate of emission and GWP. For example, although the rates of methane emissions are considerably lower than CO_2 emissions, the GWP of methane is much higher, so it still represents a significant contributor to global climate change (see Table 3.1).

TABLE 3.1 Global Warming Potential (GWP) of Greenhouse Gases

GREENHOUSE GAS	ATMOSPHERIC LIFETIME (YEARS)	% OF TOTAL EMISSIONS	20-YEAR GWP	100-YEAR GWP
Carbon dioxide	20–200[a]	77	1	1
Methane	12	14	86	34
Nitrous oxide	114	8	268	298
HFC-134a (hydrofluorocarbon)	13	<1	3,790	1,550
CFC-11 (chlorofluorocarbon)	45	<1	7,020	5,350
Carbon tetrafluoride (CF4)	50,000	<1	4,950	7,350
HFC-23 (hydrofluorocarbon)	270	<1	12,000	14,800
Sulfur hexafluoride	3,200	<1	16,300	22,800

[a]Between 65% and 80% of CO_2 released into the air dissolves into the ocean over a period of 20–200 years. The rest is removed by slower processes that take up to several hundreds of thousands of years, including chemical weathering and rock formation.

Most GHGs have both natural and anthropogenic sources. The 2007 "Fourth Assessment Report" and subsequent reports compiled by the Intergovernmental Panel on Climate Change (IPCC)[4] noted that anthropogenic emissions of GHGs have been the primary cause of increases in their atmospheric concentrations and increases in global average temperatures since the mid-20th century. The resultant climate change has already had very serious consequences, as described in Chapter 1. Here, we examine the sources of each GHG and the role that each plays in climate change.

WATER VAPOR

The most abundant atmospheric GHG and the largest contributor to climate change is water vapor. Human sources have only a small direct influence on atmospheric water vapor concentrations, which are largely controlled by rates of evaporation from surface waters. Another natural source of water vapor is volcanic activity. However, as climate warming occurs through anthropogenic increases in other GHGs, rates of evaporation will rise and atmospheric water vapor concentrations will increase, contributing further to warming. This amplifying effect is referred to as a positive feedback loop (Figure 3.3). This positive feedback effect of water vapor nearly doubles the amount of warming caused by CO_2. So if there is a 1°C change caused by CO_2, the resultant increase in atmospheric water vapor through evaporation will cause the temperature to increase another 1°C (see Figure 3.3).

CARBON DIOXIDE

After water vapor, CO_2 is the next largest contributor to climate change. CO_2 naturally cycles with O_2 through the processes of photosynthesis, respiration, and decomposition. Human activity has altered the balance of this natural cycle. Sources of atmospheric CO_2 include the burning of fossil fuels, solid waste, and trees and wood products. Another common source is the production of cement. CO_2 is removed from the atmosphere, or sequestered, when it is absorbed by terrestrial and aquatic plants, and it is dissolved and converted to bicarbonate HCO_3 in oceans. Thus, both forests and oceans serve as net carbon sinks. However, the latter results in ocean acidification, which has very negative consequences for marine organisms and ecosystems, especially coral reefs. Also, as oceans warm, their capacity to absorb CO_2 decreases. Thus, the increase in atmospheric CO_2 is directly linked to increased emissions from anthropogenic sources and losses of carbon sinks primarily through deforestation and decreases in CO_2 solubility in warming oceans.

METHANE

Globally, agricultural animal production is the primary source of methane (CH_4) emissions. Domestic livestock produce large amounts of CH_4 as part of their normal digestive process. Also, CH_4 is produced as animal manure is broken down. CH_4 is also emitted during the production and transport of coal, natural gas, and oil. A considerable amount of CH_4 is generated in domestic landfills. Because organic matter is buried in landfills, it decays in the absence of oxygen or anaerobically. This results in the production of CH_4 instead of CO_2, which is emitted through aerobic decay. Methane is also emitted by natural sources such as wetlands; however, more than 60% of CH_4 emissions are the result of the human activities described previously.

OTHER GHGs

Already discussed as criteria air pollutants, ozone and nitrous oxide are also GHGs. Synthetic halocarbons or, specifically, fluorinated gases are powerful GHGs that have no natural sources and only come from human-related activities. They are emitted through a variety of industrial processes, such as aluminum and semiconductor manufacturing. Some of them are replacements for ozone-depleting substances. Examples include hydrofluorocarbons, perfluorocarbons, and sulfur hexafluoride. These gases are typically emitted in smaller quantities, but because they are highly potent GHGs, they are sometimes referred to as high-GWP gases (see Table 3.1).

Acid Deposition

Two of the criteria air pollutants, nitrogen oxides and sulfur dioxide, which are emitted primarily through fossil fuel combustion, react in the atmosphere with water, oxygen, and other chemicals to form nitric and sulfuric acids. This results in wet or dry acid deposition. In wet deposition, also referred to as **acid rain**, the acids reach the ground in the form of rain, snow, fog, or mist. In dry deposition, the acids fall to the ground, incorporated into dry materials such as dust or smoke. There are minimal direct health effects of acid rain because the reduced pH of the water and rain is not acidic enough to harm humans. However, the air pollutants responsible for acid deposition have well-established negative health consequences, as described previously. Furthermore, acid deposition is a serious environmental problem that affects large regions near industrial and urban sites.

Acid deposition can result in extreme acidification of lakes and streams, drastically affecting their ecology. Most lakes and streams have a pH between 6 and 8, but acid precipitation has led to chronic acidification of lakes and streams with acidic pH levels well below 5. Acid rain lowers the pH of these waters by directly supplying acids from the atmosphere beyond their natural capacity to neutralize them. Also, as rain flows through soils that then drain into these surface waters, aluminum is released from them. Thus, as the pH in a lake or stream decreases, aluminum levels increase. Both low pH and increased aluminum levels are directly toxic to fish and other aquatic organisms. By the early 1980s, hundreds of lakes in New York state became highly acidic as a result of atmospheric deposition, resulting in the local extinction of numerous aquatic animals.

Acid deposition also has serious consequences for forest health. Acid deposition weakens the trees' natural defenses, making them more vulnerable to diseases. Acid rain can directly damage plant tissues

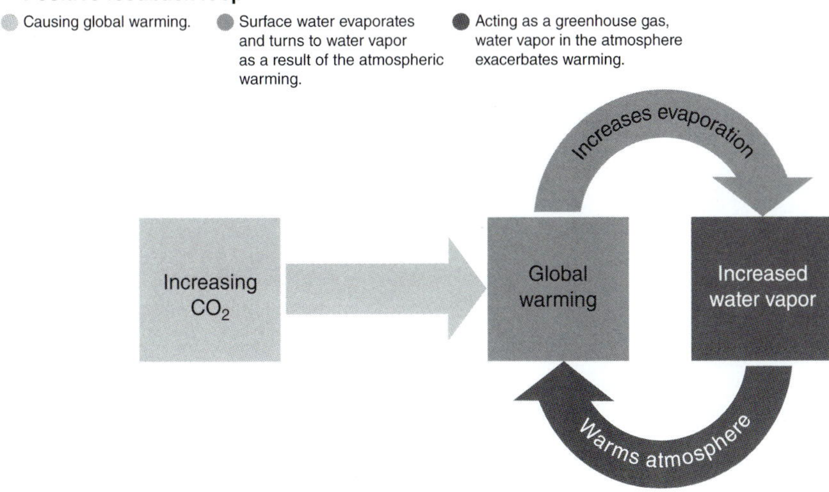

Figure 3.3 Warming caused by increased atmospheric CO_2 increases evaporation and the amount of atmospheric water vapor, another GHG. This positive feedback effect of water vapor nearly doubles the amount of warming caused by CO_2.

and leach vital calcium from them. It also leaches necessary minerals and nutrients such as calcium, magnesium, and potassium from soils before trees can absorb them. Also, acidification of soils mobilizes toxic metals such as aluminum, zinc, and mercury that are normally bound to soil particles. These metals can be toxic to plants. The combination of direct damage, nutrient deprivation, and exposure to toxins weakens trees and makes them more susceptible to disease, insect damage, cold exposure, and high winds. The result is often the death of large areas of forests of both ecological and economic importance.

Acid deposition can also negatively impact human-built objects. The acid damages stone, metal, and paint, including automobile coatings. Thus, acid deposition accelerates the deterioration of buildings and monuments, bridges, and automobiles. Their repair has cost billions of dollars. It also accelerates the corrosion of pipes, which can result in the release of lead into drinking water and the environment. Ancient monuments and buildings damaged by acid rain, such as the Parthenon in Greece, can never be replaced.

Acid deposition is a regional problem around industrial corridors such as the mid-Atlantic states in the United States and areas of Europe where there has been extensive acidification of freshwater systems and significant forest decline (Figure 3.4). For example, by the early 1980s in Germany, essentially the center of European industry that has had the highest levels of acid rain, more than 50% of forest trees were dead or dying. In the legendary Black Forest, tree mortality has reached 75%. In the northeastern United States, tens of thousands of lakes and streams have suffered from chronic acidification. Acid deposition has also been implicated in forest degradation along the eastern United States, including the death of 50% of red spruce trees in the White and Green Mountains of New England. From 1965 to 1983, on Camel's Hump Mountain in Vermont, the number of red spruce declined by 73%, and lower altitude sugar and beech

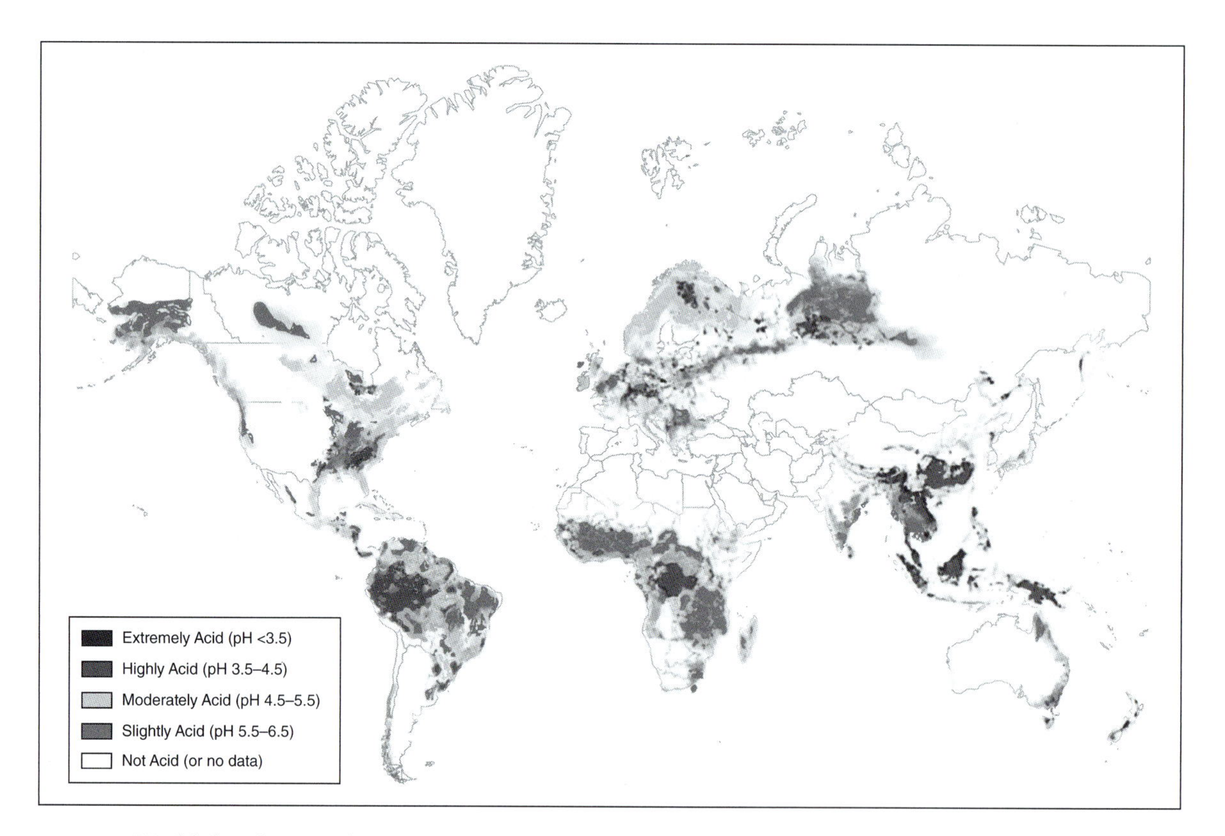

Figure 3.4 World distribution of acid soils. Adapted from Likens, G. E., Butler, T. J., & Rury, M. A. 2012. Acid Rain. In H. Anheier & M. Juergensmeyer, *Encyclopedia of Global Studies.* SAGE Publications, Inc.

tree biomass decreased by 25% due to acid mist and rain with pH ranging from 3.6 to 4.2.[5]

A number of solutions related to air pollution have begun to alleviate the problem of acid deposition in some developed countries. However, as industrialization occurs in other areas of the world, additional regions will be subjected to acid deposition. This is particularly true for large areas of Southeastern Asia, where economic growth in countries such as China and India has led to increased fossil fuel use without significant pollution controls. Many of these areas will be highly sensitive to acid deposition and will experience more drastic effects because the soils there lack the buffering capacity to neutralize the acids (see Figure 3.4).

Transboundary Air Pollution

Although most air pollutants are emitted by local or regional sources, air pollution does not stop at national borders. **Transboundary pollutants** are those that are generated in one country but are then transported by wind across borders, where their environmental and health impacts are experienced in other countries. Often, those countries are not the ones generating the emissions. Pollutants that are persistent in the atmosphere are more readily transported across boundaries than are those that have shorter atmospheric lifespans.

Transboundary air pollution was first recognized in the 1960s when scientists demonstrated the interrelationship between sulfur emissions in continental Europe and the acidification of Scandinavian lakes. A similar effect was seen in Canada, where acid precipitation generated in the northeastern United States was transported and deposited in Canada, detrimentally impacting its lakes, streams, and forests. More recently, serious air pollution events in China became a major concern for people in Japan because particulate matter was drifting across the sea in the winds from China. GHGs obviously have transboundary effect as well, but in the case of climate change, the impacts are more global. The issues of transboundary and global air pollution must be addressed at the international level and thus represent a significant challenge.

Environmental Justice and Air Pollution

Air pollutants that threaten public health and welfare often disproportionately impact specific populations, especially the poorer minority communities. Many low-income neighborhoods experience greater exposure to air pollutants, especially air toxics, because of their proximity to highways and transportation hubs, heavy industry, power generation facilities, waste incinerators and transfer stations, and area sources such as dry cleaners and auto body shops. When one community experiences more negative environmental consequences than another because of disproportionate placement of these emitters in its neighborhoods, air pollution becomes an environmental justice issue. Environmental justice will only be achieved when everyone has the same degree of protection from environmental and health hazards and has equal access to the decision-making process about where the sources of pollution are located.

There are many reasons for this environmental injustice, but they are mostly economic. Poorer, minority communities tend to lack the economic and legal resources, community organization, and power to fight placement of air polluters in their neighborhoods. The polluters generate profits in these poorer communities, and the residents see no economic benefit while being subjected to poorer air quality compared with other neighborhoods. One report showed that environmental injustice in the United States results in three times the number of asthma-related deaths among Blacks compared to Whites. Among children, this rate increases to 5:1. In some inner-city communities, one-third of all Black children have been diagnosed with asthma.[6]

A recent investigation by the Center for Public Integrity and the organization Reveal showed that nearly 8,000 US public schools lie within 500 feet of highways, truck routes, and other roads with significant traffic, exposing 4.4 million students to high levels of toxic air pollution.[7] The study showed that the concentration of air pollution is much higher near the roadways, and it enters schools' ventilation systems, thereby directly delivering it to students while in the classrooms. Such pollution can stunt lung development, affect the ability of children to learn, and increase the incidence of asthma, brain maladies, and cancer. Because many of these schools are in urban environments, this disproportionately impacts minority and lower income students. Approximately 15% of schools where more than three-fourths of the students are racial or ethnic minorities are located near a busy road. This is in comparison to only 4% of schools in which the demographics are reversed.

Although prevalent in the United States, environmental injustice with regard to air pollution is a global problem. In Australia, communities with the highest number of polluting sites, emission volume, and toxic air emissions have significantly greater proportions of indigenous populations and higher levels of socioeconomic disadvantage. Similar patterns regarding the disproportionate exposure to air pollution among the poor have been documented throughout the developed world. However, this problem is even greater in the developing world, where risks are generally higher because of the combination of the lack of investment in modern technology and weak environmental legislation in these countries. The rapid urbanization occurring in developing countries has resulted in cities with extremely poor air quality with high levels of exposure for the poorest communities.

Environmental injustice has become prevalent in Mexico as a result of the free trade policy with the United States. Such policy has resulted in a proliferation of assembly plants, or *maquiladoras,* that are typically located in the poorest neighborhoods. Residents in these neighborhoods, including thousands of schoolchildren, are exposed to high levels of dangerous diesel emissions from the steady stream of trucks that serve these plants. In addition to local sources of air pollution being disproportionately located in poorer communities, global climate change will disproportionately impact the global poor. This is because the poor tend to make their living growing food, and they live in regions that are more susceptible to flooding and in countries in which adapting to global climate change may be less economically viable.

Indoor Air Pollution

Indoor air pollution also represents a significant health problem, and it is related to environmental justice. Globally, more than 1.2 billion people lack access to modern energy sources, including electricity. These people must rely on other fuel sources for cooking, heating, and lighting, such as firewood, charcoal, biomass, or kerosene, which they burn in simple lanterns or stoves with incomplete combustion and little or no ventilation (Figure 3.5). This causes indoor air pollution, to which WHO attributes 4.3 million deaths each year in what may be today's deadliest environmental problem.[8] This number is more than the number of deaths from HIV/AIDS and malaria combined. Women and children living in severe poverty have the greatest exposures to household air pollution and suffer the most from related health consequences.

Figure 3.5 Indoor air pollution caused by burning biomass as cooking fuel without ventilation by coffee growers in Nicaragua. *Photo by R. Niesenbaum.*

Indoor air pollution is not just an issue for the global poor. Reduced indoor air quality (IAQ) is also a problem in developed countries with modern approaches to building, heating, and cooking. IAQ is influenced by indoor air pollutants and the degree of ventilation. Such indoor air pollutants include carbon monoxide from the combustion of fuels for heat generation, CO_2, VOCs, PM, second-hand smoke from tobacco products, ozone, and bacteria and molds. Efforts to increase home efficiency by decreasing air exchange with the outside have resulted in decreases in IAQ. CO serves to deprive the brain of oxygen, which can lead to nausea, unconsciousness, and death. The fact that it is colorless and odorless makes it impossible to detect without electronic sensors. VOCs, many of which can be serious health hazards, are emitted from paints, carpets and furnishings, building materials, and other household products and chemicals.

Sustainable Solutions for Air Pollution

There is much to be optimistic about with regard to solving the problem of air pollution and improving the quality of the air that we breathe. Some of the most successful environment policies have been in the area of clean air. Many of these have stimulated innovation and provided economic opportunity. There have also been advancements in technologies for reducing air pollution and cleaning already polluted air. There are cases in which stakeholders with very different perspectives have come together to solve air quality–related problems. There are also many ways that individual action can contribute to cleaner air. However, there are challenges and barriers to improving the quality of air for all people and to solving problems that are global in nature. But by employing certain approaches, these challenges and problems can be overcome.

Air pollution and its associated problems can be dealt with in essentially three ways: (1) stopping or reducing emissions of pollutants, (2) cleaning air that has already been polluted, and (3) adapting to our changed/changing environment. For these solutions to be sustainable, they must consider justice, transboundary, and public health issues while allowing for economic opportunity and growth that serves the majority of people. These three solutions can be achieved through local, national, and international policy; technical innovation; and changed behavior.

A first step in controlling pollution is to stop or reduce emissions through a combination of policy and technical innovation. Much of the world's economy is still based on energy from fossil fuels, and most air pollution results from their combustion. One obvious way to reduce emission of air pollutants is to decrease or stop our use of fossil fuels by lowering energy demands; increasing efficiency; and developing alternative, renewable energy sources. These are the focus of Chapter 6, which addresses energy and transportation. Our food system is also responsible for the emission of many air pollutants. This is considered in Chapter 5, which examines food and agriculture. Here, we consider other policy and technical approaches to emission reduction and elimination within the context of sustainability.

Policy Solutions to Reduce Air Pollution

There have been a number of policy approaches to reducing the emission of air pollutants at local, national, and international levels. Some of these policies have been quite effective, whereas others have not made much of a difference. Effective policy often requires broad public support, must address root causes rather than symptoms, and must take into account a context that is broader than the problem itself. Policy alone is not sufficient. Those that have been effective establish regulatory bodies and mechanisms of enforcement in conjunction with policy. The precautionary principle is often the basis of pollution-related policy. The principle can be traced to German national environmental law in 1976.[9] It was later applied to multilateral treaties, including the 1987 Brundtland Report and the 1992 Rio Declaration. The principle embraces the idea that where there are possible threats of serious harm or irreversible damage, scientific uncertainty about those threats should not preclude policy to limit them. However, there are many types of air pollution for which the scientific evidence for their potential to harm health and environment is equivocal but they clearly require regulation. In this section, a number of examples of air pollution policies are provided, and both their strengths and their weaknesses within the context of sustainability are considered.

THE US CLEAN AIR ACT

In response to rapidly declining air quality and associated health implications, the US Congress passed the Clean Air Act (CAA) in 1973. This Act is a federal law that was designed to protect human health and the environment from the effects of air pollution. In conjunction with the Act, the EPA was established in order to regulate emission of these pollutants. State and local governments also monitor and enforce CAA regulations, with oversight by the EPA.

The Act charges the EPA with establishing scientifically based emission standards for certain common and widespread pollutants as described previously in this chapter. These standards are based on the protection of public health and welfare with respect to point, mobile, and fugitive sources. States are then required to adopt enforceable plans to achieve and maintain these air quality standards, and they must also control emissions that drift across state lines. The EPA uses a transparent, stakeholder approach to design and implement clean air programs. Stakeholders include federal, state, local, and tribal leadership; affected industries; environmental groups; and individual communities.

The CAA has a well-established record of success. It has resulted in marked improvement in air quality (Figure 3.6) and has saved lives and improved public health by reducing hundreds of millions of cases of respiratory and cardiovascular disease. Successes of the Act include decreasing tropospheric

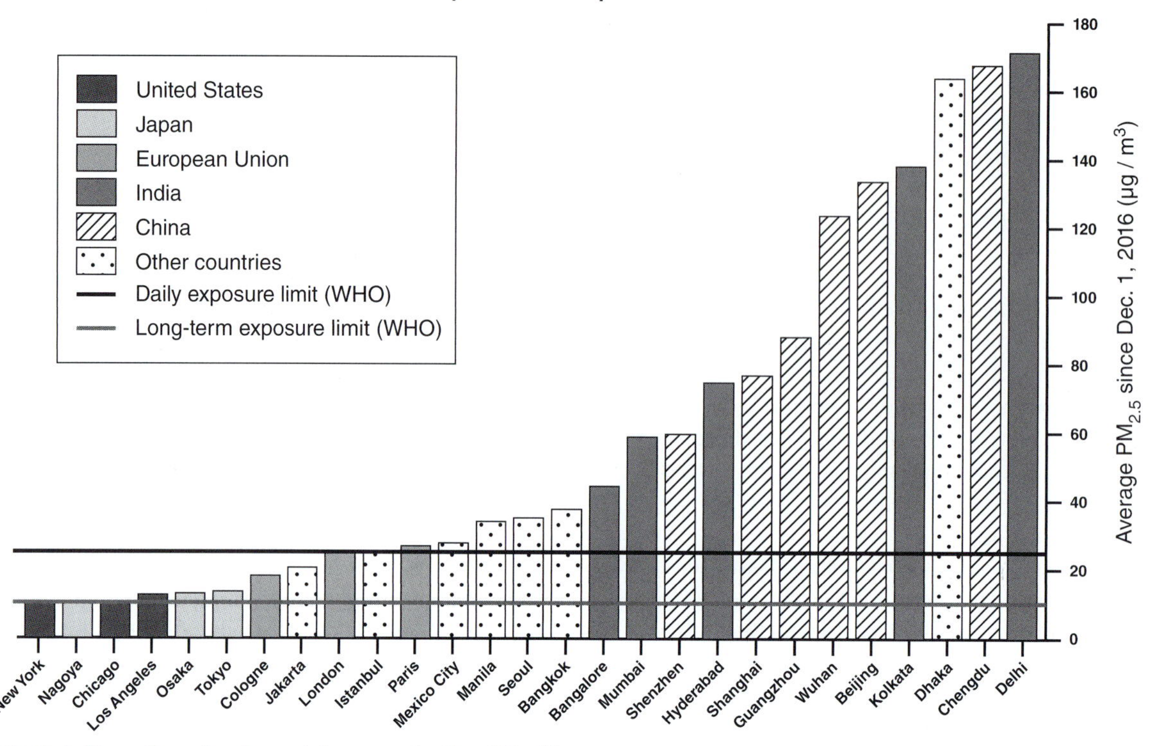

Figure 3.6 Largely due to the Clean Air Act and the EPA, US cities have the lowest amount of particulate air pollution compared to other large cities. This holds true for many other air pollutants as well. [(public domain) Adapted from BerkeleyEarth.org, http://shrinkthatfootprint.com/explain-carbon-budget, https://cdn.theconversation.com/files/59356/area14mp/fwn4wwz4-1411008433.png].

ozone levels by more than 25% since 1980, reducing mercury emissions by 45% since 1990, and significant reductions in the main pollutants that contribute to acid rain. Some might argue that the success of the EPA in improving air quality is conflated with the industrial decline that occurred in most US cities during the same period. However, improvements have moved well beyond industrial release. For example, one of the greatest EPA success stories was the phaseout of leaded gas in the United States. Tetraethyl lead was a gasoline additive that significantly improved engine performance but was phased out beginning in the 1970s because of its neurotoxic effects and mechanical-related issues. Since then, lead air pollution in the United States has decreased by 92%. Internationally, leaded gasoline remains available in only a handful of countries.

Evidence suggests that the CAA, while significantly reducing air pollution and improving public health, has stimulated rather than halted US economic growth. By creating new market opportunities in the area of emissions reduction technology, the Act stimulated innovation and new business in the development of cleaner technologies. One study shows that the direct economic benefits of regulation far exceeded compliance costs, and it predicts that they will continue to do so well into the future. In addition, there have been a number of indirect economic benefits of improved air quality. Cleaner air has resulted in fewer air pollution-related illnesses, which has reduced medical expenditures and absenteeism among American workers. Over the long term, these two indirect benefits alone will more than offset the expenditures for pollution control.[10]

The CAA, with its success in reducing air pollution, improving public health, stimulating economic growth and job creation, and its use of a stakeholder approach, meets many of the criteria of sustainability. One area in which it could have done better is in promoting environmental justice. Neither the CAA nor other federal regulations have provisions for environmental justice. This has placed the burden on states and communities to take action, but they often have little legal recourse to do so. Typically, action only occurs if a community comes together to address the addition of potential new sources of pollution. One successful example of this is California's banning of new school construction on land within 500 feet of freeways to protect children from exposure to air pollution. In addition, California's South Coast Air Quality Management District earmarked settlements from polluting companies and other funds to pay for the installation of air filtration systems at 80 schools with predominantly low-income, minority students that are near freeways or other pollution sources.

More recently, the EPA has made environmental justice a priority. This stems from President Bill Clinton's 1994 executive order, "Federal Actions to Address Environmental Justice in Minority Populations and Low-Income Populations." Although this did not result in specific policy, this executive order required that all federal agencies, including the EPA, make environmental justice a part of their mission to the greatest extent feasible and permitted by law.

THE CLEAN AIR ACT AND CLIMATE CHANGE

Another area in which the EPA and the CAA have been weak is with regard to regulating GHGs to prevent climate change. This changed in 2007 when the US Supreme Court ruled that GHGs are air pollutants and would subsequently be subject to EPA regulation under the CAA if a thorough scientific investigation showed that they endanger the public's health and welfare. In 2009, the EPA released its scientific findings, concluding that GHG emissions present a danger to public health.

This recognition of the danger of GHGs by the EPA led to President Obama's 2014 Climate Action Plan, which is a portfolio of cabinet-level energy and climate policies that were a cornerstone of his second term. The argued basis for this plan is that inaction on climate change could come with a $150 billion price tag. This cost will increase 40% with each decade of deferred action. The central objective of the plan, which was developed using a backcasting approach, is to reduce CO_2 emissions 30% by 2030 compared to the levels in 2005. This will be achieved by direct regulation of GHG emissions, increasing energy efficiency standards for appliances, making renewable energy production on federal lands a top priority, and leveraging natural gas to reduce GHG emissions.

Besides the usual political rhetoric of climate change deniers, there are some valid criticisms of this plan. Some argue that it may be too little too late. Others express concern about the plan's recommendation to use natural gas, which when combusted emits less CO_2 than other fossil fuels. However, its extraction employs environmentally risky hydrofracturing techniques. In this case, a systems approach to policy analysis could allow for a complete examination of the trade-offs between a proposed environmental benefit and resultant detriment. Others have argued that the plan is costly and will slow economic growth. However, as previously discussed, clean air regulation drives innovation and stimulates economic growth. Also, increases in efficiency and increased energy independence will reduce costs.

With the 2016 US presidential election, renewed calls for decreased regulation and the promotion of business interests over environmental protection threaten to weaken or even eliminate the CAA and the EPA. Given their prior success, the obvious importance of air quality, and accelerating climate change, this would prove to be a grave error. Individual scientists and activists have been playing a major role in ensuring that this does not happen or, if it does, that any such regulatory changes will later be reversed as the political climate shifts.

OTHER NATIONAL POLICIES

The United States is certainly not the only country in which policies have been put in place to limit the emission of air pollutants. China passed its first major environmental protection law in 1989, which imposed fines on air polluters. However, since then, China has transformed into a global manufacturing hub with expanding automobile use. It is now one of the largest consumers of energy and the world's largest carbon emitter (Figure 3.7). Air quality in most cities in China has seriously declined. For example, Beijing's air quality in 2013 failed to meet government standards on 52% of the days due to smog and pollution. This decline in air quality has had serious public health consequences, and it has become a leading cause of social unrest. One of the reasons for the policy failure is that fines were not high enough to encourage the installation and operation of pollution controls. It was simply more cost-effective to pay the fines and continue to pollute. Also, there was little monitoring and enforcement activity.

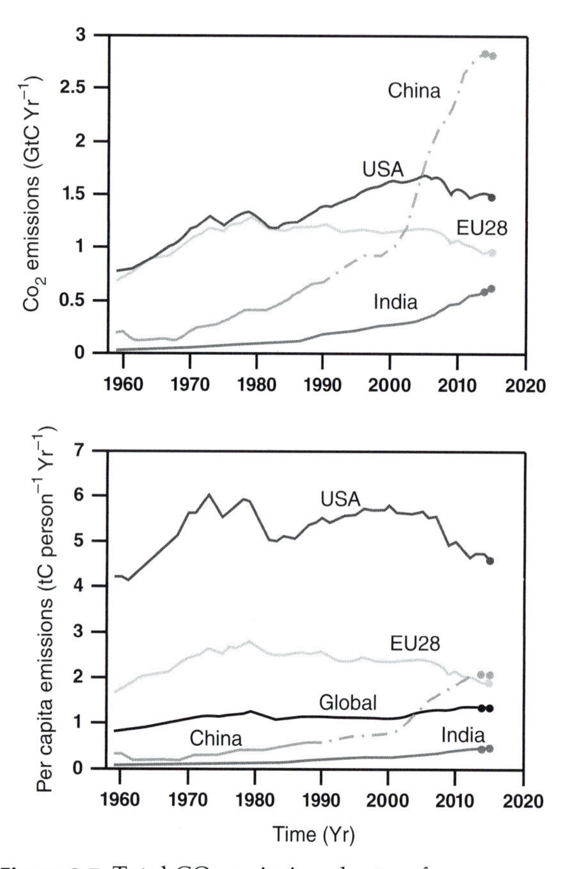

Figure 3.7 Total CO_2 emissions by top four global carbon emitters by year (above). In 2015, this represented 59% of global emissions. Per capita CO_2 emissions by top four emitters in comparison to global average (below). *Sources: The Global Carbon Budget 2016 and the US Department of Energy Carbon Dioxide Analysis Center, http://www.globalcarbonproject.org/ and https://www.carbonbrief.org/what-global-co2-emissions-2016-mean-climate-change*

In 2014, China moved to address the environmental damage that has been a by-product of its massive economic growth, and it has done so by amending the 1989 law that makes punishment for polluters much harsher and provides for increased monitoring. It allows for consecutive daily fines on polluters if they do not make improvements, and it offers channels for whistle-blowers to make environment-related appeals. For the first time, non-government groups can file lawsuits for environmental damage. Company managers can be detained if they have not done an environmental impact assessment or if they ignore orders to stop construction or continue to pollute after being asked to stop. It is too soon to determine whether the 2014 amended policy has been effective, but there is much hope for improvement. Although China has been a top global investor in clean energy, it will still be reliant on coal for most of its electricity through 2030, when its power needs will have more than doubled.[11] Policy that will encourage transition from coal to renewable energy will be needed and is being developed in China.

As less developed countries have experienced economic growth, industrialization, and urbanization, many of them have had to implement air quality standards and policies that regulate emissions. Examples include the development and implementation of modern air pollution policies in Malaysia, Thailand, and India. African nations have tended to lag behind in this regard. To address this, in 2012 the Environmental Ministries of Ghana and Nigeria, with support from the Climate and Clean Air Coalition, the African Climate Policy Centre, and UNEP, hosted a three-day event to examine the critical need for national action to address and mitigate air pollution. Other participating countries included Benin, Côte d'Ivoire, Ethiopia, Gabon, Malawi, Senegal, Togo, and Zimbabwe.

Europe has historically had air quality issues. In London, the shopping district on Oxford Street has been dubiously recognized as the most polluted street in the world. In 2013, the average NO_2 level on this shopping strip was 135 micrograms per cubic meter of air ($\mu g/m^3$), with spikes as high as 463 $\mu g/m^3$.[12] This is more than three times higher than what has been deemed safe for human exposure. The source of these high levels is diesel fumes from buses and automobiles combined with the narrow street and tall buildings that trap the fumes. This is despite a 1995 environmental act for the United Kingdom that regulates air quality. This problem has motivated the designation of locations with high amounts of air pollution as air quality management areas, for which local authorities are required to develop emissions reduction plans and make them available to the public. However, because the bulk of the air pollution in this case is from transportation, reducing emissions through this sort of policy has proved challenging. Mobile sources of air pollution are difficult to regulate and require different approaches compared to fixed sources, such as implementing and enforcing automobile emission standards like those in the United States and particularly the state of California.

Like the US Climate Plan of 2014, a number of other nations have developed their own policies to limit GHG emissions in order to address climate change. In reality, the United States has been late to the table regarding climate policy. In a book assessing climate policy,[13] Vivian Thomas shows that Germany has long been a leader in climate change mitigation and renewable energy. From 1990 to 2010, Germany reduced its GHG emissions by 26%, while its per capita gross domestic product (GDP) increased by 36%. During the same time period, US GDP grew slightly less, while GHG emissions declined by only 2%. Other leading nations in climate change and GHG emission reduction are Brazil, Denmark, Portugal, France, and Sweden.

INTERNATIONAL POLICY

National policy alone may be insufficient to address air pollution given its transboundary and global nature. This is particularly true for acid rain, ozone depletion, and global climate change, all of which have impacts that transcend national borders. International policy can be difficult both to design and to implement, especially when the interests and economic status of the stakeholders vary so widely.

When countries are already aligned for political and economic reasons, international regulation of air pollution may be easier to achieve. This is true for the European Union (EU), which has formed its own multination environmental agency. The EU's Seventh Environment Action Program aims to achieve levels of air quality that minimize risks to human health and is considered by some to have the most extensive environmental laws of any international organization.

Operating much like the EPA, the EU policies aim to reduce exposure to air pollution by setting limits and target values for air quality for all its member nations. In late 2013, the European Commission proposed a new clean air policy package for Europe that aims to achieve full compliance with existing air quality legislation by 2020 and further improve Europe's air quality by 2030. The anticipated exit of Britain from the EU with the "Brexit" vote in 2016, and similar rumblings from other EU nations, may put the future of this policy at risk.

Ozone depletion is another environmental issue that transcends national boundaries. The 1987 Montreal Protocol on Substances that Deplete the Ozone Layer was designed to reduce the production and use of ozone-depleting substances worldwide to reduce their abundance so as to halt and reverse the depletion of the Earth's protective ozone layer. Central to this was the phaseout of ozone-depleting CFCs used primarily as refrigerants and aerosol propellants. This phaseout was in direct response to an enlarging upper atmosphere ozone hole.

Implementing the Montreal Protocol proved to be a challenge for two reasons. First, a number of poorer nations were just coming online with CFC technologies and lacked the economic resources to transition to new technologies. This was addressed by employing a delayed implementation strategy for poorer nations and establishing a multilateral fund to offer economic assistance to developing country parties making the transition. Second, CFCs are part of a class of valuable chlorinated chemicals that represented a billion-dollar-a-year product for the world's dominant producer, DuPont. Typically, when policy contradicts the interests of a large global corporation such as DuPont, it has been a challenge to implement the policy. However, in this case, it was in the interest of both environmentalists and DuPont to ban CFCs (Box 3.1).

Adherence to the Montreal Protocol was successful in decreasing the atmospheric concentrations of the most threatening CFCs and related chlorinated hydrocarbons, and by 2010, the stratospheric ozone layer was significantly restored. Because of this, the Montreal Protocol has often been called the most successful international environmental agreement to date. However, this success can only be viewed within

BOX 3.1

Stakeholders and Collaborators: Coalitions of the "Green and the Greedy" in the Banning of CFCs

One of the greatest environmental success stories was the ban of CFCs as part of the 1987 Montreal Protocol. The objective of this agreement was to reduce the production, use, and resultant release of ozone-depleting substances in order to stop their accumulation in the upper atmosphere. This became an urgent priority after the discovery that as CFCs used as refrigerants and aerosol propellants accumulate in the atmosphere, they cause the degradation of the stratospheric ozone layer. This layer is responsible for filtering harmful UV radiation. Even more significant was the discovery of a large ozone hole over the Antarctic in 1985, which accelerated the interest in halting the use of CFCs to protect the ozone.

One of the greatest obstacles to passing environmental policy has been the perception that the policy will have an associated economic cost. These costs often contradict the interests of powerful corporations, which in turn use their influence to shape or stop the policy from being enacted. In this case, CFCs were generating billions of dollars in revenue per year from the global market for their primary producer and patent holder, DuPont. It seemed unlikely that the Montreal Protocol would be adopted if it contradicted the profit objectives of this large, multinational corporation. Given this obstacle, how did the worldwide phaseout and eventual ban of CFCs occur? According to an article by James Maxwell and Forest Briscoe in *Business Strategy and Environment*,[1] it turns out that the ban was actually in DuPont's economic best interest. Because of this, DuPont chose to play a major role in influencing the international regulatory regime for ozone-depleting substances.

DuPont was the dominant producer of CFCs, which it marketed as Freon. Primarily used for refrigeration, as a propellant in aerosol sprays, and for other industrial uses, Freon was one of DuPont's most successful products. The discovery of the large Antarctic ozone hole was a tangible and visual environmental and public health concern that motivated consumers to seek alternatives to CFC-propelled aerosol spray products. The marketplace thus drove the phaseout of CFCs as aerosol propellants. The resultant market decline and subsequent regulatory developments for CFC aerosols had serious consequences for DuPont's CFC business, costing the company nearly 20% of its global market share. This market change motivated DuPont to develop and shift to alternatives, which it now believed would demand higher prices than CFCs. The new products could be marketed as specialty rather than commodity chemicals that would be characterized by lower production volume and higher prices and profit margins as a result of reduced competition. Moreover, DuPont's patent on CFCs was due to expire, which could further diminish its profit potential.

According to Maxwell and Briscoe, DuPont was able to capitalize on the ban for a variety of reasons. First, by the early 1980s, a number of chemical alternatives had been identified. Second, the corporate leadership at DuPont recognized that it would be more beneficial to be an early leader in the transition rather than opposing what appeared to be an inevitable change. This allowed the company to take advantage of heterogeneity in the industry, in which it stood to gain while others would lose. Third, the cost of the transition would be shared among DuPont and the end-users who would have to make the infrastructural and product design changes required by the new chemicals. Finally, DuPont's new "corporate environmentalism" improved its relationship with the government and enhanced its reputation with the growing environmentally aware public.

Thus, regarding long-term economic interests, the ban on CFCs and eventual abandonment of a multibillion-dollar product would serve DuPont well. As a result, the company ultimately supported the Montreal Protocol, which mandated a 50% reduction by 2000. A more cynical environmentalist view of this would be that environmental policy only gets passed when it is in the best interest of a large multinational corporation. However, a more pragmatic, sustainability perspective would seek some lessons from this and explore methods to encourage potential industry winners to support environmental policy. The international community adopted the Montreal Protocol because there was market-driven political will and clear economic incentives for a major corporate player. Exploring other ways that the "greedy" and the "greens" might collaborate and the conditions by which a mutually beneficial end result might occur will be important to consider in our effort to develop and implement new environmental policy.

1 Maxwell, J., and F. Briscoe. 1997. "There's Money in the Air: The CFC Ban and DuPont's Regulatory Strategy." *Business Strategy and the Environment* 6: 276–86.

the narrow sense of stratospheric ozone layer protection and restoration. This is because the replacements for CFCs, HCFCs, are potent GHGs (see Table 3.1). This demonstrates that different environmental problems are linked and must be considered in concert with each other by employing a systems approach. There now must be increased efforts to link ozone protection to climate protection.

INTERNATIONAL POLICY ON GHGs AND CLIMATE CHANGE

Because of the transboundary impact of GHGs and the global nature of climate change, international policy will be an essential component in controlling them. A first step in addressing climate change from an international perspective was the creation of the IPCC by the World Meteorological Organization and UNEP in 1988. The mission of the IPCC is to prepare, based on available scientific information, assessments on all aspects of climate change and its impacts with a view to formulating realistic response strategies.

The work of the IPCC led to the UN Conference on Environment and Development, or the Earth Summit, held in Rio de Janeiro in 1992. At this summit, the UN Framework Convention on Climate Change was ratified by all of the UN member states. The treaty provided a framework for negotiating specific international treaties or protocols that could set limits on GHG emissions. The treaty itself contained no limits or mechanisms for enforcement. These types of policies were developed in subsequent negotiations.

THE KYOTO PROTOCOL

The first international agreement to establish binding GHG emissions reduction targets was the Kyoto Protocol, adopted in 1997. The legally binding treaty required industrialized nations to reduce their yearly emissions of GHGs by varying amounts, averaging 5.2%, by 2012 compared to 1990 levels. This would be the equivalent of a 29% cut in the values that would have otherwise occurred without those reductions. Although the United States, Australia, and some other countries signed the treaty, they refused to ratify it unless China and other developing nations agreed to limit their emissions. China had objected to doing this before the developed world acted. The United States pulled out in 2005, and the protocol did not become international law until more than halfway through the period established for reduction. Both the lack of participation by the United States and the delay in establishing the law undermined the proposed GHG limits. Some countries and regions, including the EU, were on track by 2011 to meet or exceed their Kyoto goals. However, many other countries were not even close, especially the two largest emitters, the United States and China. The amount by which these two countries exceeded the GHG emission limits nullified all of the reductions made by other countries during the Kyoto period (see Figure 3.7).

BEYOND KYOTO

Subsequent negotiations have occurred in Copenhagen (2009), Cancun (2010), Durban (2011), Doha (2012), and Lima (2014). These imposed further limits on GHG emissions for developed nations and began to recognize the issues faced by developing countries by integrating GHG limits with plans for sustainable development. There was recognition that a key element of sustainability is that social and economic development and poverty eradication are priorities for development, but they should be integrated with low-emission development strategies and actions to reduce current emissions. It was also recognized that in most cases, poorer nations would lack the resources to limit, mitigate, and adapt to climate change, and they stood to be the most impacted by it. To address this, developed countries committed considerable financial resources to support emissions reductions and adaptation in developing countries.

One issue with the Kyoto Protocol is that it established limits based on a percentage cut in the values that would have otherwise occurred if emission rates remained the same rather than using a backcasting approach based on a meaningful objective such as stopping climate change. The later protocols took the latter approach by establishing limits to hold global warming below 1.5–2°C relative to preindustrial averages. This would be achieved not just through emissions limits but also would include mitigation approaches and opportunities for carbon offsets and emission trading (Box 3.2). Later protocols also established a framework for national and international monitoring for both developed and developing countries, but they were more like calls for action than actual policy frameworks.

There have been numerous criticisms of these international protocols. The primary one is that they have not achieved their stated goals of reducing CO_2

BOX 3.2

Policy Solutions: Buying and Selling the Right to Pollute Through "Cap and Trade"

Emissions trading is a market-based approach that provides economic incentives to companies for reducing their emissions of air pollutants. It is also referred to as cap and trade, and it basically provides opportunities to buy and sell the right to pollute. With cap and trade, the government does not mandate specific mechanisms or technologies that corporations must implement to reduce their emissions, many of which can be quite costly. Rather, a limit or cap on the quantity of emissions of specific pollutants is imposed. In this way, each company has a given amount that it is allowed to emit per year, essentially a pollution allowance. Each company can then choose how to use its allowance. One approach is to restrict output through pollution control so as not to exceed the cap. But allowances can be traded on the open market, so a company can further reduce its emissions and sell its allowances to another company. Alternatively, the company that exceeds its allowance can purchase the right to pollute from a company that has excess allowances. Thus, those that can most cheaply limit their pollutants are motivated to do so because of the opportunity to sell their allowances, and for companies whose reduction is too costly, there is the opportunity to purchase those allowances rather than reduce emissions.

Although suggested by economists since the early 20th century, a cap and trade policy was not put in place until 1990 in the United States under President George H. W. Bush, when it was incorporated into the Clean Air Act. This was in response to the issue of acid rain caused by excessive sulfur dioxide release by US power plants. There was also pressure from Canada, which suffered from the transboundary impacts of US emissions. After 10 years of failing to implement a "command and control" approach requiring power plants to install scrubbers to remove sulfur dioxide and implementing pollution taxes, this market-based solution seemed more palatable to the free-market conservatives in power at the time. It seems to have worked: by 1995, when cap and trade was in full effect, acid rain emissions declined by 3 million tons.[1]

Although the political history of cap and trade reveals a novel approach that brought together conservatives and environmentalists in the late 1980s, implementing a similar policy to reduce CO_2 emissions has proven more challenging. Emissions trading was incorporated into the Kyoto Protocol, setting caps for the six major GHGs for industrialized nations. In addition to being able to sell allowances when nations emit less than their quota, GHG removal through reforestation and stopping deforestation can be used or traded to count toward emissions reductions. This has created an international carbon trading market that includes a number of industrial nations and the European Union. However, the greatest GHG emitters, the United States and China, are not participating.

The major benefit of emissions trading is that it allows for pollution reduction at the lowest cost to society. It is often preferred by governments and corporations and is thus easier to pass than taxes, fines, and command and control requirements to stop emissions. Cap and trade also requires polluters to pay for the acknowledged costs of pollution, allowing internalization of what has typically been an externality. Finally, the emissions trading scheme that grew out of the Kyoto Protocol has resulted in factories and power plants switching from more GHG-intensive fuels such as coal to those such as natural gas or renewables that emit less and help keep the factories within their limits.

Criticisms of emissions trading are that in order to deal with global climate change, more radical reductions will be required and these cannot be achieved through currently established limits. Also, carbon trading schemes tend to reward the heaviest polluters when they are granted enough carbon credits to match their historic production. Carbon trading does not stimulate the needed long-term structural change in industry and energy use because purchasing offsets is currently less expensive than making those changes.

1 Coniff, R. 2009. "The Political History of Cap and Trade." *Smithsonian Magazine*, August, https://www.smithsonianmag.com/science-nature/the-political-history-of-cap-and-trade-34711212/

emissions. Others have argued that the stated goals, even if they are achieved, are insufficient to mitigate the impacts of global climate change. The challenge in achieving significant results through this international process is due to the fact that the UN Framework Convention on Climate Change is a multilateral body, an inefficient system for enhancing international policy. Because the framework system is governed by consensus of its 190 member countries, it is easy for small groups or even individual countries to block progress. National commitments that spur innovation, focus on more specific approaches, and bring all of the stakeholders—including businesses—to the table may turn out to be more effective and sustainable.

THE 2015 PARIS CLIMATE CONFERENCE

There is increasing optimism that we can solve the climate crisis. Even without the ratification of prior agreements, many countries are moving forward with efforts to reduce GHG emissions. The transition from fossil fuels to renewable energy is occurring at a rate that has far exceeded expectations. The most recent UN Climate Change Conference, also known as COP 21, held in Paris in 2015 established binding commitments for all countries and a framework by which to pursue measures to reduce emissions with the aim of keeping global warming below 2°C. To achieve this, the agreement calls for the goal of zero net anthropogenic GHG emissions to be reached during the second half of the 21st century.

A key to the success of COP 21 would be the commitment of both the United States and China, the two largest emitters, to agree to limit their emissions. They did so in 2014 when President Obama and General Secretary Xi Jinping agreed to limit GHG emissions, thus paving the way for the success of COP 21. On April 22, 2016 (Earth Day), 174 countries signed the agreement in New York and began adopting it within their own legal systems through ratification, acceptance, approval, or accession.

Meeting the COP 21 objectives of moving to a low-carbon future and limiting average global temperature rise to 2°C will require a variety of approaches, including decarbonizing electricity production by transitioning to renewable energy, improving resource efficiency, and the use of emissions trading schemes (ETS), which is essentially adopting cap and trade (see Box 3.2) policies but on an international scale. Currently, 35 countries are employing ETS to achieve carbon reduction targets.

Another often cited approach for meeting COP 21 emissions targets is the adoption of a carbon tax. A **carbon tax** is a fee imposed on users of fossil fuels that motivates the switch to carbon-free energy. A typical carbon tax rate is from \$5 to \$10 per ton of CO_2, but some have adopted higher rates. Some proposed carbon taxes are considered revenue neutral when the taxes are used to pay down deficits, create jobs, or invest in new carbon-free technologies. Carbon taxes are garnering public support, even in the United States.[14] However, changing political climates may limit their adoption. Carbon taxes are being proposed and enacted in a number of countries, including Ireland, Australia, Chile, and Sweden, and in cities such as Singapore and Boulder, Colorado.

Similar to a carbon tax is a **carbon fee-and-dividend approach**. The difference between a tax and a fee is that a tax has the primary purpose of raising revenue for the government. By contrast, a fee recovers the cost of providing a service from a beneficiary, which can then be passed on as a dividend to end-users. The fee is based on per ton CO_2 equivalent emissions from burning fossil fuels imposed upstream at the mine, well, or port of entry. This fee in part accounts for the true costs or externalities of fossil fuel emissions. Because the fee is passed on to consumers, they are able to make a true cost comparison of various energy sources when making purchasing decisions. Ultimately, the fees would be returned to consumers as a monthly dividend. Both import fees on products from countries without a carbon fee and rebates for exporters would discourage the relocation of companies to locations where they can emit more CO_2. Such a fee-and-dividend program could reduce US CO_2 emissions 50% below 1990 levels and add 2.8 million jobs to the economy by 2037.[15]

An alternative to carbon taxes and fees that might be more palatable in the current political and economic climate is a tax cut for zero-emissions utility plants. These are referred to as **zero-emissions tax credits (ZEECs)**. Because electricity is already taxed, a tax cut for reducing emissions will reduce utility bills. ZEECs will also spur investment and incentivize innovation in order to reduce the carbon footprint of energy generation. ZEECs are seen as a more effective way to reduce carbon emissions because tax cuts are more readily enacted than new taxes.

Despite the strong scientific, economic and social arguments for meeting the emissions reduction plan of the Paris Agreement as well as the various ways in which it could be accomplished, in June of 2017 President Donald Trump announced that the US would cease participation in it. Criticism of the withdrawal was substantial, spanning political and religious affiliations, coming from business leaders, scientists, environmentalists and citizens in the US and around the world. However, according to the agreement, the earliest possible date that the US can withdrawal is after the November 2020 presidential election. The White House later indicated that the US would honor this and remains obligated to maintaining its commitments under the agreement until that date. This time frame offers voters a significant opportunity to determine the future commitment of the US to this agreement. Also, in response to the announcement of the withdrawal, governors from 16 US states and Puerto Rico formed the US Climate Alliance in order to advance the objectives of the Paris Agreement even if the withdrawal still goes into effect. Similar commitments have been expressed by local and state governments and businesses.

Innovative Solutions to Reduce Emissions

In response to emission limits as part of local, national, and international policy, there have been a number of technological advances in the area of emissions reduction. A suite of technologies has been developed in order to offer increased efficiency and to limit air pollution well below current emission regulations, contributing to a cleaner and healthier environment. These technologies have been quite effective in reducing emissions of criteria pollutants to such a degree that acid deposition has been greatly reduced in industrial areas where it was once a crisis. Other pollutants, particularly GHGs, have been more challenging to remove from emissions. The ultimate goal should be and is to develop near-zero-emissions factories and power plants.

FILTRATION

One class of clean-air technology that directly removes pollutants from exhaust gases from power plants, incinerators, and other industries includes electrostatic precipitators and complex filtration devices. An **electrostatic precipitator** is a filtration device that effectively removes particulates and aerosols from a flowing gas using the force of an induced electrostatic charge. High voltage is applied to the gas to be emitted using discharge wires, thereby charging the particles, which then adhere to charged collecting plates, thus removing them from the gas. This is done in a way that minimally impedes the flow of the gases. The plates are then cleaned, and the particulates are collected and disposed of in a way that does not release them into the air (Figure 3.8).

Direct filtration devices include **baghouse filters**, which use fabric to collect particulate matter. The addition of different types of fabric coatings helps capture other pollutants, including other gases and metals. Baghouse filters are affordable and effective,

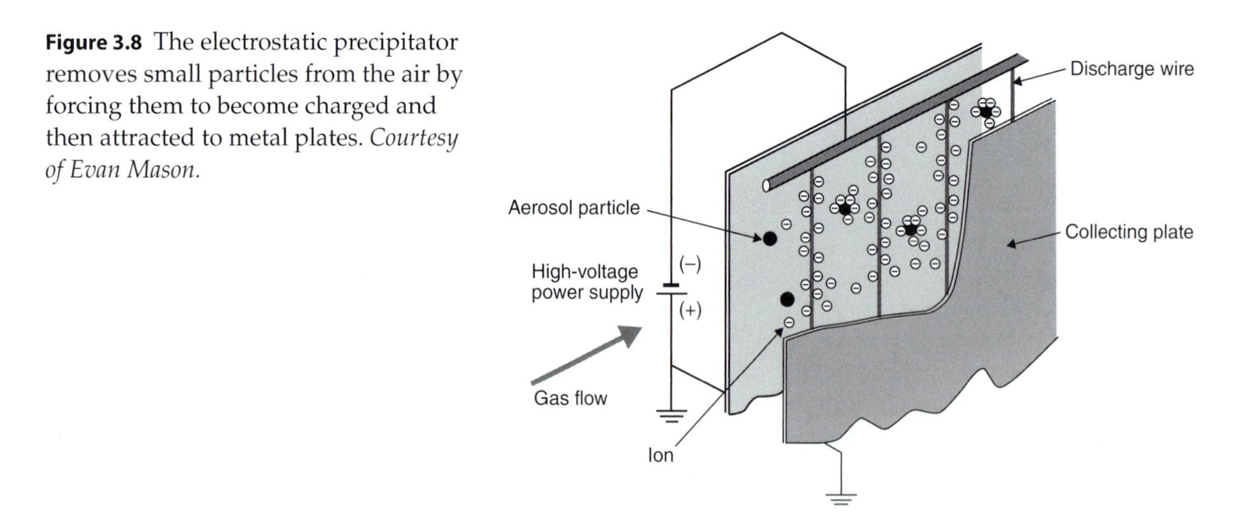

Figure 3.8 The electrostatic precipitator removes small particles from the air by forcing them to become charged and then attracted to metal plates. *Courtesy of Evan Mason.*

but drawbacks include reduction in gas flow and the need for frequent maintenance.

SCRUBBERS

Another common type of pollution control device is the scrubber system. Scrubbers are a diverse group of air pollution control devices that can be used to remove particulates and gases from industrial exhaust streams. Typically, they use liquid to wash unwanted pollutants from a gas stream. Alternatively, dry scrubbers inject a dry reagent or slurry into a dirty exhaust stream to remove acid gases.

Scrubbers have been effectively used to remove NO*x* and SO*x* gases that are responsible for acid rain and were limited by the US cap and trade policy (see Box 3.2). Scrubbers have also been used to remove mercury from emissions at power plants and other industrial facilities. Another approach to removing NO*x* is through selective catalytic reduction (SCR). With SCR, a gaseous reducing agent such as anhydrous ammonia, aqueous ammonia, or urea is added to a stream of flue or exhaust gas and is adsorbed onto a catalyst. This is an extremely effective way to remove more than 90% of NO*x* from emissions. However, one downside is that CO_2 is a reaction product when urea is used as the reducing agent.

DUST CONTROL

A great deal of particulate matter is generated as dust during construction and in agriculture. Many municipalities require that companies and property owners control dust emissions. Dust can be effectively controlled at construction sites through physical barriers, site traffic control, water application, earth-moving management, and soil stabilization techniques. In agriculture, soil conservation practices such as no-till agriculture and the use of cover crops can reduce dust emission while preserving soil quality.

CO_2 CAPTURE

There have been major technical advances in cleaning industrial emissions of criteria pollutants, particulates, contributors to ground-level ozone and smog, VOCs, and metals such as mercury. However, the technology for removing CO_2 has lagged behind, and it remains largely theoretical and cost-prohibitive despite the current urgency to reduce CO_2 emissions. The working concept is that emissions would be pumped into a scrubber where the CO_2 comes into contact and is captured by ion exchange resin. Once captured, the carbon is removed from the resin using very high temperatures and either reused or stored. One possible use is to pump the collected CO_2 into commercial greenhouses to increase plant production. There is also potential in growing algae with the collected carbon, which could then be used as food or fertilizer or to make biodiesel.

There has also been much recent investment in developing ways to store captured carbon in geological formations deep inside the Earth. Captured carbon can be pumped into oil wells, allowing for increased extraction efficiency particularly toward the end of the life of the wells. The emptied well can then be capped, thereby storing the captured CO_2 indefinitely. Dealing with a captured gas remains a major cost factor and impediment to CO_2 removal. A recent development that may lead to a cost-effective solution to this is the discovery that the compound guanidine reacts with CO_2 to form crystals containing carbonate.[16]

REDUCING MOBILE SOURCE EMISSIONS

In addition to point source emissions, mobile sources such as trucks, cars, buses, and other forms of transportation fueled by fossil fuels are also a major source of air pollution. There are a number of approaches to reducing these mobile sources of air pollution. The first is to burn less fuel by making vehicles more fuel efficient. The auto industry has made great advances in this area. A second approach is to produce cleaner fuels. The phaseout of leaded gasoline was a major first step toward this goal and has been quite successful in nearly eliminating lead from vehicle emissions. The removal of sulfur and light hydrocarbons from fuels has also controlled engine emissions.

There have also been advances in engine technology to reduce the amount of pollutants in the exhaust. The advent of the **catalytic converter** for automobiles was a significant advancement (Figure 3.9). Located in the exhaust pipe, the catalytic converter catalyzes or speeds up the conversion of hydrocarbons, carbon monoxide, and NO*x* into less harmful gases. The catalysts consist of platinum, palladium, and rhodium. As the gases from the engine are passed over the catalyst, oxidation or reduction reactions take place on its surface, converting the pollutants into harmless gases. One exception is the production of CO_2, which is converted from the more toxic CO but still the major GHG. In addition to the catalytic converter, other engine

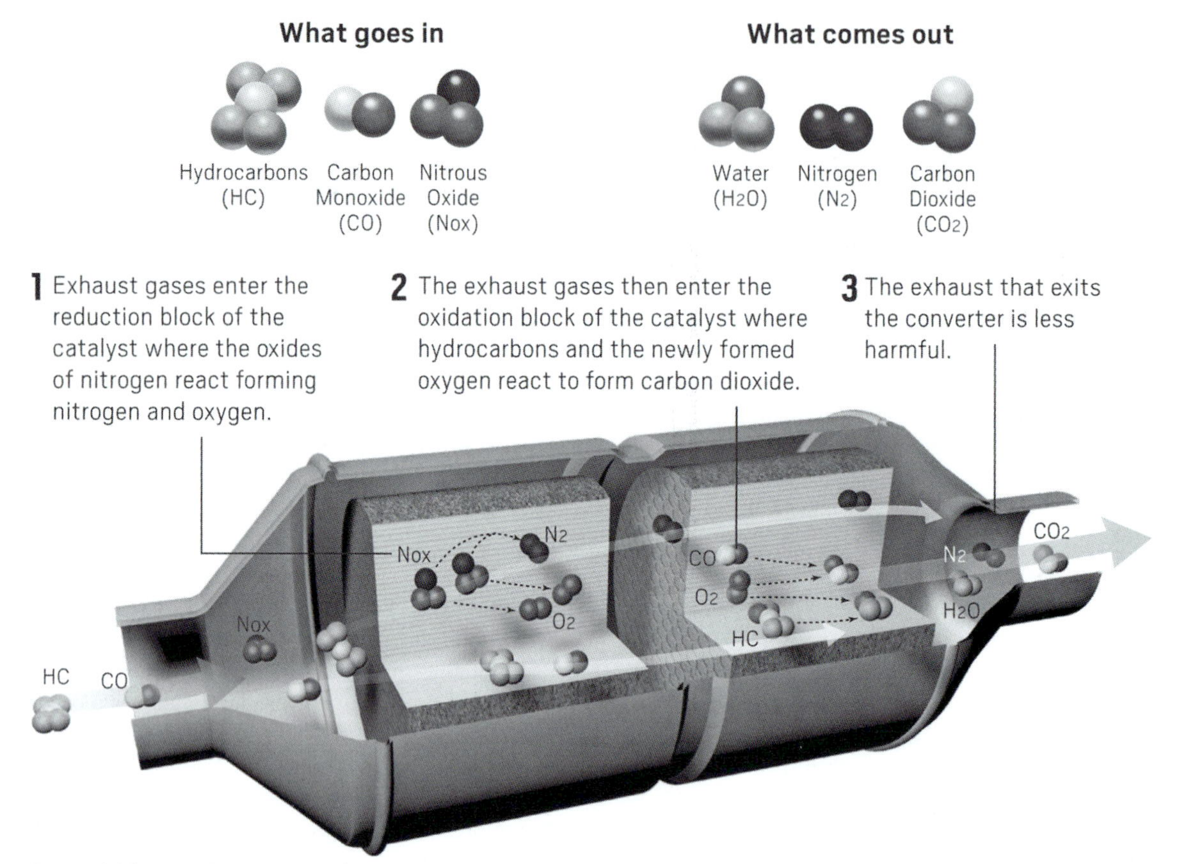

Figure 3.9 The catalytic converter converts hydrocarbons, carbon monoxide, and NOx into less harmful gases by using a combination of platinum, palladium, and rhodium as catalysts. *Source: California Air Resources Board. Jeff Goertzen/Staff.*

redesigns that adjust air-to-fuel ratios, ignition timing, turbulence in the combustion chamber, and exhaust gas recirculation have also helped curb emissions.

The development of alternative fuels, zero-emissions vehicles, and public transportation systems is perhaps the most effective way to limit mobile source emissions. Alternative fuels such as natural gas, methanol, ethanol, and hydrogen burn cleaner than gasoline and diesel and have lower carbon content, resulting in reduced CO_2 emissions. Zero-emissions vehicles would be most effective in reducing mobile sources of emissions. Electric and hydrogen fuel cell cars completely eliminate tailpipe emissions of pollutants, and gasoline–electric hybrid vehicles significantly reduce them. Norway was one of the first countries to incentivize the development of these vehicles and aims to have all zero-emissions vehicles by 2025. Other countries are following suit, including Austria, China, Denmark, Germany, Ireland, Japan, the Netherlands, Portugal, Korea, and Spain, all of which have set official targets for electric car sales. The United States does not have a federal policy, but at least eight states have set out goals for transitioning to all electric, zero-emissions vehicles. The major car manufacturers are now poised to meet this new demand. For example, General Motors reports that it will put two new, fully electric models on the market in 2018 and then at least 18 more by 2023. This is another example in which government policy and consumer demand can drive innovation and the marketplace.

Efficient public transportation, particularly when it incorporates low-emissions vehicles, can be extremely effective at decreasing air pollution, but only when it reduces the total number of vehicles on the road. Finally, personal actions such as using public transportation, carpooling, cycling, fuel-efficient driving techniques, and reductions in idling are also effective solutions (Box 3.3). We consider these further in Chapter 6.

Solutions that Clean Polluted Air

As discussed previously, both policy and technology can be very effective at limiting emissions, resulting in better air quality. However, depending on the location, emissions may still be an issue. This is particularly true for CO_2, for which policy has been less successful and technology for removing it from emissions is lagging behind and is currently cost-prohibitive. Thus, there is a need to develop ways to improve air quality by mitigating or cleaning the air once pollution has occurred. Here, we explore a variety of ways that this can be achieved.

VEGETATION

One obvious way that vegetation can mitigate air pollution is through the uptake and storage of CO_2 through photosynthesis. Thus, increases in atmospheric CO_2 can be limited by increasing the photosynthetic capacity of the planet. This can be achieved through halting deforestation, conducting reforestation, and managing forest and agricultural systems so that carbon uptake and storage are maximized. At issue is that when plants die and decompose, or if they are burned, the carbon they had removed is returned to the atmosphere. Thus, reforestation with longer lived trees will sequester and more gradually release carbon. Maintaining the bulk of plant carbon that we use in a stored form, such as wood or fiber, will also help in the reduction of CO_2. This will require more efficient harvest and processing techniques. Preservation of wetlands and peat bogs, which are important carbon reservoirs, will also help limit atmospheric CO_2.

Reforestation can play a major role in carbon sequestration. As trees grow, they remove CO_2 from the atmosphere so that it can no longer contribute to global warming. However, this is probably most effective in the tropics. Nadine Unger, a professor at Yale University's School of Forestry and Environmental Studies, makes the case that reforestation in cooler climates actually may make global warming worse. She argues that the dark color of trees causes them to absorb more of the sun's energy and thus serves to raise the planet's temperature, counteracting any benefit from CO_2 uptake by those trees. They also release climate-warming VOCs.[17] She concludes that whereas planting trees in the tropics will effectively prevent global warming, doing so in cooler regions would be less effective. There have been a number of criticisms of her work. Most important, when we apply systems thinking to this argument by taking into account other ecological benefits of planting trees, the value of reforestation becomes more evident.

One leader in using tropical reforestation to limit climate change has been the Costa Rican government. Costa Rica generates revenue by taxing domestic fossil fuel use and tourism and also through the sale of carbon offsets to industry as part of a domestic cap and trade system (see Box 3.2). It is also exploring ways to sell carbon offsets to foreign countries as part of the Kyoto Protocol, and it has sold offsets to foreign companies such as the Superior Nut Company of Cambridge, Massachusetts, that choose to offset their own CO_2 emissions. This has generated substantial funds that support payments to farmers and landowners for preserving forest and replanting deforested areas on private land. These reforested areas provide a variety of ecosystem services, such as watershed protection, flood control, biodiversity preservation, and the maintenance of scenic beauty. They also remove significant amounts of carbon from the atmosphere, which far exceeds the quantity that Costa Rica itself generates. Reforestation projects also provide jobs, and shade crops such as coffee and certain fruit can still be grown within forest interiors, providing food and additional opportunities for human communities in these forests. Costa Rica's market-based reforestation–carbon offset project meets many of the criteria for sustainability, and it has served as a model for other nations, particularly in the developing tropics.

BOX 3.3

Individual Action: Stop Idling!

Many drivers leave their automobile running while they are waiting to pick up a passenger, while they run into a store, or as a means to warm up their engine. In fact, Americans spend upwards of 16 minutes per day idling their vehicles. Large trucks, buses, and freight trains often leave their diesel engines running even longer to maintain power systems and engine temperature. When idling, these engines continue to burn fuel and release air pollutants. In the United States, passenger cars, light trucks, medium-duty trucks, and heavy-duty vehicles consume more than 6 billion gallons of diesel fuel and gasoline per year while idling. Emissions from idling include large quantities of NO_x and VOCs, which increase urban smog, and significant amounts of climate-changing CO_2. In the United States, idling passenger cars release 40,000 tons of CO_2 per day.

Figure 3.3.1 To reduce emissions, anti-idling laws have been passed in many municipalities. *Courtesy of metamorworks/Shutterstock.*

This is a case in which a small action by individuals could result in significant change. Voluntary idling is a daily action that seems negligible but when taken cumulatively can have major impacts on air quality and reducing CO_2 emissions. For example, reducing idling by 5 minutes per day in a car with an eight-cylinder engine would result in preventing the emission of 440 pounds of CO_2—or more than 150,000 pounds per year. A study of Chicago drivers found that if all passenger vehicles reduced their idling by 5 minutes per day, VOC emissions would be reduced by 6.5 tons per day and NO_x emissions would be reduced by 1.89 tons per day.[1] These numbers are even greater for larger commercial vehicles. Some have argued that cars operate best when warmed up through idling and that restarting a car uses more gasoline and is harder on the engine and battery compared to idling. This is not true, particularly for today's electronic engines with fuel injection and catalytic converters. A study by the US Department of Energy showed that idling for more than 10 seconds uses more fuel and emits more CO_2 than engine restarting.[2]

So one sustainable solution is for individual drivers to reduce or eliminate their idling. This can be achieved if drivers turn off their engine for waits greater than 10 seconds. Also, rather than leaving engines idle to warm them up, it is recommended that this be done while driving. Rather than through idling, today's engines are best warmed up by easing into the drive and avoiding excessive engine revving. Instead of sitting in line at a drive-through, parking and entering an establishment will significantly reduce idling time. Schools are places where much idling occurs as buses and parents wait to discharge and receive children. This has been shown to decrease air quality in the immediate vicinity of schoolchildren. As a result, many municipalities have restricted idling on and near school properties (Figure 3.3.1). Advocating for new anti-idling laws and their enforcement for both personal and commercial vehicles will also further the cause to reduce emissions and conserve fuel. New technologies for commercial vehicles, such as direct-fired heating and battery-powered air conditioners and refrigeration, should make these laws easier to implement.

1 Shipchandler, R., J. Janssen, and G. Miller. 2008. "Estimating Smog Precursor Emissions from Idling Vehicles in the Chicago Metropolitan Area." Urbana–Champaign: Illinois Sustainable Technology Center, Institute of Natural Resources Sustainability, University of Illinois at Urbana–Champaign.

2 Gaines, L., E. Rask, and G. Keller. 2013. "Which Is Greener: Idle, Stop, or Restart?" Argonne, IL: Argonne National Laboratory, US Department of Energy.

TREES AND OTHER POLLUTANTS

In addition to the obvious benefit of carbon sequestration through photosynthesis, plants also help mitigate other pollutants. When planted along roads, vegetation can trap and hold particulate material. Depending on conditions and which tree species are planted, trees can either increase or decrease ozone air pollution. Trees emit a substantial amount of VOCs, which can contribute to the formation of ozone and smog. However, when nitrogen oxide concentrations are low, VOCs may actually remove ozone. VOC emission rates by trees vary by species. Plants with the greatest relative effect on lowering ozone, such as cherry, linden, and honey locust, should be preferentially planted in urban environments. The cooling effect of shade trees can reduce temperature-dependent processes responsible for ozone production.

In addition to the removal of particulates, trees can absorb a variety of gaseous pollutants through the stomata, or pores, in their leaves. A study published in *Environmental Pollution* showed that in the United States, removal of air pollutants by trees resulting in a less than 1% improvement in air quality would have a substantial positive impact. The study estimated the annual health care savings associated with this reduced pollution at approximately $7 billion. This includes saving more than 850 human lives and preventing 670,000 cases of acute respiratory symptoms each year in the United States.[18]

One example of where reforestation efforts are being used to abate air pollution and create jobs is in India. Six cities in India rank among the worst in the world for air quality. India also has a high unemployment rate among the youth population aged 10–24 years. To address both of these problems, India's government is hiring as many as 300,000 youth to plant 2 billion trees along the nation's highways. This program will help reduce particulate matter and smog in urban areas, help combat climate change, provide economic opportunity for unemployed youth, and beautify the country's roadways. This commitment to improving air quality and providing economic opportunity for the underserved is a model sustainable solution.

AGRICULTURE

A number of agricultural methods can enhance carbon removal from the atmosphere, including the use of photosynthesizing cover crops between planting seasons. An essential part of this is to further sequester the captured carbon in the soil by incorporating the cover into it rather than harvesting it. This can also be done with unused portions of crops, such as stalks and roots. By increasing the carbon soil pool in this manner, a significant amount of atmospheric CO_2 is sequestered. It also improves soil quality and can increase yield with fewer chemical inputs.

USING TECHNOLOGY TO CLEAN THE AIR

A number of technologies are being developed that could potentially be used to clean polluted air. Much of this development is occurring in China, where smog and poor air quality are a chronic problem. A recent study claimed that polluted air is the fourth largest threat to the health of Chinese people and that in 2007, air pollution caused 350,000–400,000 premature deaths in that country.[19] In response to what some refer to as China's "airpocalypse," the Chinese government has invested 982 billion Yuan ($162 billion) to fight pollution.

Among the innovative technologies being developed in China is the pumping of super-chilled liquid nitrogen into the atmosphere as an anti-smog agent. Preliminary research shows that this results in the formation of crystals on small particles, causing them to fall to the ground. Another approach being developed is the use of rooftop sprinklers to spray water into the atmosphere to wash particulate matter to the ground in the same way that rain cleanses the air. Seeding clouds to make rain has also been proposed as a possible solution.

Another proposed idea is the development of a smog-sucking electrostatic vacuum cleaner. Such a cleaner would use high-voltage, low-amp electricity to create an electrostatic field that would essentially filter out particles. Prototypes used in parking garages have effectively removed 99% of the particles in the size range that is most detrimental to human health. The scalability and capture and disposal of removed particulates, however, are important considerations in the development of these technologies.

Throughout the world, a variety of smog-reducing technologies are being integrated into innovative building design. The façade of the Palazzo Italia in Milan will be built with air-purifying biodynamic cement, which removes air pollutants and converts them into inert salts. Similarly, the Manuel Gea González Hospital in Mexico City has a "smog-eating" façade that uses a titanium dioxide coating that reacts with ambient UV light to neutralize elements of air pollution, breaking them down to less noxious compounds such as water. Other projects have incorporated titanium dioxide nanoparticles in billboards and even clothing. It is predicted that one person wearing such catalytic clothing could remove 5 or 6 g of nitrogen dioxide from the air per day, and two pairs of jeans could clean up the nitrogen dioxide emitted from one car.

In terms of GHGs and global climate change, there is promising technological innovation in the area of CO_2 removal from the atmosphere. It would use the same anionic exchange resin that is being developed for CO_2 scrubbers, but it would be used to capture CO_2 from ambient air. This approach could serve to reduce atmospheric CO_2 levels and slow or even reverse the process of climate change. Captured CO_2 could be used to meet industrial CO_2 demands such as food production, refrigeration, and the manufacture of dry ice. These uses could potentially offset the costs of the technology. In addition, CO_2 storage in geological formations will be an essential component for carbon management.

Mitigating Indoor Air Pollution

As previously discussed, indoor air pollution is a deadly environmental problem in developing countries and to a certain extent in developed countries. Because so many of the global poor are impacted by indoor air pollution, it is an environmental justice issue. For the billions who lack access to energy, the solution is the development of cleaner burning technologies and fuels that can then be made available to these poorer populations. This has proven challenging. Cleaner burning cooking stoves with better ventilation have been developed, but many communities have been reluctant to adopt them. In the short term, these newer stoves have been effective at reducing smoke inhalation and related diseases such as childhood pneumonia. However, households stopped using them after a while because of needed maintenance and repairs.

This is clearly a case in which it might be more effective to work with local communities to design and implement their own approach to clean cooking and heating rather than imposing on them a technology that seems foreign to them. Clean energy approaches such as solar cookers and heaters (Box 3.4), as well as the development of infrastructure for the production and delivery of renewable energy, will help solve this grave public health issue and at the same time reduce GHG emissions.

As discussed previously with regard to outside air pollution, plants can be quite effective in improving IAQ in modern households. Potted houseplants can remove VOCs and CO_2 and can reduce airborne microbes. Energy-efficient heating, cooling, and ventilation systems that also remove indoor air pollutants can help maintain healthy indoor environments. Complete exchange of indoor air with that from the outside can be done in a way that conserves heating and cooling energy through countercurrent exchange and heat wheel systems.

Second-hand tobacco smoke is a serious threat to IAQ that causes lung cancer, bronchitis, and pneumonia. The problem of second-hand smoke can best be solved by eliminating indoor smoking either as a household policy or through legislation for public spaces. The latter is gaining momentum and has been quite successful at national, state, and local levels.

Another common indoor air pollutant is radioactive radon gas that seeps into households from geological sources through foundations and basements. Radon gas is a silent killer, and it is estimated that it is responsible for as many as 14,000 lung cancer deaths each year.[20] The presence of radon is easily detected and can be mitigated through the installation of radon exhaust systems. In many cases, radon testing is part of the home inspection process prior to purchase. Testing and mitigation should be required in the certification process for rental properties as well. This is particularly important in older, poorer neighborhoods where renters may be unaware of the potential for such exposure.

Adapting to Our Impacted Environment

In cases in which reducing emissions and cleaning air that has already been polluted have not been effective, it will be necessary to adapt to our changed environment. This is and will continue to be a necessary strategy with regard to climate change. This is because climate change, its associated consequences, and necessary accommodations have already begun. These changes will continue to occur because of unwillingness, particularly among industrialized countries, to significantly curb GHG emissions. One reason behind this is our short-sightedness in relation to current costs of implementing current and new technologies that could halt or even reverse the process of climate change. However, in the long term, the costs of making change now are likely to be much lower than the total economic, social, and environmental costs of climate change. Science policy experts argue that at this point, adapting to climate change would be a more effective means of dealing with global warming than reducing emissions of greenhouse gases.[21] Part of their argument is that even if emissions are stabilized relatively soon, which is not likely, climate change and its effects will last many years.

BOX 3.4

Innovators and Entrepreneurs: SolSource—Cooking Anything Under the Sun

Household air pollution from solid fuel used for cooking inside the home is one of the deadliest environmental problems today, accounting for as many as 4 million deaths per year. The concentration of air pollutants inside a dwelling located in otherwise pristine areas has been shown to be more than 10 times greater than that on the streets of Beijing, China. Not only do the fuels that people use pollute but also their collection, primarily by women, takes hours each day, and they often cost families a large proportion of whatever income they earn.

Founded by Caitlin Powers and Scot Frank, One Earth Designs works to remedy this situation by providing a means for people to use a free source of energy for cooking that creates no indoor air pollution. That source is the sun. Their revolutionary sun-powered grill, SolSource (Figure 3.4.1), requires no fuel, heats up quickly, is easy to clean, and most important, is emissions-free. SolSource grew out of their work with rural Himalayan families to help find a solution to meet their energy needs. After numerous prototypes with help from contributions through programs such as Kickstarter, a funding platform for creative projects, they created a solar-powered stove that significantly improves the quality of life for thousands of families who rely on them for their daily needs. SolSource saves them time and money, reduces their exposure to harmful stove pollution, and preserves the environment in terms of both air quality and reducing deforestation due to the harvest of wood for fuel.

One Earth Designs operates as a hybrid non-profit/for-profit business that serves to meet the needs of all customers equally and to develop products for emerging market customers. Their approach allows them to meet the needs of their customers rather than the needs of intermediary funders. Instead of giving their products away as a donation, they

Figure 3.4.1 A parabolic solar cooker with segmented construction similar to the SolsSource produced by One Earth Designs. *http://www.atlascuisinesolaire.com*

continued

continued

work with local governments and non-governmental organizations to offer subsidies for their products. They also offer financing options such as installments or rental models that enable people from all walks of life to purchase SolSource. They view this as a way to empower and respect their customers, offering them a true sense of ownership and care for product than if it had been given to them as charity.

The ultimate goal of One Earth Designs is to empower the 3 billion rising consumers in emerging markets to attain a higher standard of living using sustainable practices and products such as SolSource. This truly is a sustainable solution. It represents an innovation that integrates principles of social, environmental, and financial longevity. In addition to alleviating a deadly environmental problem, the company works toward creating a positive impact in every stage of its product's life cycle. It does this by selecting high-quality sustainable raw materials, working closely with responsible manufacturers, and maximizing product durability and recyclability.

Because global action may not stop climate change, the world needs to explore how to live with it by adapting to that change. The development of methods of adaptation is an urgent priority because those means are needed now. This is made evident by recent events caused by global warming, such as devastating floods, extensive droughts and other major weather events, and rising sea levels. It is vital that we find ways to live with scarcer water, higher peak temperatures, higher sea levels, and weather patterns—all of which threaten our current agricultural practice and have the potential to cause catastrophic disasters. Solutions being developed include the breeding of new heat- and drought-tolerant crop varieties, water conservation and improved water supply, and increased emergency preparedness, including response to weather-related catastrophic events and emerging public health issues. Flood control including flood barriers, the damning of glacial lakes, tethering icebergs, and community relocation must also be explored. Detailed ways that we must adapt our food systems are presented in Chapter 5.

Developing countries will need assistance in order to adapt to climate change because it is predicted that they are the most likely to be impacted by it and lack the resources to adapt. People and communities in developing countries are too poor to adapt on their own, and if emissions caused by the consumption of the rich impose adaptation costs on the poor, justice demands support from industrialized nations. Recognizing this, the UN Framework Convention on Climate Change recommended the creation of a fund of $1 billion per year to assist developing countries in combating the consequences of climate change. However, implementation of this fund remains to be seen. Finally, note that climate change mitigation and adaptation are not mutually exclusive, and the most effective solutions will be those that accomplish both. For example, farming techniques that improve the capacity of soil to store moisture may also increase its capacity to store carbon.

Asssessing the Sustainability of Clean Air Action

Clean air is a vital priority. To determine if policies and approaches to maintaining excellent air quality represent sustainable solutions, we must consider them within the context of our definition. This must include consideration of environmental, economic, and social elements as well as the principles by which these solutions have been developed. Simple quantitative indicators can reveal how effective our approaches have been developed and employed to determine success of policy and other efforts. Measures of community sustainability often include such indicators of air quality and associated measures of health.

Nationally in the United States, the EPA reports on very similar kinds of indicators. For example, since the implementation of the CAA, US emissions of the six criteria pollutants have declined an average of 72% while GDP has grown 219%. There have been concurrent declines in air quality–related health problems. These indicators suggest that policies such as the US CAA improve environment and public health without compromising economic growth. In fact, it has been argued that environmental regulation has actually stimulated economic growth and development.

The transboundary and global nature of pollution has constrained our ability to both assess and identify the most effective approaches to sustaining excellent air quality. Improvements in the United States mean nothing when global production of CO_2 impacts the rest of the world. Moreover, industrialized nations disproportionately contribute to air pollution while their poorest residents and the developing world suffer the bulk of the consequences. The only real sustainable solutions to air pollution must address these justice issues on the ground at the level of individual households and communities with policies that address this sort of disparity. Finally, given the essential nature of clean air, the precautionary approach should be applied in developing clean air policy and technical solutions. Even when we are uncertain about the potential dangers of air pollution, mitigating it should be a priority because the savings by not dealing with the problem in the present will most likely be surpassed by future costs.

Chapter Summary

- The layer of atmosphere at the surface of the Earth, the troposphere, contains the air that we breathe. In addition to consisting of oxygen, nitrogen, and carbon dioxide, air is made up of trace gases, water vapor, and particulate matter. The composition and quality of the air in the Earth's atmosphere are beginning to decline through both natural and anthropogenic causes.
- There are two types of anthropogenic air pollutants: primary air pollutants and secondary air pollutants. The former describes a hazardous substance that is released into the air directly from a source, whereas the latter defines a pollutant that is not harmful until it interacts with other chemicals in the atmosphere. Air pollution can originate from stationary points, mobile, and fugitive sources.
- Greenhouse gases such as water vapor, CO_2, methane, nitrous oxide, and ozone are causing global climate change. GHGs in the upper atmosphere warm the Earth by absorbing and re-radiating heat that is generated when light reaches the surface of the Earth and is reflected as heat. GHG emissions through human activity since the Industrial Revolution have and continue to contribute to climate warming.
- The US EPA has set limits based on human health and environmental effects of the six most common pollutants: carbon monoxide, nitrogen oxides, sulfur dioxide, particulate matter, ozone, and lead. Other pollutants of concern include VOCs and HAPs, which include air toxics and heavy metals.
- Acid precipitation, a serious environmental problem that can devastate entire ecosystems, forms when nitrogen dioxide and sulfur dioxide react in the atmosphere with other chemicals including water and oxygen to form nitric and sulfuric acid. When these acids fall to the Earth, they can severely alter the pH of bodies of water and degrade forest health.
- Transboundary air pollutants are those that are generated in one country and blown across boundaries, thus impacting another country. GHGs released from one country influence the climate of the entire world. Air pollution and global climate change disproportionately impact poorer communities. Sustainable solutions for air pollution must include policy and technology that address the transboundary and global nature of air pollution, and they must ensure environmental justice.
- The US Clean Air Act of 1973 established policies to protect humans and the environment from potential harm caused by air pollution. Since passage of the Act, air quality in the United States has improved dramatically and public health has also improved while stimulating economic growth. The Act was amended in 2007 to include the regulation of GHGs. Similar environmental policies in China, Europe, and developing countries have been implemented with varying success in maintaining air quality and limiting GHG emissions.
- Given the transboundary and global nature of air pollution, international policy is essential for limiting air pollution and climate change. Examples of such international efforts include the Montreal Protocol to limit ozone-depleting substances and international policy on GHGs and climate change. The latter included the establishment of the IPCC in 1998, which was charged with assessing climate change and establishing a plan for halting its progress. A number of international protocols aimed at reducing GHG emissions were established but have had limited success.
- A number of technical innovations have been developed to reduce emissions of anthropogenic air pollutants. These include scrubbers, filters, and catalytic converters, many of which are extremely

effective. There has been less success in developing ways to remove climate-changing CO_2 from emissions.

- There are a variety of solutions for cleaning air that has already been polluted, including the use of trees and vegetation, alternative agricultural practice, and smog-reducing technologies.
- Indoor air quality is an often-overlooked problem that can have serious health implications for humans in both the developed and the developing world. A variety of solutions, such as energy-conserving air exchange systems and improved cooking and heating equipment, can help improve indoor air quality.
- Given that the impacts of global climate change are already occurring and are most likely to worsen in the coming years because of the lack of effective mitigation, it will be necessary to adapt to these changes. This will be particularly challenging in developing countries because they are most likely to bear the brunt of the effects of climate change and lack the resources to adapt.
- Assessment of clean air action requires the development of quantitative indicators for emissions reduction and improvements in air quality and public health. The transboundary and global nature of pollution has constrained our ability to both assess and identify the most effective approaches to sustaining excellent air quality. Effective clean air regulation has been shown to stimulate innovation and economies. Sustainable solutions to air pollution must address disparities between developed and developing countries and the environment.

Digging Deeper: Questions and Issues for Further Thought

1. Give examples of how systems thinking has better helped us understand the complexities of air pollution and potential solutions.
2. Recently, US politicians have made economic arguments for the reduction or elimination of clean air regulations and the ability of the EPA to enforce them. How would you counter those arguments?
3. How does environmental justice relate to air pollution, and how can sustainable solutions best address justice issues?
4. Give examples of how policy and other approaches to reducing air pollution including climate-changing GHGs have or can stimulate innovation and economic growth.
5. Give examples of nature-based solutions that can mitigate air pollution.

Reaping What You Sow: Opportunities for Action

1. BreezoMeter is a Web-based application that offers accurate, hyper-local air quality data in real time. Visit https://breezometer.com and enter a specific location of interest into the Air Quality Index Map. Monitor the air quality at that location on a weekly basis. Assess the air quality, and make some recommendations for improvement if needed.
2. Locate your US representatives at http://www.house.gov. Draft them a letter that makes a reasoned case for maintaining the Clean Air Act and EPA.
3. Identify three non-profit organizations that focus on clean air as part of their mission. What are their approaches, strategies, and measures of success? Which one would you be most willing to support?

References and Further Reading

1. World Health Organization. 2014. "7 Million Premature Deaths Annually Linked to Air Pollution." Accessed February 28, 2017. http://www.who.int/mediacentre/news/releases/2014/air-pollution/en
2. Blaustein, A. R., J. M. Kiesecker, D. P. Chivers, and R. G. Anthony. 1995. "Ambient UV-B Radiation Causes Deformities in Amphibian Embryos." *Proceedings of the National Academy of Sciences of the USA* 92: 11049–11052. doi:10.1073/pnas.92.24.11049. PMID 9391095
3. Brown, S. K., M. R. Sim, M. J. Abramson, and C. N. Gray. 1994. "Concentrations of Volatile Organic Compounds in Indoor Air: A Review." *Indoor Air* 4 (2): 123–34.
4. Pachauri, R. K., and A. Reisinger, eds. 2007. *Climate Change 2007: Synthesis Report. Contribution of Working Groups I, II, and III to the Fourth Assessment Report of the Intergovernmental Panel on Climate Change*. Geneva: Intergovernmental Panel on Climate Change.

5. Vogelmann, H. W., T. D. Perkins, G. J. Badger, and R. M. Klein. 1988. "A 21-Year Record of Forest Decline on Camels Hump, Vermont, USA." *European Journal of Forest Pathology* 18 (3–4): 240–49. https://doi.org/10.1111/j.1439-0329.1988.tb00923.x
6. Rauh, V. A., P. J. Landrigan, and L. Claudio. 2008. "Housing and Health: Interaction of Poverty and Environmental Exposures." *Annals of the New York Academy of Sciences* 1136: 276–88.
7. The Center for Public Integrity. 2017. "The Invisible Hazard Afflicting Thousands of Schools." Accessed February 21, 2017. https://www.publicintegrity.org/2017/02/17/20716/invisible-hazard-afflicting-thousands-schools
8. World Health Organization, "7 Million Premature Deaths Annually."
9. Stevens, M. 2002. "The Precautionary Principle in the International Arena." *Sustainable Development Law and Policy*, Spring/Summer: 13–15.
10. US EPA, Office of Air and Radiation. 2001. "The Benefits and Costs of the Clean Air Act from 1990 to 2020: Final Report." Washington, DC: US EPA.
11. Pew Charitable Trusts, Environmental Initiative. 2013. "Who's Winning the Clean Energy Race?" Philadelphia: Pew Charitable Trusts.
12. London Air Quality Network. "Statistics Maps." Accessed March 1, 2017. http://www.londonair.org.uk/london/asp/news.asp?NewsId=OxfordStHighNO2
13. Thomas, V. 2014. *Sophisticated Interdependence in Climate Policy*. London: Anthem Press.
14. Amdur, David, Barry G. Rabe, and Christopher P. Borick. 2014. "Public Views on a Carbon Tax Depend on the Proposed Use of Revenue." *Issues in Energy and Environmental Policy*, No. 13, July.
15. Nystrom, S., and P. Luckow. 2014. "The Economic, Climate, Fiscal, Power, and Demographic Impact of a National Fee-and-Dividend Carbon Tax." Report prepared by Regional Economic Models, Inc., and Synapse, Inc., for Citizens' Climate Lobby: 36.
16. Seipp, C. A., N. J. Williams, M. K. Kidder, and R. Custelcean. 2017. "CO_2 Capture from Ambient Air by Crystallization with a Guanidine Sorbent." *Angewandte Chemie International Edition* 56 (4): 1042–45. https://doi.org/10.1002/anie.201610916.
17. Unger, N. 2014. "Human Land-Use-Driven Reduction of Forest Volatiles Cools Global Climate." *Nature Climate Change* 4 (10): 907–10. https://doi.org/10.1038/nclimate2347.
18. Nowak, D. J., S. Hirabyashi, A. Bodine, and E. Greenfield. 2014. "Trees and Forest Effects on Air Quality and Human Health in the United States." *Environmental Pollution* 193: 119–29.
19. Chen, Zhu, Jin-Nan Wang, Guo-Xia Ma, and Yan-Shen Zhang. 2013. "China Tackles the Health Effects of Air Pollution." *Lancet* 382 (9909): 1959–60. doi:10.1016/S0140-6736(13)62064-4
20. Ford, E. S., A. E. Kelly, S. M. Teutsch, S. B. Thacker, and P. L. Garbe. 1999. "Radon and Lung Cancer: A Cost-Effectiveness Analysis." *American Journal of Public Health* 89 (3): 351–57. doi:10.2105/AJPH.89.3.351
21. Arizona State University. 2007. "Adaptation to Global Climate Change Is an Essential Response to a Warming Planet." *ScienceDaily*. Accessed September 17, 2014. http://www.sciencedaily.com/releases/2007/02/070207171745.htm

CHAPTER 4

WATER: USING OUR PLANET TO QUENCH OUR THIRST

We are what we drink, and we are water. All living things are made up of mostly water. Our human bodies are approximately 60% water. Our lungs are 90% water, our brains are 70% water, and our blood is more than 80% water. Simply stated, there is no life without water (Figure 4.1). Each day, our bodies must replace 2.4 liters (approximately 2.5 quarts) of water, which we get either from the food we eat or by directly drinking it. Second only to oxygen, the amount of time that we are able to live without water is short—approximately a week depending on temperature and other conditions.

In addition to the essential nature of water for the survival of all organisms, water is a fundamental part of all functioning ecosystems. We also need it to grow our food, to clean ourselves and our spaces, for the sanitary disposal and treatment of our bodily wastes, and for most industrial processes. Water is perhaps the most vital resource on our planet, and perhaps the most limiting and most threatened. Less than 1% of the Earth's water is liquid fresh water, and it must be shared among the more than 7 billion people of the world.

PLANTING A SEED: THINKING BEFORE YOU READ

1. What constitutes a basic human right? Do you think access to safe, clean drinking water is a basic human right? Why or why not?
2. What ecosystem services do aquatic systems provide? What value do they offer? How can they be protected?
3. When considering improving access to water and sewer services in your community, who are the major stakeholders?
4. What are the advantages and disadvantages of drinking bottled water versus tap water?

Water is life's mater and matrix, mother and medium. There is no life without water.
—Albert Szent-Györgyi

Figure 4.1 A Khmu girl drinking water in Xieng Khouang, Laos. *Credit: Eric Lafforgue/age fotostock.*

As discussed in Chapter 1, water is considered a renewable resource. However, due to overexploitation and contamination, rates of depletion often exceed rates of renewal. Depending on location and climate, water for drinking and agriculture can be very limited. Climate change further limits our supply of water through increased drought and contamination of fresh water with salt, a process called **salinization** that occurs as sea levels rise and with increased rates of evaporation. Related to the availability of clean water is the lack of sanitation and hygiene, particularly in developing countries. This results in the direct contamination of drinking water by pathogens that are a major source of mortality, especially among children. Thus, increasing access to clean water, sanitation, and hygiene must be a major priority where these needs are not being met.

In this chapter, we examine the sustainability of water as a resource. We start by exploring some basic **hydrology**, or how water is moved and stored on the Earth. Then we consider how access to water is becoming limited due to overdepletion, pollution, and other forces. Next, we investigate how human activity has threatened aquatic ecosystems and the services they perform. Finally, we explore how policy, innovation and technology, and ecological approaches can serve as solutions to these problems. This will reveal that despite great challenges, progress is being made in the development of these solutions, and there is significant potential to sustain our current and future needs for clean water.

Hydrology: The Storage and Movement of Water

Hydrology is the study of the movement, distribution, and quality of water on Earth. As mentioned in Chapter 1, 97.5% of the Earth's water is in the form of undrinkable salt water in the oceans. Of the remaining 2.5% of fresh water, approximately 69% is frozen in glaciers and ice caps, 30% occurs as **groundwater** below the Earth's surface within layers of permeable rock referred to as **aquifers**, and approximately 1% occurs as surface water, in the atmosphere, or in living organisms (Figure 4.2).

Water cycles between these stored sources by way of the hydrological cycle (Figure 4.3). Heat from the sun moves water to the atmosphere as vapor through **evaporation**, or the conversion of water from the liquid to its gaseous state, and **sublimation**, the conversion of solid ice and snow directly into water vapor. The sun also drives **transpiration**, the process in which plants absorb water through their roots and then return it to the atmosphere in the form of water vapor as it evaporates through pores, or **stomata**, in their leaves. Other processes that drive the cycling of water include condensation, precipitation, deposition, run-off, infiltration, melting, and groundwater flow (see Figure 4.3).

Less than 1% of the water on Earth is potentially accessible fresh water. Although this represents a large volume of water, much of it is threatened by overdepletion pollution, and drought. In the following section, we closely examine these threats and other factors that limit access to useable water. Then, given the vital nature of water for our own survival, we explore ways to conserve and provide fresh drinkable water and to protect and restore freshwater ecosystems.

Limited Access to a Vital Resource

Water, a vital resource for drinking and growing food, is not unlimited. As of 2017, approximately 663 million people lacked access to improved water sources, and 2.4 billion people had no access to basic sanitation services, such as toilets or latrines.[1] As a result, at least 1.8 billion people have been using sources of drinking water contaminated with human feces. This results in more than 500,000 deaths each year from preventable

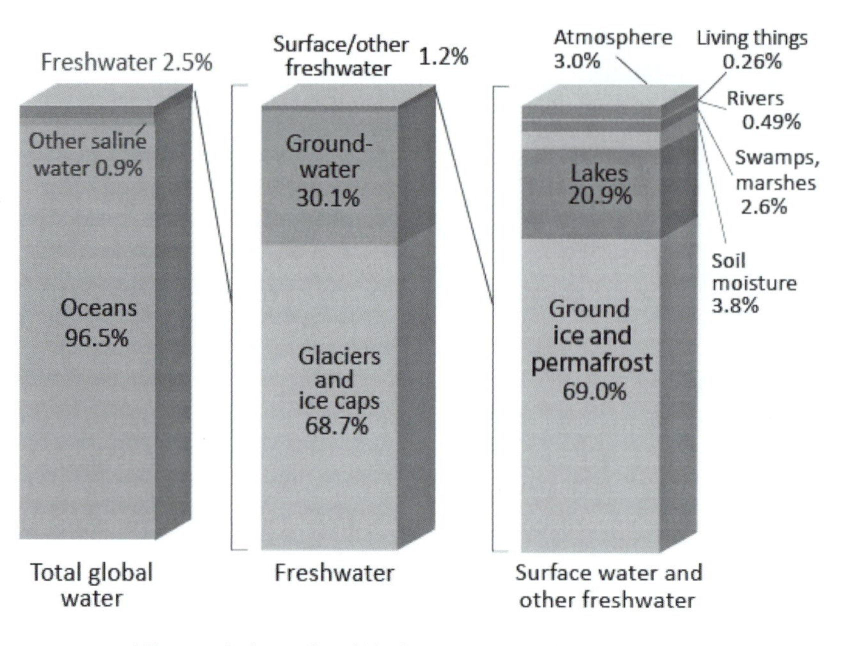

Figure 4.2 The distribution of water on Earth. Although water is abundant on Earth, most of it is undrinkable salt water. Less than 1% is readily accessible fresh water (Adapted from USGS, https://water.usgs.gov/edu/earthwherewater.html)

water- and sanitation-related diseases such as dysentery, cholera, typhoid, and polio.[2]

Water scarcity already affects every continent, where it is impacting our ability to grow food. Nearly 1.2 billion people, or almost one-fifth of the world's population, live in areas where water is scarce, and 500 million others are approaching this situation. Another 1.6 billion people, or almost one-fourth of the world's population, face water shortages due to the lack of necessary infrastructure to take water from rivers and aquifers; however, water is being extracted, primarily for irrigation, at rates that exceed the capacity of these systems to recharge.

In this section, we examine how and why over-depletion of water occurs and limits availability. We consider how population growth and inequities in consumption, urbanization, agriculture, energy, industrial processes, and global climate change are contributing to the exhaustion of this resource and how it is impacted by pollution. It is also shown that limited access to water is directly related to poverty, and we discuss a growing trend in the commodification of water, including the privatization of municipal water supplies and profiteering through the expansion of the bottled water market. We then consider water scarcity as a rising cause of international tension and conflict.

Population Growth and Inequities in Consumption

In 2014, Earth's human population exceeded 7.2 billion. It is expected to soar to 9.6 billion by 2050 and 10.9 billion by 2100. Most of this growth will take place in less developed, lower income nations, mostly in sub-Saharan Africa and Asia, where water scarcity is already a serious issue. As a result of this population growth, it is predicted that total global water consumption is likely to increase 10-fold by the end of this century.

Population growth directly impacts water availability through increased demand and indirectly through further land degradation, deforestation, and contamination and pollution of water supplies. Increased population necessitates a need for more agriculture and often results in increased urbanization and industrialization, all of which further deplete and contaminate water supplies.

Although population sizes are much more stable in developing countries, per capita rates of consumption are much higher. As described in Chapter 1, the IPAT model integrates population size and rates of consumption to determine impacts on natural resources. Water consumption can be represented as the per capita **water footprint**, or how much water it takes to meet the needs of an average person (Box 4.1). The water footprint can help explain water scarcity and reflects consumption inequities between developed and developing countries. For example, in the United States, the average American family uses approximately 300 gallons of water per day. An American taking a five-minute shower uses more water than the average person in a developing country uses for an entire day.

Meeting the current US demand for water is not sustainable. As a result of individual use coupled with agriculture and industrial processes, at least 36 of the 50 US states regularly face water scarcity. A 2014 assessment by NASA[3] showed that groundwater reserves in the US Southwest are severely low, and prospects for their long-term viability are bleak because persistent drought continues to prevent recharging. One of the more extreme examples of how US populations are depleting water resources is Lake Mead, which is shared by Nevada and Arizona. The lake, which currently supplies water to 22 million people per day, is predicted to be completely depleted by 2021.

Urbanization

It is projected that the urban share of the global population will exceed 60% by 2030. In Asia and Africa, the population of people living in cities will double

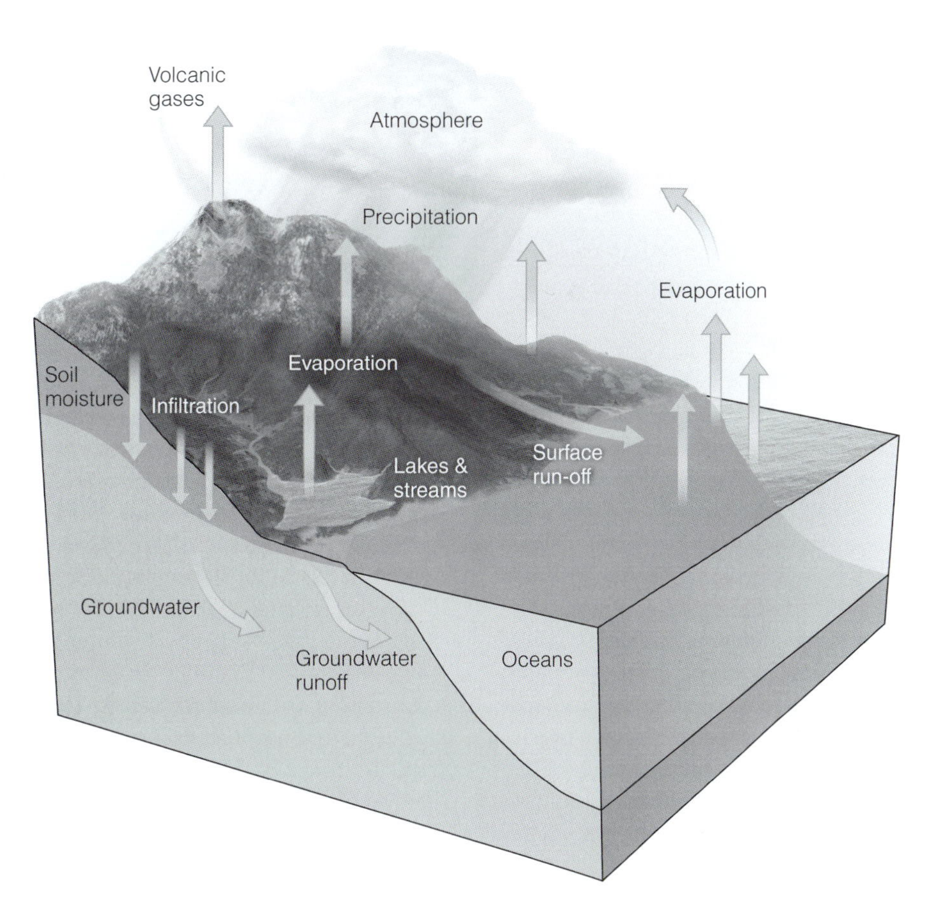

Figure 4.3 The hydrologic cycle showing how water is stored and circulates among different reservoirs.
Hellenic Center for Marine Research–Institute of Inland Waters, http://prewec.ath.hcmr.gr/overview.html

BOX 4.1

Individual Action: Reducing Your Water Footprint

The water footprint (WF) is an indicator of freshwater use that takes into account both direct and indirect water use by an individual, community, business, or nation. It is the volume of water used or degraded for a given task or for the production of a given quantity of some product or crop. WF goes beyond direct individual consumption to include a complete life cycle analysis that considers all of the water used or degraded in all products from their manufacture, use, and maintenance to their final disposal. Thus, for example, when an individual consumes a pint of beer, their WF does not merely increase by a pint. Rather, it increases by that pint plus the amount of water that went into the production of that pint, including the agricultural ingredients, which ranges from 120 to 360 liters per pint of beer depending on location.[1]

World water use has been increasing rapidly in the past 100 years, and individual use is significantly greater in the industrial world, especially in the United States, where the per capita WF is 2,842 m^3, well above twice the global average. Overdepletion and degradation of global water supplies through increasing WFs has resulted in a global water scarcity crisis that will continue to worsen.

One action that individuals can take to deal with the issues of water scarcity and depletion is to reduce their own WF. There are two approaches to decreasing WF: directly reducing water consumption or choosing and consuming products that have lower WFs. One can reduce one's direct WF by using higher water-use efficiency plumbing and appliances. For example, switching to water-conserving dual-flush toilets can save up to 69,000 liters of water per year for a family of four. Checking for leaks is another way to reduce direct water consumption. A leaky faucet can waste more than 11,000 liters of water per year or approximately 3,000 gallons—the amount of water a typical person uses in 180 showers.

Figure 4.1.1 A retrofitted sink for your toilet lid that delivers clean water with every flush. It then refills the tank with the gray water that is produced. *Courtesy of Source Sink by Environmental Designworks LLC, http://realtoiletreviews.com/the-sink-positive-toilet-sink-the-perfect-way-to-add-a-sink-to-that-tiny-house-cabin-or-workshop/*

Other ways to reduce one's direct WF are discussed in this chapter, including xeriscaping, rain barrel water collection, and the use of gray water reclamation and purification systems. One example of a simple way to capture and use gray water is a toilet-top sink that provides fresh water for handwashing and drains the gray water into the toilet tank for flushing (Figure 4.1.1). This provides a daily water savings of up to 2 gallons or nearly 10 liters per person.

1 SABMiller, GTZ, and World Wildlife Fund. 2010. *Water Futures: Working Together for a Secure Water Future.* Woking, UK/Goldalming, UK: SABMiller/WWF-UK.

Many municipalities in the United States where there are severe water shortages offer rebates or incentives for these types of household improvements.

In addition to reducing one's direct use, individual WF can be drastically reduced by choosing products and foods that require less water during their life cycle. Because the largest consumer of water is agriculture, consuming approximately 70% of the water globally and 80% in the United States, an obvious way to lower one's WF is to choose foods that require less water to produce. Eating fewer animal products is an obvious choice. Industrial-scale meat production uses 10 times more water than that used to produce the same amount of calories from most vegetable crops.[2] By skipping meat one day a week, an individual can reduce his or her WF by up to 25,000 gallons per year. Also, WF can be lowered by choosing grass-fed beef produced at smaller scales, which is less water consumptive than large-scale factory farmed beef.

Other product choices, such as choosing artificial fibers over cotton or—better yet—the purchase of second-hand clothing, will make a difference as well. Avoiding bottled water will reduce one's WF because it takes five times more water to produce the bottle and the water than the actual volume of water purchased. Finally, because the production of energy accounts for approximately 30% of water extractions in the United States, using less energy and choosing suppliers that use less water-intensive renewable sources of energy can also lower one's WF.

A number of resources are available to guide individuals and organizations in lowering their WFs. The Water Footprint Network (http://www.waterfootprint.org) is an excellent source of information and offers a number of tools and recommendations for reducing WF. Its website provides a WF calculator for various products and foods and a country-specific calculator for individual WF. The latter is based on detailed data regarding domestic water use, food consumption, outdoor water use, and industrial goods consumption. It does not, however, take into account types and amount of energy consumed by individuals. One additional tool could either be water-use labeling or an easy smartphone application that would indicate WFs for specific products. As the individual movement for reducing WF grows, so will that for institutions, communities, and nations. This will be an essential part of dealing with our growing water scarcity issue.

2 Mekonnen, M. M., and A. Y. Hoekstra. 2012. "A Global Assessment of the Water Footprint of Farm Animal Products." *Ecosystems* 15 (3): 401–5.

to 2.4 billion by 2030. This intensive urbanization will strain already limited water supplies both through increased consumption and through contamination. Of the 60 million people added to the world's towns and cities every year, most live in informal settlements, or slums, with no sanitation facilities, threatening the safety of the drinking water in these communities. Water-consuming industrial processes will also increase with urbanization, further limiting the availability of water for drinking and agriculture.

Agriculture

The need to feed growing populations places enormous stress on water supplies. The agricultural sector is the largest user of water in the world, representing approximately 70% of global consumption (Figure 4.4). As a result of increases in population size and the loss of farmers and farmland due to urbanization, more intensive, irrigation-dependent agriculture is required in order to raise yield.

As the global rate of meat consumption increases, there will be greater demand for water. The amount of water required to produce 1 kilocalorie from beef is more than 10 times that required for most vegetable, fruit, and cereal crops and twice that on a per gram of protein basis.[4] Nearly 2,500 gallons of water are required to produce 1 pound (0.45 kg); that equates to 2.5 million gallons (9.5 million liters) per head of cattle. Other water-intensive crops include rice, cotton, alfalfa, oil crops, and most nuts.

In California, nearly 10% of water consumed each year, a stunning 1.1 trillion gallons (4.2 trillion liters), is for almond production. Almond milk is marketed as a more healthful and sustainable alternative to cow's milk. However, whereas it takes approximately 2,000 gallons of water to produce 1 gallon of cow's milk, it takes approximately 2,500 gallons of water for each gallon of almond milk. The demand for almonds and almond milk is intensifying the water stress in a state that is already facing extreme, long-term drought.

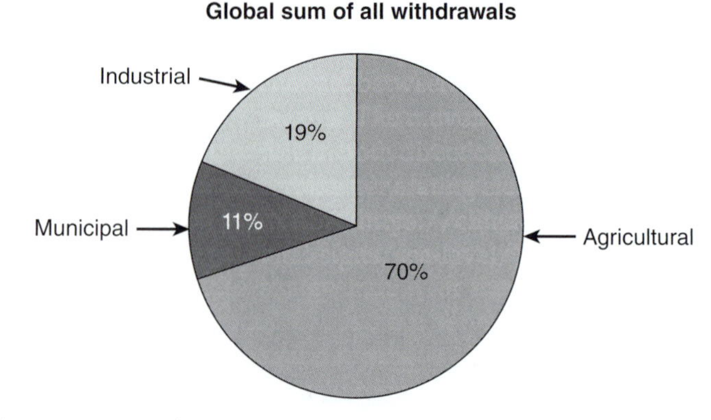

Figure 4.4 Global water use by sector: agricultural including irrigation, livestock, aquaculture, and crop-based biofuels; municipal use including domestic; and industrial water withdrawal including the energy sector except biofuels. (FAO. 2016. AQUASTAT website. Food and Agriculture Organization of the United Nations (FAO) http://www.fao.org/nr/water/aquastat/main/index.stm)

The production of corn and other crops for use as biofuels further exacerbates water shortages where this is being done.

In less developed countries, many of which already face acute water shortages, water-intensive agriculture is the predominant part of the economy. This is especially true in more arid regions of the world. In the Middle East, limited water availability has always been a critical issue. This is worsened by increases in agricultural irrigation. For example, in Saudi Arabia, 90% of groundwater supply is used for irrigation of crops. Approximately half of all the irrigated land on Earth is in China, India, Pakistan, and the United States, and all are experiencing water shortages. Sub-Saharan Africa accounts for only 4% of the total irrigated land because of the lack of infrastructure for irrigation. As this infrastructure comes online in order to meet growing regional agricultural needs, the depletion of water resources will intensify in that area of the world.

It is important to note that the intense agricultural demand for water and irrigation described previously varies both by geographical location and by specific farming techniques. For example, pasture-raised beef on non-irrigated landscapes uses much less water than concentrated feeding operations in industrial meat production. In some areas of the Midwestern United States, corn can be grown without irrigation, whereas in the West, highly consumptive flood irrigation is required. This kind of variation can provide context in the development of sustainable solutions with regard to agriculture and water consumption.

Energy

Although energy is the focus of Chapter 6, it is important to examine the relationship between water consumption and energy production in what is referred to as the **water–energy nexus**. The energy sector consumes approximately 150 trillion gallons (580 billion m^3) of water per year. According to the World Bank Group and the International Energy Agency, much of this water consumption occurs during the extraction and refinement of fossil fuels as described in the following examples.[5] In the extraction of coal, between 800 and 3,000 gallons of water are used for each ton that is mined. In the United States, the extraction of coal consumes 70–260 million gallons per day. Conventional crude oil production can also be water intensive, especially in enhanced oil recovery, in which a reservoir is flooded with water or steam to increase oil flow to the surface. This is problematic because many oil-producing regions are located in areas of water scarcity. Refining crude oil into gasoline is also water intensive, requiring 1–2.5 gallons of water to produce 1 gallon of gasoline. In the United States, that equates to 1–2 billion gallons of water to refine nearly 800 million gallons of petroleum products every day. Conventional natural gas production uses very little water, and water used for natural gas drilling is matched by water recovered from the natural gas itself.

As conventional sources of fossil fuels are depleted, the extraction of unconventional fossil fuels, such as shale gas, oil sands, coal bed methane, oil from shale, and enhanced oil recovery, is on the rise. These methods of extraction are extremely water intensive. The recent growth of shale gas extraction through hydraulic fracturing (fracking) has brought the conflict between energy extraction and water depletion to the forefront. Water is by far the largest component of fracking fluids. An initial drilling operation at a single well can consume up to 600,000 gallons of fracking fluids, and over its lifetime an average well may require up to an additional 8 million gallons of water for full operation. Much of that water is trucked in from other sites and is unrecoverable once used.

The production of renewable biofuels is also water intensive, and in some cases more so than unconventional fossil fuel extraction. On average, it takes 3.5–6 gallons of water to produce 1 gallon of ethanol from corn (maize), which is more than twice the amount of water needed to produce 1 gallon of gasoline. Increases in corn production to meet the growing demand of biofuels are being met by cultivating areas where irrigation is needed. This water is generally drawn from Midwestern US water supplies that are already under stress due to intensive food-based agriculture and drought. Because of their projected growth, it is anticipated that by 2030, as much as 8% of US fresh water might go toward biofuels.[6]

In addition to the extraction and production of fuels, the use of those fuels in the generation of electricity is also water intensive. For example, most electricity is generated by thermoelectric power plants in which fossil fuels or nuclear energy are used to create steam to turn turbines that generate electricity. These plants use millions of gallons of water to cool and condense the steam coming from the turbines. In the United States, thermoelectric power plants account for nearly half of total water withdrawals, exceeding that used for agriculture. It is anticipated that essential carbon capture and sequestration technologies as well as clean coal facilities will use even more water. Most renewable sources of energy, such as solar, wind, and geothermal, use considerably less water than fossil fuels and nuclear power. Hydropower uses large quantities of water, but the majority of the water is passed through the turbines with negligible loss. Reservoirs created by large hydroelectric projects do lose water through evaporation, and they often reduce flow and availability of water in regions that are below the dams.

Industrial Processes

In addition to the generation of energy, industrial water consumption is a major drain on the world's limited water supply (see Figure 4.4). The textile industry is one of the largest consumers of water. For example, it takes 2,900 gallons of water to produce a single pair of jeans in processing and dyeing the fabric.[7] That does not include the water used for growing the cotton that comprises the jeans. Automotive manufacturers also consume large quantities of water. It takes nearly 40,000 gallons of water to produce the average domestic car, including the tires.[8] Other water-intensive industries include mining and metal refining, pharmaceutical production, paper production, commercial beverage production, and electronics (see Box 4.1).

Global Climate Change

Water scarcity is already a problem in many countries today due to the factors described previously. However, the effects of global climate change could place millions more people at risk of water scarcity through resultant changes in rainfall and evaporation, putting 40% more people at risk of losing access to water. If the world warms just 2°C by 2100, which now seems all but unavoidable, up to one-fifth of the global population could suffer severe water shortages. This impact will vary geographically. The Mediterranean, the Middle East, the southern United States, and southern China will see a pronounced decrease in water availability, whereas southern India, western China, and areas of eastern Africa could see an increase.[9]

Water Pollution

In addition to the overdepletion of water, access to clean water is further limited by contamination and pollution. Water pollutants originate from various human activities and can contaminate surface waters, groundwater aquifers, and other freshwater, brackish, and marine ecosystems. Pollutants are often classified as either **point sources**, which are localized, stationary sources of pollutants, or **non-point sources** that are more diffuse. Examples of point sources are outlet pipes from factories, power plants, mining operations, and sewage treatment plants. Non-point sources, which are often more difficult to monitor, include run-off from farms, storm water drainage, and deposition of pollutants from the atmosphere. Specific types of point and non-point sources of pollution are described next.

PATHOGENIC POLLUTANTS

Water pollutants can be further classified as biological pathogens or chemical contaminants. Biological pathogens enter the water as run-off from animal wastes from feed lots or farms or from contamination from human excrement. The excrement is often contaminated with pathogenic bacteria, viruses, and other parasites. Examples include typhoid fever, cholera, hepatitis A, and a variety of diarrhea-causing pathogens. In the developing world, there is limited access to running water, and more than 2.5 billion people do not have access to a toilet or latrine. As such, contamination of water supplies is common, and water-related diseases have become a serious public health issue.

Pathogenic pollution from sewage is not limited to developing countries. In developed countries, sewage treatment infrastructure is aging and is often insufficient. This results in the release of untreated sewage into the environment. This is particularly true for combined sewer systems that collect rainwater run-off, domestic sewage, and industrial wastewater in the same pipe. During periods of heavy rainfall or snow melt, the sewer system or treatment plant may not be able to handle the larger amount of wastewater. This creates run-off that pollutes surface waters and aquifers with contaminated sewage, which can result in illness.

CHEMICAL POLLUTION

Toxic chemicals and other hazardous materials used in industry, agriculture, and transportation are a serious threat to water quality. The US Environmental Protection Agency (EPA) has identified 582 toxic chemicals that are produced, manufactured, and stored in locations throughout the United States. Similar to what was done for air pollutants, the EPA has designated approximately 150 of these chemicals as criteria water pollutants. For each criteria pollutant, the EPA has established recommended water quality criteria aimed at protecting both aquatic life and the health of humans consuming or exposed to those pollutants in drinking and surface water. These criteria are provided on the EPA website.[10]

Water supplies can be contaminated by chemical pollutants through storm water run-off that contains pollutants from fields, lawns, and roads and other paved sites that is then drained into aquatic systems that often lead to municipal water supplies. The impact from storm water run-off is a more significant problem in urban and developed areas, where there is a greater percentage of impervious surfaces. Storm water run-off and the pollutants it carries represent a significant contribution to the recharge of surface waters. For example, in New York state, storm water run-off is identified as a major source of contaminants in 37% of all water bodies. Examples of these contaminants are fertilizers, pesticides, bacteria from animal wastes, oil and grease from automobiles, sediment from construction, and litter such as plastic bottles.

Groundwater has been assumed to be a relatively pristine source of water compared to surface water supplies. This is because many organic compounds used in the past, such as DDT and chlordane, have low solubility in water and a strong tendency to chemically attach to soil particles so they tend not to run into aquifers. However, in the past decade, a variety of synthetic organic compounds, mostly pesticides, have been discovered in groundwater at concentrations far exceeding those in surface water supplies. This is due to the development of more persistent carbamate pesticides, including aldicarb, carbofuran, and oxamyl, which are more soluble in water and more readily leach from soils into groundwater. For example, aldicarb is commonly detected in wells near areas of application.[11] Long-term exposure to even low concentrations of carbamate pesticides has a number of health consequences, including cancer, liver and kidney damage, and mutation.[12]

Chemical pollutants can also enter drinking water by way of aging distribution infrastructure. Older distribution systems are often composed of lead-containing pipes, fixtures, and solder. The lead can leach into the water supply, causing extremely elevated levels in drinking water, especially where water is more corrosive due to either high acidity or low mineral content.

A well-known case of citywide lead contamination of drinking water occurred in Flint, Michigan, starting in 2014. In an effort to reduce costs, the economically struggling city switched from purchasing treated Lake Huron water from Detroit, as it had done for 50 years, to treating water from the Flint River. The water from the Flint River was much more corrosive, and because the city did not treat the water with corrosion inhibitors, lead from aging pipes began to leach into the water supply. As a result, between 6,000 and 12,000 children were exposed to drinking water with high levels of lead, doubling the number of children with elevated blood lead levels from 2013 to 2015. This exposure resulted in health problems that included impaired cognition, behavioral disorders, hearing problems, and delayed puberty.[13] The Flint case brought national attention to this type of lead contamination, and subsequent testing revealed that nearly 20% of the water systems in the United States exceeded the EPA's criteria for safe levels of lead.

Some toxic chemicals can also enter into the groundwater and drinking supply through natural processes. For example, arsenic can enter the groundwater as aquifers are recharged by water filtering through naturally occurring arsenic-rich rocks. Arsenic contamination of groundwater is a global problem affecting more than 137 million people in more than 70 countries. Arsenic is a well-established hazard to human health, known to cause bladder, lung, and skin cancer, and it may also cause kidney and liver cancer. Arsenic also damages the central and peripheral nervous and circulatory systems, and it may cause birth defects and reproductive problems.

The presence of arsenic in drinking water has garnered much attention in Bangladesh, where it widely occurs in well water. Prior to the 1970s, the primary source of drinking water in Bangladesh was fecal-contaminated surface waters. This resulted in high levels of water-related illness and accompanying high rates of child and infant mortality. In response to this, UNICEF and the World Bank advocated the construction of millions of deep groundwater tube wells. This was successful in reducing infant mortality and diarrheal illness by as much as 50%. However, of the more than 8 million wells constructed, approximately one in five is now contaminated with arsenic from natural sources. The level of arsenic in these wells far exceeds drinking water standards. More than 30 million people in Bangladesh have had chronic arsenic exposure, and this has been directly linked to increased rates of lung cancer.[14] Some have criticized the aid organizations, despite their good intentions, for failure to recognize and mitigate this problem because they did not effectively consult and work with local stakeholders and scientists.

THERMAL POLLUTION

Thermal pollution is the degradation of water quality by any process that changes water temperature. A common cause of thermal pollution is the use of water as a coolant by power plants and industrial manufacturers. After the cooling process, the water is returned to the environment at a higher temperature, thereby warming the body of water. The amount of oxygen that can be dissolved in water decreases as temperatures rise. Both the warming and the concomitant decline in oxygen availability have negative consequences for the ecosystem, including declines in biodiversity and the creation of environments that favor alien aquatic species. Thermal pollution is not thought to directly affect human health, except in cases in which it may cause toxic algae blooms that, when they accumulate in shellfish, are often toxic to humans. Breathing the airborne toxins from these algae may aggravate asthma. There can also be economic consequences if local fisheries are impacted.

Water and Poverty

The global poor lack access to clean drinking water and effective sanitation. The suggested minimum requirement for drinking water is 20 liters per day. More than 700 million people in the developing world are unable to meet this requirement, whereas average per capita water use in wealthier Europe and the United States is between 200 and 600 liters per day.

People in less developed countries increasingly lack access to water both because of scarcity and because of water pollution, including contamination from pathogens from poor sanitation. Decreased access to water also limits their ability to produce their own food. In addition, there are other social and economic impacts of limited water supplies of clean water on the poor. In places such as sub-Saharan Africa where water sources are often miles from villages, community members are forced to spend hours each day simply finding and transporting water instead of attending school or working.

High rates of waterborne illness also limit the ability to work and attend school. For example, the United Nations (UN) estimates that globally 443 million school days are lost due to water-related illness each day. In sub-Saharan Africa alone, people spend 40 billion hours per year collecting water instead of working. That is the equivalent of a whole year's worth of labor by France's entire workforce. Women and girls are disproportionately affected because young girls are often forced to abandon their education in search of clean water. This has been shown to perpetuate female disempowerment. Former UN Secretary General Ban Ki-moon has stated that safe drinking water and adequate sanitation are crucial for poverty reduction and for sustainable development, and they must be on the forefront of any development agenda.

The Commodification of Water

Most people assume or consider that access to clean drinking water is a basic human right, but until recently it has not been formerly recognized as such. The right to water is not specifically mentioned in the Universal Declaration of Human Rights. However, it has been recognized that without access to water, other basic human rights, such as the right to a standard of living adequate for health and well-being, cannot be met. In 2006, the Human Rights Council and then, in 2010, the UN General Assembly, in a landmark resolution, declared access to water a human right. Having recognized safe and clean drinking water and sanitation as a human right, pressure is now on local and national authorities to provide better infrastructure so that this right can be universally met.

PRIVATIZATION OF PUBLIC WATER SUPPLIES

Despite the recognition that access to water is a basic human right, there have been disturbing trends toward the commodification, privatization, and commercialization of water that place limits on that right. One of these has been a trend toward privatization of what were once public water supplies in developed cities and municipalities. A report produced by Food & Water Watch[15] showed that with the economic downturn of the mid-2000s, many local governments throughout the United States have begun to sell or lease their public water and sewer utilities to generate funding to fill budget deficits. The deterioration of state and local budgets from 2010 to 2012 led to the privatization of nearly 200 US public systems that serve more than 13 million people. Privatization of public water supply systems is also common in Europe. A recent European Union (EU) directive encouraging water privatization in countries such as Greece and Portugal as a condition to continue receiving aid funds has raised concerns and objections that it could lead to forced privatization.

This privatization of municipal water supplies in the United States and other industrialized nations has been problematic for a variety reasons. The two primary concerns with regard to privatization are increases in water rates and loss of public control of a vital resource. In privatization, a private water company pays a large upfront payment as a purchase price or concession fee to a municipality so that it can balance budget deficits. Private companies must recoup that initial investment and then ultimately turn a significant profit. This is typically achieved through excessive rate increases. According to Food & Water Watch, the average annual rate increase for the privatized water utilities was 15%, with some municipalities raising their rates well over 30% per year. Because privatization generates revenues through rate increases rather than raising taxes, low-income families often end up being saddled with this burden disproportionately more than the wealthy because their households often use more water. Thus, in many ways, privatization is like a regressive tax that benefits the wealthy at the expense of lower income families.

By selling or leasing a water system, a local government surrenders control over a vital public resource. Because a private water company is more concerned about profits than public perception and future votes, decisions are based on money savings and often result in reduced services and water quality. Private water companies often enter into deals with developers facilitating suburban sprawl to increase the number of customers. In addition, these privatization deals are difficult to undo even when a municipality or its residents become dissatisfied with the arrangement. Proponents of privatization argue that it has led to improvements in the efficiency and service quality of utilities.

Privatization has not been limited to the industrialized North. As in the case of Greece and Portugal described previously, foreign aid to developing countries is often contingent on water privatization. In numerous cases, large-scale water privatization has been stipulated as a condition for assistance from international financial aid institutions such as the World Bank and the Inter-American Development Bank. This approach to promoting privatization and free market capitalism is referred to as **neoliberalism**.

One example of this neoliberal approach to international financing occurred in the late 1990s in Bolivia, where two cities were forced to turn over their water utilities in concessions to private corporations as part of a condition of a World Bank loan of $20 million. In one case, the city Cochabamba lost control of its water utility to a subsidiary of the large US Bechtel Corporation, the sole bidder for the concession. In the first years of the concession, Bechtel raised rates by an average of 35% and in some cases as much as 200%. This far exceeded the budget of the city's poor and left many without access to water. Licenses were even required for individuals to collect rainwater from their roofs, and people were charged for water taken from their own wells. This resulted in demonstrations and uprisings that eventually led the Bolivian government to declare martial law and ultimately to the rejection of these concessions. The neoliberal approach of linking foreign aid and loans to the privatization of water like that in Bolivia has been a global problem that has favored private water companies at the expense of the right of people to have reasonable access to water.

BOTTLED WATER

Another way that private corporations are profiteering from water has been through the expansion of the bottled water market, which has more than tripled during the past 10 years. In 2017, the global bottled water market reached more than $200 billion. More than 75% of people in the United States drink bottled water, and one in five people drink only bottled water. In 2017, this amounted to more than 100 billion gallons. Much of the water that is bottled is merely tap water from municipal systems or local wells without any functional additives or purification. However, when this water is sold in bottles, it can be up to 1,000 times more expensive than when consumers take it straight from the tap themselves. Furthermore, blind taste tests consistently reveal no preference for bottled water over tap water. Where bottled water is most often consumed, it is typically no safer than tap water. Bottled water is less regulated than tap water, which is subject to more stringent federal safety regulations. These regulations require rigorous testing of tap water safety, and these test results must be made available to the public. Although bottled water is marketed for its purity, a report by the Environmental Working Group[16] showed that in laboratory tests of the 10 most popular brands, the water contained 38 chemical pollutants altogether, with an average of 8 contaminants in each brand. Many of these contaminants were well above acceptable limits imposed on tap water.

In addition to being expensive for the consumer and less healthy than tap water, there are substantial environmental costs associated with bottled water. Bottled water wastes fossil fuels in production and transport. The manufacture and transport of a 1-liter bottle of water consume approximately 1 liter of fossil fuel and emit 562 g of greenhouse gases. In the United States, it takes more than 1.5 million barrels of oil to meet the demands of the bottled water industry. This amount of oil far exceeds the amount needed to power 100,000 cars for a year. Bottled water production is also water intensive; it takes approximately 25 liters of water to produce and transport a 1-liter bottle of water. Rather than being recycled, approximately 75% of the empty plastic bottles end up in landfills, lakes, streams, and oceans, where it may take up to 1,000 years for them to degrade.

There are also equity issues associated with bottled water. The large beverage companies extract water from municipal or underground sources that local people depend on for drinking water at prices that are often lower than those for the average consumer. The companies then sell that water for huge profits while leaving water sources overdepleted. This was seen in California, where severe droughts have increased in frequency, but the Nestle Corporation continues to extract water for bottling. A number of California towns are now out of water. For example, in the city of East Potterville, which is located in one of the United States' most productive agricultural regions, more than 500 wells are completely dry. To make things worse, Nestle is capitalizing on this disaster to which it contributed by selling back the water to Californians at inflated prices. The depletion of water resources by beverage bottling companies is common practice throughout the world. Moreover, a major shift to bottled water could undermine funding for tap water protection, raising serious equity issues for the poor who cannot afford to drink bottled water.

Water Scarcity, Conflicts, and International Security

The combination of water scarcity and growing demand has resulted in individual countries trying to find ways to maximize their extraction and use of this vital resource. This is further complicated by the transboundary nature of river basins and aquifers. The UN reports that globally there are 276 transboundary river basins and more than 200 aquifers that lie beneath multiple countries.[17] As water sources have become depleted, competition between nations that share common sources of water has intensified. This in turn is creating international tension and the potential for conflict between nations. In an edited book on water resource conflicts and international security,[18] the case is made that water supply and infrastructure are increasingly likely to be the objectives of military action.

A prime example of this is in the Middle East, where the Jordan River basin is shared by Jordan, Israel, the West Bank, Lebanon, Syria, and Turkey (Figure 4.5). As a result, water scarcity has become what many consider to be the most destabilizing factor in the Middle East. Former Egyptian president Anwar al Sadat has said that the only reason Egypt would ever go to war again would be for water.

Other regions of tension over water include Pakistan, India, and Bangladesh, where water has already has been an objective of military action. Similar tensions over water exist in the Volta River Basin in West Africa and between China and India. Because of shortage and these conflicts, water is often referred to as "the new oil." However, in many ways, the growing crises over water are much more serious than those for oil because unlike oil, there are no alternatives to water and our survival absolutely depends on it.

Impacting Aquatic Ecosystems

We have seen that aquatic ecosystems play an important role in the hydrologic cycle. They are a source of water for human consumption and hygiene, for agriculture and food production, and for industry and energy production. In addition, aquatic systems are a direct source of food, including fish, shellfish, plants,

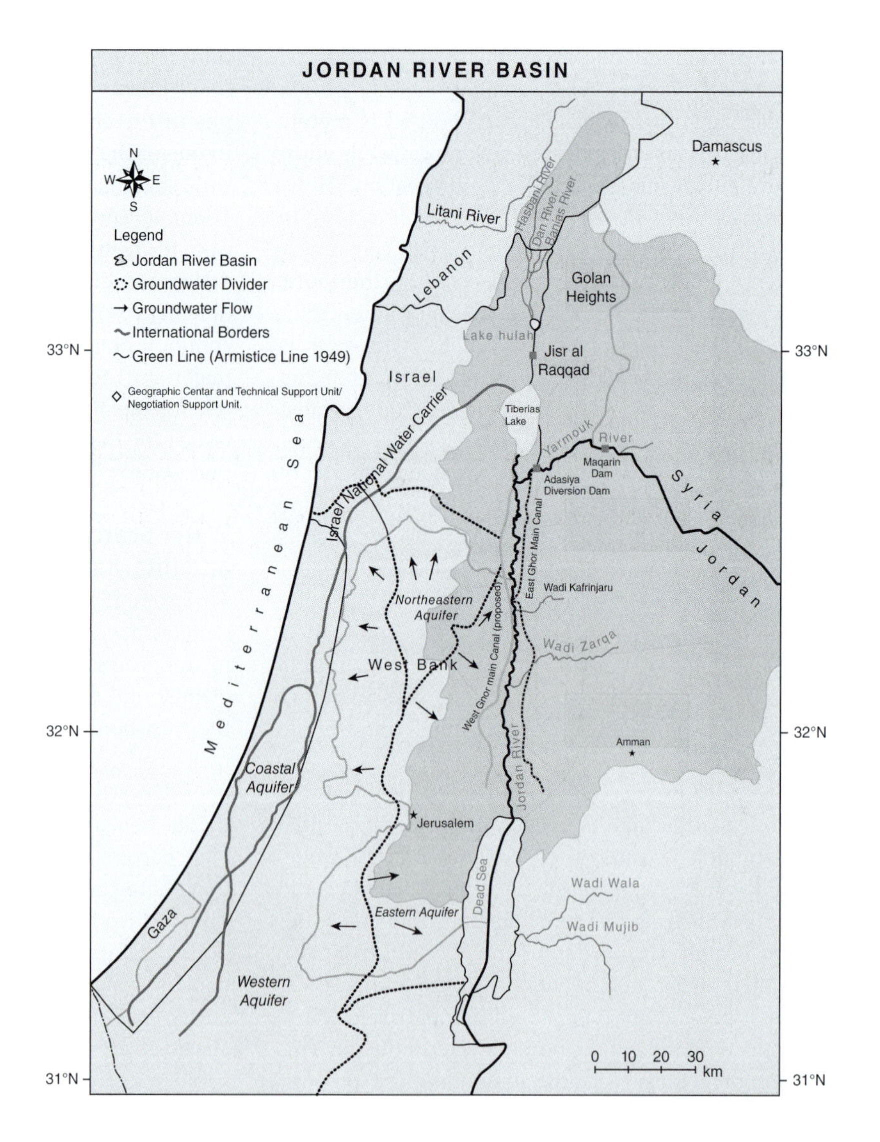

Figure 4.5 The Jordan River Basin shared by Jordon, Israel, the West Bank, Lebanon, and Syria where water scarcity is seen as a risk to international security. (https://waterinventory.org/sites/waterinventory.org/files/chapters/chapter-06-jordan-river-basin-web.pdf UN-ESCWA and BGR (United Nations Economic and Social Commission for Western Asia; Bundesanstalt für Geowissenschaften und Rohstoffe. 2013. Inventory of Shared Water Resources in Western Asia. Beirut. Chapter 6 Page 177)

and algae. Aquatic ecosystems are also important ecologically because they provide habitat for high levels of biodiversity and are major contributors to global primary productivity. They also provide a number of ecosystem services that benefit humans, such as flood regulation, filtration of run-off, as well as serving as a major source of recreation and aesthetic beauty. As discussed in this section, these aquatic ecosystems are being severely impacted by human activities.

Acidification, Eutrophication, and Invasive Species

Chapter 3 discussed how air pollution has directly impacted aquatic systems through acidification and eutrophication. These acidic materials and limiting nutrients can also enter aquatic systems from point and non-point terrestrial sources. For example, nitrogen run-off from agricultural fields or releases from sewage treatment plants can cause eutrophication of aquatic ecosystems and associated ecological consequences. Freshwater ecosystems and estuaries throughout the world are also vulnerable to invasive species, as discussed in Chapter 1.

Sedimentation

Another negative impact on aquatic ecosystems is sedimentation from erosion as a result of agriculture, forestry, and construction activity or from direct run-off from mining and manufacturing processes. **Sediment** consists of particles of either mineral or organic origin that range from micron-sized clay particles to silt, sand, gravel, rock, and boulders. **Erosion** is the movement or displacement of sediment by the action of wind, water, gravity, or ice. **Sedimentation** is the natural process by which these materials settle from solution and are carried to the bottom of a body of water, forming a solid layer. Sedimentation impacts aquatic environments by smothering bottom-dwelling or **benthic organisms** such as corals, shellfish, or stands of plants such as mangroves. For example, erosion from the extensive banana plantations on the Caribbean slopes of Central America has resulted in large amounts of sedimentation in the coastal Caribbean coral reefs, causing substantial mortality of those reefs. Sediments can also harm swimming, or **pelagic**, organisms by damaging their gills, degrading egg-laying sites, and compromising their ability to see prey and predators.

Dams

Another major impact on river ecosystems has been the construction of dams. Dams have been built for many reasons, including flood control, to drive mills and other types of industrial machines, for fire control ponds, to supply drinking and domestic water, for hydropower generation, for crop irrigation, and for aesthetic and recreational reasons. Worldwide, there are currently approximately 40,000 large dams (>15 m high), with many more under construction. China has approximately 19,000 large dams. The United States is the second most dammed country with approximately 5,500 large dams, followed by Russia, India, and Japan. The number of small dams is unknown but is likely in the millions.

Dams that obstruct the world's rivers have numerous direct impacts on the biological, chemical, and physical properties of river environments. They can completely change the patterns of circulation in rivers, and water quality both up- and downstream from the dams can be negatively impacted, which can have serious consequences for biological organisms.

Dams also negatively impact fish living in rivers by altering the structure of the riverbed, which is where fish spawn and obtain invertebrates for food. They also act as barriers to fish that move up and down rivers, such as sturgeon, shad, salmon, and trout. Many of these species are important economically. To remedy this, many dams have been constructed with fish ladders (Figure 4.6). These are structures on or around dams that facilitate the movement of fish up and over them by enabling the fish to move up a series of relatively low steps into the waters above. However, a number of studies have shown that fish ladders are largely ineffective in that less than 3% of fish make it past all the dams in these rivers to their historical spawning areas, and some species, such as sturgeon, never use the fish ladders.[19]

Dams can also have negative social implications. The construction of large dams often displaces human communities—between 40 and 80 million people thus far, mostly in China and India. Typically, the majority of people evicted are impoverished farmers and indigenous people. Many believe that this displacement results in further economic impoverishment, cultural decline, high rates of sickness and death, and major psychological stress. In some cases, people receive no or negligible compensation for their losses. On the other hand, Harvard University water resource expert

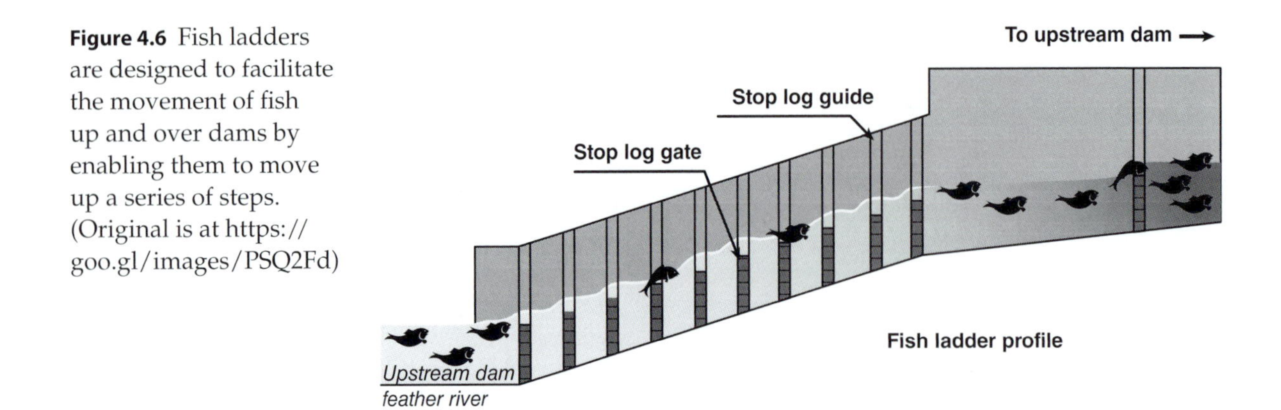

Figure 4.6 Fish ladders are designed to facilitate the movement of fish up and over dams by enabling them to move up a series of steps. (Original is at https://goo.gl/images/PSQ2Fd)

John Briscoe has argued the opposite. He has seen that in Bangladesh and Mozambique, people displaced by dams appreciated new schools, increased access to loans and credit, and increased economic opportunity with the building of new roads and bridges associated with the dam project.[20] Whether large dam construction projects contribute to or detract from sustainability is complicated, and this assessment requires a systems approach. They are expensive to construct, damage river ecosystems, reduce flow and water availability downriver, and displace poorer communities. However, they are also a renewable source of carbon-neutral energy, can increase access to water for drinking and agriculture, and in some cases may improve the well-being of displaced communities.

Loss of Wetlands

Another impact on aquatic ecosystems is the alteration and outright destruction of wetlands. A **wetland** is a land area where water covers the soil or is present either at or near the surface of the soil all year or for varying periods of time during the year. Wetlands may support both aquatic and terrestrial species. The prolonged presence of water creates conditions that favor the growth of specially adapted plants, creating a distinct ecosystem characterized by unique vegetation. Like a sponge, wetlands soak up rainfall and slowly release it over time.

Freshwater wetlands include swamps, marshes, flooded forests, bogs, and fens. Coastal and estuarine wetlands include deltas, tidal mudflats, mangrove forests, and salt marshes. Wetlands are important ecologically in that they are biologically diverse and provide habitat and food for aquatic and terrestrial animals. It has been estimated that freshwater wetlands are home to more than 40% of all the world's species, including 12% of all animal species. Many of them are **endemic**, meaning that they occur only in a particular region.

Wetlands perform a number of ecosystem services that benefit humans. They recharge groundwater supplies, filter pollution, provide flood control, and offer shoreline and storm protection. They also provide spawning and nursery habitat for many commercially important fish. For example, in Brazil, the Várzea flooded forests that cover approximately 180,000 km^2 of the Amazon basin are a breeding ground for more than 200 species of fish. Wetlands are used for rice production and provide wood fuel. People rely on wetlands as sources of medicines, peat, and other usable materials. They also offer recreational and cultural value.

Despite their recognized value, wetlands are being destroyed at an alarming rate. Half of the world's wetlands have disappeared since 1900. Rates of wetland destruction are among the highest in East Asia, where 1.6% of existing wetland area is destroyed per year. Wetlands are destroyed as they are drained to be used for commercial development or to provide dry land to house expanding coastal populations. Increasingly more wetlands are being drained for expanding agriculture. Wetlands are also impacted by dams, pollution, overextraction of minerals and peat, siltation, and invasive species. The rise of coastal waters and increased storm activity with global climate change do and will continue to threaten wetlands.

In developing countries, urbanization results in the expansion of slums onto wetlands. For example, one-sixth of the area of Kampala, the capital city of Uganda, is covered by wetlands. As Uganda experiences intense urbanization pressure, slums and industrial and agricultural sites are rapidly expanding onto these wetlands. The National Environmental

Management Agency of Uganda[21] reports that 75% of Kampala's wetlands have been significantly impacted by human activity, and up to 14% are seriously degraded. Less than 10% of Kampala residents are served by a public sewer, and more than 33,000 persons discharge their household wastes into these wetlands. Similar situations are occurring throughout the world.

Watershed Management

When considering the protection of aquatic resources, it is important to do so within the context of the watershed. A **watershed** is an area of land bounded by topographic features where any water that drains from it goes to a shared destination, such as a large river, lake, or ocean. Watersheds capture precipitation, filter and store water, and determine its release. A watershed, therefore, is a catchment or drainage basin that divides the landscape into hydrologically defined areas and includes all of the land, lakes, streams, reservoirs, wetlands, and aquifers in that area. Because activities such as agriculture and deforestation in a watershed can impact all the water resources within it, they have become a priority consideration for the management and conservation for a particular region.

A specific example of a major watershed is the Mau Forest, which is Kenya's largest closed-canopy forest ecosystem and the most important water catchment in the African Rift Valley and western Kenya. Its rivers act as arteries carrying the Mau's waters throughout western Kenya and eventually to its most populated rural areas near Lake Victoria. These rivers support agriculture, hydropower, urban water supply, tourism, rural livelihoods, and wildlife habitats throughout much of Kenya. However, many areas of the Mau Forest Complex have been deforested or degraded and are negatively impacting downstream water resources. This is a common pattern seen throughout the world, reaffirming the need for watershed management, conservation, and restoration to become a global priority.

Policies to Sustain Water

Our access to clean, safe water and the health of aquatic ecosystems are far from sustainable. Current needs are not being met, and the potential for future generations to meet their needs for water is even bleaker, especially for the global poor. Our most serious public health problems are centered on water. All of our food and energy systems are heavily dependent on water. Climate change has and continues to negatively impact water resources. Solutions must be developed that can sustain this vital resource. These must include policy and innovative technical solutions that will increase our ability to obtain clean, safe water and protect aquatic resources.

This section presents examples of national, international, and global policy approaches for protecting and providing clean water. It also examines the role of non-governmental organizations (NGOs) and public–private partnerships in these efforts. We briefly consider changes in agricultural and energy policy within the context of water use. We also explore how international conflicts over water can be resolved and how to deal with issues surrounding the commodification of water. We then explore new technologies and innovations that will make water resources more sustainable and improve access to water, sanitation, and hygiene (often referred to as WASH) and discuss ways to protect and restore aquatic ecosystems.

The Clean Water Act

One example of a policy aimed at protecting water resources is the US Clean Water Act (CWA), which is often touted as one of the greatest successes in environmental law. The basis of the CWA was enacted in 1948 and was called the Federal Water Pollution Control Act, but the Act was significantly reorganized and expanded as the CWA in 1972. Prior to its expansion in 1972, many US waterways were so polluted that they were considered ecologically dead. It had been commonplace to discharge raw sewage and pollution into harbors and rivers. Bacterial levels in both the Hudson River (New York) and the Charles River (Massachusetts) often exceeded 200 times the safe limit. The Cuyahoga River that runs through Cleveland, Ohio, was so polluted that it caught fire at least 13 times (Figure 4.7). Unfortunately, this was the norm in the United States, where only one-third of the nation's waters were safe for fishing and swimming, and wetland losses were estimated at approximately 460,000 acres per year.

Supported by both Democrats and Republicans and signed into law by President Richard Nixon, the CWA would regulate the discharge of pollutants into waters throughout the country. The CWA made it illegal to release pollution, including raw sewage, from point sources into US waters, and it provided billions of dollars in grants to construct and upgrade publicly

Figure 4.7 Photo from 1952 Cuyahoga River fire caused by the river's severe pollution, showing firemen on railroad bridge at left battling the blaze on the river below. *Cleveland State University, Michael Schwartz Library, Special Collections.*

owned sewage treatment plants nationwide. In recognition that wetlands filter and purify water as it flows through them, the CWA had a provision for their protection and regulated their development.

The CWA has been remarkably successful. Since its inception, pollution has been kept out of US rivers, and the number of waters that meet clean water goals nationwide has doubled. This has had direct benefits for drinking water quality, public health, recreation, and wildlife. Because of the CWA, rivers no longer burn, but policy could be improved to widen its focus and address new threats to water quality. The CWA focuses only on point sources of pollution, so run-off from farms, irrigation ditches, and storm water are not regulated. Pollution from households also remains unregulated, and household contaminants such as chemical cleaners, insecticides, and pharmaceuticals appear in most of the streams that have been sampled. Unconventional methods such as fracking of shale gas and oil are also not regulated by the CWA. Because up to 3,700 spills of chemical-laden water occur at fracking sites each year,[22] this technology should have consistent federal monitoring and regulation in order to protect US drinking water.

The CWA is responsible for protecting navigable waterways and their tributaries, and it gave the EPA the authority to regulate activities in wetlands. However, protection for approximately 60% of the nation's non-navigable streams and millions of acres of wetlands has been challenging since Supreme Court decisions in 2001 and 2006 weakened the CWA's ability to do so. This is problematic because 117 million people get their drinking water from small streams that lost their protection. In 2015, the Clean Waters Rule was passed to clarify which streams and wetlands are to be protected by the CWA. Under this rule, streams and specific water features, including wetlands, from which people either directly depend on for drinking water or that impact downstream waters are now protected. Prior to developing and enacting this rule, the EPA and the Army Corps of Engineers held hundreds of meetings with stakeholders throughout the country, reviewed more than 1 million public comments, and listened carefully to perspectives from all sides. However, increasingly politicians are working to dismantle this rule, but this may be a legally complex process.

International and Global Policy

Although many individual nations have clean water policies, sustaining clean water and sanitation requires both regional and global approaches. Because of its established political structure, the EU has been able to take a regional approach to simplify the regulatory environment in the EU for the water sector. The EU recently adopted two new legal instruments setting strict rules on the treatment of wastewater and the use of nitrates in agriculture. It also adopted a Water Framework Directive in order to standardize and improve the efficiency of water protection legislation for all EU nations. Because of this, water resources are to be managed across national boundaries, following a coordinated approach within river watersheds.

Water resources are most seriously impacted and threatened in the developing world, where the ability to develop and implement clean water policy has been economically constrained. This, coupled with the transboundary nature of water resources, makes the provision of clean, safe water a global issue that must be managed from that perspective. This can best be achieved through the work of global organizations focused on the basic human right of access to safe, clean water.

The World Health Organization (WHO) produces international norms on water quality and human health as the basis for regulation and standard setting in developing and developed countries worldwide. It provides scientifically based water quality management guidelines and support to strengthen the capacity of member states to make sustainable improvements in water quality management. This is done using a stakeholder approach that facilitates cooperation between the relevant authorities, businesses, and the public. The US Agency for International Development (USAID) has also developed a global water policy that has equitably and affordably brought safe water and sanitation to more than 50 million people.

The UN Millennium Development Goals (MDGs) also included goals for safe standards of drinking water and sanitation. Under Goal 7, to ensure environmental sustainability, the objective was to halve the proportion of people without sustainable access to safe drinking water and basic sanitation by 2015. This objective was met five years ahead of schedule. By the end of 2011, 89% of the world population had access to improved drinking water sources, and 55% had a piped supply on premises. However, according to a joint WHO/UNICEF report,[23] an estimated 768 million people were still left without improved sources of drinking water. The report also reveals considerably less success in the area of sanitation, with 2.5 billion people still lacking access to basic sanitation such as toilets and latrines. This is viewed as one of the most under-realized goals of the MDGs. The newly implemented Sustainable Development Goals (SDGs) aim to address these deficiencies, with the Goal 6 target to achieve, by 2030, universal and equitable access to safe drinking water and sanitation and hygiene for all and to end open defecation, paying special attention to the needs of women and girls and those in vulnerable situations. There is also increased emphasis on reducing pollution, increasing water-use efficiency, and improved water resource and ecosystem management.

Although there is strong national and international political will to solve the global water problem, the major limitation has been the funding needed for new capital investment in infrastructure and for recurrent expenses of operations and maintenance of that infrastructure. The burden of this funding has primarily been placed on the central governments of the nations in which these improvements are made, either through user fees or through domestic taxes. The problem is that many low- to middle-income nations do not domestically raise the sufficient funds needed to meet the recommendations and goals of the international community, especially in the area of sanitation. Often, there is insufficient funding for the operation and maintenance of water and sanitation infrastructure, thereby undermining the sustainability of these projects. For example, regional spending in Africa has been 25% of what is required to provide safe drinking water.

Because of the inability of poorer nations to meet the financial demands associated with needed water and sanitation improvements, they must rely on external sources of funding. External funding has come from the World Bank, the Asian and African Development banks, and from foreign aid from developed countries. Funding for improved water and sanitation has also come from USAID and UNICEF. For example, they recently established a partnership to invest $14 million to improve water, sanitation, and hygiene services for chronically vulnerable and displaced populations in South Sudan. However, according to a report by WHO, total funding for these types of programs has been insufficient.[24]

In response to the global water and sanitation crisis, the US Congress passed the Senator Paul Simon Water for the Poor Act of 2005. It was then signed into law by President George W. Bush. The goal of this bill was to involve the US Department of State in improving water resources management, optimizing the productivity of water, and mitigating tensions associated with transboundary waters. However, the act was never fully implemented or funded. In 2012, new legislation, the Senator Paul Simon Water for the World Act, was introduced in both the House and the Senate to facilitate the implementation of the original Water for the Poor Act, but this bipartisan act has yet to be passed.

Non-Governmental Organizations and Private–Public Partnerships

Because of poor economic conditions and limited foreign aid, central governments in water-stressed nations have lacked the resources to meet the global recommendations for improving access to water and sanitation; thus, the non-profit NGO sector has provided help. One example is the Center for Strategic and International Studies, a US non-partisan, non-profit organization focused on the governance challenges in ensuring the sustainable use and stewardship of water in diverse world contexts. Its work includes analyses of the effectiveness of US foreign assistance regarding water, the role of the private sector in extending access to water supplies, and the integration of water and sanitation into global health, gender, security, and environment agendas.

There are hundreds of non-profit organizations dedicated to promoting and providing WASH. Well-known examples are CARE and WaterAid. These are international NGOs that focus on empowering people to make decisions about their own water and sanitation systems and water management. They do this by influencing policy at national and international levels and by providing direct aid and training to develop WASH infrastructure and technologies for the most vulnerable and marginalized populations.

Another NGO, Water For People, brings together local entrepreneurs, civil society, governments, and communities to establish creative, collaborative solutions that allow people to build and maintain their own reliable, safe water systems. The International Water Management Institute is a non-profit, scientific research organization focusing on the sustainable use of water and land resources in developing countries. It is headquartered in Colombo, Sri Lanka, with regional offices throughout Asia and Africa. Another, SPLASH, increases access to hygiene by utilizing technology and supply chains already in use by major food and hotel chains in Nepal, China, India, Thailand, Bangladesh, Cambodia, and Vietnam. By building local businesses to create safe water projects that incorporate approaches already in use in each country, SPLASH has been able to sustain access to WASH at the local level.

Another approach to improving WASH in developing countries is through private–public partnerships. One of many examples is a recent collaboration among the Peace Corps, USAID, and the Coca-Cola Company to improve local capacity to deliver sustainable water supply, sanitation, and hygiene services for the reduction of waterborne disease throughout the world. The focus of the project is to strengthen training and materials for Peace Corps volunteers and their in-country partners so that they can disseminate information and best practices among community stakeholders regarding WASH. Some might question Coca-Cola's involvement in this project given its history of extracting hundreds of billions of liters of water globally each year, often from drought-stricken regions. But this is another case in which public perception of a corporation might influence its role in sustaining the very resources it overexploits. Both through giving hundreds of millions of dollars to projects such as this collaboration and through the development of new sustainable technologies that conserve and replenish water supplies, Coke hopes to untarnish its reputation without losing market share.

Agriculture and Energy Policy

Because water consumption is so tightly linked to agriculture and energy, it is important that policies and independent agencies that deal with each of these issues recognize and address these interactions. The Common Agricultural Policy and the Water Framework Directive in Europe accomplish this by bringing together decision-makers, stakeholders, and scientists from fields related to agriculture, energy, and water. USAID has also developed a global water policy by convening a range of stakeholders and university researchers in a monthly water working group that focused on the interdependency of food, water, and energy. This initiative helped bring water and sanitation to more than 50 million people, and it assisted governments and private firms to manage and distribute water more equitably and affordably.

Current farm policy in the United States and other industrialized countries that subsidize large, industrial farms that are fossil fuel, chemically, and irrigation intensive should consider favoring approaches that conserve and protect water resources. Examples include subsidies for implementing water-efficient irrigation techniques, less use of toxic pesticides and fertilizers, and those that favor smaller, local farms (see Chapter 5). Current US agricultural and energy policies that favor water-intensive biofuels should also be reconsidered in light of water issues (see Chapter 6).

New energy policy initiatives are also considering water conservation issues. This includes a shift away from subsidizing the fossil fuel industry and support of the development of less water-intensive renewables. Policies such as New York state's 2014 moratorium on the water-consumptive and contaminating practice of fracking is an example of an energy policy that protects water and human health. The US Department of Energy has prepared a report on the nexus of water and energy that frames an integrated approach for policy and technical development around these two issues.[25] Energy policy, including its relation to water resources, is discussed more completely in Chapter 6.

Water Dispute Resolution

Despite the complex problem of shared transboundary water resources, historically these disputes have been handled diplomatically. The history of international water treaties dates back to 2500 BC when the two Sumerian city-states of Lagash and Umma negotiated a treaty ending a water dispute along the Tigris River. Since 805 AD, there have been more than 3,600 treaties related to international water resources. Nations continue to value these agreements because they make international relations over water more stable and predictable. According to the UN Department of Economic and Social Affairs, in the past 50 years there have been 150 treaties signed regarding shared water resources, in contrast to the 37 acute disputes that resulted in violence.

Agreements on shared water resources have been negotiated and maintained even as conflicts have persisted over other issues. For example, Cambodia, Laos, Thailand, and Vietnam have been able to cooperate since 1957 within the framework of the Mekong River Commission, and they had technical exchanges regarding this even throughout the Vietnam War. Since 1955, Israel and Jordan have held regular talks on the sharing of the Jordan River, even as they were recently in a legal state of war. Similar examples exist between India and Pakistan and in the Nile River basin, which is home to 160 million people and shared among 10 countries.

As the pressure on water intensifies, so does the potential for destabilization and conflict over transboundary sources of water. To address this, there have been a number of UN initiatives to minimize conflicts and disputes over water. Among these is the UN General Assembly's 2011 resolution referred to as the Law of Transboundary Aquifers. The resolution encourages nations to make appropriate bilateral or regional arrangements for the proper management of their transboundary water resources and to cooperate in controlling the pollution of shared aquifers. It also provides some key guiding principles on the equitable and reasonable utilization of international water sources and the obligation of nations to protect shared transboundary waters from pollution.

This type of international cooperation is an essential component of sustainable problem solving. However, there is consensus among experts that much more work needs to be done to reduce the potential for international conflict over water. New and revised agreements need to be more concrete and must also include specific conflict resolution mechanisms in case disputes occur. In order to be successful, treaties should incorporate detailed monitoring provisions and enforcement mechanisms to which all parties can agree. Given the increases in variation in the abundance of water and changing needs for water over time, there must be flexibility in water allocation provisions. Finally, successful water conflict management will require well-funded third-party support trusted by all of the stakeholders.[26]

The future success of limiting disputes over transboundary water supplies will require agreements that included detailed recommendations for identifying clear yet flexible water allocations and water quality standards that account for short-term hydrological events such as droughts, longer term variation in water resources, and social changes on either side of the boundary. It also established measures for enforcement and conflict resolution in case disputes erupt. The concrete, detailed nature of this international agreement and its recent successes offer hope for our ability to handle water disputes diplomatically even as pressures on this essential resource rise.

Stop the Commodification and Privatization of Water

Part of the effort to meet what has become the globally recognized human right to water includes stopping and reversing the rising trend in the privatization of public water supplies and to stop making it a condition for international aid. This process of returning control of water from private for-profit companies to local municipalities, referred to as **remunicipalization**, has been shown to maintain the access, affordability, and quality of water for the average citizen.

Globally, there has been success in remunicipalization of water. From 2000 to 2015, 235 cities from 37 countries remunicipalized their water services, benefitting 100 million people.[27] In other cases, attempts at privatization have been halted through public outcry and even uprisings such as those in Bolivia described previously in this chapter. Some municipalities have leased their water to non-profit water authorities rather than to privately held companies. Although not without controversy, this has allowed continued local oversight and a certain amount of rate control while providing the municipalities with needed economic relief.

Stopping and reversing the growth of the bottled water industry is another essential component of sustaining water resources. One way is to educate the public about the negative aspects of extracting and depleting public water supplies and then re-selling that water for large profit so that individuals can then choose not to purchase it. Another is to work with institutions so that they ban or limit the sale or provision of bottled water. Although this is sometimes challenging to achieve, it is a growing trend in higher education and other places that offer food and beverage services (Box 4.2). Finally, bottling companies should be charged higher fees for their right to extract public water supplies for private sales so that they are responsible for the external costs of overextraction.

Innovative and Technical Solutions

Policy alone cannot solve the global water crisis or issues related to WASH. In combination with new policy, we need new infrastructure and technical innovations that will conserve, generate, and provide equitable access to clean water and improvements in sanitation and hygiene. In this section, we examine a variety of technical and innovative solutions that can enhance our ability to do this. We also discuss examples of how changes in behavior at the individual and community levels, in addition to ecological approaches, can help sustain our water resources. Finally, we address how to most effectively stimulate innovation in these areas and how to make technology transfer and program implementation more sustainable.

Generating Fresh Water

Given increasing demand and reduced availability of fresh water, solutions that effectively produce it are sorely needed. There has been much recent innovation in the development of technologies that produce, harvest, clean, or reclaim used water, making it available for consumption and other uses by humans. These include desalination, condensation, rainwater capture, filtration, and gray water reclamation. Challenges that are being addressed include the cost of and the dependency on fossil fuels of many of these solutions.

DESALINATION

Because greater than 97% of the Earth's water is in the form of undrinkable salt water, one obvious solution is to develop technologies to remove the salt from the water through the process of desalination. Used for thousands of years, early methods of desalination removed salt either by boiling water to evaporate fresh water away from the salt or by using clay filters to trap salt, but usually on a very small scale. The same basic approaches are employed today, but technological advances have allowed for greater freshwater production.

Most desalination plants incorporate thermal technologies such as distillation, which involves boiling the water to generate salt-free water vapor that can be condensed and collected, leaving the salt behind. Another approach is the use of membrane filtration

BOX 4.2

Stakeholders and Collaborators: Beyond Bottled Water Bans

Despite having access to cheap, safe, and highly regulated public water, Americans buy approximately 43 billion single-serve plastic water bottles annually. The millions of gallons of oil used to produce and transport this product contribute to climate change. The environmental harms continue as people throw away the majority of plastic bottles, polluting our land and waterways. Due to the unnecessary nature of the product and the waste it creates, environmental groups at dozens of US colleges and universities have tried to ban bottled water.

Some movements have successfully prohibited sales on campuses,[1] but many have failed due to opposition from student bodies, administrators, and beverage companies. The competing interests of the different stakeholders—for example, environmental groups and beverage companies—make policy changes such as the banning of consumer products difficult within our capitalist society. Consumer bans, like bottled water boycotts, challenge the principles of individualism and free choice that pervade our culture and economy. In this context, stakeholders' positions appear to be incommensurate, and sustainable solutions seem impossible.

Are there ways to promote sustainability through stakeholder collaboration? Is the problem that stakeholders' interests are incongruous or that we need more creative initiatives? One example that has offered another way to promote sustainability and overcome some stakeholder conflict over bottled water is the "Just Tap It" campaign at Muhlenberg College. Similar to many institutions, a bottled water ban initiated by the college's environmental group was rejected by the college's governing bodies (Figure 4.2.1). Student government representatives opposed the policy because they believed it obstructed students' consumer rights, would inconvenience students because the existing water fountains could not accommodate water bottles, and that the tap water was unappealing because of its high mineral content. Also, the beverage provider claimed that forbidding bottled water would violate its contract with the college, which worried administrators. Finally, college officials were unwilling to introduce an unpopular policy even though a ban would reduce the institution's waste and carbon footprint.

Rather than give up, the environmental group considered all stakeholders' concerns and drafted a new proposal to reduce bottled water purchases. After negotiating with the different stakeholders, the college initiated a "Just Tap It" program, which included installing attractive filtered water fountains with bottle refilling docks; removing bottled water from meal deals, thus requiring students to pay out of pocket for bottled water; reducing product advertising; moving water vending machines to the back of the campus convenience store; providing new students with free, refillable water bottles; and publicizing the changes on campus. Overall, the campaign decreased the convenience and availability of bottled water while increasing people's access to appealing and alternative water sources.

Figure 4.2.1 Environmental groups at dozens of US colleges and universities protest to have bottled water banned on campus. *Photo by R. Niesenbaum.*

1 For a list of campuses with existing or ongoing campaigns against bottled water, see https://www.banthebottle.net/map-of-campaigns.

continued

continued

Within two years, the college's bottled water sales declined by 92%, and the "Just Tap It" program diverted approximately 1.5 million bottles from landfills. The resounding success of the program illustrates that it is possible to institutionalize sustainability and collaborate with stakeholders to achieve environmentally desirable outcomes.

—Contributed by Krista Bywater, PhD
World Vision, Washington, DC

technologies that are permeable to water but not to salts. Pressure-driven membrane technologies such as reverse osmosis work by applying pressure to the saltwater side of the membrane, forcing fresh water through it while leaving behind the salt in a concentrated brine solution. This is referred to as **reverse osmosis** because fresh water normally moves across a membrane to dilute concentrated solutions in a process referred to as **osmosis**. Through the application of pressure to the saltwater side, this process is reversed. The passage of fresh water through a membrane can also be driven electronically in processes such as electrodialysis, in which electrodes are used to attract the salt ions.

Desalination has been used successfully in the provision of water for drinking and for irrigation. In Israel, five of the world's largest seawater reverse-osmosis plants produce more than 600,000 m^3 of water each year and provide 75% of household water. There have been other success stories with desalination throughout the world. As of June 2015, there were 18,426 desalination plants operating worldwide, producing 86.8 million m^3 per day, providing water for 300 million people.[28]

Despite the success of desalination and its potential for contributing to our supply of fresh water, economic, environmental, and even health concerns have been raised, limiting its more extensive development. Desalination plants are expensive to construct, energy intensive, and can only serve coastal areas with access to salt water because long-distance transport infrastructure is lacking and expensive. Oil-rich areas in the dry Middle East and North Africa that also have access to seawater rely heavily on desalination because they have the energy and economic capability to make it happen. However, this technology has yet to serve poorer, land-locked nations.

One of the major issues with desalination is that with current technology, it is an energy-intensive industry that relies almost entirely on fossil fuels. The solution to this is to decouple desalination from fossil fuels and to use renewables such as solar, wind, or geothermal to power the process. Another issue is that the technology is expensive and not cost-effective for most nations. However, with greater investment in technological advances, costs will be reduced and desalination will soon begin to serve wider populations, including water-stressed poorer nations. One report indicated that between 2010 and 2016, $88 billion in total capital investment was channeled into this industry, and it suggested that widespread renewable-powered desalination can become a reality by 2050.[29]

Another method, solar distillation, involves placing a clear lid over a saltwater pond, which uses solar heat to cause evaporation and then traps that water vapor as it condenses on the lid. This approach is more affordable and has reduced energy costs, but it requires large amounts of land. Recently, chemists at the University of Texas and the University of Marburg devised a method of using a small electrical field that removes the salt from seawater. This new "water chip" method is much simpler and consumes less energy than other forms of desalination.

An environmental issue with desalination is that fish and other marine organisms can be killed during the seawater extraction process. Screening inlet pipes can help reduce these deaths. Another major environmental challenge of desalination is the disposal of the highly concentrated salt brine that remains from the process. This brine can kill marine organisms near disposal sites. This problem can be solved by first filtering salt from the effluent and then using special diffusers to dilute the brine as it is discharged. More recently, a link between regular consumption of desalinated water and heart disease has been established due to the lack of magnesium in the water.[30] A solution to this problem is to add magnesium to the desalinated water, although this increases costs.

DRAWING WATER FROM AIR

In addition to desalination, another, perhaps more effective, way to generate water is through the condensation of water from ambient air. This is done by using a compressor to circulate refrigerant through a condenser and an evaporator coil that cools the air that is drawn over it. When the air comes in contact with the chilled coil, the water in it condenses, which can then be collected and stored for human consumption. This technique has been combined with wind turbines, which both power the compressor and draw the air over it. These wind–water turbines can provide at least 1,000 liters of fresh, clean water per day, even in areas where humidity is low. Because these systems require no underground plumbing or connection to an electrical grid, they can be placed in isolated areas. These innovative designs have the potential to provide clean drinking water to hundreds of millions of people without the use of chemicals or fossil fuels. The only drawback is the cost of these systems, which range from $400,000 to $1 million. However, from a sustainability perspective, they may be a solution worth pursuing by aid organizations, local and national governments, and through community microfinancing programs.

Another recent innovation that is being developed by the company SunGlacier extracts water from air using a solar-powered Peltier element. When the Peltier element is charged, it becomes hot on one side and cold on the other side. As air comes to the cooler side, water is condensed and can be collected. This technology is still under development, but SunGlacier invites the public to innovate its design, which is available online as open-source technology. This collaborative approach is expected to make SunGlacier's groundbreaking, low-maintenance design accessible to the greatest number of people possible.

RAINWATER CAPTURE

Although highly variable across time and space, approximately 100,000 km^3 of rain falls on the Earth's terrestrial surface each year. Much of this enters surface water or percolates into the groundwater. However, more than 1 million km^2 of the Earth's surface is paved, and the water that hits that surface typically does not percolate into the ground where it can recharge aquifers but, rather, enters the water waste stream as storm water run-off.

Rainwater capture is a simple solution that could help deal with water shortages both for irrigation and for drinking. It can also reduce groundwater and surface water withdrawals and relieve some of the pressure on aging drainage and sewage infrastructure. It is a simple technology that can be developed at different scales. To employ this technology, a rainwater catchment system is required, which could be as simple as a roof, open cistern, shallow infiltration well, or paved surface with drainage. Depending on what the water will be used for, it can then be purified using simple UV radiation and carbon filtration systems and stored in tanks.

Rainwater capture systems have been implemented worldwide on a variety of scales. In Tokyo, rainwater harvesting is promoted to mitigate water shortages, control floods, and secure water for emergencies. In Indonesia, where groundwater is becoming scarce in large urban areas due to reduced water infiltration, the government introduced a regulation requiring that all buildings have an infiltration well. It is estimated that this has reduced the country's water deficit by 16%. In Iran, a double-roof system engineered for arid environments was designed to collect and store rainwater and to promote natural cooling (Figure 4.8). Rainwater collection is becoming widespread in Africa, with numerous projects throughout the continent but especially in Kenya and Botswana. In the United States, rainwater capture systems are marketed on a variety of scales, but they have become particularly popular for home gardening and small farm irrigation, including rain barrel systems that are linked to downspouts.

Another way to recapture water that is being lost as it flows into combined waste water and storm water systems is to promote approaches to building and development that allow the water to infiltrate the ground surface and recharge aquifers. Techniques that promote infiltration and storage of water in the soil column, such as infiltration trenches and basins, permeable pavements, soil amendments, and reducing impermeable surfaces, have been incorporated into new and existing residential and commercial developments and local building code. "Green parking lots" that incorporate pervious concrete, porous asphalt, and pavers allow rainwater that normally runs off the parking surface and surrounding building roofs to infiltrate underlying soil. Another approach is the use of rain gardens, which are depressions planted with native plants that collect rainwater run-off from impervious urban areas such as roofs, driveways, walkways, parking lots, and compacted lawn areas. These approaches not only encourage aquifer recharge but also can reduce local flooding.

Another water capture system is mist collection, using nets in foggy areas to collect water for drinking or irrigation. Water droplets attach to the netting and

Figure 4.8 Designed for use in arid countries like Iran, the double-roof system, which includes a domed roof beneath a bowl-shaped catchment area, is designed to allow even the smallest quantities of rain flow down the roof and eventually coalesce into bigger drops before they evaporate. With this system, an estimated 28 m^3 of water could be harvested with just 90 m^2 of a concave roof surface. Also, stacking a concave roof on a convex roof promotes natural cooling through shade and wind movement between the two roofs. *Courtesy BMDesign Studio.*

run down into gutters beneath the nets into storage tanks or wells. One square meter of netting can provide 5 liters of water per day. Mountainous areas with seasonal fog are the most appropriate for mist collection, especially those with reduced surface waters and deep, inaccessible aquifers. Examples of these areas include costal mountain areas of Asia, southern Africa, and Central and South America.

Gray Water Reclamation

Gray water is residential and commercial waste water other than toilet wastes; it includes water from sinks, showers, tubs, and washing machines. Although it does not contain fecal material, there are other contaminants, including dirt, food, grease, hair, cleaning products, and microbes. One sustainable solution to conserve water is through the reclamation and use of gray water. This requires piping that separates sewage lines from gray water lines and can incorporate natural treatment processes depending on use (Figure 4.9). Gray water can then be used for irrigation and toilet flushing, reducing freshwater extraction from rivers and aquifers and promoting groundwater recharge.

One well-known example of a gray water reclamation system is the "Eco-Machine" designed by eco-pioneer John Todd that uses an ecosystem approach to cleaning gray water. This technique uses microbes in a pretreatment to take up nutrients and, in particular, convert ammonia into nitrates. The water is then run through simulated wetland systems, in which plants and microbes further consume nitrates and phosphate, eliminate odor, and clarify water. Water can be further filtered and stored in oxygenated tanks prior to use. This approach has been incorporated into Oberlin College's Adam Joseph Lewis Center for Environmental Studies, where all gray water is processed and reused. The system is operated by a group of approximately 10 students and provides opportunities for them to explore issues of waste water, wetland ecology, microbiology, and plant dynamics. Gray water reclamation systems are currently being developed for groups of houses and have been shown to significantly reduce a building's water footprint (see Box 4.1).

Technical Solutions in Agriculture

In addition to changes in farm policy and food choice that promote water conservation, there are a number of technical solutions to reducing the amount of water needed for food production. One example, **smart irrigation** allows farmers to irrigate with much more precision and less waste. Using a combination of water sensors and valve-controlled irrigation, water can be released based on how much water is available and can be absorbed by growing crops. **Drip irrigation** is another approach to water conservation in agriculture. A drip or micro-irrigation system delivers water directly to the root zone of a plant, where it seeps slowly into the soil one drop at a time. Almost no water is lost through surface run-off or evaporation, and a well-designed drip system uses as much as 50% less water than other methods of watering, such as spray systems. Similarly, subsurface irrigation systems save water and reduce evaporation. Another approach is an optimization strategy of **deficit irrigation**, in which irrigation is applied only during drought-sensitive growth stages of a crop. This results in a slight drop in yield but conserves large amounts of water for alternative uses. All of these irrigation techniques can be integrated with the rainwater capture and storage systems described previously.

Other approaches to land management, planting techniques, and crop choice can also save water.

Figure 4.9 Rain capture combined with gray water reclamation and storage systems reduce freshwater extraction from aquifers and surface waters. (CPM Grey_water Harvesting via https://www.buildingservicesindex.co.uk/entry/44900/CPM-Group/Oasis-grey-water-and-rain-water-harvesting-system/)

For example, **swales** or depressions created in the land used in permaculture capture water, reduce run-off, recharge groundwater, and capture water for agriculture in sloped landscapes. Other approaches that conserve water involve the construction of small pits, ditches, or canals that capture water around each crop. For example, the Zai planting pits used in Africa are hand-dug holes around each plant that serve to both trap water and increase soil fertility. Organic farming, permaculture, rice intensification systems, rotational grazing systems, and the development of drought-tolerant crops and livestock breeds also offer opportunities for water conservation and adaptation to increased drought brought on by climate change. These and other approaches are discussed in more detail in Chapter 5.

Industrial Solutions

After agriculture, industry is the largest user of water, consuming 5–10% of global water withdrawals. Thus, innovative changes that can reduce this consumption are vital. Water conservation by the industrial sector can be achieved through a combination of changing industrial processes and behavior, increasing internal reuse and recycling of water, and modifying and/or replacing equipment with more water-efficient equipment. Changes in behavior can be made both through economic incentives from water providers, such as discounted rate structures with lower levels of consumption, and through employee training programs. Reuse and recycling of water can be done using gray water reclamation, as described previously.

A variety of industrial processes have been re-engineered in order to make them more water efficient. The automotive industry has taken the lead in this regard by replacing water-intensive machining processes with dry machining processes, altering the painting process, and recapturing and reusing water. In this manner, car companies such as Toyota and Ford have reduced water consumption in many of their plants by as much as 70%. Another method, referred to as **short interval control**, increases water use efficiency by regularly collecting short-term data on major losses and water recovery through each stage of production. Using this approach, beverage companies such as Coca-Cola and MillerCoors have reduced water use by 35%. As this approach is implemented across all bottling plants, hundreds of billions of gallons of water could be saved annually. Given that conserving water is a money saver for these and other corporations, there is much motivation for developing innovative processes that conserve water.

Reducing Household Water Consumption

In the United States, each person uses 80–100 gallons of water per day, which equates to a daily total of nearly 3.2 billion gallons just for domestic use. The largest water-consuming activities at home are bathing and showering, toilet flushing, clothes and dish washing, and outdoor lawn and garden watering (Figure 4.10). Small leaks, especially those from dripping faucets or running toilets, also consume a great deal of water, sometimes up to 10% of daily water consumption. Given that we know how water is being consumed, a number of water-saving solutions that can be implemented in the household should become obvious.

Loss through leaks can be greatly reduced through the use of water meters that detect and alert the home owner of even small leaks. Such smart meter systems provide immediate notification when consumption surpasses established hourly or daily benchmarks and can alert the homeowner to remedy the situation so that the leak can be stopped. Many municipalities use these kinds of smart meters as their standard water meter, benefiting both the consumer and the water conservation efforts of the municipality.

Other effective ways of conserving water include low-flush toilets, low-flow shower heads and faucet aerators, and pressure reduction valves. A new technology for clothes washing has been developed that uses almost no water. The British company Xeros has developed a washing machine that uses thousands of nylon beads with a molecular structure that binds with detergent and attracts dirt. The beads can be used for hundreds of washes before being recycled without ever being released into the environment. Such a machine will use 70% less water and 50% less energy than conventional machines.

Gardening and watering consume a significant amount of water, particularly in locations where plantings do not match the climate, such as grass lawns in desert environments. Water savings can be achieved both by selecting plantings that use less water and by use of more appropriate irrigation techniques. The careful design of landscapes, also known as **xeriscaping**, is an innovative, comprehensive approach to landscaping that incorporates plant choice, soil management, and irrigation techniques that reduces or even eliminates supplemental water use.

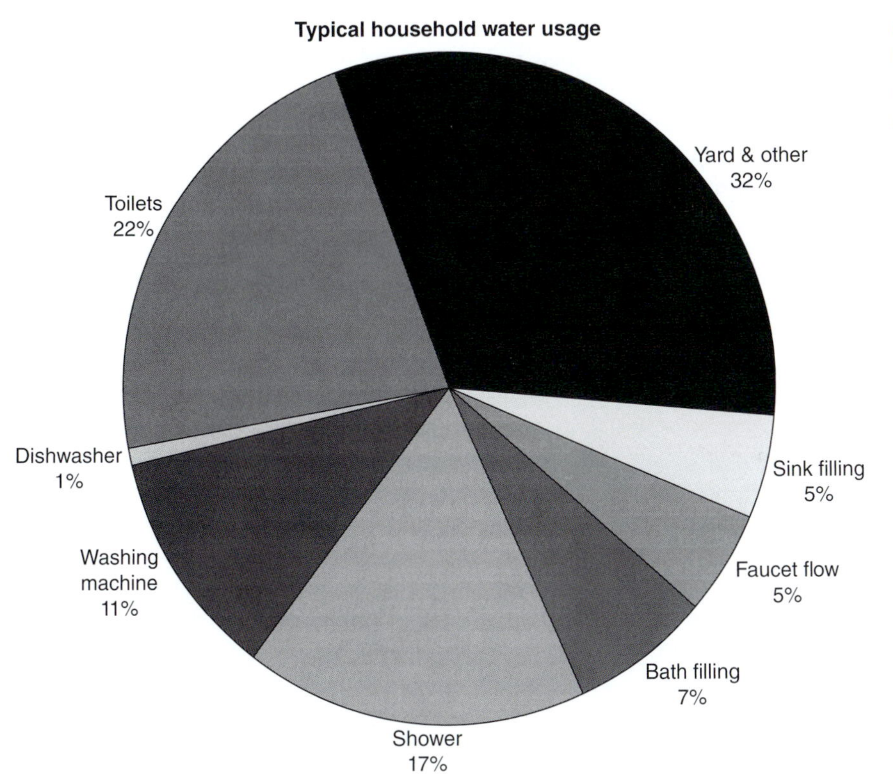

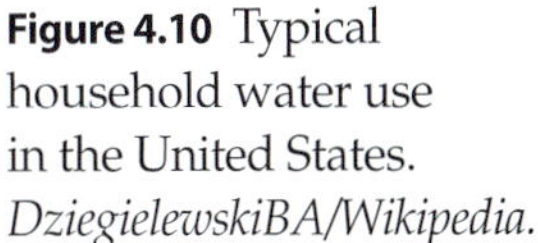

Figure 4.10 Typical household water use in the United States. *DziegielewskiBA/Wikipedia.*

Examples include scheduling irrigation for early morning or evening hours; the use of cycle irrigation, which provides the right amount of water at the right time and place for optimal growth; and the use of low-precipitation-rate sprinklers, bubbler/soaker systems, and drip irrigation systems. These approaches incorporated with previously discussed rainwater capture and gray water reclamation systems can significantly reduce the water burden of landscape.

Changes in behavior can also reduce water consumption. For example, water consumption can be reduced significantly by running the dishwasher and clothes washer only when they are full; reducing the amount of time faucets are open when washing hands and dishes, brushing teeth, and shaving; and taking shorter showers. These individual actions coupled with choosing products that require less water to produce are an essential part of reducing the individual and household water footprint (see Box 4.1).

Improving Water, Sanitation, and Hygiene

Improving WASH has been one of the highest priorities for international aid organizations, international and global policy efforts, and the work of numerous NGOs and private–public partnerships. Drastic improvements have been made in increasing access to drinkable water. However, improvements in sanitation are lagging behind, with 2.5 billion people still lacking access to basic sanitation such as toilets and latrines. As a result, waterborne disease remains one of the leading causes of death and illness in the world. Policy has been effective at setting guidelines, norms, and goals for water quality and human health and creating legal instruments to protect water. However, meeting these global policy objectives requires the development of a variety of innovative technologies and approaches.

Numerous innovations have been developed that could improve WASH. Some of them are quite simple and inexpensive; others require significant initial and ongoing investment but have the potential to serve the most people. In many cases, simple innovations offer both solutions and entrepreneurship opportunities. As important and as effective as these innovations might be, certain conditions are essential to implement improvements to water supply and make sanitation services sustainable.

First, there must be a demand for such improvements from the users. This is because a top-down approach to technology transfer, especially from governments and foreign organizations, is rarely successful. Second, the scale of the project needs to be appropriate for the implementation site. Third, the ongoing costs of maintenance and operation that typically far exceed initial costs have to be affordable to users who must be willing to pay for them. Finally, once in place, local communities and individuals need to assume ownership and responsibility for these improvements, and they must be trained how to do so. This must include implementation of the program or new technology and the establishment of community boards that are responsible for the sustainable operation and funding of the given operation or program. An example is described in Box 4.3. Not meeting any of the previously discussed conditions represents a barrier to sustainability. However, if they are met, many of the programs, innovations, and technologies described here could sustainably and significantly improve WASH globally.

BOX 4.3

Policy Solutions: Community-Based Tariffs for Water and Sanitation

When controlled by local municipalities, billing customers for their actual water use based on accurate metered measurement has been found to contribute directly to water conservation and to be an effective way to support infrastructure for water delivery and sanitation. At least this has been the case in industrialized nations. This approach is beginning to be applied in less developed countries—for example, in the small community of Gonzolo Moldonado located in the Negrito district of northern Honduras. Its water system failed, causing residents to rely on unsanitary surface sources of water. In a collaborative effort that included the local government, community members, and the NGO Water For People, the community's water system was rehabilitated. But more important, a solid foundation was put in place to make this a sustainable success. That foundation consisted of a pay-for-use system that was controlled by an established community water board. Although there was strong opposition, water meters similar to those used in the United States and a consumption-based payment system were put in place to support the water treatment and distribution system in the long term. A local water board was established that runs an affordable tariff system based on metered use. As a result, all 65 households in this community now have reliable, safe drinking water 24 hours a day; water is conserved; and there are funds to maintain the new infrastructure.

In another small Honduran community, Las Crucitas, rather than handing out pour flush toilets that use small amounts of water, families were given loans to purchase or construct their own. These sanitation loans are atypical in developing countries such as Honduras, where government toilet handout programs are the norm. Handout-based sanitation programs result in waste, poor or ineffective use, and lack of maintenance. But when families and communities take ownership of and are the focus of the solution, the effects have been shown to be more sustainable. The small, low-interest loans with payback plans that are affordable for community members originate from the community-based water boards with some government and NGO support. Unlike handouts, the loan program has given families ownership and care over their own in-house sanitation. The loans cover the cost of a pour flush toilet bowl, cement for a slab and pit cover, and a PVC pipe to carry flushed water from the toilet to the pit. The end product for families is a pour flush toilet with no odor. Because they are integrated into the structure of their existing home, families properly use and maintain the systems.

These communities in Honduras represent case studies that show that when self-determination, ownership, and responsibility are integrated into community-based projects, solutions are more sustainable. They also show how support from public–private partnerships can be leveraged to support everyone in the community. One challenge of this type of project is that as they expand, individual communities will need to begin to cooperate so that capital maintenance funds for regional rural water systems can be established. This may require the establishment of regional associations of individual community boards and additional capital from the central government and NGOs.

INCREASING ACCESS TO WATER

The UN Water Global Analysis and Assessment of Sanitation and Drinking-Water defines reasonable access as the availability of at least 20 liters of drinkable water per capita per day from a source within 1 km of a user's dwelling. Currently, millions of people do not have such reasonable access to clean water. Most are women and children who walk several hours each day to collect household water with heavy 20-liter (5-gallon) buckets balanced on their heads. A simple innovative solution that allows for effective water collection and transport are water rollers such as those designed and distributed by the Hippo Water Roller Project (Figure 4.11). This social enterprise is a sustainable for-profit business that is able to make the Hippo Roller affordable in rural communities through social investment, community fundraising, and individual contributions. In so doing, it has distributed approximately 51,000 water hippos, directly benefiting 500,000 people in more than 20 countries while sustaining itself as a business. Social enterprise as a sustainable solution is discussed further in Chapters 9 and 10.

Improved access to water can also occur through in-house connections or shared community sources. Such sources can be through rainwater collection, the drilling of new wells, and the use of cost-effective pumps to extract and distribute the water. Knowledge of the underlying geology is an essential component so that problems such as naturally occurring arsenic contamination like that seen in the wells of Bangladesh can either be avoided or mitigated. Wells need to be properly located away from sources of human or animal waste including garbage, drilled correctly, lined, and sealed at the surface to protect them from surface contamination.

Figure 4.11 The Hippo Water Roller requires much less effort to collect water and is able to carry more water per trip (5x that of a single bucket!). This invention improves health and also allows more time for education or income-generating activities. This is a perfect example of how a simple invention can make a large difference in the lives of many. *https://www.hipporoller.org*

Hand and alternatively powered pumps are needed to extract the water. One creative approach to this has been harnessing the energy of children using playground equipment to pump water. These so-called "joy pumps" were developed in the eco-village of Gaviotas located in Vichada, Colombia, and have since spread globally. Another is a bicycle-powered pump that was conceptualized in a small village in India—an innovation that is currently being designed and sold by Xylam Water Solutions in Vadodara, India. As previously mentioned, these wells and pumps have both initial and ongoing costs, but when factoring the time lost from work while searching for water or due to water-related illness, these costs are affordable even in very poor communities. It has been argued that drinking water wells and supply pumps can do more to extend the average life expectancy and improve the quality of life in developing countries than all other public health efforts combined. When the community makes decisions about placement, construction, use, and payment for ongoing costs, it may be one of the most sustainable solutions.

WATER PURIFICATION

Whether drawn from wells or surface waters, purification is an important step in maintaining the quality of the water and the health of the users. In most rural and poverty situations, home or point-of-use treatment is typically easiest and most cost-effective and has been shown to reduce diarrhea episodes by 50%. Boiling water is a common and effective means of sanitizing water, but in many cases, it introduces other health risks associated with indoor air pollution from burning wood to heat the water (see Chapter 3).

Other purification techniques include the use of chlorination tablets or solutions, solar disinfection, and the use of ceramic and other types of filters. A recent innovation is a foam filter made with spent coffee grounds that removes heavy metal ions such as lead and mercury from contaminated water, thus reducing waste while cleaning water.[31] A simple handheld device, the LifeStraw, is another example of an entrepreneurial solution that has been sold for profit in the developed world to support its free distribution and use in the developing world (Box 4.4).

HUMAN WASTE MANAGEMENT

Open defecation and urination are an enormous problem and the only option for more than 1 billion people. This practice contaminates the water supply, and it is the reason why disease and waterborne illness are such major global killers. It disproportionately impacts children, resulting in more than 2.5 deaths every minute of children younger than age 5 years. It also disproportionately impacts girls and women. Because of a lack of facilities, menstruating girls and women must miss school and work. Women who lack private sanitation options risk sexual harassment or assault and are often constrained to go out for urination and defecation only during the hours of darkness, which has safety and health implications. Improved sanitation is clearly one of the greatest and most important challenges that can save the lives of children and emancipate women through relatively simple solutions.

The first step in the improvement of sanitation is to move the process of waste elimination from the outside into an enclosure that is culturally acceptable and that ensures privacy and dignity. This can be done in the form of simple public latrines and home latrines and toilets. Opportunities to clean the body and the hands after using the facility are also essential. It has been shown that simple handwashing with soap after the use of a latrine or other facility can reduce the incidence of diarrhea by up to 42%. This requires both the availability of soap and education to effect behavioral change.

The next step is to provide an acceptable way to eliminate waste and then ensure that the waste material is properly contained and disinfected. Simple latrine toilets with seating, portable toilets, and permanently mounted composting or pour flush toilets are the norm in most WASH programs. Containment can be as simple as a bag (see Box 4.4), bucket, hole in the ground, or more complex septic and treatment systems. The simplest way to disinfect the waste is to allow naturally occurring microbes to break down or compost the feces. In some cases, the addition of bio-additives—which are combinations of microbes, enzymes, nutrients, and chemicals—is used to reduce the volume, eliminate odor, and speed up the decomposition process.

A number of new technologies to disinfect and process fecal material are in development. One example is the development of waterless toilets that capture waste and then disinfect it using an electrochemical process. The solid dry feces are then used as fuel to generate electricity to power that process. Through the support of the Gates Foundation, prototypes are being tested in Gujarat, India, with the goal of having such systems cost less than five cents per day to operate.

Also supported by the Gates Foundation, Janicki Bioenergy, an engineering firm based near Seattle, Washington, is developing a waste processor that can generate both clean drinking water and useable energy. The machine, which is referred to as the Omniprocessor, boils incoming sludge, removing all the liquid and capturing it as water vapor that can then be condensed and processed, making it suitable to drink. The remaining solid waste is incinerated, and the heat that is produced is used to power the machine. Excess energy can then be sold and transferred into the power grid. This self-sustaining machine will soon be launched in a pilot project in Dakar, Senegal, and will eventually be sold to local entrepreneurs and communities to generate income through waste processing and water and energy production.

These income-generating approaches embody our definition of sustainability, and micro, small, and medium enterprises are emerging as important players in WASH service delivery for the poor. Private–public partnerships have been more effective at professionalizing service delivery compared to voluntary, solely community-focused approaches. One example of this social business approach to providing sanitation services is being implemented by the non-profit organization SOIL (Sustainable Organic Integrated Livelihoods) in Haiti. Customers pay an affordable monthly fee for a locally made, safe, dignified toilet to be placed in their home. The fee also pays for collection of full waste receptacles and replacing them with empty sanitized ones. These fees also help SOIL earn revenue that is critical for operating public toilets. Because of private ownership in combination with paid services, the impact of this program appears to be more lasting than the simple construction of latrines by aid agencies.

BOX 4.4

Innovators and Entrepreneurs: Two Simple Solutions for WASH

Given that improving WASH is such a vital global priority, simple and accessible innovations that increase access to clean water or offer better opportunities for sanitation and hygiene are needed. It has been shown that affordable point-of-use solutions are the most sustainable. One such example that has significantly increased access to drinkable water is the LifeStraw water filter designed and manufactured by the Swiss Company Vestergaard. The LifeStraw is a personal point-of-use water purification tool that is used like a conventional drinking straw. Water is drawn through a tube containing a microfiber-based filter that removes virtually all bacteria and protozoan parasites that can contaminate water, and it reduces turbidity by filtering particulate matter. In its lifetime, this simple tool can filter 1,000 liters, meeting one person's annual needs. Vestergaard also makes higher volume point-of-use water purifier filters that employ the same technology for households, schools, and community use (Figure 4.4.1).

Vestergaard, a for profit-company, refers to its business model as "humanitarian entrepreneurship," with the goal of inventing life-changing health innovations and then creatively finding ways to make them available to the global poor who need them the most. This is important given that the greatest barrier to large-scale safe water provision is the lack

Figure 4.4.1 Vestergaard higher volume point-of-use water purifier filters are used to provide families and communities long-term access to safe drinking water. Pictured here, the author provides such a filter to a family in need in the community of Mérida on Omotepe Island in Nicaragua. *Courtesy R. Niesenbaum.*

continued

continued

of financing. One way that the company achieves this goal is to market LifeStraw products to consumers in developed countries and use a portion of the proceeds to provide in-school purifiers in developing countries. It also partners with a variety of aid organizations that raise money to provide individual- and community-sized filters for free as part of public health campaigns or in response to complex emergencies such as the 2010 earthquake in Haiti. Vestergaard is also using carbon credits it earns because its filters eliminate the need to boil water to pay for further distribution of its products. This is the first program to directly link carbon credits with safe drinking water, and it serves as a model for deriving income from market-based emission reductions that can also result in significant public health outcomes.

Another company with a similar philosophy but focused on providing universal access to hygienic and dignified sanitation is PeePoople AB. It does this by providing an affordable and readily implemented alternative to more expensive infrastructure-based solutions that require complex investments and institutional changes. Its product is the Peepoo toilet, which is a slim, biodegradable bag with an inner layer that unfolds to form a wide funnel that serves as a toilet that is designed to be used once while sitting, squatting, or standing (Figure 4.4.2). It is easy to store, handle, and use. The Peepoo is intended to be used a single time, by one person, whenever and wherever needed. It is always clean and can be used in complete privacy.

The Peepoo is self-sanitizing. Each bag contains urea, a non-hazardous chemical that, when it comes into contact with feces or urine, is broken down and produces ammonia carbonates, which essentially neutralize any disease-causing microorganisms, bacteria, viruses, and parasites. This process rapidly transforms human waste into safe, pathogen-free fertilizer that can be used to enrich depleted soil and quickly improve food security. The bags are made of biodegradable plastic that disintegrates into carbon dioxide, water, and biomass.

It is anticipated that due to the scarcity and expense of fertilizers in less developed countries, fertilizer in the form of used Peepoos will be used in household or community gardens or sold profitably to local farmers. Because the Peepoo

Figure 4.4.2 Peepoo is a personal, single-use, self-sanitizing, fully biodegradable receptacle for human waste that prevents feces from contaminating the immediate area as well as the surrounding ecosystem. After use, Peepoo turns into valuable fertilizer that can improve livelihoods and increase food security. *Peepoo, https://commons.wikimedia.org/wiki/File:Peepoo_Camilla_Wirseen_Peepoople(1)_(6082004235).jpg or http://bhekisisa.org/article/2016-05-03-kiberas-flying-toilets-flushed-out-by-peepoo-bags*

is simple and inexpensive to produce, it can be sold to groups with the weakest purchasing power at a very low price. In addition, under disaster relief conditions, it can be readily purchased and provided by international aid organizations.

The humanitarian yet for-profit companies that make LifeStraw and Peepoo are able to provide increased access to water and sanitation in a long-term, economically sustainable manner. They also provide the user with a sense of ownership and responsibility for these simple, innovative approaches to WASH—a factor that has been shown to make such efforts more successful.

Climate Change

One of the major consequences of current and future climate change will be severe water shortages. In anticipation of climate-induced water shortages, it is important that we increase our resiliency by developing ways to adapt to these shortages. An examples of adaptation is redesigning our global food distribution system to meet the needs of water-scarce regions. Another way is to begin to engineer the redistribution of fresh water over space and time. This would include the construction of water pipelines and reservoirs. One Icelandic company has proposed a new global water shipment industry and infrastructure that would deliver glacial spring water by sea to areas of water scarcity.

Ecological Approaches to Sustain Aquatic Resources

Aquatic ecosystems serve as reservoirs in the hydrologic cycle, perform ecosystem services such as flood regulation and water purification, provide drinking water, and support biodiversity and ecological processes. As such, it is important to consider specific ways to protect, restore, and minimize our impact on them. In this section, we briefly address this by examining techniques in watershed management, wetlands conservation and restoration, the use of riparian buffers, and dam removal to protect and restore aquatic systems.

Watershed Management

Watersheds and their resources should be managed in a way that enhances ecosystem functions that affect the plant, animal, and human communities within a watershed boundary. This can include management of water supply, water quality, drainage, storm water run-off, water rights, and the overall planning and utilization of watersheds. The two largest threats to our aquatic resources are habitat degradation and non-point pollution; these may best be managed through a watershed approach. This approach is a decision-making process that reflects a common strategy for all stakeholders within that watershed.

The watershed approach incorporates three main elements. The first is to identify and prioritize water quality problems in the watershed. The second is to increase public involvement and coordination with other agencies in watershed management and decision-making, including land use. The third is consistent monitoring and evaluation of any management activities. The overall goal of the approach is to manage the watershed in a way that maintains or improves the quality of water that originates in that watershed.

One new approach to watershed management has been to develop markets for their services and to bring benefits to poor people living in those watersheds so that their livelihoods are enhanced. The latter can be achieved by having downstream beneficiaries of upstream land and water pay for that service. Paying for ecosystem services or externalities, and enhancing the well-being of those who manage watersheds well, is consistent with our model of sustainability.

Wetlands Restoration

Although there are some policies in place to protect wetlands, in many cases wetlands have been so degraded that they can no longer perform their ecosystem functions. Their rate of destruction and alteration by direct human activity and climate change impacts continue to occur at alarming rates. Wetlands restoration uses a number of techniques to accelerate their recovery in a way that returns and sustains their health and function. The process is complicated but involves restoring a site's ability to retain water by undoing the altering effect such as plugging drains or removing constructed dykes, canals, and levees.

Once the natural hydrology has been restored, the site should be replanted with native wetland species and managed for continued protection and improvement.

Riparian Buffers

One effective technique for maintaining water quality in streams, rivers, and other surface waters is the maintenance of **riparian buffers zones** or vegetated areas next to water resources that protect them from the impact of adjacent land uses and play a key role in maintaining water quality. As land has been deforested and developed for agriculture and other human uses, riparian buffers have been in decline. This has negatively impacted many aquatic resources, and as a result, the conservation and restoration of riparian buffers has become a very common conservation practice aimed at increasing water quality.

Riparian buffers maintain higher water quality by stabilizing stream banks, preventing erosion, and trapping sediment. They also filter out eutrophying nutrients, pesticides, and other pollutants carried in run-off, and they maintain cooler stream temperature through shading. The vegetation serves as valuable wildlife habitat, absorbs CO_2, and contributes organic material to the aquatic ecosystem. The recommended width of the buffer depends on the ecological objectives, surrounding land use, and slope and soil type. Typically, buffers range from a minimum of 3 km up to 100 km on each side of the stream.

Encouraging private land owners to plant and maintain riparian buffers is becoming common practice throughout the world. An example is in the agriculturally rich areas of Costa Rica, where farmers are offered financial incentives to plant and maintain riparian buffers to protect small tributaries that supply drinking water to downstream communities. In the United States, the planting of riparian buffers on public lands and parks has been met with some controversy because some stakeholders object to their impact on the aesthetics and their direct access to streams for recreation. This could be remedied by incorporating broad stakeholder participation and education in buffer restoration projects.

Dam Removal

Because dams have large negative impacts on water quality and aquatic ecology, there have been many efforts to remove them from river systems, especially when they have become obsolete. Typically, communities working in partnership with non-profit organizations and state and federal agencies spearhead dam removal projects. The non-profit group American Rivers reports that in the United States, 72 dams were removed in 2016, adding to the nearly 1,500 dams removed since 1912. The group has also documented significant recovery of the riparian systems after dam removal. There have been objections to dam removal based on their historical value and the aesthetics of the small waterfalls and the loss of recreational ponds and lakes that the dams create.

The largest dam removal project in the United States was completed in 2014 along the Elwha River on the Olympic Peninsula in Washington state. In the late 1800s, the growing need for lumber motivated the construction of the Elwha and Glines Canyon dams on the river. The dams supported the timber industry and fueled regional growth. But they also blocked the migration of salmon upstream, disrupted the flow of sediment and wood downstream, and flooded the historic homelands and cultural sites of the Lower Elwha Klallam Tribe. In 1992, Congress passed the Elwha River Ecosystem and Fisheries Restoration Act, authorizing dam removal to restore the altered ecosystem. After two decades of planning the largest dam removal in the United States, both were removed, allowing the Elwha River to once again flow freely and thereby restoring this important ecosystem.

Assessing Sustainable Solutions to the Water Crisis

Most of the problems and solutions presented in this chapter can be assessed with quantitative indicators. This is because the problems themselves and specific objectives to solve them consider rates of improvement over time. Examples include the percentage increase over time in access to clean water and sanitation, rates of habitat destruction or restoration, improvements in water quality, changes in water footprint, drought frequency, changes in the number of deaths related to water quality, the number of dam removal projects, and increases or decreases in the number of conflicts over water. This section presents some specific programs of assessment. We examine their strengths and weaknesses. Although there have been considerable successes in some areas, we identify others in which challenges remain. Finally, we identify key features that improve the sustainability of outcomes, based on lessons learned in this chapter.

Assessing Government Policies

Government policies to protect water resources or increase access to WASH need to be continually assessed. One example of a successful policy is the US Clean Water Act, indicated by quantifiable improvements in water quality over time. The major reason for this success is that the Act provided for billions of dollars in grants for infrastructure and funded permitting and enforcement programs. Despite its success, challenges remain. The CWA's lack of regulation of non-point and household sources, and recently fracking, is limiting. The added protection of the Clean Water Rule that extends federal jurisdiction over smaller streams and waterways has recently become politically vulnerable.

Policy solutions are even more challenging in low- to middle-income nations. Poorer nations most often cannot raise the funds domestically because of the inability of citizens to pay additional taxes or user fees to meet the goals of global policymakers. An assessment of the effectiveness of Uganda's environmental policies[32] precisely reflects this situation. The Ugandan government formulated a number of policies to regulate land use and impacts on the environment. The assessment revealed a glaring gap between the existence of laws and policies and the reality of implementation on the ground. Despite these regulations, there has been large-scale destruction of wetlands, encroachment on watersheds, and depletion of water resources. This is simply because funding is lacking to effectively enforce the laws and policies and maintain infrastructure. The assessment revealed that another reason for failure was limited community participation in policy development and implementation. Community and stakeholder involvement and increased funding through NGOs and private–public partnership, including large corporate foundations, will be required to achieve policy success in protecting and improving water resources and WASH.

There has been a well-developed approach to assessing MDGs. The goal that related to water—to reduce by half the proportion of the population without sustainable access to safe drinking water and basic sanitation by 2015—was assessed through the Joint Monitoring Program for Water Supply and Sanitation (JMP) involving WHO and UNICEF. The JMP has produced specific indicators that illustrate major successes and challenges. The target of halving the proportion of people without access to improved sources of water was met five years ahead of schedule. This occurred concurrently with a decline in the annual number of deaths due to water-related illness, from 3.4 million to an estimated 842,000 since the start of the MDG program in 2000. Despite this major progress, the assessment reveals challenges that remain, including that 748 million people remain without access to an improved source of drinking water and 2.5 billion people in developing countries still lack access to improved sanitation facilities.

There have been some criticisms about assessment of the MDGs. First, the discrepancy between the target for access to clean water and that for improved sanitation could in part be explained by the use of different benchmarks in assessing improvements in each. For drinking water, the benchmark is community-level access, and for sanitation, the benchmark is household access. Thus, for example, a pit latrine shared between houses would not count toward the sanitation target.

Also, reporting global data can mask stark differences between countries or even regions within countries. For example, data on global success do not reveal the lack of progress in countries such as Somalia, Ethiopia, and Papua New Guinea in achieving water-related targets. This variation may be due to economic disparities because MDGs are funded both by foreign aid and by domestic revenues. Some countries cannot generate the revenue needed to pay for their portion of the implementation. This is worsened by recent decreases in foreign aid. In addition, geographical variability in success results from differences between countries in their ability to reform their policies to successfully implement projects to meet targets. These criticisms are being taken into account as indicators for the SDGs are developed.

Water Footprint as an Indicator

In Box 4.1, it was shown how calculating one's own water footprint can be a way to assess both direct and indirect water use by an individual, community, business, or nation. As global water scarcity has worsened, a variety of water footprint concepts and tools have been developed to support better management of water resources. Water footprint assessments help raise awareness of issues of water use and scarcity for individuals and decision-makers in industry and government.[33] Water footprint has also been effectively used by businesses to assess risks relating to water scarcity and pollution and to motivate aid agencies to work with stakeholders to improve access to water. Water footprint has been less successful at enhancing public policy, and it has been criticized for not considering potential impacts on food production and for insufficient consideration of regional water scarcity.

Ecological Improvement

There are a variety of tools for monitoring and assessing habitat loss and restoration. One of them, **geographical information systems (GIS)**, is a computer system that can capture, analyze, and present all types of spatial or geographical data both visually and quantitatively. GIS technology is an effective tool for assessing and mapping changes in habitat destruction or restoration, including wetlands, riparian buffers, and other water resources. It can also incorporate biodiversity, human population, and socioeconomic data to assess current and future pressures on ecosystems. This can be done by overlaying spatial data to delineate and visualize relationships between ecosystems and human pressures to help identify prioritized sites for conservation and restoration.

Recipes for Greater Success

In this chapter, we have discussed some incredible challenges regarding the state of water as a resource and efforts to help alleviate them. There have been many successes and improvements, but expanding human pressure, the impacts of climate change, and economic limitations will further threaten the sustainability of water. In this chapter, we featured a number of solutions that many believe could ease the global water crisis by 2050.

Throughout this chapter, we have discussed recipes both for failure and for success in addressing the global water crisis, and some common themes have emerged that seem to be essential for success. For success, solutions must include a combination of measures that both increase the supply and reduce the demand for water. These measures need funding for implementation, and this has been a challenge. Many enacted policies appear promising but are often made ineffective because the burden of funding the implementation of the policy recommendations on the ground is not met by the policymakers. Local, state, or national governments often lack the funding that is required for new infrastructure, monitoring and enforcement, and other policy recommendations. For example, replacing infrastructure that contributes lead to drinking water in impoverished cities such as Flint, Michigan, will require large amounts of external support.

Given the economic and consumption inequities between the developed and the developing world, much of this funding should come from wealthier nations. This can occur through support from the governments of wealthier nations, international aid agencies, and private–public partnerships including contributions from corporations that have larger water footprints and foundations with a focus on environment and humanitarian relief. Foreign aid that places conditions on assistance, such as privatizing public water systems, can have disastrous results. Agencies should collaborate with local stakeholders and experts to avoid and mitigate unexpected consequences of their efforts, such as the naturally occurring arsenic found in wells as part of a solution in Bangladesh.

In addition to foreign support, domestic fees or taxes that are affordable to end-users should also contribute to initial and ongoing costs of any project. When this is done in a participatory transparent manner, projects are more sustainable (see Box 4.3). It has also been shown that handout-based programs result in waste, and top-down approaches to technology transfer from governments and foreign organizations are rarely successful. But when families and communities express a need for a particular solution and take ownership of it, the result is more sustainable. Social enterprise approaches that provide reasonably priced solutions while generating profit, such as the Hippo Water Roller, and open-source approaches to innovation, such as that of SunGlacier's Peltier element, have increased potential for sustainability.

Water links disparate areas that are managed by very different agencies. Better communication among different agencies must occur to solve the water crisis. Examples of this include the USAID global water policy that was developed by convening a working group that focused on the interdependency of food, water, and energy; and the US Department of Energy report on the nexus of water and energy that frames an integrated approach for policy and technical development around these two issues.

As we set priorities regarding water, we must also perform real cost analysis. Things such as the dollar value of the benefits provided by aquatic ecosystems that have been estimated to be more than $7 trillion a year should be factored into decision-making about land and water use. The economic impact of work and school hours lost while searching for water or due to water-related illness should be factored into cost analysis of WASH improvements. Finally, the connections between poverty, gender issues, social justice, and the basic human right to water need to be central to our decision-making processes about supporting and implementing appropriate solutions. We are already seeing great success as we begin to solve the global water crisis. Existing and new policies and putting technical solutions in the hands of the users while considering these recipes for success offer much hope that our water needs will be met well into the future.

Chapter Summary

- The hydrologic cycle is driven by the sun, which circulates water among the different reservoirs. Although 75% of the Earth's surface is covered by water, only 0.77% of Earth's 1.4 billion km^3 of water is potentially accessible fresh water.
- Water scarcity is driven by population growth, increased urbanization, and use in the agriculture and energy sectors and industry. The abundance and availability of fresh water are further threatened by global climate change and water pollution. Lack of sanitation and sewage treatment and open defecation expose millions to pathogens and are major public health problems in the developing world.
- Access to clean water is directly related to poverty and is further impacted by the commodification of water. This includes the privatization of public water supplies and the expansion of the bottled water market.
- The combination of the growing demand for water and its scarcity and the transboundary nature of the resource have implications for international security and will continue to be a source of international tension and objectives for military action.
- Aquatic ecosystems that provide a variety of services and have much value are being impacted by human activity. These impacts include eutrophication, sedimentation, the construction of dams, habitat destruction, and poor watershed management.
- Despite these threats to this vital resource, there are many solutions that could potentially ease the global water crisis by 2050. These include national, international, and global policies designed to protect and increase access to water resources. The US CWA is an example of a policy that has been somewhat successful, and the UN Millennium Development Goals related to water have also had some success.
- One of the greatest limitations to implementing water-related policy, particularly in poorer nations, has been insufficient funding. Funding by NGOs and private–public partnerships, including contributions from large corporations and foundations, have helped alleviate this issue.
- Other policy changes related to agriculture, energy, water dispute resolution, and stopping the trend in the commodification of water are also an essential part of lessening the global water crisis. There have been some trends in each of these areas, but greater emphasis on the connections between agricultural policy and energy policy in relation to water use is needed.
- A number of innovative technical solutions that include methods for generating water, harvesting rainwater, recycling used water, technical innovations in agriculture and industry, and ways to reduce household water consumption have shown great potential in alleviating water stress.
- Improving access to clean water, sanitation, and hygiene (WASH) has become a high priority given its huge impact on public health in developing countries. Solutions range from simple (see Box 4.4) to more complex technology and infrastructure. Approaches that allow families and communities to be invested in the improvement are more sustainable (see Box 4.3).
- Dealing with population growth and either mitigating or adapting to climate change will be crucial in dealing with the global water crisis, as will ecological approaches such as watershed management and restoration.
- Assessment of sustainable solutions to the global water crisis must use quantitative indicators and common benchmarks and be viewed critically.

A variety of tools, such as water footprint calculations, a technology application framework, and GIS, can aid in efforts to assess implemented solutions.
- The most important elements of a sustainable project aimed at meeting the challenges of the global water crisis are community buy-in and capacity building, in addition to ongoing revenue generation either by the community or by a humanitarian-based business. The revenue can then be used to service and expand improvements.

Digging Deeper: Questions and Issues for Further Thought

1. What have been some of the successes of the CWA? Despite these successes, there is an increasing trend to want to dismantle or weaken this policy. What are the reasons for this trend, and how can it be opposed?
2. In an effort to restore stream quality in a popular urban park, it has been recommend that a dam be removed and that riparian buffer zones be planted. Who are the stakeholders that should be consulted before initiating this project, and what issues might each have?
3. In this chapter, you have read about a number of innovative, social enterprise solutions to problems related to water. Try to identify other examples of this type of solution.
4. Whether large dam construction projects contribute to or detract from sustainability is complicated. Employing a systems thinking approach, map out all of the potential environmental, social, and economic consequences of a large dam project indicating whether they are negative or positive and how they are related.
5. Give examples of "nature-based solutions" that protect clean water.

Reaping What You Sow: Opportunities for Action

1. Go to http://waterfootprint.org and calculate your water footprint. Develop an action plan to reduce it.
2. Design a campaign for reducing bottled water use at your school or place of work.

References and Further Reading

1. United Nations. 2015. "17 Goals to Transform Our World. Goal 6: Ensure Access to Water and Sanitation for All." Accessed March 8, 2017. http://www.un.org/sustainabledevelopment/water-and-sanitation.
2. World Health Organization. 2016. "Drinking-Water." Accessed March 8, 2017. http://www.who.int/mediacentre/factsheets/fs391/en
3. NASA. 2014. "Satellite Study Reveals Parched U.S. West Using Up Underground Water." July 24. Accessed July 24, 2014. http://www.nasa.gov/press/2014/july/satellite-study-reveals-parched-us-west-using-up-underground-water/index.html#.U9FLxhaUKA0
4. Mekonnen, M. M., and A. Y. Hoekstra. 2012. "A Global Assessment of the Water Footprint of Farm Animal Products." *Ecosystems* 15 (3): 401–5.
5. Rodriguez, Diego J., Anna Delgado, Pat DeLaquil, and Antonia Sohns. 2013. "Thirsty Energy." Water Papers. Washington, DC: World Bank.
6. Mielke, E., L. D. Anadon, and V. Narayanamurti. 2010. "Water Consumption of Energy Resource Extraction, Processing, and Conversion." Cambridge, MA: Belfer Center for Science and International Affairs.
7. Goldman, R. L. 2010. "Ecosystem Services: How People Benefit from Nature." *Environment* 52 (5): 15–23.
8. US Geological Survey. n.d. "Water Resources of the United States." Accessed March 8, 2017. https://www2.usgs.gov/water
9. Schewe, Jacob, Jens Heinke, Dieter Gerten, Ingjerd Haddeland, Nigel W. Arnell, Douglas B. Clark, Rutger Dankers, et al. 2014. "Multimodel Assessment of Water Scarcity Under Climate Change." *Proceedings of the National Academy of Sciences of the USA* 111 (9): 3245–50. doi:10.1073/pnas.1222460110
10. US EPA. "National Recommended Water Quality Criteria." http://water.epa.gov/scitech/swguidance/standards/criteria/current/index.cfm
11. Jones, R. L., and R. Allen. 2007. "Summary of Potable Well Monitoring Conducted for Aldicarb and Its Metabolites in the United States in 2005. *Environmental Toxicology and Chemistry* 26 (7): 1355–60.
12. World Health Organization. 1991. "Environmental Health Criteria 121: Aldicarb." pp. 11–13. Geneva, International Programme on Chemical Safety.

13. Hanna-Attisha, M., J. LaChance, R. C. Sadler, and A. Champney Schnepp. 2016. "Elevated Blood Lead Levels in Children Associated with the Flint Drinking Water Crisis: A Spatial Analysis of Risk and Public Health Response." *American Journal of Public Health* 106 (2): 283–90.
14. Argos, Maria, Faruque Parvez, Mahfuzar Rahman, Muhammad Rakibuz-Zaman, Alauddin Ahmed, Samar Kumar Hore, Tariqul Islam, et al. 2014. "Arsenic and Lung Disease Mortality in Bangladeshi Adults." *Epidemiology* 25 (4): 536–43. doi:10.1097/EDE.0000000000000106
15. Trends in Water Privatization: The Post-Recession Economy and the Fight for Public Water in the United States Food & Water Watch. 2010. +. Washington, DC, Food & Water Watch.
16. Environmental Working Group. 2014. "Bottled Water Quality Investigation." Accessed November 20, 2015. http://www.ewg.org/research/bottled-water-quality-investigation
17. United Nations Department of Public Information. 2004. "Water Without Borders: Backgrounder." http://www.un.org/waterforlifedecade/pdf/waterborders.pdf; United Nations Development Programme (UNDP). 2006. Human Development Report 2006. Beyond Scarcity: Power, Poverty and the Global Water Crisis. Chapter 6. New York: UNDP.
18. Vajpeyi, Dhirendra K. 2011. *Water Resource Conflicts and International Security: A Global Perspective.* Lanham, MD: Lexington Books.
19. Brown, J. J., K. E. Limburg, J. R. Waldman, K. Stephenson, E. P. Glenn, F. Juanes, and A. Jordaan. 2013. "Fish and Hydropower on the U.S. Atlantic Coast: Failed Fisheries Policies from Half-Way Technologies." *Conservation Letters* 6: 280–86.
20. Briscoe, John. 2005. "Water as an Economic Good." In *Cost–Benefit Analysis and Water Resources Management*, edited by Roy Brouwer and David Pearce, 46–70. Cheltenham, UK: Elgar.
21. Muramira, T., and L. Emerton. 1999. "Uganda Biodiversity: Economic Assessment." Kampala, Uganda: National Environment Management Authority.
22. Patterson, L. et. al. 2017. "Unconventional Oil and Gas Spills: Risks and Mitigation Priorities. "*Environmental Science & Technology* 51(5): 2563–2573.
23. World Health Organization and UNICEF. 2013. *Progress on Sanitation and Drinking Water.* Geneva: World Health Organization.
24. World Health Organization. 2012. *UN Water Global Analysis and Assessment of Sanitation and Drinking-Water: The Challenge of Extending and Sustaining Services.* Geneva: World Health Organization.
25. US Department of Energy. 2014. "The Water–Energy Nexus: Challenges and Opportunities." Washington, DC: US Department of Energy.
26. United Nations Development Programme (UNDP). 20065. Human Development Report 2006. Beyond Scarcity Power, Poverty and the Global Water Crisis. Chapter 6. New York: UNDP.
27. Kishimoto, S., Lobina, E. and Petitjean, O., 2015. Our public water future: The global experience with remunicipalisation. Transnational Institute (TNI)/Public Services International Research Unit (PSIRU)/Multinationals Observatory/Municipal Services Project (MSP)/European Federation of Public Service Unions (EPSU).
28. International Desalination Association. n.d. "Desalination by the Numbers." Accessed October 20, 2017. http://idadesal.org/desalination-101/desalination-by-the-numbers
29. Sood, Aditya, and Vladimir Smakhtin. 2014. "Can Desalination and Clean Energy Combined Help to Alleviate Global Water Scarcity?" *Journal of the American Water Resources Association* 50 (5): 1111–23. doi:10.1111/jawr.12174
30. Shlezinger, M., Y. Amitai, I. Goldenberg, and M. Shechter. 2016. "Desalinated Seawater Supply and All-Cause Mortality in Hospitalized Acute Myocardial Infarction Patients from the Acute Coronary Syndrome Israeli Survey 2002–2013." *International Journal of Cardiology* 220: 544–50.
31. Chavan, A. A., J. Pinto, I. Liakos, I. S. Bayer, S. Lauciello, A. Athanassiou, and D. Fragouli. 2016. "Spent Coffee Bioelastomeric Composite Foams for the Removal of Pb^{2+} and Hg^{2+} from Water." *ACS Sustainable Chemistry & Engineering* 4 (10): 5495–5502.
32. Rwakakamba, Twesigye Morrison. 2009. "How Effective Are Uganda's Environmental Policies?" *Mountain Research and Development* 29 (2): 121–27. doi:10.1659/mrd.1092
33. Egan, Matthew. 2013. "Water Footprint: Help or Hindrance?" *Social and Environmental Accountability Journal* 33 (2): 113–14. doi:10.1080/09691 60X.2013.820396

CHAPTER 5

FOOD AND AGRICULTURE: USING OUR PLANET TO SUSTAIN OURSELVES

Food is an obvious thing to consider within the context of sustainability because it is the very thing that sustains us and keeps us alive and healthy. Food cannot be taken for granted. We are forced to make choices about what and how much to eat multiple times on a given day. It is something we work for, search for, or cultivate ourselves. It is central to our cultures, economies, family connections, and community welfare.

PLANTING A SEED: THINKING BEFORE YOU READ

1. What percentage of household income do you think your family spends on food?
2. Why is it often more expensive to buy a salad than to buy a beef hamburger at a fast-food restaurant?
3. Would you eat food that has been developed through genetic modification? Why or why not?
4. What kind of consumer choices best contribute to a sustainable food system? Which of them are you willing to make?

If we don't get sustainability in agriculture first, sustainability will not happen.
—Wes Jackson

We currently meet the basic nutritional requirements of only approximately 85% of the world's people. Approximately 1 billion people lack adequate nutrition, and this number will likely rise as populations continue to grow. This is true particularly in the poorest regions of the world. The cause of this problem is threefold. First, we are not producing enough food to meet the needs of current consumption. Second, there are global inequities in per capita consumption that contribute to our inability to meet the basic nutritional needs of all people (Figure 5.1). The recommended daily caloric intake for a moderately active male adult is between 2,200 and 2,800 kcal. These numbers are lower for women, children, and those with sedentary lifestyles. In the United States, average caloric consumption is approximately 3,600 kcal/day. In other areas of the world, especially in developing countries, average daily per capita consumption is less than 2,000 kcal/day, well below what is recommended. Third, approximately one-third of all food produced globally is lost as waste. These three issues indicate that the hunger problem is not merely due to insufficient food production but also due to unequal access, distribution, and consumption of food as well as excessive food waste. Solutions to our hunger problem must include sustainable methods that will increase production, ways to encourage and support a more healthy and equitable distribution of consumption, and reductions in food waste.

Hunger and malnutrition occur not only in developing countries but also in developed countries in which large proportions of impoverished people live in areas designated as **food deserts**. These are urban neighborhoods and rural towns without ready access to fresh, healthy, and affordable food. Our inadequacy to meet the food needs of our current and growing populations is made worse by food production systems that deplete our capacity to provide food through environmental degradation and overextraction of resources.

Food is directly linked to all three sectors of our sustainability model. Food production has impacts on the environment, including soil, water, and air quality. It results in natural habitat loss including deforestation, pollution, and contributes to climate change. Food is central to the global economy, and individuals and families often spend a significant portion of their personal income on it. Numerous social issues center

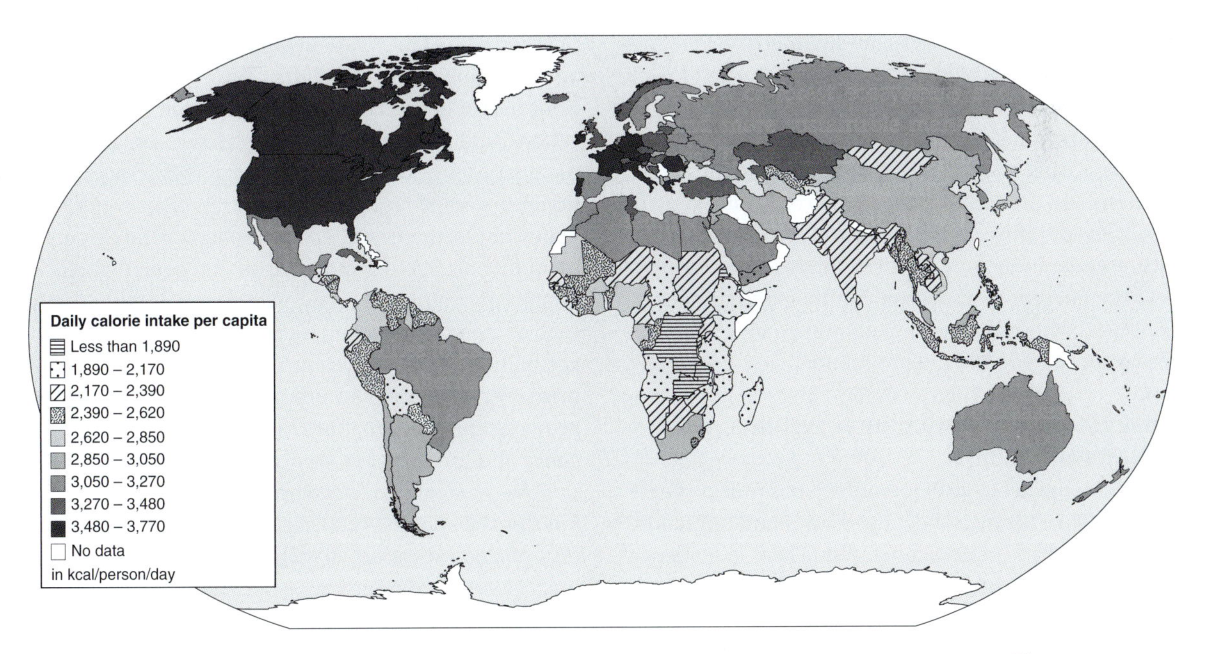

Figure 5.1 Daily per capita caloric intake across the globe reflecting consumption inequities. The average recommended daily caloric intake for a moderately active male is 2,500 kcal. (FAO, "Food Balance," http://chartsbin.com/view/15619)

on food, including social inequity, public health, shifts from community to global food production, and negative treatment of workers in food production systems.

A sustainable food system should provide healthy food to meet the current needs of all people through enhanced production and distribution. It should maintain healthy ecosystems and manage resources in order to meet the dietary needs of subsequent generations. It should encourage local production and distribution in a way that increases availability, accessibility, and affordability of nutritious, healthy foods. The production of food should be done in a manner that protects producers, consumers, and communities.

In this chapter, we examine our food systems through this lens of sustainability. We do this by first reflecting on the history of agriculture in what has ultimately resulted in a large-scale industrial agricultural system. We next examine the environmental, social, and human health aspects of the current food system. We then discuss and assess potential solutions that will increase our ability to make the food system more sustainable.

A Brief History of Agriculture

In this section, we trace the history of agriculture from its origins through our present-day food system. We discuss how agricultural policy and the "Green Revolution" of the 1950s and 1960s increased our ability to produce affordable food, but that affordability comes at a large, unaccounted cost through detriment to the environment and human health, corporate control of our food, abuse of workers, and the potential for future generations to provide food. We then explore how the current food system can be made more sustainable through changes in policy and new and innovative approaches to food production.

The Evolution of Agricultural Systems as Natural Systems

Humans existed on Earth for more than 2 million years without agriculture. They lived in small, nomadic groups that had limited geography. There is evidence that disease and malnutrition were prevalent, and it has been suggested that the only way that these small groups could sustain themselves given the limitations of food and other resources was through population control, including the use of infanticide. Preserved bones and seeds uncovered from prehistoric **middens** or garbage heaps reflect what was an omnivorous diet. Humans from this era were hunter–gatherers or foragers rather than food producers.[1]

The first archeological evidence of agriculture is approximately 10,000 years old, representing a time period that comprises less than 0.5% of human existence. The transition from forager to farmer perhaps first began as simple **forest gardening** in which, once collected, forest food plants were cultivated within their natural environment. This then developed into farming where the environment was more drastically altered, and the process of domestication of previously foraged plants and hunted animals allowed for increased food production and availability.

There are a number of explanations for the transition of human groups from nomadic foragers into more settled agrarian communities. One argument is that the changing climate and overexploitation by hunter–gatherers resulted in food shortages, which provided strong pressure to develop forest gardens into more productive systems. It is believed and it makes sense that those plants that sprouted up in garbage heaps in human settlements were the first to be planted and then domesticated. Others have argued that the development of religious rituals including animal sacrifice paralleled the domestication of animals and the plants on which they might feed and that agriculture is really a product of our cultural evolutions as a species. There is evidence that the domestication and farming of grains began and expanded for the purposes of making beer, which served as a nutritional beverage and a pathogen-free source of water. For example, there is extensive archeological evidence, including 6,000-year-old recipes for beer written in Sumerian from ancient Mesopotamia.[2] Most likely, each of these factors contributed to the development of agriculture, but clearly hunting and gathering as the primary source of food production could not sustain growing populations nor compete against the emergence and growth of more productive agriculture.

As agriculture developed, humans began to modify the species that they raised through the process of domestication, artificial selection, and breeding. They also began to alter their environment, optimizing conditions for growth, increasing production, and protecting their crops from predation and herbivory. This occurred independently in a variety of regions throughout the world referred to as primary centers of domestication. For example, wheat was domesticated in the Middle East, whereas maize (corn) and potatoes were domesticated in the Americas.

Farms were small and remained so until farming gradually shifted from manual cultivation to heavier animal-driven cultivation and eventually to fuel-driven mechanization. These advances gradually increased the amount of surface area that could be cultivated per individual worker, allowing for the diversification of the workforce. Because food acquisition was no longer the work of all, further development of culture and eventually leisure occurred. However, until less than 100 years ago, farms remained relatively small, were maintained and supported by families and communities, and provided locally produced, healthy food. Thus, many of the criteria for sustainability were met except for perhaps the most important: the ability to produce enough food for the exponentially growing human population.

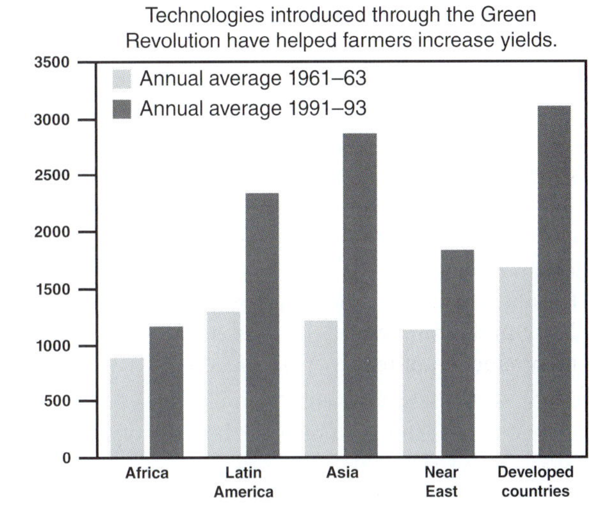

Figure 5.2 Crop yield (tonnes/ha) by region before and after the Green Revolution showing significant increases in most of the world, but less so in Africa. (Data from FAO.)

The Green Revolution and the Development of Industrial Food Production

The Green Revolution or the development of fossil-fuel/chemically based, industrial agriculture grew out of the post-war economy of the 1950s with apparent noble goals. These included the strengthening of world agriculture by creating high-yielding and disease-resistant varieties through conventional plant breeding and also the development of chemical, industrial, technological approaches to agriculture. Associated technical innovation allowed for the expansion of irrigation and the subsequent spread of farming to previously unarable lands. The quest for innovation led to the further development and public funding of agricultural and technical educational institutions known as land grant universities.

In many ways, the Green Revolution was a success. Improved high-yielding grain varieties spread throughout much of the world, as did the fertilizers, pesticides, and irrigation needed to reap their full potential. Fertilizer applications in developing countries increased by 360% from 1970 to 1990, pesticide use increased by 7% or 8% per year, and the amount of land under irrigation increased by one-third. **Yield** (production per unit area) nearly doubled during that same period, suggesting the Green Revolution offered us the possibility for sustaining food production to meet the needs of our planet's growing population (Figure 5.2). In addition, particularly in the United States, food costs remained low and relatively stable. However, a closer analysis reveals many environmental, social, and health issues that are related to the industrialized food system that when fully considered suggest that the result of the Green Revolution is anything but sustainable.

One main problem associated with modern agriculture is that the increase in production of cheap food does not consider the externalities or indirect costs of food production. These are the costs associated with the negative health as well as environmental and social problems that are not met by the food system and are incurred by other components of the economy. In other words, a full consideration of the true costs of food reveals the myth of cheap, affordable food. As we consider the real costs of food and the problems with our current food system, we can begin to identify solutions that may increase the sustainability of food production systems.

Problems with the Green Revolution and the Real Cost of Cheap Food

Although the apparent intent of the Green Revolution was to increase food production, end world hunger, and make food more affordable, there were a number of limitations to achieving these goals—and there were negative consequences that resulted from the activities it promoted. It is true that food was made more available and less expensive, but this lower cost did not include externalities such as environmental impacts, unequal access to new technologies, and the social consequences of a food system that favored a larger, more industrial form of agriculture. We explore these further here.

UNEQUAL ACCESS

High-yielding crops that were developed during the Green Revolution are dependent on fertilizer, herbicide, and pesticide applications; increased water usage; mechanized farming; and an increased dependency on fossil fuels. To reap the benefits of the potential increase in yield, farmers needed access to new information and technology, and they had to meet the cost of increased mechanization and energy inputs for modern agriculture. This technology was dependent on larger scale production of individual crops or monocultures. In more developed countries including the United States, poorer farmers with small land area and diverse crop production could not meet these demands or increases in costs. Because of these costs, the Green Revolution resulted in the transformation of agriculture from collectives of small, diverse family farms to larger, industrial farms specializing in single crops (Figure 5.3).

The goals of the Green Revolution were less obtainable in the poorest areas of the world where hunger and malnutrition were on the rise. These new and more costly technologies and approaches were essentially unavailable to subsistence farmers in less developed countries. This is one reason why the poorest countries, particularly those on the African continent, saw little or no increase in yield after the Green Revolution (see Figure 5.2), and it helps explain the geographical inequity in caloric consumption that occur today (see Figure 5.1).

BIG FOOD AND THE FACTORY FARM

The Green Revolution, in combination with farm policy and the expansion of processed and fast foods, has led to a corporate or "big food" economy in the United States. In the 1970s under the leadership of Secretary of Agriculture Earl Butz, farm policy was designed to support agribusiness. The goal was to re-engineer the American food system so that farms could provide cheaper and consistent raw materials for manufacturing processed foods and for the rapidly expanding processed, fast-food, and soft drink industries led by Coca Cola, Nestle, General Mills, Tyson, Cargill, and Archer Daniels Midland (Box 5.1). Rather than providing farmers with security through loans, subsidies were provided when prices were depressed. These subsidies supported the expansion of large, regional farms at the expense of small, local farms (see Figure 5.3). The focus was on the cheapest way to produce highly caloric food primarily in the form of corn, corn by-products, and animals that consume corn. This led to greater development of the factory

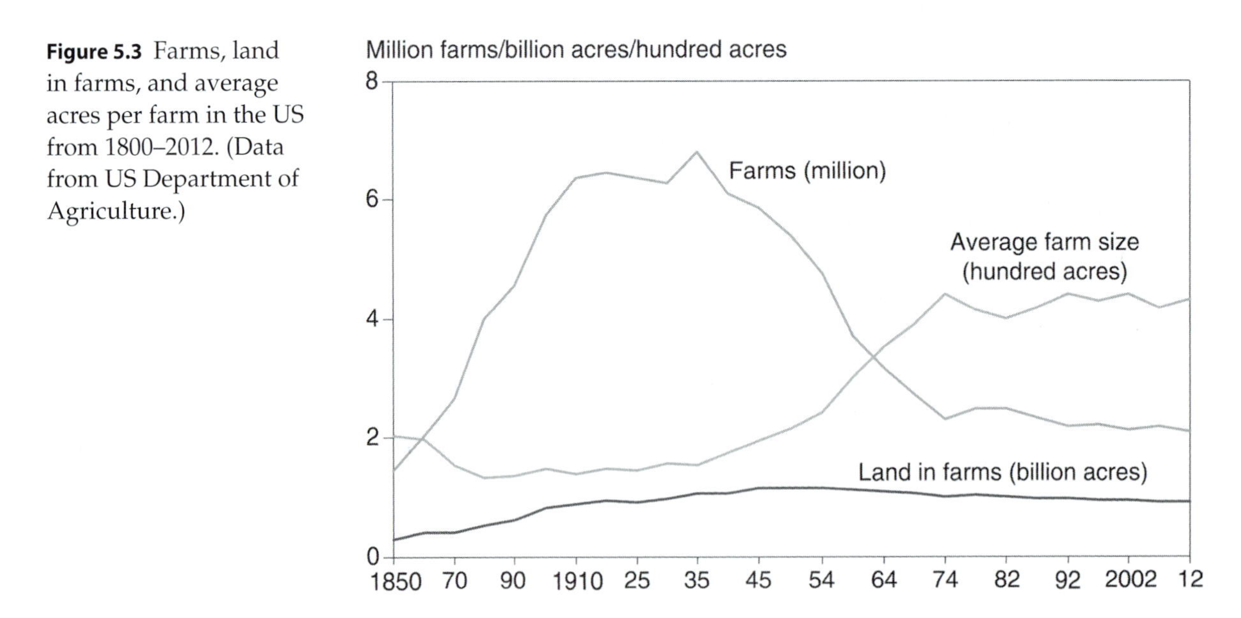

Figure 5.3 Farms, land in farms, and average acres per farm in the US from 1800–2012. (Data from US Department of Agriculture.)

BOX 5.1

Policy Solutions: Envisioning New Farm Policy

Since the early 1970s, US farm policy has undermined the goals of sustainability. Developed by US Department of Agriculture Secretary Earl Butz, this policy served to undo the prior New Deal agricultural policies that were established to protect the interest of farmers that encouraged restraint and managed supply. The Butz policy was established to serve expanding agribusiness that needed a steady and consistent supply of cheap corn and soy as inputs for its processed and fast-food industries. This shift from supply management to subsidies that maximized cheap output favored large, industrial farms that are fossil fuel and chemically intensive monocultures. As a result, federal dollars are being used to support an agricultural system that is damaging to the environment and supports a food system that contradicts federal nutritional regulations (Figure 5.1.1). Thus, a fast-food hamburger and a beverage consisting mostly of high-fructose corn syrup are less expensive than healthier options such as fresh salads and vegetables, and many attribute some of the blame for the US obesity epidemic on farm policy.

More recent farm policies, such as the Organic Foods Production Act of 1990 and the Farmers Assistant Provision of the 2013 Appropriations Bill, favor big farms and corporations at the expense of small farms and consumers. The latter, also known as the "Monsanto Protection Act," effectively bars federal courts from being able to halt the sale or planting of genetically modified organism seed independent of any health issues that may arise concerning them in the future. The provision was drafted in part by individuals with ties to Monsanto who added to the bill without review from Congress' Agricultural or Judiciary committees. Unfortunately, global agricultural policy mirrors what is happening in the United States. Global agricultural subsidies are approaching $5 billion, nearly all of which goes to industrialized countries, with the bulk of it benefiting large-scale industrial farming, thus negatively affecting poor countries and limiting crop diversity.[1]

Is it possible to fix US farm policy, or for that matter global farm policy, so that farmers, consumers, and the environment are protected while allowing for increased production of a diversity of healthy foods? What would such a

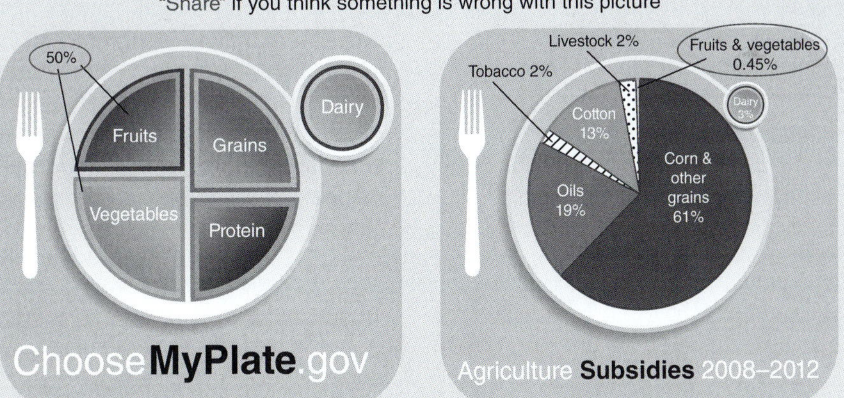

Figure 5.1.1 Federal subsidies for food production and federal nutritional recommendations are inconsistent. (Adapted from Physicians Committee for Responsible Medicine.)

1 Biron, C. L. 2014. "Global Agricultural Subsidies Near $500b, Favoring Large-Scale Producers." MPN News. https://www.mintpressnews.com/global-agricultural-subsidies-near-500b-favoring-large-scale-producers/187275

continued

continued

policy look like? Should a farm bill go beyond the goal of feeding people for the least amount of money? Marion Nestle, Paulette Goddard Professor of Nutrition, Food Studies, and Public Health at New York University, and her students have suggested a number of policy elements for a more sustainable farm bill.[2] They include the following:

1. Protect and support farmers, especially those who employ sustainable practices and growers of nutritious fruits and vegetables. Worker protection, risk reduction, and farmer training should be central to this.
2. Support human health by linking agricultural and nutritional policy to reduce the disconnection between the two (see Figure 5.1.1). The 2008 farm bill explicitly prohibits any government support for the production of nutritious vegetables and fruits. A new bill should provide incentives for the production beyond the commodity crops of corn and soy so that the bulk of our plants grown are no longer used to feed only cars and cows. Federally funded supplemental nutritional programs are already a part of the farm bill and could be restructured to encourage the purchase of fresh, healthy foods over processed foods.
3. Require subsidy recipients to engage in environmentally sound practices such as soil conservation and efforts to reduce emissions and chemical inputs. The planting of riparian buffer zones or strips of forest between fields and streams that could help protect those aquatic resources should be supported, as should restoration and preservation of uncultivated lands.

Could or would the US government implement such a policy shift? Opponents would argue that the government should not meddle in the business of food production. Perhaps, as Marion Nestle suggests, it is reasonable for the federal government to play a role in ensuring adequate food production for its people at an affordable price given its essential nature.

There are cases of policy implementations in other countries that have helped promote many aspects of sustainability. One example is Cuba, where a shift in policy facilitated the transition from large state-owned commodity farms to smaller scale, organic production. This was a necessary response to the food crisis that resulted from the collapse of the market for these commodities, the collapse of the Soviet Union, and trade embargos of the 1990s. In 1992, the Cuban government broke up the state farms and turned them over to the workers. These worker-owned enterprises helped ensure food security while increasing economic opportunity. Other policies encouraged urban farmers to produce diversified, healthy products and promoted the conversion of vacant lots in Havana into small farms and grazing areas. These policies have resulted in the creation of more than 350,000 jobs and the city of Havana self-sufficiently providing healthy food to its more than 2 million residents.[3] Stories such as this provide hope that policy shifts can make food systems more sustainable. Although the case in Cuba was a response to crisis, the public health and environmental problems in the United States may have also reached a state of crisis.

2 Nestle, M. 2014. "Utopian Dream: A New Farm Bill." *Dissent Magazine.* http://www.dissentmagazine.org/article/utopian-dream-a-new-farm-bill

3 Koont, S., 2008. A Cuban success story: urban agriculture. *Review of Radical Political Economics, 40*(3): 285-291.

farm and the industrialization of meat and poultry production. The end result is that a small group of multinational corporations now control the majority of the food that is consumed in the United States. For example, the top four beef companies now control more than 80% of the market and all of the farmers that supply them with feed.

The rapid expansion of the factory farm in the United States has led to increases in productivity and the advance of cheap food, but at what cost? This mass-produced, processed food is not as healthy. Its reliance on high-fructose corn syrup has contributed to the diabetes epidemic, especially among less advantaged communities. To maximize production, a steady stream of hormones and antibiotics are used in the production of animal products, thus presenting additional health issues. Meat processing plants are mechanized assembly lines that create a number of social and risk issues for workers. Environmental impacts of these large farms are massive, and the processes and approaches are anything but renewable. All of this in conjunction with the decline of the family farm have led to a less healthy population and a people who are culturally disconnected from their food.

CHEMICAL FERTILIZERS

The Green Revolution has resulted in significant negative environmental impacts, some of which may be irreparable particularly in the short term. Chemical-intensive, mechanized agriculture has degraded soil quality in ways that have depleted its organic material and soil nutrients and resulted in increased soil salinity. Agriculture became a significant polluter primarily through the run-off of excessive chemical inputs. The excessive application of chemical fertilizers results in nutrient leeching and run-off, which can lead to eutrophication and anoxia in aquatic systems, as described in Chapters 3 and 4.

Throughout the world, the process of eutrophication can be seen in small aquatic systems near farms. In Asia, 54% of lakes are eutrophic, whereas 53% of lakes in Europe, 48% in North America, and 41% in South America are eutrophic. In Africa, where the expansion of the Green Revolution has been limited, only 28% of lakes are eutrophic, suggesting that intensive agriculture is contributing to this. More extensive anoxic or hypoxic events through eutrophication have impacted larger coastal and marine systems, many of which provide significant economic opportunity through the harvest of aquatic foods and through tourism. Major anoxic events related to agricultural run-off have occurred in Chesapeake Bay, Long Island Sound, the Great Barrier Reef, the Gulf of Mexico, and the Caspian, Baltic, Caribbean, and Mediterranean seas.[3] Human control and alteration of the nitrogen cycle through modern agriculture have contributed to our exceeding the planetary boundary of the nitrogen cycle.

PESTICIDES

In addition to the use of chemical fertilizers, the Green Revolution has resulted in the extensive use of chemical pesticides. These synthetic chemicals have been designed and are used to control insect pests, weeds, fungal pathogens, and other biological threats to crops. Now that agriculture has developed into large-scale monoculture production, their use has become an essential part of maintaining high yields, requiring multiple applications each season. This is because the loss of crop diversity and natural predators of plant pests favor their uncontrolled proliferation. Pesticide use has numerous environmental and public health implications. Annually, more than 1 billion pounds of pesticides are used in the United States, and approximately 5.6 billion pounds are used worldwide. More than 95% of sprayed insecticides and herbicides reach a destination other than their target species, including air, water, bottom sediment, and natural and human communities.[4]

Insecticides are indiscriminate in the insects that they kill. As a result, beneficial insects, including natural insect predators, decomposers, and pollinators, are killed. Pesticide use has been attributed to the current ecological crisis of pollinator decline, and their presence has been shown to disrupt the sensory cues of important crop and native plant pollinators.

It is well established that streams, aquifers, and other aquatic systems have experienced extensive pesticide pollution worldwide. Pesticides **bioaccumulate** in the environment as they are stored in the fatty tissues of organisms and become more concentrated at higher trophic or food chain levels. They have significantly impacted bird, fish, and other ecological populations and communities. Amphibians are particularly sensitive to pesticides, and the global decline in amphibian populations and diversity has been attributed, at least in part, to pesticide use and exposure. Some pesticides contribute to global warming and ozone depletion. Consistent use of pesticides has resulted in the evolution of pest resistance, rendering these chemical controls less effective and requiring even greater application.

Most pesticides are acutely toxic, and concentrated exposure by workers can result in immediate illness and mortality. Acute effects include skin irritations, birth defects, blood and nerve disorders, endocrine disruption, and death. In many developing countries, protective equipment to control pesticide exposure is non-existent or its use is limited. As a consequence, it has been estimated that as many as 25 million agricultural workers globally experience unintentional pesticide poisonings each year.[5] Long-term or chronic exposure also has consequences: Many pesticides have been shown to be linked to cancer in humans, including soft tissue sarcoma, malignant lymphoma, leukemia, and cancers of the lung, ovary, and breast (Table 5.1).

TABLE 5.1 Commonly Used Pesticides and Their Acute and Long-Term Health Effects

PESTICIDE	TYPE	ACUTE HEALTH EFFECTS	LONG-TERM HEALTH EFFECTS
2,4-D	Herbicide	Irritation and inflammation of eyes and skin, hives, nausea, vomiting, throat irritation, headache, dizziness, coughing, difficulty breathing	Genetic damage, cancer causing, leukemia
Aldicarb	Insecticide	Nausea, vomiting, diarrhea, sweating, muscle twitching, slow heartbeat, seizures, loss of consciousness, shock, stillbirth	Probable carcinogen, colon cancer
Atrazine	Herbicide	Eye injury, endocrine disruption, reduced immunity	An endocrine disruptor; increased risk of birth defects, infertility, and cancer
Carbaryl	Insecticide	Stinging eyes, wheezing, sweating, and nausea	Melanoma, reduced sperm count
Chlorpyrifos	Insecticide	Respiratory paralysis	Increased risk of children born with lower IQs and potential for ADHD
Diazinon	Insecticide	Headache, nausea, dizziness, tearing, sweating, blurred vision, and memory problems	All cancers, leukemia, lung, Hodgkin's lymphoma
Dichloropropene	Soil fumigant	Chest pains, breathing difficulties, skin rashes, irritation of the eyes and respiratory tract, liver and kidney damage, cardiac arrhythmias	Lung damage, carcinogen
Glyphosate	Herbicide	Eye irritation, burning eyes, blurred vision, skin rashes, burning or itchy skin, nausea, sore throat, asthma and difficulty breathing, headache, lethargy, nose bleeds, dizziness	Increased risks of the cancer, non-Hodgkin's lymphoma, miscarriages, and attention deficit disorder
Malathion	Insecticide	Headache, nausea and vomiting, burning eyes, difficulty breathing, lethargy	Potential carcinogen, sperm and chromosomal damage, thyroid damage, hormone disruption and birth defects
Metam sodium	Soil fumigant	Burns, eye irritation, difficulty breathing, nausea	Immune reduction, liver damage, carcinogen
Permethrin	Insecticide	Tremors, elevated body temperature, increased aggressive behavior, disruption of learning	Neurotoxin, carcinogen

Source: Pesticide fact sheets from the Northwest Center for Alternatives to Pesticides, Eugene, OR.

Exposure to pesticide residues by the consumer may also have long-term health implications, but this has been disputed. Pesticide residues are found on commercially sold fruits and vegetables and in processed foods for which these serve as ingredients. There is added concern for the health effects of these residues for children and pregnant women and the fear that even at very low dosages, long-term-exposure puts these consumers at risk. Organophosphate pesticide exposure during pregnancy has been shown to impact the fetus' developing nervous system; negatively impacts subsequent learning, reasoning, and memory; and cause other attention and behavioral problems. The US Food and Drug Administration (FDA) monitors and enforces tolerances for pesticide residues in raw agricultural commodities and processed foods for domestically produced and imported foods. However, there is often no monitoring or enforcement in other areas of the world, where consumer exposure to residues may constitute a significant public health issue.

HABITAT LOSS

The expansion of agriculture has resulted in natural habitat loss. According to the UN Food and Agriculture Organization (FAO), at 5 billion hectares (ha) there is more land under agricultural cultivation than currently covered by forest and woodlands.[6] Increases in irrigation have caused the depletion of aquifers and have resulted in the loss of natural wetland habitat. The loss of natural habitats as land comes into cultivation reduces ecosystem function and the services that ecosystems provide. As farms consolidate and smaller local farms are eliminated, they are not returned to their natural state but, rather, developed for industrial or residential use.

LOSS OF CROP AND GENETIC DIVERSITY

There are approximately 50,000 known edible plant species. Of these, 250–300 are cultivated as food. Despite the diverse potential of farmable foods, no more than 20 species are major crops that constitute more than 50% of the human diet. These include rice, wheat, maize (corn), soy, sorghum, barley, beans, and several root and fruit crops. The food system in the United States is dominated by corn and soy. Both represent the bulk of calories consumed either directly as ingredients in processed foods or indirectly as animals that are fed exclusively on these crops. This is the result of the development of agribusiness that emphasizes the production and trade of food as **commodities** or basic raw materials.

Among other crops, the numbers of varieties and genetic diversity have been driven down by a need for higher yield and product consistency. For example, according to the International Potato Center, more than 4,000 varieties of potato have been historically cultivated in the Andean highlands of Peru, Bolivia, and Ecuador, selected over centuries for their taste, texture, shape, and color (Figure 5.4). However, today only a few varieties of potato are in major cultivation, typically in monoculture and often produced through

Figure 5.4 A few of the many varieties of potatoes historically cultivated in the Andean highlands. © *International Potato Center (CIP).*

clonal propagation, thus limiting both varietal and genetic diversity. The same is true for many other crops. There is major concern that the loss of this agrobiodiversity reduces the resiliency of the food system and threatens food security through increased vulnerability to insects, disease, and changing climate.[7]

GLOBAL AGRICULTURE AND SEASONAL FOODS

With the advent of improved, refrigerated transportation and free trade agreements, we have entered into an era of global agriculture. One apparent benefit of this is that there are fewer seasonal restrictions on food availability. Less than 20 years ago, specific fruits and vegetables were only seasonally available. In the United States, we ate apples in the fall, citrus in the winter, berries in the spring, and peaches, plums, and melons in the summer. A trip to the supermarket today reveals that all of these are now available year-round because they can be cultivated in Central and South America and shipped northward.

Imported fruits and vegetables are affordable, but mainly through the exploitation of cheap labor and lax environmental constraints abroad. The banana is an example of a food that has historically been a major part of the US diet and that has been consistently available year-round. This is because of the development of large monoculture banana plantations in the tropical Americas and the Caribbean in the late 1800s. This industry has been notorious for its establishment of monopolies, exploitation of workers, and environmental damage, including extensive deforestation, pesticide use, and soil erosion. Banana production has been a motive for war and political change in so-called "Banana Republics."[8] With the expansion of global agriculture, the exploitation of foreign workers in agriculture, and of immigrant and migrant workers in the United States, is yet another hidden cost of what appear to be relatively cheap fruits and vegetables.

The global food economy and industrial agriculture have resulted in greater transportation costs and use of fossil fuels. A typical plate of food in the United States accumulates 1,500 miles (2,400 km) from source to table. A small container of yogurt sold in Germany contains ingredients from four different countries that traveled more than 600 miles (1,000 km), and even more puzzling, the global trade of food has resulted in Britain importing as much milk as it exported within a single year.[9] This suggests the global food system is more about trading commodities for profit and less about providing healthy, affordable food to communities.

Global agribusiness in many ways can undermine the sustainability of our food system. People are generally unaware that most of what they spend on food goes to corporate middlemen, not farmers. In the United States, for example, distributors, marketers, and input suppliers take 91 cents out of every food dollar, whereas farmers keep only 9 cents.[10] As global corporations take over food marketing, small shopkeepers are also being squeezed out: In the 1990s alone, approximately 1,000 independent food shops—grocers, bakers, butchers, and fishmongers—closed in the United Kingdom each year.[11]

Global agriculture is dependent on and reinforces the large-scale, monoculture, highly mechanized, and chemically intensive methods described previously. The increased transportation and use of fossil fuels and related carbon dioxide (CO_2) release, the possibility for exploitation of workers and the environment, and the political complications of importing and exporting food are also significant consequences of global agriculture. Food that is transported long distances often requires more preservatives or is less fresh. This, coupled with the other consequences of industrial agriculture that are now central to the global food systems, has major public health implications. The globalization of food has increased risks to food security by driving down crop diversity and eliminating traditional crops from the market. Moreover, globalization has led to a disconnect between people and their food that has made our food system less resilient due to a loss of locally adapted food systems and a breakdown of small farming communities worldwide.

MEAT AND ANIMAL PRODUCT CONSUMPTION

Historically, in most areas of the world meat consumption has not been a significant issue. For people in developing countries, the majority of calories and protein have been obtained from plant foods, with less than 5% of their calories obtained from animal products. These numbers have been significantly higher in the wealthier North, where up to 20% of a higher caloric intake is met with animal products. However, the amount of meat consumed in developing countries during the past decade increased three times as much as it did in developed countries and continues to rise (Figure 5.5). This increase in consumption is related to development and rising affluence in these countries.

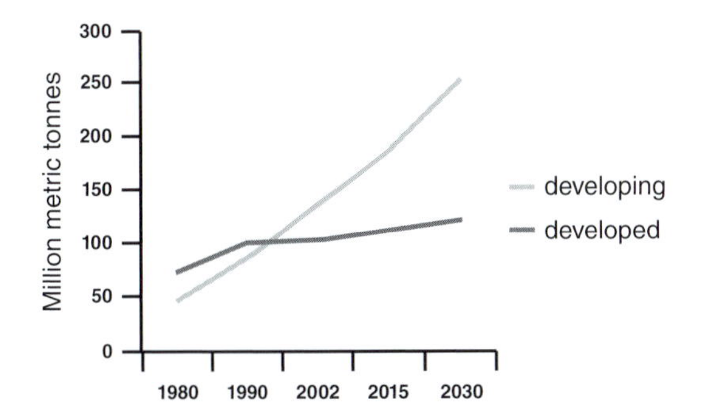

Figure 5.5 Gains in meat consumption in developing countries outpaced that of developed countries. *Source: FAO Statistics and international Livestock Research Institute.*

On a per calorie basis, the production of meat and animal products uses more resources and requires more land compared to crop production. This is because of **trophic inefficiency**, which is the inefficiency of conversion of plant calories into meat calories (Figure 5.6). Because it is land intensive, it has been a major cause of global deforestation. Once deforested, that land is even further degraded through overgrazing. Raising animals, particularly on an industrial scale, is extremely energy and water intensive and creates pollution. Meat production contributes to climate change through energy use and methane production by animals, and it is significantly more energy intensive in the United States than in the rest of the world.

The consumption of meat and animal products in the United States is a major component of the factory farm agro-economy that exploits both the environment and lower wage, immigrant workers. The fast-food and processed food industries, which depend on industrial beef, often import their meat from developing tropical countries. This has resulted in large-scale deforestation to create extensive cattle ranches that produce beef mostly for export.

There are also serious health implications from eating meat. Death rates from bowel cancer and heart disease are directly related to per capita meat consumption rates. There is also evidence that the antibiotics used in industrial livestock production are passed through to consumers. Exposure to these antibiotics results in the development of antibiotic-resistant bacteria in humans and has been shown to be a contributor to obesity, particularly in children.[12]

DEPLETION OF FISH AND SEAFOOD

Until recently, fish and other types of seafood have been the last mass-marketed food supplied to consumers by hunter–gatherers. Consumer demand for these products has been rapidly increasing on an annual basis. As a result, we are experiencing a global collapse of natural fisheries, with more than 75–90% of stocks being depleted through overfishing. Because of environmental inputs, exacerbated by the process of biomagnification, many natural-caught fish contain toxins, including mercury and polychlorinated biphenyls (PCBs). Such fish have been deemed unhealthy to consume, especially by pregnant mothers.

The supply of what was once a natural source of healthy protein is no longer sustainable. As a result, fish farming or aquaculture is the fastest growing segment of agriculture worldwide. Aquaculture now supplies more than 30% of all fish consumed. However, there are often environmental and food safety issues with current approaches to aquaculture. In large-scale net/pen or open pond aquaculture, tremendous numbers of fish are raised in systems that have free exchange with the environment. Impacts include nutrient release and eutrophication and escape of non-native species into surrounding natural aquatic systems. The high density of fish often requires the addition of antibiotics. Because these systems are open, fish are exposed to any pollutants or pathogens that may be in the surrounding environment.

Most large-scale fish farms use corn as a primary feed ingredient. As a result, these fish are less nutritious and lack the omega-3 fatty acids that are one of the more important healthy aspects of consuming fish. Moreover, the use of corn-based feeds furthers the larger corn-based industrial agriculture described previously.

GENETICALLY MODIFIED ORGANISMS IN MODERN AGRICULTURE

What may be viewed as a second revolution in agriculture is the advent of **genetically modified organisms (GMOs)**. Since the early times of domestication, humans have been genetically manipulating plants and animals through artificial selection, hybridization, backcrossing, and selective mutation. GMOs are another type of genetic modification in which a segment of DNA is extracted from one organism and spliced into a recipient organism's preexisting DNA. Like other aspects of the Green Revolution, the use of

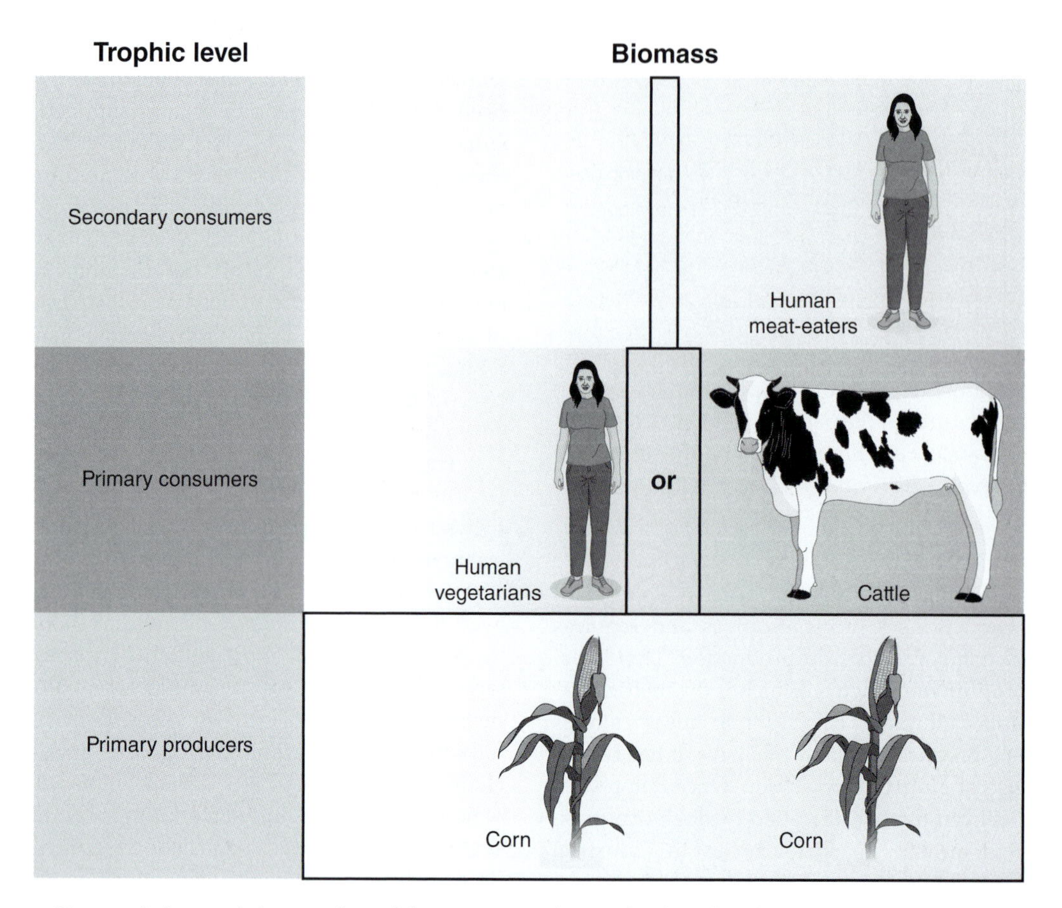

Figure 5.6 Energy is lost as it is transferred from one trophic or feeding level to the next. Thus, the same amount of plant production can support less biomass at each additional level as indicated by the pyramid. With a plant-based diet, people eat at a lower trophic level, allowing for greater ecological efficiency, and as a result, less plant crop biomass production and associated resources can feed more people.

GMOs offered great promise and potential for making our food system more sustainable. The potential benefits included increases in yield on smaller amounts of land, increased nutritional quality of food, and a decreased dependency on pesticides. This truly has been a revolution, as an estimated 80% of crops grown in the United States are GMOs. These numbers are higher for the major crops, corn and soy. Despite intent, GMOs have not turned out to be the panacea for food sustainability.

The most vocal critics of GMOs focus on potential health threats of GMO foods. Concerns center on the possible development of unknown toxic and allergenic components in GMO foods. Another concern is the possibility for transgenic DNA fragments to be absorbed by the tissues of animals or people consuming genetically modified foods. These fears are probably misplaced, and there is very little documented evidence of specific health issues related to the toxicity of GMO foods. In addition, GMO fragments have been shown to degrade in the animal and human gastrointestinal tract, eliminating the possibility for transfer. There have been more documented cases of death and illness due to pesticide poisoning or exposure on any given day than have ever been attributed to the consumption of GMOs. However, there is a scarcity of sound data, and there has been essentially no research on the potential hazards of GMO foods. The FDA does not test the safety of GMOs, nor does it require independent pre-market safety testing. Instead, it places the responsibility on the producer or manufacturer to ensure the safety of the food.

More research on the health aspects of GMOs is needed, but corporations such as Monsanto prohibit open research on their products. In 2013, what has become known as the "Monsanto Protection Act" was signed into law as an attachment to a larger bill that permitted the United States to exceed its debt limit. The bill states that even if future research shows that GMOs or genetically engineered seeds cause significant health problems, the federal courts no longer have any power to stop their spread, use, or sales. This policy serves to protect corporations such as Monsanto from litigation suits over any subsequent discovery of negative health effects of GMOs. Despite the lack of current evidence of negative health consequences of GMO foods, more than 26 nations have bans or partial bans on GMOs. The United States is not one of these. Major corporations, however, are now responding to market pressures and creating GMO-free brands, and they have begun to make these available even in the United States.

There are many serious issues related to the use of GMOs independent of potential negative health consequences that currently limit their potential to making our food system more sustainable. First, yields per unit area of land have not significantly increased as promised. Rather, GMOs have allowed certain crops to thrive on previously unarable land, thereby increasing the loss of habitat that was essentially protected because food could not be grown on it. Second, in many cases, GMOs have resulted in the increased use of toxic chemistry. The best known case of this is the engineering of plants to withstand the herbicide glyphosate or Roundup. Roundup is a herbicide or weed killer. Prior to engineering of so-called "Roundup Ready plants," herbicides could only be applied prior to planting because as plant killers, they would kill the crop. With Roundup Ready plants, farmers can now control weeds while growing their crops through extensive application of this chemical throughout the year. The increased use of this chemical has negatively impacted aquatic systems; reduced biodiversity, especially among amphibian species; increased pest resistance to Roundup; and has potential human health effects, including links to cancer. The developing resistance of these GMOs to Roundup has caused farmers to begin to increase their use of Roundup and additional pesticides. Roundup Ready soy has been shown to have higher residue concentrations of glyphosate and its principal breakdown products compared to soy without the Roundup Ready gene.[13] Monsanto both engineered the plants and is the seller of Roundup, which is one of the company's largest moneymakers. In this case, the GMO did more for the sustainability of Monsanto's profits and less toward a sustainable food system.

Another concern about the use GMOs is the transfer of genes from GMO crops to native plants and the potential for impacting them negatively and perhaps causing them to be herbicide resistant. The incorporation of a gene from the soil bacteria *Bacillus thuringiensis* (*Bt*) into corn and other crops has raised questions about potential impacts on biodiversity. *Bt* is a naturally occurring soil bacterium that produces an insecticidal toxin. The bacteria themselves have been applied to crops, even as a technique in organic agriculture, as a way to decrease pest damage. Genes from *Bt* can be inserted into crop plants to make them capable of producing the same insecticidal toxin and therefore resistant to certain pests. There are no known adverse human health effects associated with *Bt* corn. However, *Bt* corn can adversely affect non-target insects if they are closely related to the target pest. One example is when the caterpillars of the Monarch butterfly are exposed to *Bt* pollen. If the caterpillar's host plants are near agricultural fields and come in contact with the pollen, the caterpillars suffer increased mortality.[14] These adverse effects are considered minor relative to those associated with the alternative of blanket insecticide applications.

GMO genetic markers have been found in the seeds of crops of farmers not planting GMOs but whose farms neighbor fields where they have been planted. This shows the potential for passive cross-contamination of GMO genes. Worse is that the GMO companies are now suing farmers who have been subject to this passive cross-contamination for "using" their GMOs without a license. Courts have ruled in favor in Monsanto on this issue. Judgments have stated that it does not matter how a farmer's seed or plants become contaminated with patented genes. Any passive contamination—whether through cross-pollination, pollen blowing in the wind, by bees, by direct seed movement, or seed transportation—results in the grower's seeds or plants becoming Monsanto's property. Even the slightest contamination is a violation of patent law, and farmers can be sued. Now small, non-GMO farmers are being forced to purchase insurance for this potential liability that is not under their control.

Because of the private investment in the development of GMOs, seed companies such as Monsanto have engineered their seeds to protect their intellectual property rights. The use of genetic restriction technologies or "terminator seeds" that render the harvested crop seed sterile has been viewed as highly controversial. The potential benefit is that it could encourage more investment in crop biotechnology and plant breeding research that may benefit all farmers. However, this technology puts to a halt seed planting, exchange, diffusion, and on-farm breeding that have been important for small farms and especially subsistent farmers in developing countries. This results in the loss of local genetic varieties that are adapted to specific conditions. Also, because subsistent farmers lack the resources to purchase seeds every season, this excludes them from access to the potential benefits of GMOs. Moreover, farmers in developing countries are often required by aid policies to purchase terminator seed, thereby forcing dependence on these seed companies. Making farmers dependent on these seed purchases could be detrimental to their livelihoods and undermine regional food security. It is believed that as this terminator technology accelerates, corporate concentration and control of agribusiness will spread and undermine public sector research.

The development of GMOs and policies related to them has been more about the control of intellectual property of the food crops on which we depend than about the sustainability of food system. The promised increases in yield with GMOs have mostly come from traditional plant breeding, not biotechnology. Genetic engineering has done more to support large-scale, mono-crop agribusiness; marginalize small farmers; and concentrate power over our food security among a few biotech corporations. This has resulted in a model of farming that undermines the sustainability of cash-poor farmers and does not serve the world's most hungry people.

Monsanto was acquired by German pharmaceutical company Bayer in 2018. As a way to avoid the negative reputation associated with the previously described controversies, the new parent company replaced the name Monsanto with its own. Although the company formerly known as Monsanto will operate much the same as it always has, Bayer executives have stated that they would like to deepen the dialog with society regarding this new segment of their business. The extent to which this leads to more sustainable use of GMOs remains to be seen.

FOSSIL FUEL DEPENDENCE AND CLIMATE CHANGE

Our current food system is fossil fuel intensive. Annually, 19% of the total fossil fuels consumed in the United States is by the industrialized food system. This is similar to the amount of energy consumed by automobiles each year in the United States. US conventional crop production consumes approximately 1,000 liters (260 gallons) of fossil fuel per hectare per year.[15] This includes energy inputs for mechanization and chemical inputs. GMOs have served to increase our use of these chemical inputs.

Meat and animal product consumption increases energy demands because although the same energy inputs are required for the production of feed, the efficiency of conversion of that energy into consumable meat is low (see Figure 5.6). In addition, there are non-feed energy costs associated with maintaining and preserving animal products. Transportation associated with global agriculture raises these numbers significantly. The rate of fossil fuel consumption in agriculture is rapidly increasing in countries such as India and China.

Obviously, the increased reliance of our food system on fossil fuels results in increased carbon emissions and subsequent influence on global climate change, and it increases dependency on foreign fuel sources. The loss of forested habitat associated with its conversion to large farms eliminates that carbon sink and releases environmentally stored carbon. Energy and agricultural policy agendas are typically not considered in conjunction with each other, but they should be (see Box 5.1).

The Green Revolution and Sustainability

It is clear that the Green Revolution has transformed our food system. The goal of increasing yield and feeding the world through the use of technology, chemistry, and mechanized equipment and consolidation of farms and food processing into large agribusinesses superficially seems reasonable from an efficiency perspective. However, there have been obvious negative impacts on the environment, communities, public health, and food security. There are also social equity issues associated with the modernization of agriculture, making it less sustainable. There are social implications associated with externally imposed technology, which has been likened to a form of cultural imperialism. The advances of the Green Revolution disproportionately benefit larger, resource-rich farms compared to those that are smaller and more limited economically. It has resulted in the loss of subsistence strategies and sharecropping, and it has damaged other social structures. Agriculture has become more exploitive of both people and the environment. Some have argued that increasing food production without controlling population growth and rates of consumption is not sustainable in that it will place excessive pressure on other resources and sectors.

So how did a revolution with the good intentions of feeding the world go bad? One reason is that corporate interests fueled this revolution. The Green Revolution grew out of the post–World War military–industrial complex. The same technology that was used to make synthetic nitrogen fertilizer was used heavily in the war effort for making bombs and poison gases. One of the inventors of this technology, Fritz Haber, was awarded the Nobel Prize for improving the standards of humans through the development of synthetic fertilizer but then, ironically, was responsible for developing nitrogen-based gases and explosives used by Germany in World Wars I and II.[16] This illustrates the potential conflict between good and bad in science.

This paradox played out further as the military–industrial complex gradually shifted into the agribusiness complex. The new corporate power in the food industry was and continues to be able to influence farm policy and other regulations that favor the re-engineered food system that benefits mostly the chemical industry, factory farms, and globally transported processed foods (see Box 5.1). Corporate–government collusion still occurs today because companies such as Monsanto have powerful influence over policy that benefits them and their proprietary chemicals and GMOs but not the food system as a whole. Food quality problems, environmental degradation, hunger and malnutrition, and food inequity still exist. Although these structural problems with the food system seem insurmountable, they are solvable.

Sustainable Solutions: Food and Agriculture

A variety of innovative approaches can make food systems more sustainable. Some of these are new innovative technologies, but many are also based on traditional methods of food production. In other words, a more sustainable way to produce food in the future will in part rely on practices that are centuries or even thousands of years old. Many of these are nature-based solutions (NbS).

Organic agriculture, seed saving, and mixed-crop livestock farming are examples of innovations or alternatives that are returns to or are built on past practice. In this book, we refer to innovations adapted from prior practice as "retrovations." Next, a number of such retrovations and new innovative approaches that have great potential to make our food system more sustainable are presented. However, it is important to keep in mind that no single approach is really a sustainable solution. Ultimately, we must integrate an entire suite of them to meet our continued nutritional needs.

Organic Agriculture

Organic agriculture incorporates proactive approaches to prevent many of the problems associated with Green Revolution, industrial agriculture. Developed—or perhaps more appropriately stated, redeveloped—in the early 1900s, organic farming is in many ways a return to more traditional forms of agriculture. Organic agriculture became a topic of both scientific and popular interest in the 1950s and 1960s as publisher J. I. Rodale began to promote the methods through his publications. He also created the first experimental farm, on which organic farming could be scientifically compared to conventional approaches. Organic farming was also advanced through the work of Rachel Carson, who made the public aware of the environmental and human health implications of pesticides.

Organic farmers recognize the linkages between healthy soil, healthy food, and healthy people, and much of their focus is on the maintenance and replenishment of soil fertility and soil organic matter. Soil is a fundamental resource that is essentially a complex living system consisting of more than 10,000 microbial species in 1 gram of soil. Soil is a vulnerable resource, especially in intensive, conventional agriculture. It takes 100–400 years for nature to create a 1-cm layer of topsoil. This resource is managed by organic farmers through the use of nitrogen-fixing cover crops such as legumes, organic composting techniques, addition of minerals, and reduced tilling and mechanical manipulation of the soil. It serves to prevent erosion, organic matter loss, acidification, reduced biological activity, and other soil quality components that are depleted through mechanized, chemically intensive agriculture.

Organic farmers also eliminate the use of toxic and persistent chemical fertilizers and pesticides. This is done through the previously mentioned soil management techniques and through the encouragement and release of beneficial insects. Also, plants can be protected through the use of row cover fabrics. The increased crop diversity of many organic farms also promotes beneficial insects and spreads the risk of complete crop decimation by disease or insect pests. Applications of insecticidal soaps and oils, and naturally occurring *Bt* proteins or dead bacteria, are also approved for organic farming. One argument of the pro-GMO lobby in support of *Bt* toxin GMO crops is that the *Bt* toxin has been safely used for decades by organic farmers.

Three important questions arise when considering organic agriculture as a potential sustainable solution: (1) Are organic crops indeed healthier for the consumer? (2) Is organic farming better for the environment? and (3) Can organic agriculture generate the yields needed to feed existing and rapidly expanding populations?

HEALTH BENEFITS

Regarding health, there is a perception that organically grown foods are better for the consumer. This belief is what drives the market for organics. A controversial analysis of the existing literature on the health benefits of organic foods by Stanford University School of Medicine suggested that there is little evidence of the health benefits of organic foods.[17] However, this study was limited by the preexisting work in this area that has had a very narrow scope. Since this study, it has been shown that organic foods typically have more antioxidants compared to conventionally grown foods.

Organic foods have also been shown to contain less non-protein sources of nitrogen, which are hazardous because they are converted to nitrosamines in the human digestive track. Even the Stanford study showed that pesticide residues are significantly reduced in organic products, decreasing the risk of long-term exposure. This is supported by analyses of urine samples taken from participants on either a conventional or an organic diet. Participants on the conventional diet had higher levels of pesticide residue in their urine compared to those on an organic diet.[18] However, data are generally lacking, particularly from comparisons of the health outcomes of populations that habitually consume organically versus conventionally produced foods over the long term.

ENVIRONMENTAL BENEFITS

There is considerable evidence that organic farming is better for the environment. The removal of chemical inputs eliminates contributions to eutrophication, biomagnification, and other issues related to chemical pollution. The focus on soil regeneration maintains rather than depletes our capacity to produce food well into the future. Fossil fuel consumption is lower in organic farming and therefore results in less CO_2 production.

Long-term farming systems trials at the Rodale Experimental Farm have shown that organic farming can reduce fossil fuel consumption by up to 45% compared to conventional systems.[19] Organic grass-fed beef systems reduced fossil fuel use by 50% compared to non-organic. Organic farming also helps mitigate global climate change by increasing the carbon storage capacity of the soil. This enhanced ecosystem service occurs because leguminous cover crops used by organic farmers to enrich and stabilize the soil release phenolic compounds that help sequester the carbon there. However, the release of nitrous oxide—a much more potent greenhouse gas—by soil microbes is the same in organic and conventional agriculture.[20]

YIELD

Answering the third question, whether organic agriculture can generate the yields needed to feed existing and rapidly expanding populations, is vital in assessing organic agriculture as a sustainable solution. One of the common critiques is that organic agriculture and the availability of organically acceptable fertilizers are insufficient to meet this demand. The 30-year farming system trials at the Rodale Institute have shown that organic yields match conventional yields and outperform them in years of drought.[21]

A review of nearly 300 different comparisons of the two agricultural systems showed that in developed countries, the average yield of organic farms is approximately 92% of conventional yields.[22] This same study showed that in developing countries, organic farms produce 80% more than conventional farms. That is because the inputs for organic farming are cheaper and more readily available to organic farmers in poorer countries. Globally, agriculture today produces approximately 2,786 calories per person per day. This study also suggests that full conversion to organic agriculture could produce between 2,641 and 4,281 calories per person per day, exceeding the recommended daily consumption of 2,100–2,500 calories per person per day.[23]

COSTS OF ORGANIC FOOD

One major criticism about organic farming in the United States is that the products are more expensive and this limits both desirability and access to these products, especially for poorer people. Some argue that as the demand increases, the market will follow and prices will drop. The reality is that in the wealthiest nation in the world, Americans spend the lowest proportion of their income on food, approximately 11%, compared to elsewhere, where that number is closer to 50%. Conversion to a primarily organic diet would increase the proportion Americans spend on food by only 3–6%, still significantly lower than the proportion spent in the rest of the world. Moreover, if we consider the real or external costs of conventional farming, then the actual cost of organic food is equivalent to or less than that of conventionally produced food.

THE GROWING DEMAND

The growing demand for organic food has had some negative consequences. As many processed foods have come to incorporate organic ingredients and are marketed as such, the supply chain has lengthened and organic products now travel longer distances. This type of market growth has also fostered the entry of large-scale agribusiness into organic production. As a result, it caught the interest of the federal government and led to the Organic Foods Production Act of 1990. This shifted the development and enforcement of organic standards from third-party certification to government control. Many have argued that this shift has weakened some standards and has favored larger scale organic agriculture, as has previous farm policy. The cost of organic certification and required record keeping can be as high as 6% of a small farm's gross sales. As a result, many smaller, direct-sales farms follow organic practice but forgo participation in the federal organic certification process. These farmers now opt for less costly labels such as "Certified Naturally Grown" that are sponsored by state and local organic farming associations.

Biodynamic farming is another approach to agriculture that in addition to adhering to organic standards, bases planting on a lunar calendar. Biodynamic farmers also apply specialized preparations made from fermented manure, minerals, and herbs. Their focus is more spiritual and attempts to harmonize the vital life forces of the farm. Although some are skeptical about the benefits of this approach, studies have revealed that biodynamic farming systems generally have better soil quality and increased crop root growth compared to conventional and standard organic systems, although yields are lower.[24] In addition, biodynamic approaches result in crops with higher polyphenol content and greater antiradical activity, which offer health benefits. The market for biodynamic products is growing, particularly for wines. Biodynamic products will remain in a higher priced, specialty niche market and will not solve major food security issues. However, biodynamic farming helps preserve chemical-free farmland, raises awareness about the problems associated with industrial and conventional agriculture, and connects more people to farmers and their food.

Integrated pest management (IPM) is another way to reduce chemical inputs while preventing crop damage from insect, weed, and disease pests; it embodies some of the organic ideals, but perhaps in a more pragmatic sense. The goal of IPM is long-term prevention of pests or their damage by managing the ecosystem. This is accomplished through the use of regular pest monitoring, the use of pest-resistant plant varieties, encouragement of beneficial insects, cultivation and irrigation practices that reduce pest establishment, and physical barriers in conjunction with responsible chemical use. IPM has allowed farmers of particularly vulnerable crops to produce high-quality crops, protect the environment, and maintain farm profitability.

Hydroponics

Hydroponics is a branch of agriculture in which plants are grown without the use of soil. The nutrients that the plants normally derive from the soil are dissolved in water, and then the plant roots are either suspended in or flooded or misted with the nutrient solution. This allows farming to occur where there is no arable land. For example, many islands and inner-city, urban environments lack space or soil appropriate for agriculture. Hydroponic farming is also vital in areas where soils are completely degraded or on small islands on which sand or saline soils are the only available substrate.

Hydroponics can meet the needs of people without degrading the soil. Because the nutrient solution is recirculated, hydroponics uses less than 10% of the water needed to grow the same crop with soil-based farming. Also, because production per unit area is four times greater than that of soil farming, less land is required. This can benefit land-limited farmers and allow for expansion without the need to acquire more land.[25] Hydroponics can provide consistent crops at a variety of scales, including local and regional. When done in climate-controlled environments, the growing season can be extended throughout the year. When climate control is achieved through the use of more sustainable forms of energy such as heat pumps and solar energy, there are additional environmental and economic benefits. Energy-efficient, wavelength-specific LED lighting can produce nutrient-rich crops in 25% of the growing time of plants produced with natural lighting.

Many have debated whether or not hydroponic growing systems can be considered organic. This makes sense given that hydroponic systems are soilless and many of the principles of organic agriculture focus on soil management, conservation, and health. Many hydroponic farms are pesticide-free but are not certified organic because of their use of dissolved mineral nutrients.

Hydroponics is an area of great innovation, and new approaches are regularly being developed. One example is the **nutrient film technique (NFT)** hydroponic system (Figure 5.7) that is potentially certifiable organic. With NFT, plants come into contact with only organic-approved materials. Mineral and insoluble nutrients are located and converted to organic nutrients by way of bacterial processes in an environment separate from the plant root environment. Another is **aeroponics**, in which exposed roots are misted with microns of water at regular intervals. Because it uses 98% less water than conventional farming, this technique would be very useful in areas where water is scarce.

Aquaculture and Aquaponics

The demand for fish, seafood, and healthy protein continues to rise. Given that fisheries and large-scale aquaculture are not sustainable, alternatives need to be explored. **Recirculating aquaculture systems (RAS)** could be a sustainable solution. RAS uses closed-loop production that relies on biological filtration, in which bacteria convert ammonia produced by fish to less toxic, potentially useful forms of nitrogen. These closed systems are environmentally compatible in that they use much less water and release substantially less waste material into the environment compared to open systems. Efficient closed-loop biological filtration and solids removal maintain clean, disease-free water and can guarantee food safety. RAS maximizes

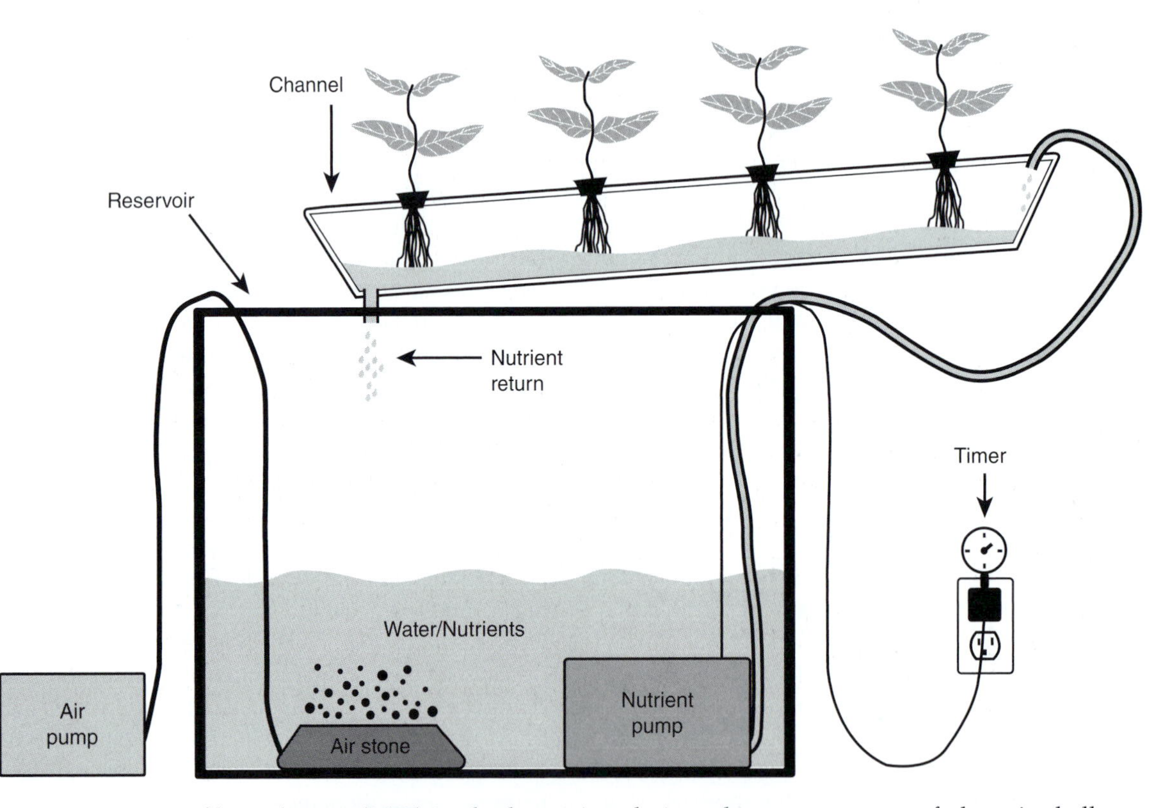

Figure 5.7 Nutrient film technique (NFT) is a hydroponic technique that exposes roots of plants in shallow channels to a recirculating film of nutrient solutions. (http://www.kratkyhydroponics.com/nft.html).

production per unit area, is expandable, and can be done indoors. It is well suited for abandoned manufacturing facilities in that occur in many cities. Urban RAS can create economic opportunities in chronically distressed urban centers and can serve both the diversity of multicultural consumers and the "locavore" food movement. RAS has been incorporated into successful business models, such as in the production of a sustainable barramundi (Box 5.2) and no-kill caviar that protects the endangered sturgeon providing this sought-after delicacy.

Aquaponics is the integration of RAS aquaculture with hydroponics. In aquaponics, filtered fish water serves as the nutrient solution for a hydroponic system for vegetable and herb production. Plants use the nitrates that the bacteria converted from ammonia wastes.

BOX 5.2

Innovators and Entrepreneurs: Josh Goldman and Australis Aquaculture

Recirculating aquaculture systems (RAS) are closed-loop production systems that continuously filter and recycle water. RAS enables large-scale fish farming that requires a small amount of water and releases little or no pollution. RAS is considered the leading edge of sustainable fish production in a market plagued by global fish stock depletion, unhealthy traditionally farmed product, and significant environmental impact.

One innovator, Josh Goldman, has been recognized as a leader in the sustainable seafood movement with his company Australis Aquaculture and the marketing of his product as The Better Fish trademarked brand (Figure 5.2.1). Beginning as a student at Hampshire College, Goldman has been an unrelenting innovator both in the bioengineering of water reuse technologies and in business strategy and marketing. A key to this has been his development of the fish, Australian barramundi, as an expanding international culinary trend. Marketed and trademarked as the Sustainable Sea Bass, the fish are produced with minimal impact. Because barramundi can synthesize their own omega-3 fatty acids, they are a healthy food and require less feed from fish and fish oils, contributing to increased sustainability. A custom feed that meets the highest standards for sustainability, health, and food safety is used. The company never uses synthetic chemicals such as hormones or colorants, and its product is free of environmental contaminants such as mercury and PCBs that frequently occur in wild-caught and traditionally farmed fish. The Australis facility in Massachusetts uses closed-containment or RAS technology pioneered by Goldman that has been green certified by Seafood Watch (see Box 5.3). The company recycles and purifies 99% of its water and donates the fish waste as fertilizer to local farmers. Australis has supported the local, rural economy by creating more than 80 jobs.

The Australis product line is sold fresh or frozen-fresh as hand-cut fillets, and it includes frozen steamable gourmet seafood entrees. As part of Goldman's plan for market expansion, Australis has moved some of its production overseas to Vietnam. Here, on what the company refers to as Eco Farms, it first hatches and nurses fish in land-based recirculating tanks and then completes the grow-out process in modern, low-density sea cages with minimal impact on the pristine waters. In terms of social welfare, the company pays living wages and offers all permanent full-time employees full benefits. All non-US facilities undergo third-party social welfare audits to ensure compliance with local laws. By moving some of his operations to the developing world, Goldman has been able to make his products more affordable and accessible to discount-oriented shoppers and has helped improve the standard of living in those foreign communities. Goldman is an innovative entrepreneur who demonstrates that for-profit businesses can promote the core values of sustainability.

Figure 5.2.1 Josh Goldman of *Australis Aquaculture.* *Copyright © 2018 Fred Collins.*

In essence, the fish, by way of bacteria, are feeding plants. The plants then in turn are cleansing the water that is returned to the fish system. The result is the production of two types of food in a closed-loop system.

Sustainability of these systems could be further improved by shortening the supply chain by way of vertical integration. **Vertical integration** is defined as the combination of two or more stages of production within a single project, operation, or company. One way to do this is to produce fish feed at the site of fish production. This could include raising worms on plant compost and the cultivation of algae and crickets. The use of fish feces on-site as compost and the biogeneration of methane are also being explored. The development of in-house fish breeding programs could eliminate the need for importing fry and/or fingerlings. Vertical integration of RAS and aquaponic systems with other industries could help close waste, resource, and energy loops.

One larger scale example of vertical integration is Plant Chicago. Plant Chicago models a systems thinking approach that integrates aquaponics with beer and kombucha tea production, energy generation, and private income-generating activities (Figure 5.8). The project has created 125 jobs in an economically distressed area. FarmedHere is an example of a vertically integrated food production project also in Chicago that produces organic produce and value-added products such as herb-based salad dressings. It has become a zero organic waste facility by converting all of its organic waste into compost. The compost is then used for urban landscaping, horticulture, and farming at other locations in the city.

Buying Local and Small

Another possible solution that has gained much recent favor in the developed world is the support and encouragement of local agriculture. In the less developed

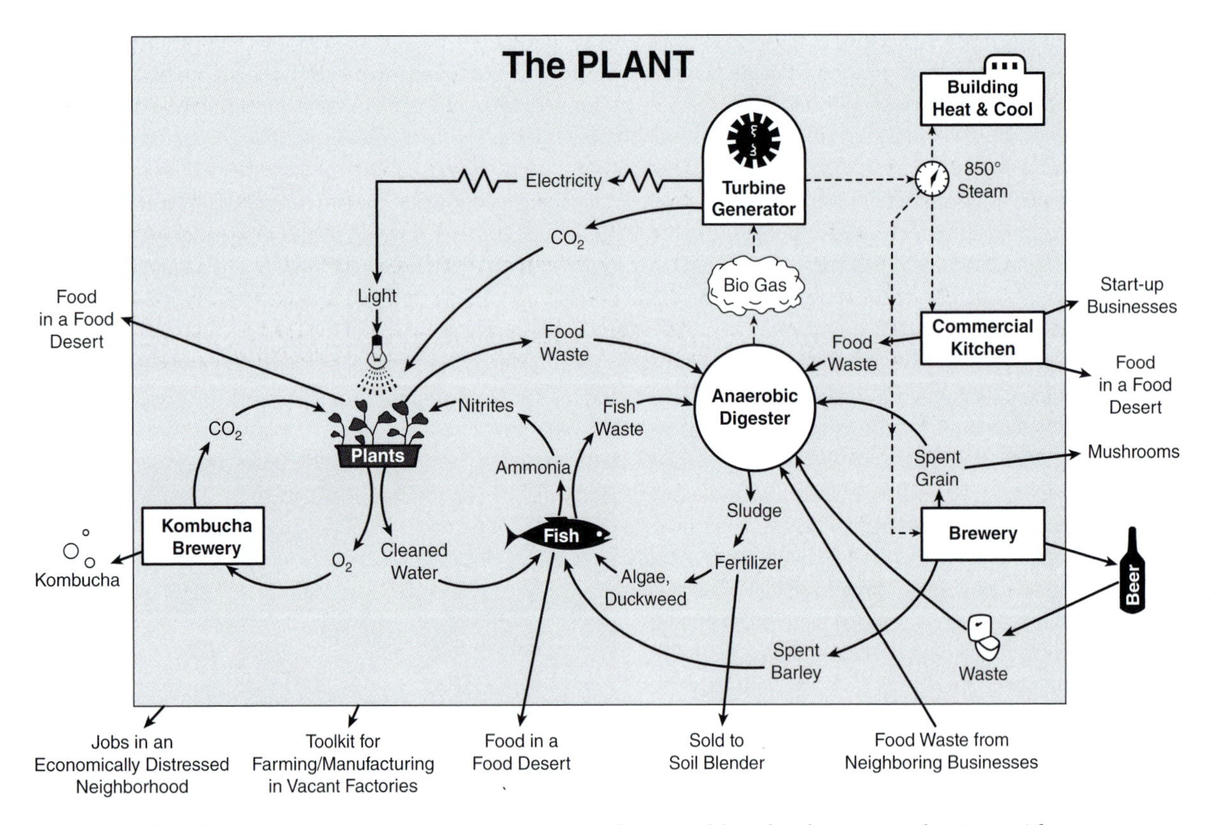

Figure 5.8 The Plant in Chicago integrates aquaponics, beer, and kombucha tea production with a net-zero energy system. This is an excellent example of using systems thinking to solve multiple problems. *Source: Matt Bergstrom, The Plant.*

world, locally produced food has always been vital given the lack of infrastructure for transport and storage and their associated costs. There is considerable evidence that small farms are more productive and contribute more to local economies than do larger farms. They reduce impact on the environment through better farming practice. Small farms tend to be more diverse, less chemically and mechanically intensive, and hence better for the environment.

Although not necessarily certified organic, most smaller operations follow similar practices. Locally sold products require less transportation, so fossil fuel use and carbon footprints are reduced. Local products tend to be picked when ripe, are fresher and better tasting, and offer greater health benefits. When imbedded within the matrix of the developing landscape, local farms serve to preserve pockets of undeveloped, more natural lands and reduce the impact of urban/suburban sprawl. The transition from a large agribusiness-type economy to a smaller system of local organic farms is possible and has been demonstrated in Cuba as a necessary consequence of changes in economic forces influenced by significant policy implementation (see Box 5.1).

A locally focused food system can be achieved through the development and support of local and regional farmers markets and food cooperatives that offer opportunities for direct sales from producer to consumer. As a result, the producer maintains a larger portion of the profit instead of corporate middlemen and chain distributors. Sales are higher for farms that engage in other entrepreneurial activities, including value-added production of farm goods, custom work, and agritourism. Producer farmers markets and **community-supported agriculture (CSA)** have become popular ways for consumers to buy local, seasonal food and products directly from farmers. Through CSAs, a farmer offers a certain number of "shares" to the public. Interested consumers purchase a share and in return receive an allotment of seasonal produce each week throughout the farming season.

Recently in the United States, the local foods or "locavore" movement has grown substantially. Many restaurants featuring "farm-to-table" meals and specialty supermarkets promote the sale and use of local foods. This has expanded to university campuses and other institutions where dining services are seeking out local producers to meet client demand. A recent poll of chefs and food enthusiasts predicted that locally produced food will become a top consumer choice, especially in higher income markets.[26] This local food movement has generated significant income for small farms. In three western Massachusetts counties, 1,960 farms cultivating 169,000 acres sold nearly $9 million of products in 2007, and this number continues to rise and spread geographically.[27]

The local food movement is confronted by a number of challenges and must overcome a variety of barriers to market entry. These challenges include meeting demands for high volumes and consistent quality and also inadequate distribution channels. These could be alleviated if small, local farms were able to capture the advantages of larger scale production systems by working as regional cooperatives. This could allow consolidation of the supply chain so that inputs could be purchased in bulk. In the case of RAS fish production systems, regional networks are being formed for group purchases of fry and fingerling, foods, and other inputs. Collaboration with regional brokers could allow consolidation of direct marketing and provide intermediaries for packing and shipping. The development of shared food transportation infrastructure, food hubs, and food business incubators would also help.

Many potential retail or institutional purchasers are not prepared to deal with unprocessed foods that come directly from farms. A solution to this is the development of shared commercial-style kitchens in which farmers can process their products for end use. This would also require farmer training programs on processing and food safety. Organizations such as Buy Fresh, Buy Local have played a role in supporting local farmers in these ways, and a number of communities and small business incubation centers such as The Plant in Chicago offer access to health-certified kitchens.

Another issue is limited seasonal availability. This can be addressed by extending the growing season through the use of greenhouses, hoop houses, and other techniques that also have the potential to incorporate sustainable energy technologies. Probably the major challenge to local agriculture is the loss of farmland. During the past 75 years, many regions have seen the number of farms decline by 80% and a loss of more than 50% of farmland. Those that have remained tend to be larger farms.

The loss of farms could be slowed through farmland preservation approaches that include agricultural protection zoning, agricultural land easements, and lower tax rates for preserved farms. Urban growth boundary programs such as those in Oregon and other states limit the development of farmland as agricultural boundaries around urban areas through legislation. The aging farming population is contributing to the further loss of farms as children often leave the farm to pursue other ways of life. New farmer training programs or so-called seed farms that help new farmers enter into local agriculture are therefore vital.

Many assume that eating local food from small farms is more ecologically sustainable and socially just. Scale theory in political and economic geography suggests that this may be a false assumption.[28] These theorists argue that there is a "local trap" that conflates the scale of the food system with desired sustainable and socially just outcomes. The local trap obscures other scalar options such as regional networks of local farmers, cooperatives, and distributors that might be more effective at achieving these outcomes. An example of such a network is the Urban Organic home delivery company that serves the tri-state area of New York, New Jersey, and Connecticut. Others have questioned whether there is sufficient local land area to support high population centers. For example, if all or even a large portion of the calories consumed by New Yorkers came from New York state, nearly all of the state's forests would need to be converted to agriculture.

One consistent complaint about the local food movement is that it is not typically accessible to poorer urban communities because of higher cost or lack of purchasing opportunities within those communities. This has resulted in the local food movement being labeled as "classist." As such, proponents of food equity have made increasing such accessibility a priority. This can be done by promoting and supporting neighborhood community gardens and urban farmers markets. Allowing and encouraging recipients of supplemental nutrition programs such as Women, Infants, and Children (WIC) or the Supplemental Nutrition Assistance Program (SNAP) to make purchases at local farmers market would benefit both producer and consumer.

The Greensgrow project in Philadelphia is an example of an effort that incorporates much of this. Its mission is to revitalize abandoned land in declining urban centers through entrepreneurial and sustainable local agriculture. This serves to beautify urban blight, is a local economic driver, and provides jobs in areas of high unemployment. The locally grown food is sold at the farm, through a CSA, and through mobile food markets that access underserved communities. These accept payment through supplemental nutritional assistance programs. They also provide training in food entrepreneurship and offer a shared community kitchen for the development of product.

Similar programs have been developed in other urban centers, including Farmscape in Los Angeles, which has installed more than 600 urban farms, and the 60-acre urban farm run by Recovery Park in Detroit. New York City has implemented a program called Health Bucks that provides food stamp clients coupons that extend their purchase power by $2 for every $5 they spend at local farmers markets. This serves to leverage federal dollars spent on food stamps to generate the local food economy while promoting the purchase of healthy food.[29]

Urban food production projects such as those described previously are effective ways to increase access to local, healthier foods for economically distressed areas while creating jobs in those communities. In addition, hydroponic, RAS, and aquaponics food production systems lend themselves well to the urban environment. Rooftop and container gardening are also options for growing foods in cities. Other innovations in urban food production include food forests and fruit maps. A food forest such as the Beacon Food Forest Project in Seattle, another example of NbS, is a gardening technique or land management system that mimics a woodland ecosystem composed mostly of edible plants. Fruit mapping projects identify and make available locations of what is referred to as "public fruit" that grows in or over public property such as streets, sidewalks, and parks. FallingFruit.org maintains a collection of maps for a variety of locales, and there is an extensive database for wild food sources in Portland, Oregon.

Urban Agriculture and Vertical Farming

The previously described innovations in urban food production can help meet current nutritional needs in the world's cities. However, by the year 2050, nearly 80% of the earth's population will reside in urban centers. It is anticipated that there will be insufficient land to meet the agricultural demand by these large urban populations. One class of solutions to address the growing urban demand for food is referred to as **vertical farming**. This includes the cultivation of plant or animal life within a skyscraper greenhouse or on vertically inclined building surfaces. According to Columbia University ecologist Dickson Despommier, a 30-story building built on one city block that is engineered to maximize year-round agricultural yield could feed tens of thousands of people who live in the surrounding area using less energy and water than food production on natural landscapes. This would be made possible through the use of LED lighting and advanced hydroponic growing techniques. Because food is produced for local consumers, transport costs and associated carbon emissions could be avoided. Vertical farming also refers to "tech-dense" indoor farms that produce food on stacked hydroponic systems in old industrial buildings or recycled shipping containers in urban settings. A number of companies such as Spread Co., Ltd., in Japan and the California company, Local Roots Farms already have such systems in operation.

Vertical farming offers great potential for providing the area required to produce local, healthier food with less environmental impact for expanding cities. It can also increase local employment opportunities. To maximize production, vertical farming will require the further development of intuitive climate controls, LED lighting systems, and artificial neural networks that directly monitor individual plants. The implementation of vertical farming should also be integrated with local renewable energy. The MIT City Farm Porject, whose motto is "food on every façade," is developing new techniques for vertical farming. The Association for Vertical Farming provides resources and training for advancing the sustainable growth and development of the vertical farming movement. An additional challenge to developing urban food production systems is that acquiring and developing land or buildings for such purposes are often constrained by archaic municipal zoning and permitting structures. Farm policy, nutrition, and urban planning are typically considered in isolation of each other. Their integration could better support the sustainability goals of a local food system (see Box 5.1).

Fair Trade

There are no local options for many of the crops on which the developed world depends. Crops such as coffee, chocolate, and bananas can only be grown in tropical climates where the poorest nations in the world are located. The production of these crops is labor intensive. Also, the going rate for them has remained globally depressed as increasingly more tropical countries, often at the encouragement of development agencies, move into their production. As a result, their supply remains relatively high. The combination of depressed prices and labor-intensive agriculture can lead to the exploitation of workers involved in their production. For example, the global wholesale price of coffee has remained below $1.00 per pound. Coffee farmers receive only approximately $0.10 a pound, which translates to approximately $1 or $2 per day for labor-intensive work, whereas the average consumer in the developed world spends more than $3.00 per day at a retail coffee shop. This is the equivalent of sweatshop labor in the field, and the resulting markup and profit for coffee retailers are massive. Similar or worse conditions occur for other agricultural products, such as cocoa, tea, honey, grains, fruits, flowers, and vegetables. In the case of West African cocoa production, much has been reported on the exploitation of children. Given the social equity component of sustainability, low wages, poor conditions, and abuse of child labor are neither acceptable nor sustainable.

Fair trade certification and marketing are a possible solution to this problem (Box 5.3). The goal of fair trade is to help producers in developing countries by guaranteeing fair wages and safe labor conditions and by banning child labor and protecting basic human rights. In addition to paying higher wages to producers, revenues from fair trade products are also used to support community development projects. The fair trade certification system also attempts to promote long-term business relationships between buyers and sellers, microfinancing for farmers, and greater transparency throughout the supply chain. Fair trade covers a growing range of products, including bananas, honey, coffee, oranges, cocoa, cotton, dried and fresh fruits and vegetables, juices, nuts and oil seeds, quinoa, rice, spices, sugar, tea, and wine. The increased cost of fair trade products is passed on to consumers who choose to pay more because they believe their purchase will support social justice and environmental protection. Businesses have increased their fair trade offerings to meet consumer demands.

Critics of fair trade argue that the extra money paid by consumers never reaches the poor farmers due to the expenses inherent in the fair trade certification, buying and selling, and marketing processes. This disadvantages the poorest producers in developing countries. Also, the fair trade system is not transparent with

BOX 5.3

Individual Action: Food Choice

According to Richard Opplenlander, author of the book *Food Choice and Sustainability*,[1] what we are choosing to eat is killing our planet. Our daily food choices increase the pace of climate change and are devastating our terrestrial, aquatic, and marine ecosystems. What we eat impacts our health, the workers who produce and process our foods, and the communities in which food is produced or production ceases to occur as a result of a changing food system. Thus, an individual action that can go a long way to promote the sustainability of our food system is choosing to purchase and consume sustainable foods.

The challenge is in making the correct determination of where our food comes from and how it was produced. The solutions offered in this chapter are a potential basis for food selection. The best way to know what you are eating is of course to grow it yourself or purchase it directly from farmers who can tell you how their food was produced. This is not always possible. Another way to make logical and sustainable food choices is to focus on labeling such as "organic," "naturally grown," "grain fed," "heirloom," and "antibiotic-free food." Certification programs such as organic, fair trade, and eco-friendly also can guide the consumer. Organizations such as the Food Alliance and Scientific Certification Systems, Inc., provide the food and agriculture industry with sustainability standards, evaluation tools, and a voluntary third-party certification program based on sustainability principles. Their websites offer a directory of certified suppliers. Organizations such as Seafood Watch offer a pocket card or smartphone application that allows the consumer to make informed decisions about seafood purchases based on ocean protection and environmental and human health.

There are other smartphone applications, such as GoodGuide, that allow one to scan the UPC code of a product and view ratings based on health, environmental impact including life cycle analysis, and social responsibility. Evaluations of hundreds of thousands of products are made by nutritionists, environmental experts, and sociologists. These ratings allow shoppers to make purchasing decisions based on sustainability criteria.

Another part of food choice is based on what *not* to purchase—for example, not purchasing meat or processed foods. Other decisions might be based on where not to make purchases. This can include chains that do not treat workers fairly or have negative impacts on local communities. Avoiding chains and large big-box discount stores are examples. This is complicated, and consumers must do their own research to make sound decisions. For example, many sustainable-minded consumers avoid stores such as Walmart. Ironically, however, this global chain is the largest seller of organic food products, so not shopping there may negatively impact organic food producers. More independently developed tools and information regarding sustainable food choice are needed to help guide those interested in green purchasing.

As food activist and author Michael Pollan suggests, the best choice that will allow one to eat more sustainably is to purchase actual ingredients and cook them. This type of diet is healthier for both the consumer and the environment, especially when intelligent choices are made in the selection of the ingredients. Related to this is the development of the Slow Food movement. This is a global movement focused on linking the pleasure of good food with a commitment to community and the environment. Started under the leadership of Carlo Petrini in the 1980s, the objectives of this global movement are to defend regional traditions and food biodiversity, build links between producers and consumers, and restore the pleasure of eating and a slower pace of life. Choosing the right ingredients and foods and then consuming them slowly can contribute to many aspects of food system sustainability.

1 Opplenlander, Richard. 2013. *Food Choice and Sustainability*. Minneapolis, MN: Langdon Street Press.

regard to the exact proportion of revenues that these farmers receive. Others argue that fair trade may help wealthier consumers feel good about themselves as they make these "ethical" purchases while doing little to alleviate global poverty. In response to these criticisms, alternatives to fair trade have emerged, such as the non-profit organization TechnoServe, which develops business solutions to poverty by linking people to information, capital, and markets.[30] The for-profit company Allegro goes beyond fair trade by paying well above the price to obtain quality product while returning 5% of its profits to charity and spending 85% in growers' communities.[31]

Permaculture

Cereal, oilseed, and legume crops such as wheat, rice, maize, and soy occupy 70% of global croplands and an equivalent proportion of human food calories. All of these are annual plants that must be replanted each year from seed. Their cultivation requires significant inputs and negatively influences water, soil, natural habitat, and global climate change. Permaculture offers an alternative to our dependence on annual crops through the development of perennial crops that are permanently in cultivation. By modeling the design of farms after natural prairies, it creates and maintains agriculturally productive ecosystems that have the diversity, stability, and resilience of natural ecosystems. The extensive root systems that develop in permaculture make the crops more competitive against weeds and more effective in nutrient and water uptake. Permaculture conserves soil, filters water, and promotes wildlife. With fewer inputs, including seed, fertilizer, and pesticide, and less use of mechanization in tilling the soil, permaculture is economically beneficial to farmers.

Permaculture can be approached by either planting annual crops into a matrix of perennial grasses or through polyculture. Polyculture attempts to imitate the diversity of natural ecosystems by growing together complementary annual crops in the same location throughout the year, thus mimicking the diversity of natural ecosystems. Although it requires more labor, it can reduce pest damage, replenish the soil, and support diverse wildlife in comparison to large monoculture farms. Another approach to permaculture is pasture cropping, which integrates grain and pasture production. It allows cereal crops to be planted directly into perennial native pastures for the benefit of both the pasture and the crop. Pasture cropping preserves the soil structure, builds biomass, and results in no loss of topsoil while maintaining fairly high crop yields by integrating grain and livestock production. It has also been shown to offer cost savings for farmers and high rates of carbon sequestration.

The goals of permaculture may be best realized through the development of new perennial crops. This has been the focus of the Land Institute in Salina, Kansas, led by the agricultural visionary Wes Jackson. The institute has focused on breeding perennial grains that do not currently exist either through domestication or through hybridization. Unfortunately, this has not been a priority for federally supported research. The National Science Foundation, the US Environmental Protection Agency, the US Department of Energy, and the US Department of Agriculture (USDA) are still focused on supporting annual crops such as maize for food and biofuels. Support of perennial crop breeding programs will be essential for the realization of regenerative permaculture.

Eating Less Meat

Given the global rise in meat consumption and the well-documented environmental and health implications associated with it, one sustainable solution would be to eat less of it. The case for this can best be made using a trophic or feeding-level efficiency argument. The metabolic conversion of plant biomass into animal biomass is inefficient, approximately 10%. Similarly, the conversion of meat into human biomass is approximately 10%. The added trophic level means that 5–10 times more plant material, typically corn or grain in the United States, is needed to feed people with meat than through direct plant consumption (see Figure 5.6). Eating less meat would mean that current crop production could sustain more people because fewer calories would be inefficiently used by cattle and other livestock. The problem is that as poverty is reduced with development, global meat consumption is on the rise. To support this, significant increases in crop production and conversion of land for that production will be needed to meet increased demands (see Figure 5.5).

A second argument is that eating less meat would significantly reduce energy and water consumption, pollution, and land degradation from unsustainable grazing practices. It would also reduce contributions to climate change through energy use and methane production by animals. While completely eliminating meat from our diets could be a powerful solution, we do not all have to become vegans to make a change. Simply de-emphasizing meat in our diet could improve the sustainability of agricultural production by increasing trophic efficiency, and it would reduce environmental impact while improving public health.

SUSTAINABLE PASTURE MANAGEMENT

There are other approaches that could make meat production more sustainable. Many of these are applications of more traditional grazing practice that is done in a way that meets the needs of grazing animals through improved soil and plant productivity. For example, mixed-crop livestock farming that is common in West Africa, which allows for animal wastes and crop residues to be used to enrich soils while animals graze, is now being applied elsewhere. These traditional techniques are most effectively applied when integrated with careful monitoring of soil condition, the use of sustainable pasture varieties, and the prevention of overgrazing. Overgrazing is best prevented through rotational grazing that maintains plants at peak growth rates.

In Colombia, where cattle occupy 80% of the agricultural area, a new approach called sustainable intensification is being encouraged. It has been found that combining grazing with tree cultivation and crops is more productive for each and increases and diversifies farmer income. It also restores degraded landscape, maintains a higher degree of ecosystem function, and may make farmland more resilient to climate change. Another sustainable approach is to redirect animal grazing from prime agricultural land to more marginal sites such as roadsides, timber plantations, or unarable hillsides. This helps reduce the loss of prime land that could be used to grow valued plant crops.

BIOGAS PRODUCTION

A major by-product of meat production is animal waste. For example, one beef cow produces as much as 11 tons of manure per year. On large-scale meat production farms, this is part of the waste stream and contributes to environmental pollution. To improve the sustainability of meat production, proper ways to deal with or even use this waste are required. On small farms and in sustainable grazing systems described previously, animal wastes can be used to enrich soils for crop production. Another possibility is to use these wastes in the production of energy. The anaerobic digestion of animal wastes to generate methane-rich gas that can be used to run electrical generators, gas-powered lights, stoves, and heaters is not new. Some date the use of this "technology" as far back as the 17th and 18th centuries—yet another example of retrovation.

Today, small farm-based biogas facilities are very common, particularly in developing countries, but they are also extensively used throughout Europe. Globally, 6–8 million family-sized, low-technology digesters are used to provide biogas for cooking and lighting fuels with varying degrees of success. Biogas production is achieved through the collection of animal wastes, including those from beef, pork, poultry, and fish systems. The wastes are placed in a sealed chamber, lagoon, or bag that provides an oxygen-depleted environment. Methanogenic bacteria feed on the manure or other animal wastes and generate useable biogas that is typically 65% methane and 35% CO_2. Another benefit is that the digestion process results in an effluent of water and solids with a significant reduction in foul odors—a benefit not to be underestimated for those who live in close proximity to their animals (Figure 5.9).

As energy demands and costs increase, interest in larger, more sophisticated methane production facilities has grown. This includes more than 160 projects in the United States and even higher numbers in China and India. Energy production per volume of biogas is approximately 60% of that for natural gas, and its production is significantly cheaper than that

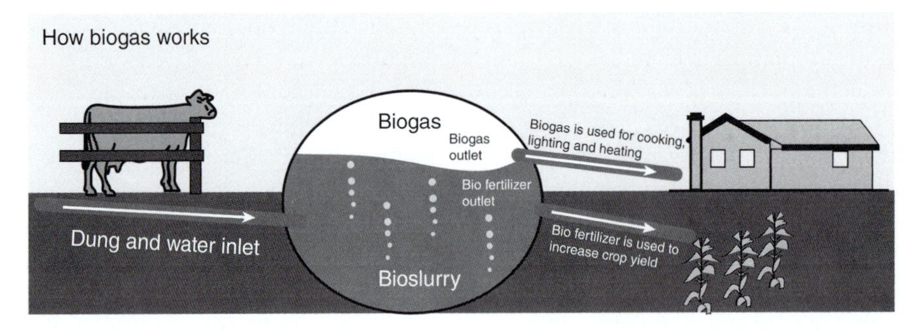

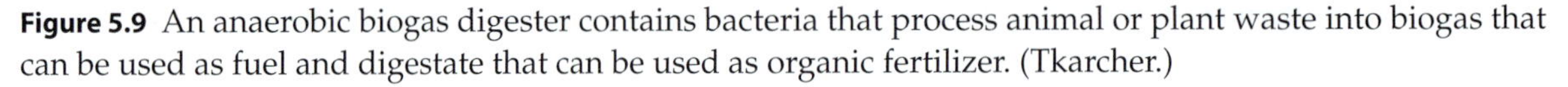
Figure 5.9 An anaerobic biogas digester contains bacteria that process animal or plant waste into biogas that can be used as fuel and digestate that can be used as organic fertilizer. (Tkarcher.)

of non-renewable, fossil fuel–based energies. One of the challenges will be to develop ways to clean biogas to meet the commercial standards of natural gas and how to deal with the remaining sludge and effluent. As meat production continues to rise, the use of biogas from animal wastes represents potential economic opportunity or savings for farmers, in addition to more effective waste management. Animal wastes in the absence of biogas production would eventually decompose and release CO_2 and methane into the atmosphere. Biogas generation is a way to capture energy from these decomposing materials. This reduces fossil fuel use and offers economic savings to farmers. In Chapter 6, we more fully examine and analyze the production, use, and sustainability of biofuels.

IN VITRO MEAT PRODUCTION

Another recent area of research is the development of meat culturing technologies or *in vitro* meat production through laboratory tissue-growth technology. This technology involves removing stem cells from animals and placing them in culture media in which they can proliferate and grow, independently from the animal. Theoretically, this process could efficiently meet global demand for meat without actually using animals, and it could have financial, health, animal welfare, and environmental advantages compared to traditional meat. In 2011, a study conducted by researchers from Oxford University and the University of Amsterdam showed that production of *in vitro* meat would require significantly less water, land, and energy compared to the conventional raising and slaughtering of animals. It also concluded that greenhouse gas emissions would be reduced by up to 96%.

Many have argued that laboratory production of meat is "not natural," but others have countered that it is no less natural than what it is intended to replace—industrial animal farming and meat processing. This is a new technology that is currently under development as part of a relatively new "Future of Food" movement.[32] As of 2012, 30 laboratories throughout the world were working on *in vitro* meat research. The limitations of in vitro meat are both technological and economic. In the short term, it is likely to remain economically impractical. In 2013, the first *in vitro* meat in the form of a beef burger was created with thousands of thin strips of cultured muscle tissue. The burger was cooked and then served at a news conference in London. Funded by an anonymous donor, the cost of this single burger was more than $300,000, indicating that laboratory-produced meat is not yet a sustainable solution.[33]

Hunting, Gathering, and Growing Your Own

One sustainable solution may be to return to our roots as hunters, gatherers, and growers. Legal, effectively managed hunting can provide individuals the opportunity to obtain healthy protein while sustaining game populations. Certainly, in the case of deer in North America, hunting is viewed as a way to manage explosive population growth that has strong negative impacts on forest health. The reason for this overpopulation is the loss of predatory keystone species such as wolves from these habitats. In most cases, due to human habitation near or in these ecosystems, reestablishment of predator populations is not feasible.

A consequence of deer overpopulation has been the decimation of forest understory plant populations. A deer exclusion study at the Smithsonian Conservation Biology Institute in Virginia has shown that the destruction of the understory habitat by deer has reduced nest sites for neotropical migratory birds and is severely impacting many species.[34] Also, hunted meat tends to be healthier for the consumer. Grain-fed, industrial meats have an abundance of intramuscular saturated fat referred to as marbling, and they lack the healthier long-chain fatty acids that are found in greater abundance in wild meat.

Because of a shared goal of habitat preservation, surprising alliances between pro-hunting agencies such as Ducks Unlimited and the National Rifle Association and local and national land conservancies have been formed. This diverse stakeholder approach to land preservation has often been successful. Critics of hunting often make animal cruelty and human safety arguments. They propose the alternative of managing populations through trap, neuter, and return programs. However, it is not clear that hunting is any less humane than the way that animals are typically slaughtered. Effective management and enforcement is essential for sustaining game resources, and it is not always feasible in poorer regions where poaching may occur. Increasingly, hunters have come to the table as stakeholders not only in wildlife conservation but also in providing food for the hungry (Box 5.4).

In many countries, illegal hunting or poaching is a serious problem and a threat to wildlife. Banning hunting is not successful when the economic

BOX 5.4

Stakeholders and Collaborators: Hunters, Conservationists, Farmers, and the Homeless

The term "hunting" holds numerous negative connotations. Many in the sustainability and conservation movements claim that hunting is immoral and an act of animal cruelty that is in direct contradiction to the goals of sustainability. However, increasingly, hunters have been considered major stakeholders who have played a role in wildlife conservation, farmland preservation, and even feeding the homeless. In the United States, hunters have played a significant role in wildlife conservation by providing financial contributions to wildlife and nature preservation efforts.

In addition to helping preserve native habitat, hunters often help manage it through effective removal of over-abundant grazers that are decimating understory ecological communities (Figure 5.4.1). Waterfowl hunters have been collaborating with farmers and paying them to set aside feed plots for ducks and geese near waterways that they inhabit. The hunters work in close collaboration with the farmers in making decisions about where and what should be planted in these feed plots. This effort has provided economic benefit to farmers and has helped preserve farmland and neighboring wetland sites.

In addition to providing a healthy source of protein for themselves and their families, a growing number of American hunters are donating part of their bounty each year to people who need it the most—the poor and the homeless. Most states have "Hunters Feeding the Hungry" programs that provide opportunities for hunters to donate their game meat to local food banks and shelters. Now in most states, thousands of hunters participate each year, donating significant quantities of meat, and donations have been increasing each year. The example of hunters in the United States shows that real sustainable problem solving might require each of us to open our minds and come to the table to develop novel collaborations that will serve the common needs and interests of diverse stakeholders such as hunters, conservationists, farmers, and the hungry.

Figure 5.4.1 Hunters can play a significant role in conservation. *Photo by Greta Bergstressor, Muhlenberg College.*

standards of poachers are so low that they have no other options for feeding their families. For example, in Zambia, researchers found that one of the key factors that made people turn to poaching was that they could not survive by farming.[35] This was because of low crop yields and a lack of markets for their crops. As a result, many farmers have turned to illegal hunting in the wildlife-rich Luangwa Valley. Because these people have no viable livelihood alternative, bans on hunting have not been effective. The only real way to prevent wildlife and habitat loss is to deal with underlying causes—poverty and hunger. To this end, a collaborative effort between the Wildlife Conservation Society and Community Markets for Conservation began training poachers in other income-generating skills, including beekeeping, carpentry, and sustainable farming methods using organic crops such as rice and peanuts. It also helps market and sell these new farm products as pro-conservation, organic goods. In return for this training and product marketing, participants have been required to surrender their guns and other equipment used for hunting. In addition to being able to produce food for themselves and their communities, this program generated $1 million in new income for more than 11,000 farmers in Zambia.[36] This is an example of a solution that is sustainable because it effectively integrates poverty reduction, food security, and wildlife conservation. It has been so successful that the approach is being applied elsewhere in Africa where rural development and wildlife conservation are at odds with each other.

Home gardening is an excellent way to provide affordable, nutritious foods in a sustainable manner. A National Garden Association study revealed that nearly 50% of people in the United States home garden. The study also revealed that $70 worth of plantings yielded $530 worth of produce, a profit or savings of more than 600%. Gardening offers other benefits, including increased physical and mental health. Gardening burns calories, and a poll by *Gardeners' World*, a UK publication, found that 80% of gardeners were satisfied with their lives versus only 67% of non-gardeners. Outreach to promote sustainable gardening techniques is essential to spread these benefits to poorer and urban communities. Community gardening and land-limited techniques such as container gardening can also serve these populations. The Rodale Institute, Feed the Future, and Kitchen Gardens International are examples of organizations that promote sustainable gardening worldwide. They work to achieve greater levels of food self-reliance through the promotion of kitchen gardening and small-scale farming, and they seek to expand the global community of people who grow their own food using best practices.

FORAGING

Although the bulk of human society transformed from hunter–gathers to an agrarian society, there are still cultures and people who forage for their own food. The !Kung of the Kalahari Desert in Botswana is one example of such a community that sustains itself in this manner. The men traditionally hunt, but the women provide 80% of the diet by collecting nuts, roots, and fruits. Elsewhere, people contribute to their diet through foraging. Examples include berry and fruit picking and foraging for fungi, mosses, and seaweed. Care must be taken because there is potential for poisoning, particularly from fungi. In the developed world, there has been a foraged food movement, with numerous gourmet restaurants featuring gathered "wild foods" such as lichens, fungi, pine needles, herbs, and berries emerging in cities throughout the world.

Food activists and authors such as Michael Pollan have made compelling arguments that the effects of not cooking run contrary to the goals of sustainability.[37] We have discussed that reliance on corporate-made, processed food has drastic health implications, weakens family and community relationships, and has negative consequences for the environment and social justice. Pollan argues that cooking is a way to regain control over one's food and is an act that individuals can take to help make the American food system healthier and more sustainable. Shifting consumer purchasing from processed foods to real ingredients can be done by focusing shopping along the perimeter of supermarkets, where these essential items are sold, and by making choices to purchase products that are in line with many of the solutions presented in this chapter (see Box 5.3).

Some have suggested that incentivizing the purchase of healthier ingredients over processed food may improve public health. A study by the Einstein Healthcare Network's Center for Urban Health Policy and Research found that cash rebates for the purchase of healthier foods resulted in a significant increase in the consumption of fresh foods and vegetables. However, it also found that when the rebate was lowered, produce purchases reverted back to almost baseline levels.[38]

Related to the purchase of ingredients instead of processed food is the growing home brew movement for the production of beer, meads, and wines. Here, ingredients are purchased rather than the processed product, allowing for creativity and freshness and resulting in a user-customized end product. In this way, beer is an excellent example of an **open-source** product—that is, one for which the technology or recipe is freely available. In many states, beer and wine are taxed at high rates. However, the ingredients are sold as food and are not taxed in those states where food is not taxed.

Obviously, home brewing remains in many indigenous populations and includes fermented beverages such as the mouth-fermented chicha of Peru and Ecuador; pakari, a fermented cassava drink from Guyana; and boza from Bulgaria. Most fermented products like these are believed to have originally been developed throughout the years by women in order to preserve a calorie source for times of scarcity and for unique tastes rather than for alcohol content.

New Food Discovery, Food Diversity, and Indigenous Crops

Thomas Jefferson stated that "the greatest service which can be rendered any country is to add a useful plant to its culture." As noted previously, we use only a small percentage of the more than 300,000 plants as a regular part of the human diet. The breadth of our diet can be increased through the discovery of new foods that have not yet been domesticated or by importing crops and foods from other regions and cultures. In the developed world, there have been a number of recent additions to our culinary repertoire, and there is potential for others, mostly from the developing world (Table 5.2). Because most of these are already foods in the country of origin, the term "new food" is slightly North-centric.

The question is whether or not adding new foods contributes to sustainability. Greater food diversity should increase resilience by spreading risk over a greater variety of crops. Because many of these foods are already cultivated in developing countries, there is potential for economic development through the expansion of the production of these current or potential crops. Also, many of these foods are highly nutritious and offer opportunities for increased public health. For example, berries from the açaí palm native to Central and South America, a traditional food in the Brazilian Amazon, are purportedly highly nutritious and have grown in popularity as a nutritional food or supplement among health-minded people. This has provided sustainable livelihoods for South American subsistence harvesters. As a renewable resource, berries are picked from trees without killing them, so their harvest does not threaten the Amazon forest. Another example, the highly nutritious, flavorful grain quinoa has been a staple crop in traditional communities in the Andes. It has entered the mainstream diet in the United States, improving nutrition for those who consume it and providing economic opportunity for indigenous populations who grow this crop.

There is legitimate concern that as the market demand for new foods increases, growers could succumb

TABLE 5.2 New Horticultural Crop Foods

COMMON NAME	SCIENTIFIC NAME	REGION OF ORIGIN	BENEFITS/USES
Açai	*Euterpe oleracea*	Central and South America	Protein rich; antioxidants
Quinoa	*Chenopodium quinoa*	Peruvian Andes	High protein, dietary fiber, and other nutrients; lacks gluten
Amaranth	*Amaranthus* spp.	Asia and the Americas	High protein, dietary fiber, and other nutrients; lacks gluten
Tamarillo	*Solanum betaceum*	Andes	Source of vitamins A and C
Naranjillo	*Solanum quitoense*	Ecuadorian Highlands	High in vitamins and nutrients
Camu-camu	*Myrciaria dubia*	Amazon rainforest	Very high vitamin C content
Freekah	*Tritium* spp. *(harvested Green)*	North Africa and Middle East	High-protein "super grain"
Guaraná	*Paullinia cupana*	South America	Stimulant
Elderberries	*Sambucus* spp.	North America	Food, flavoring, and medicinal uses
Jujubes	*Ziziphus jujuba*	Asia	Nutritious fruit

to the destructive agribusiness model of clear-cutting lands, sprawling plantations, and liberal application of pesticides and fertilizer. For both açaí and quinoa, there is evidence that as demand and subsequently price for these crops have increased, the very people who have historically relied on these nutritious foods can no longer afford them. For example, in Bolivia, the rise in price of quinoa has resulted in decreased domestic consumption of this nutritionally important, traditional food.[39] Others have argued that overall the growth of quinoa has raised the standard of living in these communities. Quinoa production has been traditionally coupled with the rearing of llamas, which contributed compost to maintain soil quality. As the size of quinoa farms has increased with the demand, llama production has declined. This, coupled with increased mechanization, threatens the sustainable potential of the soil.

These problems can be dealt with through organic and fair trade certification, which has been the practice of a primary açaí exporter, the US company Sambazon. Indigenous food protection has to be a priority in new crop foods development as it directly relates to nutrition and the protection of cultural heritage and traditional ecological knowledge. This can best be achieved through a stakeholder approach to managing new foods development and indigenous agriculture and also through appropriate protection of the intellectual property of both new and traditional foods.

Reducing Food Waste

According to FAO, given current trends in food consumption and population growth, global food production must increase 60% by 2050 in order to meet the growing demands. Yet, approximately one-third of the food produced for human consumption, or 1.3 billion tons per year, is wasted globally. The amount of food waste is similar in developed and developing countries, but the causes and occurrences along the supply chain are different. In developing countries, most food losses happen after harvest and during processing, whereas in developed countries, the majority occurs at the point of sale and at the hands of the consumer.[40]

The generation of food waste has negative implications for food security, contributes to food scarcity and hunger, and results in greater environmental impact. As it decomposes, food waste typically generates methane, which is a potent greenhouse gas. The global contribution of greenhouse gases from decomposing food waste is estimated to be more than twice the emissions of US road transportation.[41]

Reducing food waste is an important component to achieving a more sustainable food system. Given the disparities in the cause of food wastage between developed and developing countries but not in the amount of waste produced, substantial but different solutions will be required for each. In developing countries, efforts should focus on improving harvest techniques and post-harvest storage. Processing techniques and improved transportation and delivery should also be emphasized. For example, the International Rice Research Institute developed a rice storage bag that enables farmers to safely store their harvest crop for 9–12 months. The bag blocks the flow of both oxygen and water vapor, allowing more effective storage of their product and thus preventing significant losses to rice farmers, who often have post-harvest losses of up to 15%, as well as loss of nutritional quality. Using greenhouse model solar dryers is a promising strategy to reduce post-harvest losses of mangoes, a crop that is important in reducing vitamin A deficiency in West Africa and, in turn, reducing child mortality.[42]

In developed countries, the improvements should be made where food is sold, prepared, and consumed. This will require both education and legislative approaches, which are often more challenging than the technical fixes at the harvest and post-harvest levels. Restaurants and institutional food services are a major source of food waste. This is because the amount of food prepared is based on predicted consumption, and the food cannot be kept for extended periods of time. Also, in cafeterias, consumers often take more food than they can consume. The latter has been approached through trayless dining programs and portion control that do not place limits on how much one can eat but require multiple trips for additional food.

In Portugal, the Menu Dose Certa Project allows participating restaurants to adopt best environmental, nutritional, and food stocks management practices, from the purchase of foodstuffs through the preparation of meals. The goal is to support restaurants in creating menus that generate notably less food waste. Participants are certified by the project and receive marketing benefits. Changing consumer habits is particularly challenging. Rescuing food waste at the consumer level is going to require significant education and outreach efforts. The French NGO France Nature

Environment initiated such a campaign, with the goal of teaching families to reduce their food waste by shopping with a list, adjusting serving size, and by teaching them food preservation techniques.

More research is needed on the environmental impact of global food wastage. FAO is in the process of producing the first global Food Wastage Footprint as a way to quantify the impact of the food grown but that then enters the waste stream on the environment and the economy, with a view to assist decision-making along the food supply chain. The problem of food waste production and the potential ways to deal with that waste are further considered in Chapters 6 and 8, respectively.

Genetic Modification and Sustainability

Recall that one of the arguments for the development and use of GMOs was to make food production more sustainable through higher yields, increased nutritional quality of food, and decreased dependency on pesticides. Pamela Ronald, a professor of plant pathology at the University of California at Davis, has argued that by increasing yield without increased need for chemical fertilizers and pesticides, the future of organic agriculture may actually benefit from the use of GMOs. She makes the case that feeding the world in a sustainable manner may require what seems like an unlikely marriage between organic farming and beneficially engineered crops.[43]

The majority of GMOs have not increased yield and have not decreased chemical use but actually increased it, nor have they solved the global hunger problem. This is because they were not developed with these goals in mind. Rather, the GMOs on the market today were developed to advantage the biotech corporations that created them through their use on large-scale, capital-intensive industrialized farms. These corporations have not been interested in meeting the needs of the poorest people of the world because they do not see significant income potential or return on research and development investment from doing so. Thus, where increases in yield, disease resistance, and increased nutrition are needed the most, little or no benefits from GMOs have been realized. Also, the current way that GMO crops are developed gives absolute control of the intellectual property of the food crops on which we depend to these corporations. Innovation, modification, safety testing, or research of any kind on these GMOs by outside scientists is prohibited. Corporate control has also limited the application of GMOs in addressing global food and nutrition issues.

OPEN SOURCE

How can we take advantage of the potential benefits of GMOs to increase food sustainability as Pam Arnold suggests? One way is to bypass and perhaps undermine corporate control through the development of **open-source GMOs**. Like open-source software, such as Linux and Firefox, open-source GMOs would be shared and made freely available for users to make improvements. Many believe that open-source genetics would advance GMO research by shifting it away from intellectually controlled and restrictive corporations to universities, government agencies, and non-profits. This could allow for a realignment of the goals of GMO crop development to be more consistent with the goals of sustainability.

There are a number of examples of how open-source GMOs and collaborative partnerships contribute to sustainability. One is the non-profit initiative Public Intellectual Property Resource for Agriculture, which brings together intellectual property from more than 40 universities, public agencies, and non-profit institutes and makes these technologies available to developing countries throughout the world for humanitarian purposes. This type of collaboration has essentially saved the papaya through genetic modification. Papaya is a widespread, economically important crop in most areas of the world, including Brazil, Mexico, and India, where crops have been devastated by the ring spot virus. A team of public researchers from the University of Hawaii, Cornell University, and the USDA, independent of private multinational companies, created genetically resistant varieties of papaya using GM technologies. It took more than 15 years to achieve this goal, but when finally approved in 1998, disease-resistant GM papaya seeds were distributed for free to the farmers who had suffered significant crop loss and economic hardship as a result of the virus, and papaya production has rebounded.

Another example of how open-source GMO technology through public–private partnerships can alleviate global food and nutrition problems is demonstrated by the Golden Rice Initiative. This initiative was aimed at developing a rice variety rich in beta-carotene to help make up for vitamin A deficiency in many developing countries. This sort of micronutrient deficiency is referred to as "hidden hunger" because it can occur even when caloric intake is sufficient, but the diseases and physical disorders related to such deficiency represent a significant socioeconomic public health problem in developing countries. Research on golden rice started in 1982 as an initiative of the Rockefeller Foundation, eventually becoming a collaborative effort with the private multinational company Syngenta. Golden rice will eventually be shared in the public domain, spreading to several developing countries. Syngenta's only commercial interest is potential revenues from developed countries.

Another fruitful way to increase innovation and access in the development of GMOs is the development of open-source biotech methods. These processes are not patent restricted, so any company or researcher can use them to develop more sustainably meaningful GMOs. An example of this is a novel method for making transgenic plants called Transbacter, created by the Australian biotech company Cambia Technologies. The company offers a licensing agreement for this technology that can be used to make any number of modifications with great potential for innovation.

Ironically, one of the major obstacles to this approach of aligning GMOs with the goals of sustainability has been the efforts of well-intentioned, but unscientific, anti-GMO activists. Another is the long and expensive regulatory process that makes the approval of GMOs in the United States open only to the largest corporations, which in turn have eliminated any liability for their products. Open source, private–public partnerships, enforced corporate responsibility, and increased ability to conduct research on the safety of GMO foods may better permit us to meet the nutritional needs of current and future generations.

BANNING AND LABELING

More than 25 countries have totally or partially banned GMOs, and more than 60 countries require that GMO foods or products with GMO ingredients be labeled as such. These bans and labeling requirements span the developed and developing world. The United States is notably absent from these lists. Because commodity crops such as corn, rice, and soy are sold and transported as mixtures from many producers, any bulk purchase of them is likely to contain some GMO product. As such, the bans have in some cases hindered the potential for export of US corn and rice. This situation, along with consumer demand, has motivated large food producers such as Kellogg's and General Mills to develop and market a GMO-free line of products; they have even begun to do so for products sold in the United States, given market pressures. The demand for labeling of GMOs begs the question of whether consumers have the right to know exactly what they are consuming. All of our food has in some way been genetically modified, whether through molecular genetic techniques or through conventional breeding, domestication, and artificial and natural selection. This suggests that the GMO label may not be very informative. We are not aware of the ingredients of all-natural products such as fruits and vegetables, and most have not been studied for potential health effects. If the ingredients for all-natural, non-GMO blueberries were listed, consumers might be alarmed by the natural occurrence of such chemicals as methylbutyrate, benzaldehyde, and butylated hydroxyl toluene.[44] If coffee were discovered today or manufactured rather than grown, it would not likely pass FDA screening given its richness in physiologically active compounds.

Given that an important component of a sustainable food system is transparency, labeling products as containing GMO or GMO-free ingredients does make sense, and more than 90% of Americans, independent of political affiliation, are pro-labeling. The question remains as to what is the most useful information for the consumer. Knowledge about pesticide residues, plant pathogen DNA and RNA, naturally occurring plant compounds, and GMO end products such as *Bt* protein may all be important. The mere labeling of foods as GMO-free really does not provide useful information, and often it does not move beyond mere marketing. For example, GMO-free applesauce and olive oil can be found in many supermarkets, but there are neither GMO apples nor GMO olives. This is akin to promoting gluten-free water.

To move toward a sustainable food system, we need to consider consumer safety and consumers' right to know what is in their food through effective labeling. We should continue to address the full economic, social, health, and environmental implications of all foods, whether they are organically or conventionally grown, produced on a small or a large farm, locally or globally grown, and with or without GMOs.

There has been a concerted effort by corporations to control the intellectual property rights of food and to expand their markets for their conventionally bred and GMO seeds into global, rural communities. As part of this effort, they have encouraged laws that make it illegal for farmers to do what they have been doing for thousands of years—harvest seeds from their crops and save and share them for future use. An example of such a law was passed in Iraq after the war that started with the 2003 invasion by a US-led coalition. The war led to the loss of the nation's largest seed bank. As part of the economic restructuring program, Iraqis were required to destroy any remaining seeds they had saved and repurchase seeds from an authorized US supplier. They were also prohibited from saving seeds from future harvests, a practice followed by 97% of farmers prior to the war. Elsewhere, farmers are encouraged or required to purchase GMO seeds, including those with terminator technology, as part of international aid programs and through corporate-influenced government policies.

The threat of laws and contingencies associated with international development programs that require the purchase of seeds from corporate sources and thereby allow corporations to control the intellectual property rights of food has led to the development of seed saving programs. As part of what is referred to as the "seed freedom movement," these programs encourage rural farmers to harvest seeds from their crops and contribute them to a community seed bank for storage and cataloguing. In subsequent years, the farmers can return to the seed bank and make a withdrawal as needed. One of the goals of seed saving is to keep track of regionally specific, highly specialized species. These programs also protect farmers who have a poor growing season in a prior year and as a result lack seed.

One example of a successful seed saving program is Navdanya, led by the food freedom activist Vandana Shiva. A network of seed keepers and organic producers spread across 17 states in India, Navdanya has helped set up more than 100 community seed banks throughout the country and trained more than 500,000 farmers in seed sovereignty and sustainable agriculture during the past two decades. Through its learning centers, Navdanya is actively involved in the rejuvenation of indigenous knowledge and culture. It also educates about the issues surrounding GMOs and defends indigenous intellectual property rights and control over food with the aim of protecting biological and cultural diversity.

Seed saving is even catching on in the United States, where there are more than 300 seed exchange centers throughout the country, often partnering with local libraries. At these "seed libraries," patrons can take seeds at the beginning of the season and then save and return the seeds for future users. This has been in response to renewed interest in local food production and heirloom and open-pollinated varieties. This, too, has been met with government restriction. Numerous states consider seed sharing without a permit an illegal act for which one can be fined, and thus they have strictly regulated it.

For example, the library system in Mechanicsburg, Pennsylvania, was shut down by the state Department of Agriculture over concerns that shared seeds could be poisonous, incorrectly labeled, or of low quality. Many states require that informal seed exchanges have similar testing criteria for quality and germination rate as those of commercial providers. A sustainable food system will need to address government and corporate control of seed saving and further protect the rights of growers.

Closing the Gender Divide

One way to improve food security is to invest in women farmers and improve their access to child care, particularly in developing countries. In Africa, women comprise more than half of the continent's farmers. But due to a variety of constraints, the average productivity on plots managed by women is significantly lower than that on plots managed by men. FAO recommends ways to increase the success of women as farmers, including offering farmer training and strengthening landowner rights for women. Also, providing child care facilities and supporting women's access to markets and high-quality seeds and fertilizers could increase productivity. An FAO report determined that if women had the same access to resources as men under similar conditions, their yields could increase up to 30%. This could result in up to 150 million fewer people going hungry out of the current 842 million people who currently do not have sufficient access to food—not an insignificant decrease.[45]

Other global efforts to increase sustainable food production and decrease hunger are and should be women-focused. The previously discussed Navdanya seed saving movement is also women-centered in recognition of them as nurturers of both children and the Earth. As new farmers are being trained in the United States and other developed nations, participation by women has risen. This has resulted in greater economic equity and increased food production. Women play a vital role in advancing agricultural development and food security, but unfortunately, their contributions to agricultural production often go unrecognized. Organizations such as Feed the Future are well aware of this and have made supporting women in agriculture central to their mission of improving global food security.

Adapting to Climate Change

As our climate changes, we will have to adapt our food systems to ensure its resiliency and future food security. Climate change-related increases in drought, severe weather, and sea level changes are impacting global agricultural systems. At particular risk are the landlocked countries of Asia and the numerous small Pacific islands where food security is threatened by climate change as inundation from rising seawater occurs. The people of these countries are disproportionately affected by changing weather patterns.

Crops that take a longer time from planting to harvest are also at risk. For example, taro crop, a main staple food in Micronesia, requires five years of growth before harvest. As a result, large-scale climate-induced losses could result in five years of starvation. There are also economic impacts. The US has not been immune to this. In California, the largest domestic producer of fruit and vegetables in the country, the 2014 drought cost nearly $5 billion. Whether domestic or global, agricultural loss not only impacts food security but also impacts the economic well-being of families that depend on agriculture for their livelihood.

Accurate predictions of climate impacts on agriculture are challenging. Crops are known to respond in complex ways to variations in temperature and precipitation. Some models predict that increases in atmospheric CO_2 will actually increase yield because this is the source of carbon in plant productivity. However, recent CO_2 enrichment studies suggest that the positive effect may be much less because of other limiting factors on yield, such as water and nutrients.[46] Changes in pest, disease, and weed dynamics also must be considered, but the way each of these interacts with crop plants is complex. More research using a systems approach is needed, but it is already clear that to maintain food security, we must adapt to climate change now. Perhaps limited research dollars should be focused on such approaches to adaptation.

It has always been part of the farming tradition to adjust to changing conditions, in terms of both crop selection and farming practice. Farmers must respond to changing seasonal patterns by adapting their planting, harvesting, fertilizing, crop rotation, and pest management schedules. They will also need to employ other farm practices, such as integrating canopy management with fertilizer application to optimize photosynthetic capacity. Crop diversification and the development of new varieties should also be a priority. For example, food scientists at Cornell University have produced a strain of broccoli that thrives in hot environments. The Interspecific Hybridization Project of the Africa Rice Center has developed improved drought-resistant varieties. Better farm financial and risk management strategies need to be developed and implemented. Livelihood diversification on the farm and beyond should be central to these strategies. This has already been seen in sub-Saharan Africa, where small farmers have begun to increase their non-agricultural income by as much as 40%.

Adaptation can occur incrementally as described previously, but the drastic changes that are occurring may require more than minor adjustments or incremental change. These changes represent more transformational adaptation at the regional level that includes changes in social–ecological systems. In Bangladesh, more significant adaption is already occurring where the rising sea level has resulted in the salinization of rice fields, the primary contributor to the Bangladeshi diet. One response has been to develop salt-tolerant, high-yielding rice varieties, but more important, they are approaching food security through diversification. This has included the development of aquaculture in small ponds located in the rice fields spoiled by rising tides. Fish production, primarily tilapia, has grown exponentially since the mid-1980s, and reaching 225,000 metric tons in 2017.[47] Bangladesh's growing aquaculture sector has been lauded as an innovative response to changing climate, but there are concerns about environmental impacts associated with this new industry.[48]

Transformational change beyond the sort that has occurred in Bangladesh will require changes in agricultural policies. In the United States, this must include a departure from commodities-based subsidies to a more integrative approach that incorporates environmental, water, and health policy. Policy must also account for and address the actual costs of food production, ecosystems services, and increased risk with climate change for all farmers (see Box 5.1).

Assessing Food Systems Sustainability

As with any assessment of sustainability, the development of indicators must first consider scope and scale. In terms of scope, one should start with the three foci of sustainability: environmental, economic, and social components. Food security incorporates all three. Assessment of environment should include both negative impacts and management approaches that influence sustained yield for subsequent generations. Scale could range from the household to a community or region up through national and global scales. Comparative, whole-systems approaches of assessment such as organic versus conventional would also be beneficial.

To identify indicators to assess the sustainability of particular solutions, specific goals and strategies should be established. The indicators can then be selected to measure the progress of each of these. The process of developing indicators should be inclusive and transparent by including input from stakeholders with different perspectives. Indicators must be measurable, relevant, and promote learning and effective decision-making. In selecting indicators, one limitation that often emerges is a gap between what we choose to measure and available data. But this gap has the potential to serve as a basis for prioritizing research needs.

There are many examples of assessments that have developed measurable indicators of sustainability in food systems. The Vivid Picture Project designed a set of indicators to measure food system progress toward sustainability in the state of California. Reflecting 22 different goals, its indicators were designed to measure a diversity of sustainability-related components. Examples include the availability of affordable healthy foods, the quality of life of food and farming workers, business ownership opportunities for workers and communities, and farmland and natural habitat preservation.[49]

The Wallace Center, in conjunction with the Food and Society Initiative of the W. K. Kellogg Foundation, developed a similar set of indicators but for application at the US national level. The Global Food Security Index was designed by engaging a panel of experts from the academic, non-profit, and government sectors to develop food security indicators that could be used to assess food affordability, availability, and quality across a set of 107 countries. The goal of this project is to determine the risks to food security that can then be used to inform policy development and the private sector investment.[50]

Overcoming Barriers

Assessment of current programs or methods will shed light on both barriers to sustainability and recipes for success on how to overcome them. One finding that has emerged from assessment of problem solving for food systems is that there is a tendency to argue for one or a few sustainable solutions rather than examine a larger suite of objectives. For example, some local food advocates have preached that the only or primary way to achieve sustainability is by transitioning to locally based food systems. Others have argued that GMOs should be completely banned and could never contribute to a sustainable food system. However, these simple arguments are often flawed. Currently, we do not have the land resources to exclusively produce local foods near large urban centers, and seasonality is an issue. There are examples and ways in which GMOs have and could contribute to food security and increased nutrition, and with the changing climate we may have no choice but to develop new crop strains using whatever technology is available and can work within the context of sustainability.

Rather than dismissing particular approaches, we should consider how they can be made more sustainable within the context of how they are pursued by social actors independent of a particular approach and how they measure sustainable and socially just outcomes of that approach. Not all small organic farms are sustainable, especially in terms of their calorie contribution to the global food supply. Moreover, not all large farms are necessarily unsustainable. For example, a close examination of the soy megafarm in Mato Grosso, Brazil, shows that indigenous soils there are very effective at retaining applied fertilizers, an important aspect of farm sustainability.[51] Moreover, since its inception, deforestation to accommodate smaller farms has been halted, slowing the overall rate of habitat destruction.

Sustainability advocates are often narrowly focused on specific causes rather than an integrative, comprehensive approach, and this can be a real barrier to sustainability. This barrier can be overcome by moving beyond this oversimplified approach and examining aspects of the food systems in concert with each other. Instead of dismissing particular approaches, we should be asking how we can make them more sustainable. How can industrial agriculture be made more sustainable? How can new technologies be wed with traditional modes or "retrovations" in food production? Can the processed food industry be made healthier and more environmentally friendly? How can locally produced food be made affordable and accessible? Can local food production be integrated with larger food networks and producers? How can organic farming, GMOs, meat production, food waste reduction strategies, consumer actions, and health be improved? These kinds of questions should be driving our pursuit of food sustainability. There is a global food and nutrition crisis, and this will only worsen as the climate continues to change and populations continue to grow with unfettered, disproportionate rates of consumption. We need to develop an arsenal of sustainable solutions. Then we need to integrate them to meet the needs of people and to sustain the environment in a way that will allow it to continue to provide abundant healthy food for future generations.

Some elements that contribute to success have emerged in this chapter, including the use of systems thinking, nature-based solutions, and vertical integration. Transparent, private–public partnerships that use open-source approaches have made practices once regarded as unsustainable more effective at increasing accessibility to healthy food while protecting the environment. So will new food and farm policies that de-emphasize protecting the interests of a few large corporations and instead are aligned with sustainability objectives.

Chapter Summary

- The primary manner in which the majority of humans sustain themselves with food has evolved from hunting and gathering to small-scale farming through to the Green Revolution of the mid-1900s that gave us our industrialized food system of today.
- There are a number of problems associated with the Green Revolution, including unequal access, factory farming, the use of toxic chemicals, habitat loss, and decreasing crop genetic diversity.
- There are also numerous environmental, health, and social problems associated with industrial agriculture, meat consumption, depletion of fisheries, the use of genetically modified organisms, and the intensity of fossil fuel use in relation to global climate change.
- Possible sustainable solutions include organic agriculture, hydroponics, local agriculture, fair trade, permaculture, and sustainable approaches to meat production.
- Hunting and home gardens, new food discovery, seed saving, investing in women farmers, and reducing food waste are also ways to improve upon the sustainability of our food systems.
- Genetically modified foods that do not improve yield or the nutritional content of food, do not decrease toxic chemical use, result in greater environmental degradation, and concentrate the power of our food security among a few large biotech corporations make our food system less sustainable.
- Open-source GMO technology could increase the ability to conduct research on the safety of GMO foods and may better permit us to meet the nutritional needs of current and future generations.
- As the climate changes, our food system will have to adapt to changing conditions, including increased drought and rising sea levels. Doing so will depend on applying new knowledge as well as learning from existing and traditional systems and how they respond to change.
- There is no single solution to making our food system sustainable but, rather, a suite of solutions must be applied to meet current and future nutritional needs.
- The sustainability of any of the proposed solutions must be assessed by first determining the scope and scale of what is being assessed and by identifying effective indicators. Rather than dismissing individual approaches, we should consider how each could be made more sustainable and used in combination with other approaches and then incorporate elements that we have determined lead to successful solutions.

Digging Deeper: Questions and Issues for Further Thought

1. What is the connection between soil and human health? What are some approaches to agriculture that could improve both?
2. What are the advantages and disadvantages of GMOs? What approaches would allow this technology to best contribute to sustainability objectives?
3. Draw a diagram that includes all the elements of a modern irrigation system linked by arrows that indicate how the elements influence each other. Using systems thinking, modify the diagram in a way that could make it more sustainable.
4. Currently, we produce enough calories of food to provide for the needs of our population, yet hunger and malnutrition remain global problems. Why is that? What are ways that these problems could be alleviated now and in the future?
5. What are the various sustainability-related arguments for eating less meat? How can the production and consumption of meat be more sustainable?
6. Draw a diagram that shows the connections among food production, water, energy, and climate change. Then integrate solutions into your diagram that increase sustainability.

Reaping What You Sow: Opportunities for Action

1. Whether at home, work, or school, develop a way to assess the amount of food wasted on a regular basis. On average, how much is being wasted? Develop and implement a plan to either reduce or sustainably manage food waste.
2. Visit a local food market. With a budget of $50, create with a "sustainable menu" for five dinners that will each feed four adults. Discuss the considerations that went into the design of your menus and why your food choices are more sustainable than other options.
3. Develop a campaign or activities that will both educate about and increase access to local foods in your community.
4. Visit https://fallingfruit.org to better inform yourself about fruit maps. Develop one fruit map for your local community, and make it publicly available.

References and Further Reading

1. Arnold, J. E. 1996. "The Archaeology of Complex Hunter–Gatherers." *Journal of Archaeological Method and Theory* 3 (1): 77–126.
2. Damania, A. B. 2012. "Early Domestication in South Asia Based on Archaeobotanical Evidence: I. Crop Plants." *Asian Agri-History* 16 (1): 3–20.
3. Selman, M. 2007. *Eutrophication: An Overview of Status, Trends, Policies, and Strategies.* Washington, DC: World Resources Institute.
4. Alavanja, M. C. R. 2009. "Pesticides Use and Exposure Extensive Worldwide." *Reviews on Environmental Health* 24 (4): 303–9.
5. Alavanja, "Pesticides Use and Exposure Extensive Worldwide."
6. Food and Agriculture Organization, electronic files and web site. http://www.fao.org/faostat/en/
7. Thrupp, L. A. 2000. "Linking Agricultural Biodiversity and Food Security: The Valuable Role of Agrobiodiversity for Sustainable Agriculture." *International Affairs* 76 (2): 283–97.
8. Langley, L. D. 2001. *The Banana Wars: United States Intervention in the Caribbean, 1898–1934.* Lanham, MD: Rowman & Littlefield.
9. Norberg-Hodge, H., T. Merrifield, and S. Gorelick. 2002. *Bringing the Food Economy Home: Local Alternatives to Global Agribusiness.* Sterling, VA: Kumarian Press.
10. Norberg-Hodge et al., *Bringing the Food Economy Home.*
11. Norberg-Hodge et al., *Bringing the Food Economy Home.*
12. Mathew, A. G., R. Cissell, and S. Liamthong. 2007. "Antibiotic Resistance in Bacteria Associated with Food Animals: A United States Perspective of Livestock Production." *Foodborne Pathogens and Disease* 4 (2): 115–33.
13. Bøhn, T., and M. Cuhra. 2014. "How "Extreme Levels" of Roundup in Food Became the Industry Norm." Independent Science News. https://www.independentsciencenews.org/news/

how-extreme-levels-of-roundup-in-food-became-the-industry-norm

14. Losey, J. E., L. S. Rayor, and M. E. Carter. 1999. "Transgenic Pollen Harms Monarch Larvae." *Nature* 399: 214.
15. Pimentel, D., and M. H. Pimentel, eds. 2008. *Food, Energy, and Society*. 3rd ed. Boca Raton, FL: CRC Press.
16. Dunikowska, M., and L. Turko. 2011. "Fritz Haber: The Damned Scientist." *Angewandte Chemie International Edition* 50 (43): 10050–62.
17. Smith-Spangler, C., M. Brandeau, G. Hunter, et al. 2012. "Are Organic Foods Safer or Healthier Than Conventional Alternatives? A Systematic Review." *Annals of Internal Medicine* 157: 348–66.
18. Lu, C., K. Toepel, R. Irish, et al. 2006. "Organic Diets Significantly Lower Children's Dietary Exposure to Organophosphorus Pesticides." *Environmental Health Perspectives* 114**:** 260–63.
19. The Rodale Institute. 2011. "Farming System Trial: 30-Year Report." Kutztown, PA: The Rodale Institute.
20. Eagle, A. J., L. P. Olander, L. R. Henry, et al. 2011. *Greenhouse Gas Mitigation Potential of Agricultural Land Management in the United States: A Synthesis of the Literature*. 2nd ed. Durham, NC: Nicholas Institute for Environmental Policy Solutions.
21. The Rodale Institute, "Farming System Trial."
22. C. Badgley, I. Perfecto, et al. 2007. "Can Organic Agriculture Feed the World?" *Renewable Agriculture and Food Systems* 22: 80–6.
23. Badgley et al., "Can Organic Agriculture Feed the World?"
24. Reganold, J. P. 1995. "Soil Quality and Profitability of Biodynamic and Conventional Farming Systems: A Review." *American Journal of Alternative Agriculture* 10: 36–45.
25. Sardare, M. M. D., and M. S. V. Admane. 2013. "A Review on Plant Without Soil-Hydroponics. *International Journal of Research in Engineering and Technology* 2: 299–305.
26. "New Shoppers® Food & Pharmacy 2014 Food Trends Forecast: More Local, Healthier Foods Top List of Predictions." MarketWatch. Accessed January 29, 2019. https://www.marketwatch.com/press-release/new-shoppers-food-pharmacy-2014-food-trends-forecast-more-local-healthier-foods-top-list-of-predictions-2014-02-03.
27. Korman, P., and M. Christie. 2014. "Outlook 2014: CISA Underscores Farming and Food Businesses as Drivers of Western Mass. Economy." *The Republican*, February 9.
28. Pimentel and Pimentel, *Food, Energy, and Society*.
29. The Rodale Institute, "Farming System Trial."
30. Technoserve. n.d. "Business Solutions to Poverty." http://www.technoserve.org
31. Allegro Coffee. http://www.allegrocoffee.com
32. Future Food. n.d. "Meat Without Livestock." http://www.futurefood.org
33. Badgley et al., "Can Organic Agriculture Feed the World?"
34. McShea, William J. and Rappole, John H. 2000. Managing the abundance and diversity of breeding bird populations through manipulation of deer densities. *Conservation Biology (14):* 1160–70.
35. Read, P. K. 2014. "Fighting Poaching Means Fighting Poverty—Conservation Farming in Zambia." Food Tank. http://foodtank.com/news/2014/03/fighting-poaching-means-fighting-poverty-conservation-farming-in-zambia
36. Read, "Fighting Poaching Means Fighting Poverty."
37. Pollan, M. 2013. *Cooked: A Natural History of Transformation*. New York: Penguin.
38. Phipps, E. J., L. E. Braitman, S. D. Stites, et al. 2015. "Impact of a Rewards-Based Incentive Program on Promoting Fruit and Vegetable Purchases." *American Journal of Public Health* 105: 166–72.
39. Romero, S., and S. Shahriari. 2011. "Quinoa's Global Success Creates Quandary in Bolivia." *New York Times*, March 29. http://www.nytimes.com/2011/03/20/world/americas/20bolivia.html
40. Food and Agriculture Organization. 2013. "Food Wastage Footprint: Impacts on Natural Resources." http://www.fao.org/docrep/018/i3347e/i3347e.pdf
41. Food and Agriculture Organization, "Food Wastage Footprint."
42. Food and Agriculture Organization, "Food Wastage Footprint."
43. Ronald, P. C., and R. W. Adamchak. 2008. *Tomorrow's Table: Organic Farming, Genetics, and the Future of Food*. New York: Oxford University Press.
44. Dvirosky, G. 2014. "What if Natural Products Came with a List of Ingredients?" http://io9.com/what-if-natural-products-came-with-a-list-of-ingredient-1503320184

45. Ford, L. 2014. "Mind the Gap: Closing Gender Divide in African Agriculture Could Reduce Hunger." *The Guardian.* http://www.theguardian.com/global-development/2014/mar/19/women-closing-gender-divide-african-farming-reduce-hunger
46. Kirschbaum, M. U. F. 2011. "Does Enhanced Photosynthesis Enhance Growth? Lessons Learned from CO_2 Enrichment Studies." *Plant Physiology* 155: 117–24.
47. Self-sufficient in fish, meat. (2018, February 11). Retrieved July 10, 2018, from https://www.thedailystar.net/frontpage/self-sufficient-fish-meat-1532953
48. Glass, J. 2013. "Amid Perfect Storm of Climate Challenges, Can Aquaculture Net Food Security Gains in Bangladesh?" *New Security Beat*, https://www.newsecuritybeat.org/2013/10/perfect-storm-climate-challenges-aquaculture-net-food-security-gains-bangladesh/
49. The Vivid Picture Project. 2005. *Proposed Indicators for Sustainable Food Systems.* Portland, OR: Ecotrust.
50. Wallace Center at Winrock International. n.d. "Charting Growth Report." http://www.wallacecenter.org/resourcelibrary/charting-growth-report.html
51. Porder, S. 2013. "Iowa in the Amazon." *New York Times*, November 24. http://www.nytimes.com/2013/11/25/opinion/iowa-in-the-amazon.html

CHAPTER 6

ENERGY: FROM FOSSIL FUELS TO A SUSTAINABLE FUTURE

Our economy is based on fossil fuels. Oil, coal, and natural gas fuel all aspects of our lives. As previously discussed, fossil fuels are directly linked to our food, water, and air. They are the sources of energy for our transportation, industry, and the power plants that generate the electricity that runs the tools and toys of our daily lives. They provide comfort as they heat and cool our homes. They are more than just our primary source of energy; fossil fuels are also incorporated into our plastics, technology, and most products that make our lives safer, cleaner, more comfortable, and convenient. If the flow of fossil fuels were to suddenly stop, our industrial economy would come to a grinding halt, as would the developing economies that feed our supply chains and provide our labor.

PLANTING A SEED: THINKING BEFORE YOU READ

1. What are the predominate sources of energy used to generate the electricity that is provided to your local community?
2. How can it possibly cost $2 or $3 for a gallon of bottled water that is extracted locally and the same for a gallon of gasoline that is made from oil extracted from the ground, transported from halfway around the world, chemically refined, delivered to a gas station, and includes 20–50 cents in taxes?
3. In addition to direct uses of energy such as transportation and electricity, what are other energy-intensive activities?
4. What are possible barriers to the transition from fossil fuels to 100% renewable energy?
5. Why is there often a disconnect between Americans who understand that climate change is real and human caused and their commitment to act in order to mitigate it?
6. Given that the majority of Americans understand that climate change is real and human caused, why do so many of our politicians who are elected to represent them deny this truth?

What the world needs is an energy revolution.
—Kandeh K. Yumkella, former United Nations Special Representative of the Secretary-General for Sustainable Energy

Is the fossil fuel economy sustainable? We know from Chapter 1 that when the rate of extraction of a resource exceeds its rate of renewal, that resource is considered non-renewable. Fossil fuels are so named because they are made of dead or fossilized microorganisms, plants, and animals that lived millions years ago in primordial swamps and oceans. Over time, these dead organisms slowly decomposed and formed coal, oil, or natural gas depending on the type of organic debris that was present and the conditions of temperature and pressure that existed as they decomposed. In slightly more than 200 years, we have depleted the majority of those fuels that took 50–300 million years to make. It is predicted that as our consumption of them continues to rise, nearly all of the economically recoverable fossil fuels will be gone within the next 100 years. Clearly, by our definition, fossil fuels are not a renewable resource.

There are other reasons why our dependence on fossil fuels is not consistent with our criteria for sustainability. In terms of the environment, they pollute our air, water, and soil as they are extracted, processed, and burned. Fossil fuels are the major contributor of greenhouse gases (GHGs) leading to global climate change. They also negatively impact human health. From economic and social perspectives, fossil fuels also contradict the conditions of sustainability. Although fossil fuels have driven industrialized economic growth, they have not created equitable economic opportunity. The fossil fuel industry has a history of exploiting workers who have labored under dangerous conditions often at reduced pay. Their impacts, such as pollution and climate change, disproportionately affect poor people.

Not only will fossil fuels be depleted and fail to meet the needs of future generations but also they already fail to meet present-day need. More than 1.2 billion people have no access to electricity and are essentially living in darkness. The need for fossil fuels has threatened international security and has been a major source of conflict. They limit the possibility of energy independence because many countries lack fossil fuel resources within their own boundaries and must import them.

As problematic as the fossil fuel economy is, there is good news. Alternatives already exist that are affordable and obtainable. Central to achieving a sustainable energy future is the transition from a fossil fuel–based economy to one that runs on renewable energies that will not run out, will have less impact on human health and the environment, and will have the potential to create a new, more inclusive economy.

In this chapter, we briefly examine the use of fossil fuels from the perspectives of the environment, economics, the human condition, international security, and the growing energy demand and the lack of access in the developing world. Then we discuss renewable alternatives to fossil fuels, focusing on innovations and technologies that will allow us to meet our growing energy needs in a more sustainable way, and we identify those that are the most sustainable. We then explore new technologies in energy distribution and management and also increased energy efficiency as vital parts of a sustainable energy system. We also examine ways to increase access to reliable energy in remote areas of developing countries and how transportation can be made more sustainable. Finally, we explore the barriers that are inhibiting the transition to a renewable-based economy and ways to remove them. We will see that a clean and sustainable energy future is obtainable, will stimulate economic growth, and can offer a bright future for all.

Fossil Fuels

Just because a resource exists on the planet does not mean it is immediately available for use by humans. We can view the amounts of fossil fuels that remain based on this availability. The total estimated amount of a fossil fuel that exists but may or may not be technically or economically feasible to extract is referred to simply as a **resource**. A **reserve** is the quantity of that resource that is technically and economically recoverable. Reserves can be classified as proven or unproven. **Proven reserves** are those that are recoverable under existing economic and political conditions using current technology with reasonable certainty (normally at least 90% confidence). **Unproven reserves** are potentially recoverable but less likely so due to technical, economic, or other limitations. The designation of a resource as a proven reserve is based on available extraction technology and its cost, in addition to the market price of the resource. Thus, the quantity of proven reserves will change both with technical innovation and with the prevailing market.

Another term used to address the availability of fossil fuels is the **reserve-to-production ratio (RPR)**.

This is the amount of a proven reserve in an area divided by the amount of resources used in one year. This ratio is used by companies and government agencies in forecasting the future number of years a resource will be available. RPR is also used to predict potential future income and whether more exploration must be undertaken to ensure continued supply of the resource.

In this section, we discuss how coal, oil, and gas are formed, extracted, and used. We examine how much of each remains and consider how new technologies may change this. We do so through the lens of our model of sustainability focusing on environment, economics, and social aspects. More details on the impacts of fossil fuel use on air and climate change, on water depletion and pollution, and in relation to agriculture can be found in prior chapters on these specific subjects.

Coal

Fossil fuels such as coal are formed from plants, animals, and single-celled organisms called protists that were alive and died hundreds of millions of years ago. Upon death, they accumulated, decomposed, and became buried under multiple layers of mud, rock, and sand. Over time, they became buried by thousands of meters of earth and in some cases covered by oceans. Under the pressure from this burial, the decomposed organic materials formed fossil fuels. Coal was specifically formed from the dead remains of trees, ferns, and other plants from swamps covered by salt or freshwater marshes. The coal that formed beneath the salt marshes tends to be sulfur rich and when burned releases more polluting sulfur oxides.

The largest coal deposits are in North America, where it is mined in 26 of the 50 states. There are also relatively large deposits in the former Soviet Union and China, with much smaller amounts distributed throughout the world. Currently, four types of coal are mined. The largest portion of the world's coal reserves is made up of lignite, a soft, brownish-black coal that is of the lowest quality in terms of energy yield. Next in quality is sub-bituminous coal, then bituminous coal, and finally, the hardest coal that produces the greatest amount of heat when it is burned—anthracite.

Coal is what fueled the Industrial Revolution. It was the first dominant worldwide supplier of energy fueling rail transportation, electricity generation, factories, and the production of steel. Coal is often mined from deep beneath the ground along layers or seams. This is a dangerous, labor-intensive process that has led to numerous coal mining disasters and deaths from suffocation, gas poisoning, mine collapse, and gas explosions. In the United States alone, more than 100,000 coal miners were killed in accidents in the 20th century, with more than 3,200 dying in 1907 alone. Coal miners also suffer many health problems, including chronic lung diseases.

When coal is found close to the Earth's surface, it can be extracted by removing the surface layers of materials to expose and remove the coal. Surface rock and soil above the coal are removed using heavy machinery to expose seams close to the surface, by removing material covering the seams in a pattern following the contours along a ridge or around a hillside, or through complete mountaintop removal. Mountaintop removal impacts the health of people who live near those sites. In the Appalachia region of the United States, rates of birth defects and cancer are greater among people who live closer to mountaintop removal sites. Surface mining is also extremely destructive to the environment in that it completely eliminates existing vegetation, destroys the soil and wildlife habitat, and permanently scars the landscape.

Historically, social justice issues have been associated with coal mining. In addition to exposure to safety and health hazards, miners were paid low wages for long hours of dangerous and difficult work. Some miners were further exploited as they were paid in scrip that could only be spent at overpriced company stores or to rent company-owned housing. Child labor was used to separate coal from slate for minimal pay. In the United States, where most coal mining was occurring, these conditions led to an active labor movement that resulted in improvements in wages and conditions, including the development of safety equipment. However, as wages grew in the United States, the mining companies turned to new mechanized technologies, which in turn reduced the labor-intensive nature of the work. This resulted in the loss of hundreds of thousands of jobs. In other countries, coal miners were and continue to be exploited as described previously.

In the United States, coal accounts for approximately 15% of total energy consumption (Figure 6.1), but this varies by location. The US Environmental Protection Agency's (EPA) Power Profiler[1] allows one to compare the fuel mix and air emissions rates of the electricity generated in a specific region compared to the national average. Globally, coal's share of the energy mix is nearly 30% and has been steadily rising since the 1970s, with China and India accounting for nearly 70% of the demand (Figures 6.1 and 6.2).

So how much coal remains? In 2012, there were nearly 900 billion tons of proven reserves of coal, which based on RPR will last more than 100 years at current rates of production and use. However, with renewed interest in increasing coal production and consumption, coal shortages will likely occur much sooner than projected by the RPR.[2] Cheaper sources of natural gas and oil through new methods of extraction, the emergence of renewable energy, and other geopolitical issues are making coal less economically viable, yet some politicians continue to argue that coal will fuel our economy and job growth.

The dwindling amount of economically available coal should not be driving the reduction or elimination of global use. Rather, coal's significant contribution to air and water pollution should do so. Emissions include nitrogen and sulfur oxides and mercury. Burning coal emits more than twice the amount of GHGs per unit energy generated than natural gas or oil, thus playing a major role in global climate change. Mining and burning coal both use large amounts of water and pollute it with heavy metals such as arsenic and lead. These negative environmental, health, and social impacts of mining and burning coal coupled with its declining abundance should have us searching for more sustainable forms of energy.

Oil

Crude oil or petroleum is also formed by the decomposition of prehistoric organisms, but unlike coal, oil

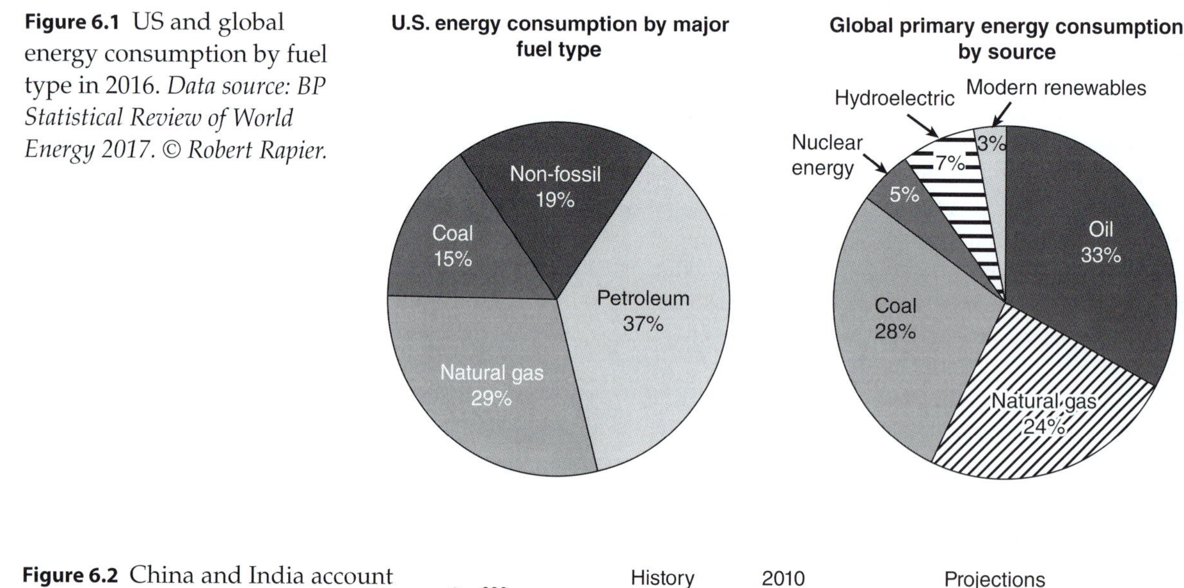

Figure 6.1 US and global energy consumption by fuel type in 2016. *Data source: BP Statistical Review of World Energy 2017. © Robert Rapier.*

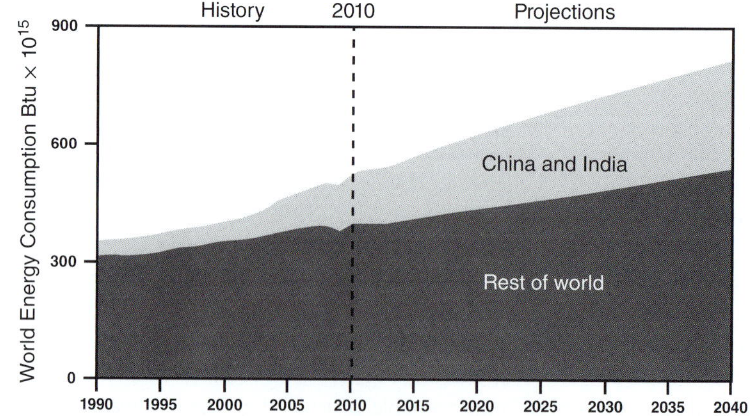

Figure 6.2 China and India account for half of the world's projected increase in energy consumption through 2040. (EIA: Energy Information Administration, International Energy Outlook.)

derives exclusively from marine and aquatic organisms that become trapped under sedimentary rock, thus forming shale. With additional sedimentation, pressure, combined with heat and bacterial activity, the organic material is converted into oil within the layers and pores of the shale. Further pressure caused the oil to flow from the shale and accumulate in large reservoirs of thicker porous rock trapped between layers of impermeable rock such as granite. In many cases, the marine environments in which the oil formed have receded, resulting in terrestrial sources of oil. Many reserves still remain beneath the ocean and require the use of offshore drilling rigs. There is a large amount of variation in the geographical distribution of oil reserves; the overwhelming majority of oil is found in the Middle East. Significantly lesser amounts occur throughout the rest of world, and many countries have essentially no available oil within their boundaries.

The conventional process of extracting oil is through wells that are drilled through the impermeable rock into the reservoir. Primary recovery relies on underground pressure and then pumps to drive crude oil to the surface. Primary recovery often taps only 10% of the oil in a deposit. Secondary recovery injects water that is separated from the oil in the initial phase of drilling back into the oil-bearing formation to bring more oil to the surface. In addition to boosting oil recovery, it also disposes of the waste water, returning it to its original location. This can bring an additional 20% of the oil to the surface. Enhanced recovery extracts even more oil by injecting steam, gases such as carbon dioxide (CO_2) or propane, or water containing soluble polymers into a secondary well. This enhances flow out of the primary well and can bring as much as 60% of the reserve to the surface.

Oil drilling can have many harmful ecological and environmental effects, primarily because it is toxic to almost all forms of life and can readily enter the environment during extraction and transportation. Entry into the environment can occur either through small undetected leaks or spills or through catastrophic oil spills such as in 1989 when the tanker Exxon Valdez spilled 10.8 million gallons of crude oil into Prince Edward Sound, Alaska, and the disastrous Deepwater Horizon oil spill in the Gulf of Mexico in 2010. The refinement of crude oil into gasoline is also a major source of water and air pollution. The use of seismic blasting in deep sea oil exploration is known to cause hearing loss in marine mammals, and it disrupts many of their essential behaviors, such as communication, feeding, and breeding. In one case, deep sea exploration by ExxonMobil near Madagascar resulted in more than 100 melon-headed whales beaching themselves.[3]

The extraction of oil is less labor intensive than that of coal, but there is substantial risk to injury, especially on offshore rigs. Although those involved in the oil extraction process tend to be paid well, there are equity and social justice–related issues associated with the extraction of oil that are quite different from those of coal. First, oil production and transportation have severely impacted the lives of indigenous people. For example, in the Ecuadorian Amazon, many of the indigenous tribes that once numbered in the thousands have been reduced to hundreds. This is a direct result of the dumping of more than 19 billion gallons of toxic waste water and the spilling of 6.8 million gallons of crude oil from a main pipeline into forests where these communities are located. According to the 1999 "Yana Curi Report,"[4] which details the impact of oil development on the health of the people of the Ecuadorian Amazon, oil exploration and extraction have resulted in higher rates of cancer and spontaneous abortion. Similar cases exist in Colombia, Nigeria, and other areas of the world. Also, polluting oil refineries are often placed in the poorest areas, representing a common form of environmental injustice. Finally, massive government subsidies to large oil corporations divert billions of dollars from impoverished social, health, and development programs.

Oil accounts for 37% of total energy consumption in the United States and 33% globally (see Figure 6.1). The United States is by far the largest consumer of oil, burning more than 20,000 barrels per day. That is nearly 4 times the next two highest consumers, China and Japan, and as much as 5–10 times the amount consumed by other countries. Since 1970, global oil consumption has steadily increased, and it is predicted to continue to do so even though it has leveled off in some industrialized countries and has actually declined in eastern Europe and Russia. This is because of the increase in demand from developing countries, especially China and India, which are projected to account for half of the world increase in energy consumption through 2040 (see Figure 6.2).

So how much oil remains? There is quite a bit of disagreement on this. Some believe **peak oil**, or the point of maximum production after which reserves begin to decline, has already occurred. Others, mostly oil companies and governments, state that it is decades away. Estimates of the RPR for oil based on current production and use range from 40 to 80 years. However, both use and production are increasing. There is an expected 2% annual rise in global oil demand, with a 3% annual decline in existing reserves. But this does not take into account new, unconventional methods of oil extraction, which are increasing the supply of oil. This includes extraction of oil from sands and shale and other techniques, most of which have greater environmental impact than more traditional sources of oil. Despite these new techniques, we will run out of oil before the next century, and if increases in consumption continue, this will likely happen sooner rather than later.

Because the majority of industrial technology and transportation is designed to be fueled by oil, we are globally dependent on it. However, the oil reserves are concentrated in certain areas, especially in the Middle East. Because these countries can control the flow and price of oil, they have disproportionate power. This power has been exerted in the form of oil embargos such as that by the Organization of Arab Petroleum Exporting Countries (OPEC) in the 1970s, which had many short-term and long-term effects on global politics and the global economy. It also had direct security implications for the United States and other industrialized nations.

Oil is a leading cause of conflict and war. Major wars have been fought over oil every decade since World War I, with smaller conflicts arising more frequently, and nearly half of wars since 1973 have been linked to oil. Wars have been fought to directly acquire another nation's oil resources. There can also be conflict over shipping lanes and international pipelines, oil-related terrorism, or conflict when oil is used to protect an aggressive leader. Wars in Iraq, Syria, Nigeria, South Sudan, Ukraine, and the East and South China seas have caused millions of people to either be displaced or slaughtered. No other resource or commodity has had as great an impact on international security. However, as discussed in Chapter 4, it is believed that rising conflicts over access to water may surpass those over oil as that resource becomes depleted.

So the argument for transitioning from an oil-based economy to renewable energy should be clear. Oil is not renewable, and it is rapidly becoming depleted. So-called market forces that keep oil relatively cheap are driven by large government subsidies and do not include externalities such as health and environmental impact including climate change. The issues of energy independence, international security, global conflict, and climate change should be motivating a much more rapid transition from oil to renewable energy.

Natural Gas

Natural gas is created during the formation of crude oil. In deeper underground rock, the increased temperature and pressure essentially cook or gasify the oil, resulting in the production of natural gas or methane. These gases then move upward into more porous rock and get trapped by impermeable layers in the same way that oil does so, sometimes with oil or in isolated pockets of pure gas. Natural gas can be found in a variety of different underground formations, including sandstone beds, coal seams, and shale formations. Natural gas is most commonly extracted by drilling a single vertical well into capped sandstone pockets. The gas readily flows upward either on its own or with simple pumping systems, and it can then be cleanly captured and transported. This method of conventional natural gas extraction has little environmental impact. Of all the fossil fuels, natural gas is the cleanest. Emissions are much lower than those of coal and oil, especially carbon. For the same amount of energy, burning natural gas releases less than 10% of the carbon released from burning oil and less than 5% of that of coal.

Because natural gas deposits are often found near oil deposits, most of the world's reserves—approximately 40%—are in the Middle East. Russia has the second highest amount of proven reserves, whereas the United States contains just over 4% of the world's natural gas reserves. In the United States, approximately 29% of the energy consumed comes from natural gas, while the global proportion is 24% (see Figure 6.1). The United States by far consumes the most natural gas per year (approximately 20% of world consumption), double that of most other industrialized countries and considerably more than most smaller and developing countries.

So how much natural gas is left? The RPR for natural gas is estimated to be approximately 70 years, which is on average slightly better than that for oil. Again, this assumes that rates of extraction and use will remain constant. However, through newer and unconventional forms of natural gas extraction, rates are increasing, as is usage, and prices are declining. There are large quantities of natural gas that are not considered proven reserves because they cannot be economically extracted using conventional methods. Rather than in porous pockets of limestone, this natural gas occurs in small cracks in underground shale. It is estimated that the United States contains the largest known shale gas reserves of this type in the world. Motivated by energy profits, the promise of new jobs, energy independence, and the cleaner nature of burning natural gas, the US shale gas industry has taken off. President Obama's 2014 Climate Action Plan called for greater extraction and use of shale gases because their contribution of CO_2 is so much lower than those of coal and oil.

The problem with this increased emphasis on shale gas is that the technique for extracting this gas, **hydrofracturing,** or fracking, is environmentally risky. This method uses horizontal drilling and then blasts open fissures in underground shale rock formations by injecting a high-pressure combination of water, chemicals, and a slurry of materials including sand and ground ceramics (Figure 6.3). This causes the gas to flow to the production well.

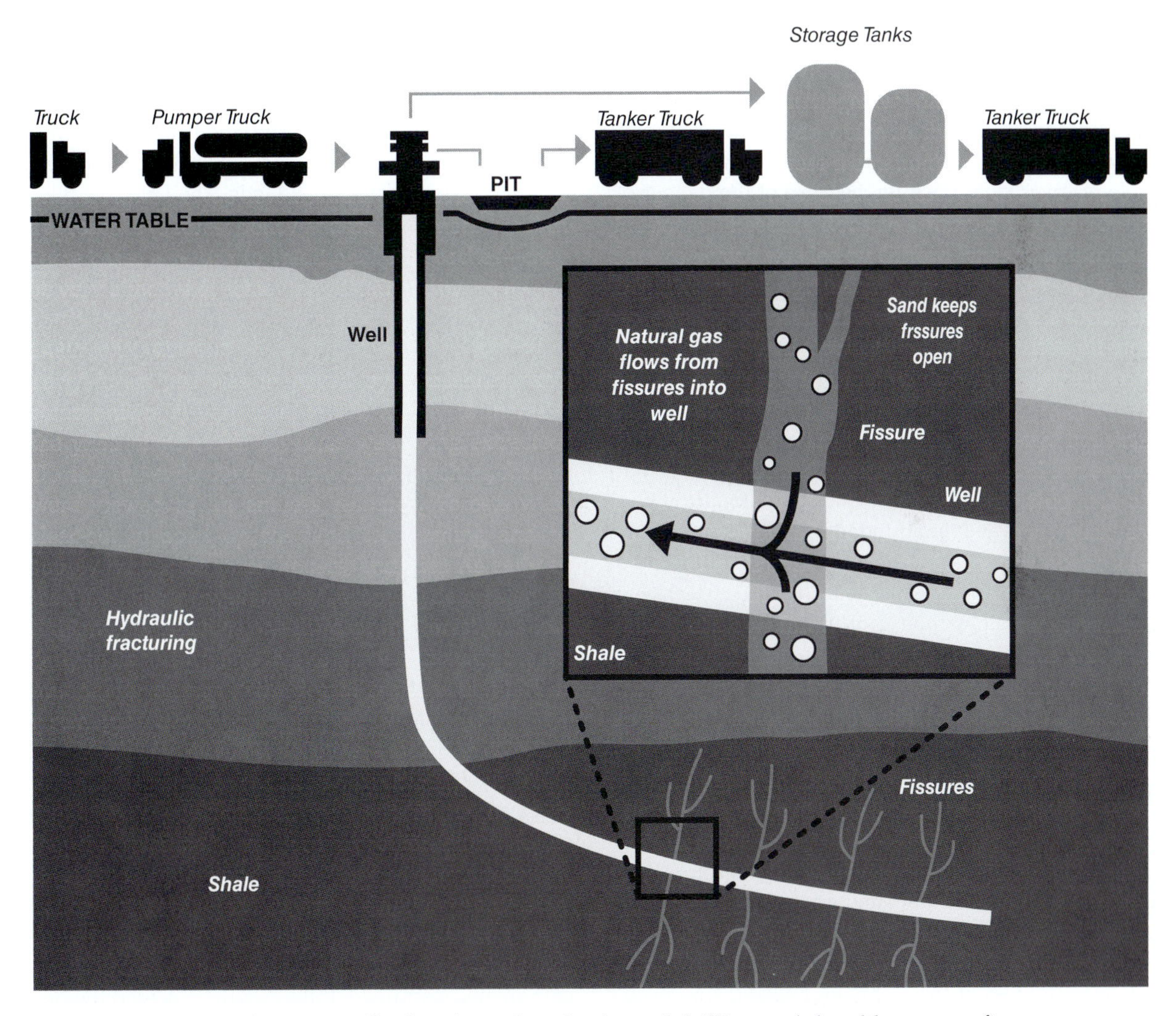

Figure 6.3 Hydraulic fracturing (fracking) employs horizontal drilling and then blasts open fissures in underground shale-rock formations by injecting a high-pressure combination of water, chemicals and a slurry of materials including sand and ground ceramics. (Denver Public Library, https://www.denverlibrary.org/blog/what-frack)

Fracking uses millions of gallons of water, which is an already limited and rapidly diminishing vital resource. In addition, fracking fluids are known to contain toxic materials, some of which are carcinogens and endocrine disruptors. These fluids can contaminate both aquifers and surface waters, many of which supply drinking water to communities in which this is occurring. Often, companies are not required to disclose what is in their proprietary fluids, and no baseline or post-fracking data are collected. As a result, the fracking companies cannot be held responsible for any environmental damage. Methane, a potent GHG that contributes to climate change, also leaks from fracking sites. Fracking is currently regulated at the state level, where fracking companies can have a strong influence. As a result, fracking policy and regulation vary greatly from state to state (Box 6.1). Despite its cleaner burning qualities and the expansion of newly recoverable reserves from fracking, because of the danger of this extraction process to environmental and human health the expansion and use of natural gas do not meet our criteria for sustainability.

Clean Coal

Burning coal is a major source of air pollution, including sulfur and nitrogen oxides. When burned, it also releases the most CO_2 per unit energy generated compared to any other fossil fuel—nearly 1.5 times that of oil and approximately 15 times that of natural gas. Because of this and the abundance of coal, particularly in the United States, there has been effort to develop what is referred to as clean coal technology. Clean coal uses gasification, carbon capture, and sequestration technologies. In gasification, the coal is not directly burned. Rather, it is heated to generate gas, which is then burned to generate electricity.

There are a number of criticisms of clean coal. First is cost. The first plant that uses this technology including carbon capture was built in Mississippi in 2014. The plant cost $5.2 billion to build, and it costs 22% more to generate electricity than what it would cost to generate the same amount of energy in a traditional coal-fired power plant. These costs will be passed on to the consumer. Even with the new plant, the effectiveness of the technology is still unproven. Others argue that even if carbon and other pollutants are captured, clean coal will never be clean because the environmental impacts of coal mining remain and because the end products from the combustion process, which are considerably hazardous to the environment, still need to be disposed of, typically in landfills. Finally, because coal is not renewable or sustainable in the long term, the return on these large investments in infrastructure may never be realized. It makes better economic sense to invest those dollars in renewable technologies that will persist long into the future.

Transporting Fossil Fuels

There are a number of issues surrounding the transport of fossil fuels, ranging from the risk of catastrophic oil spills at sea to the dangers of transporting flammable, explosive, toxic fuels by rail or truck through communities. Relatively recently, there has been an emphasis on developing pipelines to transport fossil fuels. This has attracted much controversy because eminent domain has been used to acquire private land to build the pipelines and because pipelines are often routed through natural areas or indigenous lands. The latter resulted in a major uprising as the Dakota Pipeline was to be routed through the Sioux Standing Rock Reservation in what has been referred to as #NoDAPL. Many have argued that these pipelines are vital for economic development and energy independence, but in reality, they create very few long-term jobs and much of the fuel transported in them may be exported.

Access to Energy

Modern energy services are crucial to human well-being and to a country's economic development, yet globally more than 1.2 billion people lack access to electricity and another 2.8 billion rely on indoor air-polluting wood or other biomass for cooking and heating and/or kerosene for lighting. Most of these people live in the rural areas or slums of developing countries, primarily in Africa and Asia. Developing Asia is home to the majority of the world's energy poor, where more than 600 million people do not have access to electricity and approximately 1.8 billion people still use firewood, charcoal, or kerosene to cook their food and light and heat their homes. These numbers will continue to grow as populations in these areas increase over time.

Without universal access to energy, sustainable development and poverty reduction in these areas are extremely constrained. Billions will be denied the opportunity to improve their lives as they will continue to depend on wood for heating and cooking, which destroys habitats and increases air pollution and related health problems. Time spent searching for wood

BOX 6.1

Policy Solutions: To Frack or Not to Frack?

The Marcellus shale play in the northeastern United States is one of the most robust deposits of natural gas in North America. Stretching from West Virginia northward to central New York state, the Marcellus shale deposit contains an estimated 141 trillion cubic feet of natural gas.[1] Although the deposit is a unified geographcal feature, it lies under an array of political jurisdictions, including at least some portion of nine states and one Canadian province.

With little federal intervention in the regulation of the extraction of natural gas from shale, state governments have played a primary role in overseeing drilling activity within the Marcellus shale play. This situation has created striking differences in policy approaches throughout the Marcellus region, with the most extreme example of policy variation occurring on both sides of the 306-mile border that separates the state of New York and the Commonwealth of Pennsylvania. This state border, which intersects the heart of the Marcellus shale, has become a divide between one of the most active regions of hydraulic fracturing in the United States and a counterpart where the shale play remains largely untouched. In some places along the border, New York residents can look south across the state line and see drill sites in Pennsylvania engaged in the process of releasing natural gas from the same shale formation that sits beneath their property.

Given the large economic, environmental, and social implications of the issue of shale gas and the process of hydraulic fracturing, or "fracking," it is not surprising that this matter has become a major political issue in both Pennsylvania and New York. In New York, Governor Andrew Cuomo's long-awaited decision on the fate of the state's six-year-old moratorium on hydraulic fracturing was announced in December 2014. Since 2008, New York had restricted the process of hydraulic fracturing within its borders in order to more fully ascertain potential risks from the process, leaving the state free of any shale gas extraction. After his re-election in November 2014, Cuomo announced that the moratorium on fracking would not be lifted, citing an array of environmental and health risks associated with hydraulic fracturing.[2] In essence, the state of New York has opted to remain out of the shale gas boom that many other states, including its neighbor Pennsylvania, have engaged in for more than a decade.

Conversely, Pennsylvania has been at the heart of the shale gas boom, with broad and intense expansion in the use of hydraulic fracturing since the process was first used in the Commonwealth in 2003.[3] In 2012, natural gas from shale accounted for more than 90% of the state's natural gas production.[4] The growth in shale gas drilling has led to large-scale public debates within Pennsylvania that have in turn produced a number of highly controversial policies, most notably the state's Oil and Gas Act.

The Pennsylvania legislature passed the Oil and Gas Act, or Act 13, in early 2012, thus establishing the framework for governance of unconventional gas drilling in the state. Act 13 is generally viewed as an "industry friendly" law that is permissive of high levels of hydraulic fracturing without tight controls on such matters as chemical disclosure and limits on the ability of local governments to control the locations of drilling sites.

Perhaps the most notable and controversial feature of Act 13 was the absence of a statewide extraction tax on gas produced through hydraulic fracturing. Former Governor Tom Corbett, a principal architect and strong Act 13 supporter, prioritized shale gas exploration and development throughout his term as governor. He received numerous campaign donations from those aligned with the oil and gas industry, and despite growing pressure from within his own party, he continued to oppose an extraction tax as he faced re-election in November 2014.[5] Corbett's opposition to an extraction

1 US Energy Information Agency. 2012. "Annual Energy Outlook 2012." June. http://www.eia.gov/forecasts/aeo/pdf/0383(2012).pdf

2 Kaplan, Thomas. 2014. "Citing Health Risks, Cuomo Bans Fracking in New York State." *New York Times*, December 17. https://www.nytimes.com/2014/12/18/nyregion/cuomo-to-ban-fracking-in-new-york-state-citing-health-risks.html

3 Harper, John A. 2008. "The Marcellus Shale—An Old 'New' Gas Reserve in Pennsylvania." *Pennsylvania Geology* 38: 2–13.

4 US Energy Information Agency. 2014. "Natural Gas Gross Withdrawals and Production." January. http://www.eia.gov/dnav/ng/ng_prod_sum_dcu_smi_m.htm

5 Rabe, B. G., and C. Borick. 2013. "Conventional Politics for Unconventional Drilling? Lessons from Pennsylvania's Early Move into Fracking Policy Development." *Review of Policy Research* 30: 321–40. doi:10.1111/ropr.12018

continued

continued

tax and the revocation of Act 13 provisions limiting local control of drilling sites became a liability for him in the 2014 election and helped Democratic nominee Tom Wolf solidly defeat Corbett. Since becoming governor in January 2015, Wolf has called for the adoption of a 5% tax on shale gas extraction and has banned fracking in Pennsylvania's state parks, but he has not called for a moratorium on drilling and has publicly supported the positive economic impacts of hydraulic fracturing in Pennsylvania.

The strikingly different policy approaches to a common resource present an excellent opportunity for a variety of comparative analyses. Will Pennsylvania ultimately regret its widespread use of fracking as environmental impacts from the process become more extensive and evident? Or will New York regret missing out on the economic gains from the production of a domestic energy source that were experienced across the border in Pennsylvania? With New York out of the shale gas game and Pennsylvania likely to remain highly engaged in hydraulic fracturing, it appears that we will get answers to these questions in years to come.

—Contributed by Christopher Borick, PhD
Department of Political Science, Muhlenberg College
Barry Rabe, PhD, Professor of Environmental Policy
Gerald R. Ford School of Public Policy, The University of Michigan

reduces opportunity for work and education, mostly for women and girls. Access to energy is needed for entrepreneurship and small business development that provide economic opportunity. Without electric lighting, there is reduced time for reading and education and a lack of personal security. There are no opportunities to take advantage of technologies that can be used for health care, education, food preparation and preservation, and communication and information exchange. These billions of people can be lifted out of poverty, and there can be sustainable development as these needs are met, but only through the expansion of clean, affordable renewable energy.

A Sustainable Energy Future: Renewable Technologies

It is apparent that the fossil fuel economy is not sustainable in both the near and the long term. The resources are finite and are rapidly becoming depleted as the global demand for energy increases. The United Nations (UN) estimates that this increase in energy demand will nearly double by 2050. Our use of fossil fuels negatively impacts people and the environment, and it is driving global climate change. Given all of this, the transition from fossil fuels to renewable energy not only makes sense but also is essential for a sustainable future.

Renewable energy is generally defined as energy that comes from resources that are naturally replenished on a human timescale, such as sunlight, wind, rain, tides, waves, and geothermal heat. Although they are not without environmental impact, they are much more benign than fossil fuels and produce little or no climate-changing GHGs. Renewable energy projects can be large scale and serve large industrial populations, but they are also suited to rural and remote areas and developing countries in which energy is crucial in human development—and is often lacking. Renewable energy is economically viable for several applications, and it is increasingly becoming cost-competitive with fossil fuel sources of energy. The combination of renewable energy with increased energy efficiency has the potential to allow us to completely replace fossil fuels to meet our global energy needs within a very reasonable time frame on the order of 20–50 years.

There are many types of renewable energy. The question we need to ask is which are the most affordable, effective, and have the least environmental impact. This section briefly introduces the different types of renewable energies, presents recent technological advances, and examines both their strengths and their weaknesses from economic and environmental perspectives. It is shown that a portfolio of renewables will most likely be required to meet the growing global energy demand and to curb global climate change.

Solar

The sun can be harnessed as a renewable source of energy either as thermal energy or through the direct conversion of light into electricity using **photovoltaic (PV) cells**. Solar thermal energy can be used for passive heating of buildings and water. Buildings can be designed to capture the sun's heat in the winter through proper orientation, placement of windows, and the use of heat-absorbing building materials. Solar hot water systems are also passive heat systems. They consist of a storage tank connected with pipes or tubing to a solar collector that is typically mounted on the roof. Gravity or active pumping moves water heated by the panel into the building, where it can heat the space or heat water for home use. Some solar collectors transfer solar heat directly to water moving through pipes. Others use vacuum tube collectors that consist of glass tubes with a layer of heat-absorbent coating through which water pipes run. The vacuum tube collectors are much more efficient, with heat retentions up to 97%. The transient nature of light due to nighttime and cloudiness may require a backup hot water system powered by another source. However, even with a backup system powered by fossil fuels, a single household solar hot water system can prevent up to 4.5 tons of GHG emissions annually.[5]

PV cells actively convert light into electricity at the atomic level using materials with properties that exhibit a **photoelectric effect**, or the absorption of photons of light that causes the release of electrons, thus creating current that can be used as electricity. A typical silicon PV cell is composed of a thin wafer of two layers of silicon. The one on top is treated with phosphorous, making it electron rich or negatively charged (n-type). The one on the bottom is treated with boron, making it electron poor and giving it a slightly positive charge (p-type). The junction between the two layers serves as a barrier through which electrons can only pass in one direction, from the p side to the n side. When sunlight hits the silicon exciting the electrons, they then move in that direction. This creates even more of an electrical imbalance between the layers, leaving what is referred to as electron holes on the p side. Electrons then flow from the electron-rich layer around a circuit connected to the cell in order to fill those holes. This electron flow provides current, and the more light that strikes the cell, the more electrons jump up and the more current flows (Figure 6.4).

Solar modules consist of a number of these solar cells that are electrically connected to each other. **Solar panels** consist of a number of modules, and solar arrays are composed of multiple panels that are designed based on a specific amount of current to be supplied. Solar cells only produce **direct current (DC)**, which can be used to charge batteries or power lights.

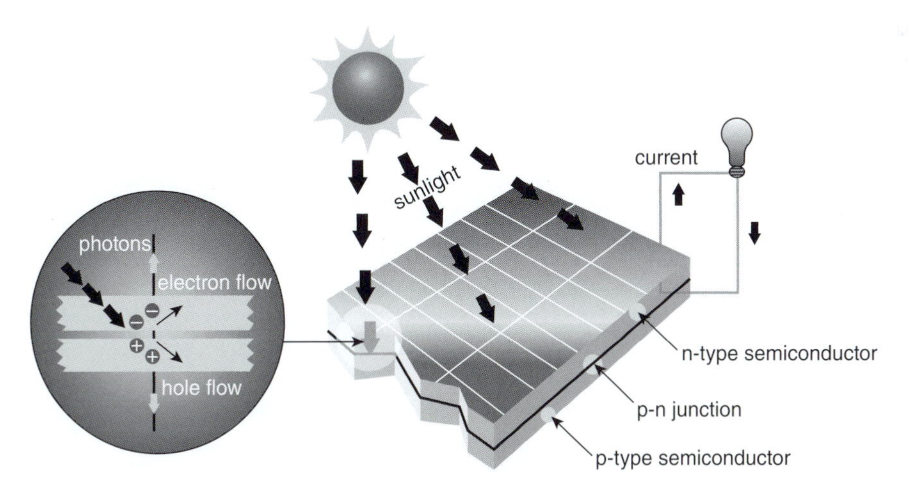

Figure 6.4 When light strikes a solar PV cell, excited electrons can only move from the p-type layer to the electron-rich, n-type layer. This leaves an electron "hole" in the p-type layer, increasing the imbalance of electrons across the p-n junction. This imbalance causes electrons to flow as current out of the n-type layer along a circuit back to the p-type layer in order to fill those holes. (https://www.electricaltechnology.org)

However, most of the electrical energy that people use is **alternating current (AC)**, so an **inverter** is required to make the conversion for more typical electricity use.

A number of novel solar technologies are in development, including heavy-duty hexagon solar glass panels that can be installed on the surface of roads and other paved areas, including parking lots and even playgrounds. They have heating elements that can melt snow and ice, and they contain light-emitting diode (LED) lighting that can signal drivers and flexibly demarcate lanes. If used to resurface roads throughout the United States, these solar roadways have the potential to produce more renewable energy than the entire nation would use. Other solar innovations include thin solar films that reduce costs and create novel opportunities for placement, including integration into building materials. Solar windows and roofing shingles have been treated with a new, see-through coating that generates electricity.

The company Rawlemon has designed a rotating glass orb that brings in energy from the sun and concentrates it onto a small surface of tiny solar panels, and it has the potential to concentrate both sunlight and moonlight up to 10,000 times (Figure 6.5). This makes its solar harvesting capability 35% more efficient than that of conventional PV panels while offering aesthetic benefits. These are just a few of the many examples in which the market potential of renewables has stimulated innovation and entrepreneurial opportunities that are generating economic growth and creating job opportunities.

In addition to the use of passive and active solar for home energy and heat production, solar power plants are currently being developed for regional supply. These plants can tie together PV panels in what is referred to as a "solar farm" or by concentrating thermal power. Concentrating solar thermal plants collect solar energy with sun-tracking mirrors called **heliostats** (Figure 6.6). They then use lenses to focus large amounts of sunlight into smaller beams that heat up air, water, synthetic oils, or salts, which then heat water to create steam that produces electricity through turbine generation. New solar plant technology is capable of generating high-pressure steam that matches the generation potential of fossil fuel power plants and at a lower cost. This point at which an alternative energy such as solar can generate the same amount of electricity as fossil fuels at equal or less cost is referred to as **grid parity**.

The major criticism of solar power has been cost. Solar hot water systems have always had a much shorter payback or a good return on investment (ROI) for homeowners. PV panels have been too expensive for significant market penetration. This is rapidly changing because of the decline in production costs as the manufacturing infrastructure increases and also with rapid market expansion as countries such as India target as much as a 500% increase in solar

Figure 6.5 Glass photovoltaic solar orb that concentrates both sunlight and moonlight up to 10,000 times, making it 35% more efficient than conventional flat solar panels. *By permission of Rawlemon.*

Figure 6.6 A solar thermal plant uses large, sun-tracking mirrors (heliostats) to focus sunlight on a receiver at the top of a tower. A heat transfer fluid consisting of molten salts that is heated in the receiver generates steam-turbine based electricity. Pictured here is the Gemsolar plant in Seville, Spain, the first of such solar plants, developed by SENER and owned by Torresol Energy. *StockStudio/Shutterstock.*

power generation by 2022. As a result, PV technology is now so inexpensive that it competes with oil and natural gas in Asia without subsidies. Even as solar subsidies expire and fossil fuel prices are in decline in the United States, investment and jobs in the solar industry are on the rise. Major oil-producing countries such as Saudi Arabia are investing billions of dollars in solar power.

Another limitation of solar power is the transient nature of the resource. The sun is not always shining. This means that surplus energy produced when the sun is shining must be stored so that it can be used when the sun is not shining. This will require new advances in energy storage and batteries, a technology sector that is rapidly developing. One new technology from Aquion Technology is a battery that operates on salt water that can provide high-performance energy storage over a wide range of scales that is safe, sustainable, and cost-effective. Tesla has developed a line of home batteries, the Powerwall, that can power an average-sized house overnight and provide backup power for a full day. Another solution is to integrate solar power with a diversity of renewable sources of energy so that temporal variation in supply and demand can be addressed. Smart grid technology, discussed later, will be an essential component of this.

Solar power has zero emissions and is completely renewable. PV cells are primarily made of silica, which has been abundant and can be extracted from the environment with little environmental impact, although there is a sense that this could be changing (see Chapter 7). One criticism of solar power is that the production of solar panels is an energy-intensive process. However, the amount of energy required to create the components is more than offset by the amount of energy they save. Furthermore, factories that construct solar panels are starting to power themselves with renewable energy. Another concern is that panel construction includes the use of rare-earth metals such as gallium and indium, which are scarce and expensive, and their extraction often has environmental, social, and public health implications, including extraction from war-torn countries as so-called "conflict minerals" (see Chapter 7). However, newer solar PV technologies are moving to more common metals that will lower prices and have fewer impacts as they are extracted.

Fabricating solar panels also requires caustic chemicals such as sodium hydroxide and hydrofluoric acid, and there have been a few cases of large-scale toxic chemical spills from PV panel factories, primarily in China. Certainly, this needs to be controlled and regulated, but these releases, although negative, pale in comparison to the toxic releases and emissions of the fossil fuel industry. Another issue is that for damaged or spent solar panels, there are currently limited recycling opportunities. These will need to be developed so that some of the metals and other materials they contain can be captured and reused. Opportunities to recycle panels will increase as the volume of recyclable materials increases. The Silicon Valley Toxics Coalition is developing sustainability standards for solar panel construction and a scorecard that rates each solar company on each of these issues.

Another issue is that some believe that solar panels negatively affect architectural, community, and natural aesthetics. This can be addressed through new technologies such as the solar road system, solar roofing panels, and solar films. In addition, there are efforts to develop new creative designs for renewable energy installations such as those promoted by the Land Art Generator Initiative, that links art with the design of sustainable infrastructure (Box 6.2). We can also adapt our sense of what is aesthetically pleasing to include solar installations.

Wind

The wind's energy is captured as it passes through the rotor blades of wind turbines, causing them to turn. The process converts the wind's kinetic energy into mechanical energy by turning the rotor blades, which are used to generate electricity by rotating the shaft of an electrical generator. As in PVs, the electricity generated is DC, so an inverter is needed to convert it to more readily useable AC current. Once converted to AC current, the electricity can be used to power a local home or facility, to charge batteries, or to provide electricity to the grid (Figure 6.7).

For wind power to be economically feasible at a particular location, average wind speeds of 6.7–7.4 meters per second or faster are needed. Wind speeds are faster at higher elevations, with taller wind turbines, and at offshore locations. A new type of wind turbine that is mounted on a latticework, fiberglass tower that can be bolted together on-site is allowing taller wind towers to be installed at higher altitude locations that have previously been inaccessible.

Although highly effective, offshore wind farms have been more expensive to install than land-based farms. However, recently developed offshore wind farms that are mounted on floating platforms rather than the seabed are more cost-effective and allow for more flexible deployment.

Wind power is completely renewable with zero emissions, and it has one of the lowest global warming potentials per unit of electrical energy generated of any energy technology. But like solar, wind has a number of limitations and concerns that need to be addressed. Early on, cost had been a factor, but like solar technology, turbine costs are declining and, in the long term, wind has proven to be very cost-effective. For example, in Denmark, where initial investment was high, wind is now the cheapest source of energy. Like the sun, wind is a transient source, so similar accommodations of storage and grid flexibility are needed.

Environmentally, the magnets of wind turbines require rare-earth elements that are mined primarily in China and that, when extracted, can have serious environmental and public health consequences on local communities. Wind turbines can also kill birds and bats that fly into the spinning blades. However, the rate of bird mortality due to collision into other manufactured structures, particularly glass buildings, far exceeds the numbers killed by turbines. The ideal location for wind farms is flat mountain ridgetops, which also often serve as migratory routes for large birds of prey, putting them at risk for increased mortality. Wind turbines may disrupt migratory or mating patterns of terrestrial animals. One solution to this has been the development of wildlife-smart wind turbines that operate at slower speeds without sacrificing energy output. Also, offshore wind farms are thought to have less impact on wildlife.

There are major aesthetic issues associated with wind farms. These issues, coupled with the noise that wind farms produce, have been a major impediment to numerous wind projects. However, aesthetic issues are subjective, and some people find wind farms pleasant or view them as symbols of energy independence and local prosperity—and in some cases wind farms have actually become tourist attractions. Perspectives on the aesthetics are changing as our culture of energy evolves, and as with solar, there is an effort to create more aesthetically pleasing designs (see Box 6.2).

Hydropower

The principle behind hydroelectric generation is similar to that of wind, except flowing water instead of wind turns the turbine in order to generate electricity. Hydroelectric generation requires the construction of a dam on a large river that drops in elevation. Water is stored at the higher elevation in a reservoir formed by the dam. Water is taken into a generation station as it is passes through the bottom of the dam, where gravity causes it to flow. This gravity-driven flow turns the

BOX 6.2

Innovators and Entrepreneurs: The Land Art Generator Initiative

The public rejection of renewable projects, whether they are small or large, can often be due to aesthetics. Solar and wind infrastructure is frequently considered unsightly, destructive to the beauty of the environment, and as having potential to lower real estate values. One innovative approach to dealing with this issue is to make renewable energy beautiful. This is the mission of the Land Art Generator Initiative (LAGI), which advances sustainable design solutions by integrating art and interdisciplinary creative processes into the conception of renewable energy infrastructure. LAGI brings together artists, architects, scientists, landscape architects, and engineers as collaborators to design and construct public art installations that uniquely combine aesthetics with utility-scale clean energy generation. The result is artwork that can continuously supply energy to the grid (Figures 6.2.1 and 6.2.2).

Figure 6.2.1 Design Submission for the 2010 Land Art Generator Initiative Design Competition by Darío Núñez Ameni and Thomas Siegl. The generator converts the kinetic energy of the swaying poles into electrical energy by way of an array of current-generating shock absorbers, which convert energy produced by the forced movement of fluid through the shock absorber cylinders. *Windstalk: A submission to the 2010 Land Art Generator Initiative competition for Dubai & Abu Dhabi Team: Darío Núñez Ameni, Thomas Siegl, Gabrielle Jesiolowski, Radhi Majmudar, and Ian Lipsky.*

continued

continued

Figure 6.2.2 Design Submission for the 2010 Land Art Generator Initiative Design Competition by Lucas Jarry, Rita Serra e Silva, Lucas Guyon, Marianne Ullmann. The generator converts wind energy into electricity by moving piezoelectric discs. *Super Cloud: A submission to the 2014 Land Art Generator Initiative competition for Copenhagen Team: Lucas Jarry, Rita Serra e Silva, Lucas Guyon, and Marianne Ullmann.*

The basis for this innovation is the recognition that art has historically created movements and generated dialog about important issues, including the problems related to sustainability. Through LAGI, artists can move beyond this by participating in solution-based art practice. LAGI achieves this through education, outreach and design competitions, and the construction of aesthetic renewable infrastructure. The ultimate goal is to provide a collaborative platform for the development of renewable technologies that are culturally and aesthetically pleasing whether they be placed in the natural environment or within communities.

For more information, visit http://landartgenerator.org

turbine to generate electricity. The water then flows into the river below the dam. The flow of water and the generation of electricity can be regulated in response to demand for electricity.

Hydropower is extremely efficient, with 90% of the water's energy converted into electricity with zero emissions. The water that fuels the electrical generation is not reduced or used up in the process and remains as clean as it was before it passed through the turbines within the dam. Because the water cycle is an endless, constantly recharging system, hydropower is considered a renewable energy.

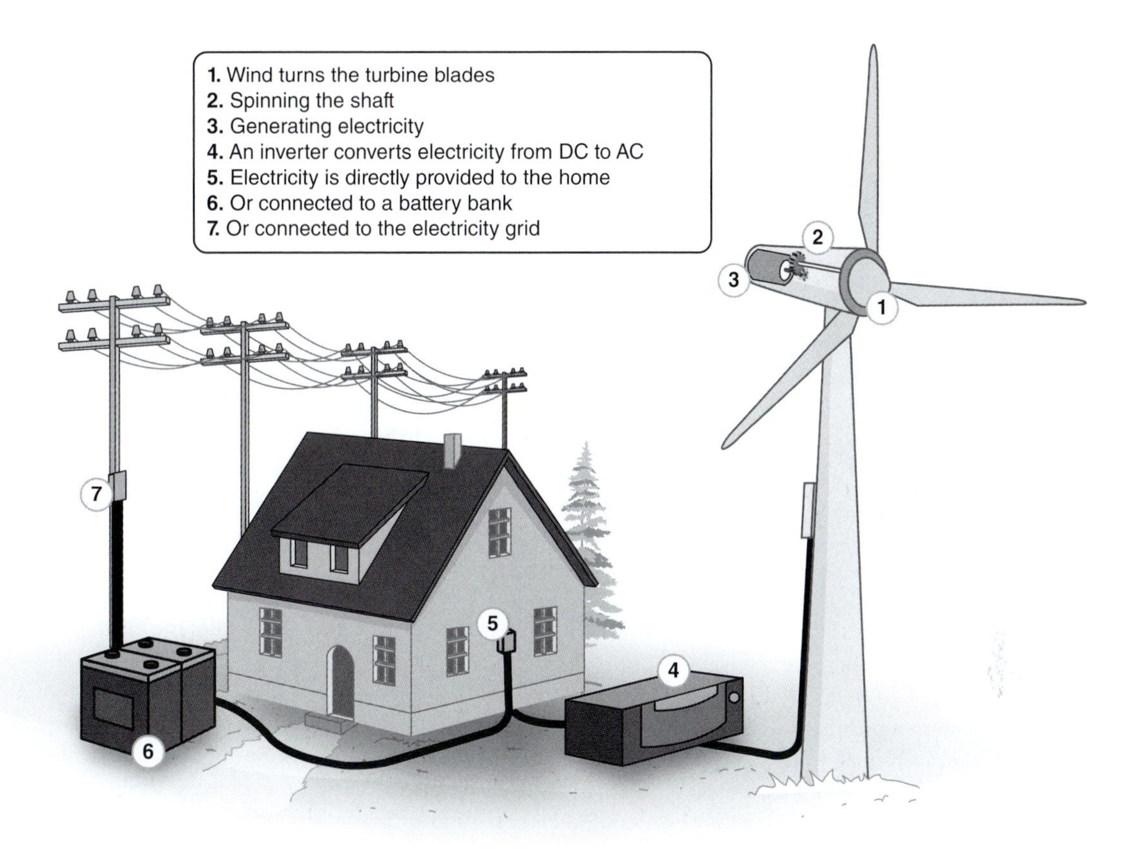

Figure 6.7 Wind energy works when wind turns the blades of a turbine, which is connected to a house, battery bank, and/or the electricity grid. (Atlasta Solar Center, https://atlastasolar.com/wind/)

Currently, hydropower accounts for the large majority (80%) of existing renewable power generation. However, globally, only approximately 20% of potentially available hydropower is being used, and its annual growth rate is less than 3% per year. Given its efficiency in clean energy production and untapped potential, the further development of hydropower is viewed as an essential part of the global renewable agenda and sustainable development. One way to do this is through smaller hydropower projects. For example, in Uttarkhand in northern India, there are 10 million people without electricity, so there has been a strong push for small-scale hydroelectric power development that includes 100 small dams. Turkey is also aggressively engaged in building micro-power facilities.

The environmental and social consequences of dams that were detailed in Chapter 4 are not insignificant, and they contradict efforts to maintain aquatic habitats and water quality. The obstruction of rivers by dams impacts the biological, chemical, and physical properties of river environments. The large reservoirs they form displace human, often indigenous, communities, which can result in economic impoverishment, cultural decline, and impacts on public health. Small hydro projects in which many small dams are placed in a single watershed could have greater cumulated environmental impacts compared to larger projects. Part of the problem is that the smaller projects are less regulated. The UN and the World Bank are beginning to focus on the impacts of smaller hydro projects and are also beginning to develop best practice and regulatory standards. Hydropower will be needed to meet the growing demand for clean, renewable energy, but in a way in which risks are minimized in order maximize the safe development of this needed resource.

Geothermal

Geothermal energy takes advantage of geological heating or cooling potential. It is clean and has minimal environmental impact. One type of geothermal energy that can be used to generate electricity is steam produced as hot volcanic material heats groundwater. In what is referred to as **hydrothermal energy**, electrical generation plants are located in areas where volcanic activity is able to heat groundwater that is close to the Earth's surface. The resultant steam is brought to the surface through wells, where it can be used to turn turbines to produce electricity. The steam is then cooled to return it to its liquid phase and returned to the environment. The use of heat exchangers during the cooling process can increase efficiency.

Hydrothermal energy is one of the cleanest sources of energy with minimal environmental impact. Unlike solar and wind, it is not intermittent and provides a constant supply of electricity. However, hydrothermal power plants are expensive to build and operate, and they can be located only where there is both volcanic activity and enough groundwater that can be sufficiently heated to power the turbines that generate power. Top locations for volcanically driven hydrothermal power plants are Central America, Kenya, Japan, Iceland, areas of Europe, and the western United States. Iceland and the Philippines have the largest proportion of their national energy production from geothermal—approximately 30%.

A new technology may be able to increase the geographical applicability of hydrothermal energy. A startup company called AltaRock is developing a new way to create a hydrothermal source of power where none naturally occurs. This is done by drilling deep into the ground and injecting cold water to shear or fracture the hot rocks that exist in the absence of naturally occurring water to produce hot water or steam. This is a promising technology that has the potential to expand and lower the cost of geothermal electric generation, although there has not yet been a complete assessment of environmental impact. Cornell University has proposed developing deep geothermal energy for the production of all its campus power and heat. The proposed system would include multiple production wells reaching approximately 16,400 feet (nearly 5 km) below the ground surface, where the rock is 120°–140°C. Water will be circulated in a closed loop through that deep hot rock reservoir and return to the surface to generate electricity and supply heat to the campus. This project has the potential to eliminate an estimated 110,000 metric tons of carbon from Cornell's annual footprint and will serve as a new, scalable model for using this sustainable energy source.

Another form of geothermal energy takes advantage of the thermal retention capabilities of the Earth through the use of a **geothermal heat pump** (Figure 6.8). A geothermal heat pump transfers heat to or from the ground to provide central heating for cooling of a home or other building. These pumps work by pumping water or antifreeze through a network of underground pipes. To heat a building, water is pumped up from the ground to bring up heat from the Earth. To cool a building, water is pumped down through the pipes to carry heat away from the building where the relatively cooler ground absorbs the heat, cooling the water that is then circulated back up to the building. The pump is able to provide both heating and cooling because the temperature beneath the surface of the Earth is consistently moderate between 8°C and 16°C year-round.

The EPA has called geothermal heat pumps the most energy-efficient, environmentally clean, and cost-effective space conditioning systems available. These systems are 70% less expensive than heating a home using traditional fossil fuel furnaces. Their only environmental drawback is that their use of refrigerants and the electricity that drives the pumps are currently most likely produced by fossil fuels. However, these impacts have or will be mitigated because nontoxic, non-ozone-depleting, biodegradable refrigerants have been developed, and the grid is increasingly being powered by renewables.

Biological Fuels

There are a number of potentially renewable fuels that can be produced biologically, including ethanol derived from crops, biodiesel from oils and fats, biologically generated methane, and wood-based fuels. Currently, the production of biofuels from crops such as corn is expanding globally, with the United States and Brazil as the largest producers. These are referred to as **first-generation biofuels** because the crop is grown specifically and exclusively for the production of ethanol that is used as an additive to or replacement for gasoline. The proliferation of crop production for ethanol is motivated by the desire for energy independence and security and also to reduce the price of gasoline. More than 60 countries have biofuel targets or mandated concentrations in gasoline.

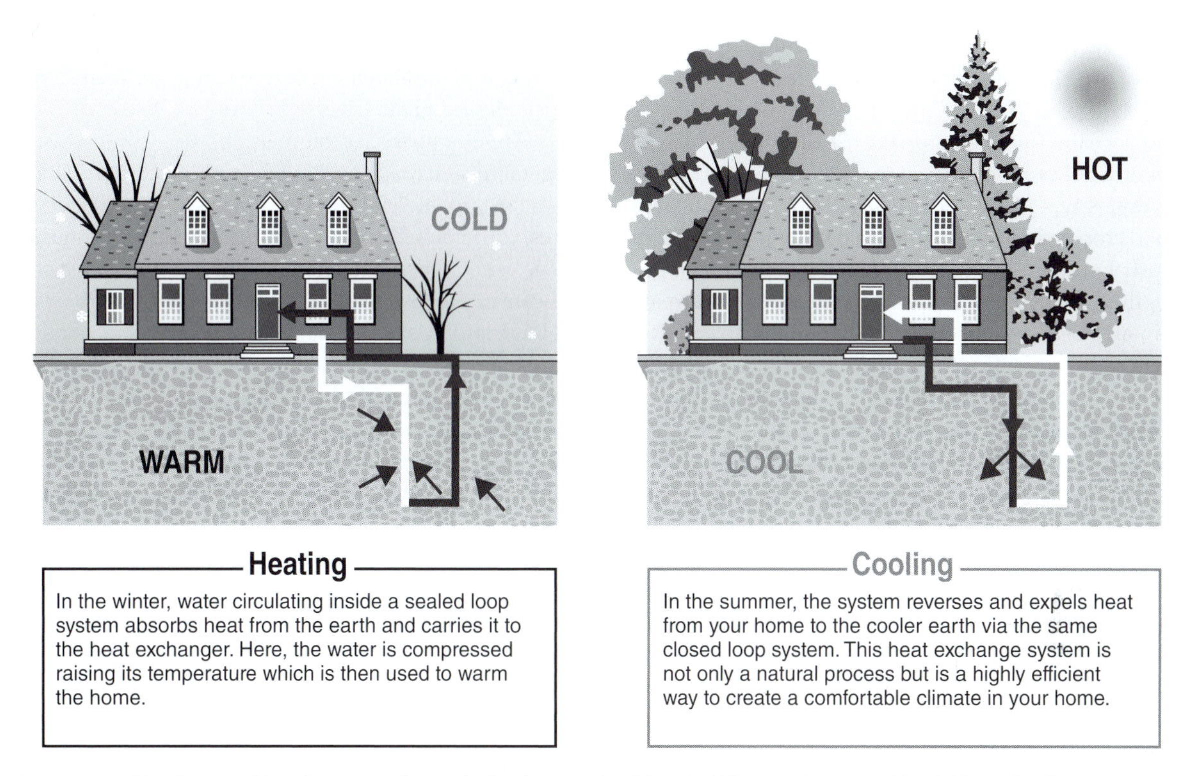

Figure 6.8 Geothermal heating uses the relatively constant heat of the earth to cool buildings in the summer and heat them in the winter. (https://goo.gl/images/bwiFg9)

Because the crops theoretically take up more atmospheric CO_2 than is released when the ethanol is burned, these biofuels are considered renewable and nearly carbon neutral. Even the Intergovernmental Panel on Climate Change reports that biofuels have direct GHG emissions that are typically 30–90% lower than those for gasoline or diesel.[6] However, these numbers do not take into account indirect GHG emissions, which include those typically associated with industrial agriculture (see Chapter 5), the conversion of natural habitat to field to meet the demand for increased cropland, and the energy used to produce ethanol from corn.

There are other negative implications from the expansion of industrial agriculture for the production of ethanol, including further soil degradation through mechanical and chemically intensive industrial agriculture and loss of biodiversity as more natural habitat is converted to monoculture cropland. To meet the growing demand of biofuels, areas where irrigation is needed are increasingly being cultivated. The increase in irrigation demand coupled with water pollution through the excessive use of fertilizers and pesticides in biofuel crop production is impacting this already stressed vital resource.

There are also a number of ethical and social justice issues related to the production of first-generation biofuels. One of these is the effect it has on food prices and food security. For example, corn ethanol production in the United States contributed to a significant increase in food prices as a large portion of the US arable land where food was once grown was converted to corn production for biofuels. In Mexico, the rising demand of corn for export to the United States for biofuel production has caused corn prices to exceed local affordability. Because corn is a staple food in Mexico, this has led to increased hunger, resulting in some highly publicized riots. Also, as the value of crops used for ethanol production rises, so does the potential for worker abuse. For example, in Brazil, there have been reports of slave-like conditions in mills where sugar cane is processed for ethanol production. There have also been reports of poor working conditions and informal child labor in the processing of crops for biofuels.

One way to reduce the environmental and social impact of biofuels is to stop the use of crops exclusively for ethanol production and to use cellulose and carbohydrates that are in the unused residues produced in food crop production. Other waste from the lumber industry and by-products from food and beverage processing, all of which normally enter the waste stream, can also be used to produce ethanol, as can cellulose materials in municipal solid wastes. These so-called **second-generation biofuels** are more sustainable because they are non-food competitive, more efficient in terms of net life cycle GHG emission reductions, and do not require water and land beyond what is needed for food or forestry production. The potential sustainability of waste-derived ethanol has led to a number of innovations and economic opportunities. Advanced Biofuels Corporation, a US company, has developed a process and platform for second-generation cellulosic ethanol production that is low cost, energy efficient, and scalable. Its business model is to deploy and franchise these operations across global economies in an effort to provide economically, socially, and environmentally positive sources of fuel.

Biodiesel is a form of diesel fuel manufactured from vegetable oils, animal fats, or recycled restaurant greases. It is a renewable, clean-burning diesel replacement that is theoretically carbon neutral. Oils can be extracted from plants grown as crops, such as soybeans, rapeseed, coconut, sunflowers, palms, or seeds from the woody plant *Jatropha curcas*. However, growing these plants as crops or on plantations creates many of the same problems as crop-based ethanol production. A more sustainable approach is to use oils from cultivated algae or waste oils to produce biodiesel. Algae can be grown in closed-loop systems and photo-bioreactors that produce nearly 30 times the energy per unit area compared to crop-based fuels.

Biodiesel is made through a chemical process called **transesterification** that splits the oil into two parts: esters, which are the diesel, and glycerin, which can be used to make soap and other beauty products. Depending on the engine, cars can be run on pure biodiesel or a blend of biodiesel and petroleum-based diesel. Cars that normally run on diesel can also be converted to run directly on filtered waste fryer oil, which is often given away by restaurants because they typically have to pay for its disposal. The conversion requires a kit and parts that are now readily available from companies such as Veggie Garage, Green Conversions, and PlantDrive.

Are ethanol-based biofuels, biodiesel, and waste vegetable oil used as fuel in converted cars renewable and sustainable? Their use produces less particulates, carbon monoxide, and sulfur dioxides, although more research is needed on emissions from waste oil-converted vehicles. They produce significantly less GHGs compared to petroleum products, and the plants used to produce the fuels capture carbon as they grow. Unlike conventional diesel, biodiesel is non-toxic, biodegradable, and spills do not require emergency response clean-up activities. With regard to sustainability, the major issue is the source of the inputs. When crops are grown only to produce bio-fuels, indirect CO_2 production and other environmental and social factors make them less sustainable. When the fuels are made from carbon-capturing materials that normally enter the waste stream or from closed-loop algae systems, the potential for sustainability is much greater. The possibility for greater energy independence, energy accessibility, and innovation and economic opportunity from the biofuel industry is also quite large.

Biogenic methane or **biogas** is generated in landfills, sewage treatment plants, and from the breakdown of animal wastes. Methane is a potent GHG when directly released into the atmosphere, and capturing and burning methane releases CO_2, a much less potent GHG. Biogenic methane can be captured from landfills. Anaerobic digesters can be used to generate methane at sewage treatment plants and also at agricultural sites at which large amounts of animal wastes are produced, where it can be captured and used as fuel. Small-scale biogas production using polyethylene biodigesters has been one way to bring energy to rural animal farmers, and it is being promoted throughout the developing world. This type of methane is a renewable resource, but its products of combustion are GHGs, although in a less potent form than that which would be released if they were not captured and used.

Wood and wood products are both traditional and contemporary biologically produced fuels that can be sustainable depending on how they are obtained and used. Wood is a renewable resource in that trees can replace themselves, especially if managed in a sustainable way that balances the rate of extraction with the rate of replacement (see Chapter 7). In most of the developing world, wood or charcoal derived from wood is the only source of fuel for cooking and heat, where more than 85% of all the wood consumed annually is used as fuel. Growing demand has resulted in both the destruction of forests and, because this wood is burned indoors, one of the most deadly environmental problems today, in which millions of people die each year from indoor air pollution. As wood becomes increasingly depleted, more time must be spent, mostly by women and children, searching for it. In this context, wood is far from a sustainable source of fuel.

In the United States, where more than 2.6 million households heat their homes with wood, there is potential for this biologically derived form of energy to meet our criteria for sustainability. Most household wood stoves are now highly efficient and clean burning. Also, a report from Duke University scientists notes that advanced wood-combustion technologies can be used to cleanly burn wood to generate electricity and potentially provide more power in the United States than we currently get from hydroelectric sources.[7] Wood production can be sustainably managed as a carbon-neutral, renewable resource. Perpetual wood, or wood cut from trees without killing them, is said to further increase carbon sequestration. Wood pellets composed of compressed sawdust from milling and manufacturing options that are typically disposed of as waste can also be burned cleanly and efficiently. As long as it can be managed sustainably and burned efficiently and cleanly, wood could be an important part of the renewable portfolio, especially in areas where forest management is already occurring. Others argue that growing wood for fuel damages forest ecosystems and results in a net atmospheric carbon gain.

Ocean Energy

Oceans have tremendous potential for providing clean, renewable energy. One example is using the rise and fall of ocean tides to generate electricity. Tides are the alternate rising and falling of the ocean, usually twice in each lunar day at a particular location, due to the combined effects of gravitational forces exerted by the moon, sun, and rotation of the Earth. The extent of tides, their frequency, and hence their potential for energy generation vary geographically and temporally with the time of day and with the lunar cycle. There are three types of tidal energy generator. Stream generators are turbines placed in a narrow channel between two land masses. The movement of the tide in both directions turns the turbine, generating electricity (Figure 6.9). Another generator type uses a large dam or barrage constructed across tidal rivers or estuaries that captures water at high tide. At low tide, the water can be released through a gate in a controlled way to turn a turbine. The third type of tidal energy generator functions like a barrage but creates a lagoon along a natural coastline. Again, water is collected at high tide and released through a turbine at low tide.

The potential for tidal energy is great, especially in areas with large tidal ranges, such as the Bay of Fundy on the northeastern coasts of the United States and Canada, where the tide can rise and fall as much as 16 meters twice per day. Tidal energy projects are also in development along the coasts of Chile, Alaska, and the United Kingdom. The advantages of tidal energy are that it is renewable, emission-free, and very efficient. Tidal energy also offers predictable output and could potentially serve as a barrier to storm surge and rising sea levels due to climate change. Tidal power plants along the coast can be integrated with energy-demanding desalination plants in the generation of fresh water (see Chapter 4).

Negative aspects of tidal power include cost as well as the limited number of sites with high potential for energy production. Because tidal energy is in its infancy, not much is known about its potential environmental impact, which could include the disruption in movement and increased mortality of large marine animals and fish migration as well as the potential for turbines to kill these animals. They may also impact coastal areas and associated wildlife through changes in sedimentation, salinity, and regular tidal patterns. However, these impacts will be more local than global. Effects on human activity include restricting and impacting fisheries and shipping channels. The Ocean Power Renewable Company, a leader in developing these technologies, is working with local communities and diverse stakeholders to allay these kinds of concerns.

Like tidal energy, wave power captures the energy of moving water to generate electricity. However, instead of the energy from the rise and fall of tides, the energy of waves generated by wind passing over the surface of the ocean is used. Wave energy is produced by placing different types of floats, buoys, or pitching

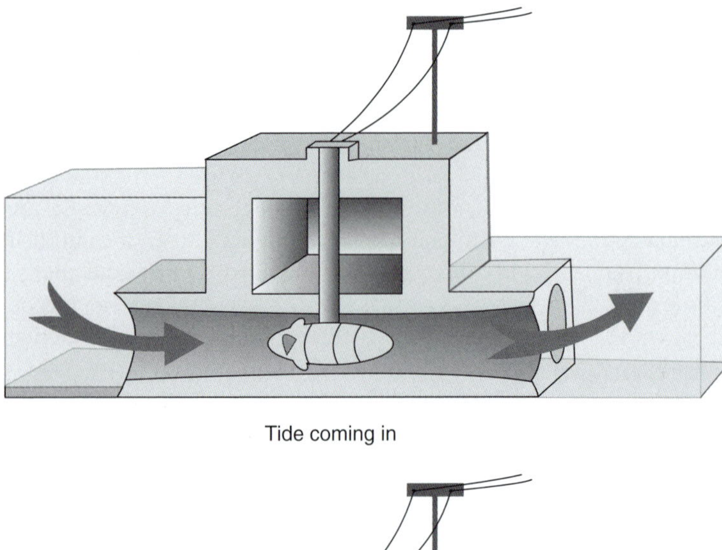

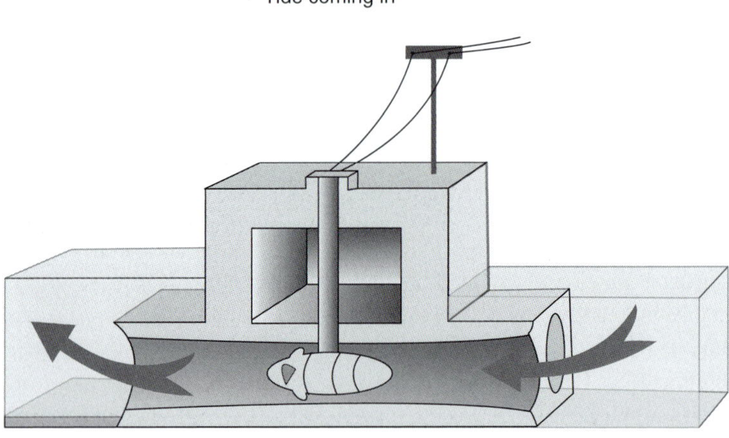

Figure 6.9 To capture tidal energy a turbine is placed in a narrow channel between two land masses. The movement of the tide in both directions turns the turbine generating electricity. (NASA, https://climate.nasa.gov/vital-signs/sea-level/)

devices to generate electricity using the rise and fall of ocean swells to drive hydraulic pumps that generate the electricity, which can be transmitted to the shore via undersea cables. Another type uses oscillating water column devices to generate electricity at the shore using the rise and fall of water within a cylindrical shaft. The rising water drives air out of the top of the shaft, powering an air-driven turbine. The development of wave energy has been limited to small-scale wave farms off the coasts of the United States, the United Kingdom, Australia, and Portugal. Another form of ocean energy that is being developed is referred to as ocean energy thermal conversion, which will use the ocean's natural thermal gradient to generate power.

Many have argued that energy from ocean tides, waves, and thermal gradients could be a major contributor to our global portfolio of renewable energy. So why have these technologies lagged behind others such as wind and solar by as much as three decades? The answer is that the technology is not as advanced, and there has not been a consensus on the optimal design for each with regard to energy generation and environmental impact. Moreover, offshore deployment and maintenance of large equipment is expensive, and the equipment must be able to withstand the force of the waves and tides that power the generators as well as resist the corrosive saltwater environment. Despite these limitations, a number of companies see the potential of ocean-based energy and are moving forward with their research, development, and eventual deployment of commercial-scale tidal and wave energy technology.

Fuel Cells

A fuel cell essentially converts chemical energy into electrical energy. Fuel cells operate much like batteries, but they can provide DC electricity continuously and indefinitely as long as the chemical inputs are provided. There are many different types of fuel cells, but hydrogen fuel cells are the most common. In a hydrogen fuel cell, electrical current is generated through the combination of hydrogen and oxygen, producing water as the only by-product (Figure 6.10). Stationary fuel cells can be used to generate electricity for commercial, industrial, and residential sites. Hydrogen fuel cells can be used to power electric automobiles, buses, and other forms of transportation. Because they produce both electricity and heat, fuel cells are ideal for the co-generation of heat and power.

An energy source that can produce electricity and emit only water and oxygen seems like the ideal renewable, carbon-neutral energy source. However, pure hydrogen gas does not occur naturally on Earth as gas, nor is it naturally renewable. It must be made using an electrolyzer that converts water to hydrogen and oxygen. This process takes a considerable amount of energy—at least twice the amount of input to make the equivalent amount of energy as electricity. Currently, that energy used to produce hydrogen primarily comes from fossil fuels. Thus, although it is clean burning, hydrogen is not a sustainable energy when

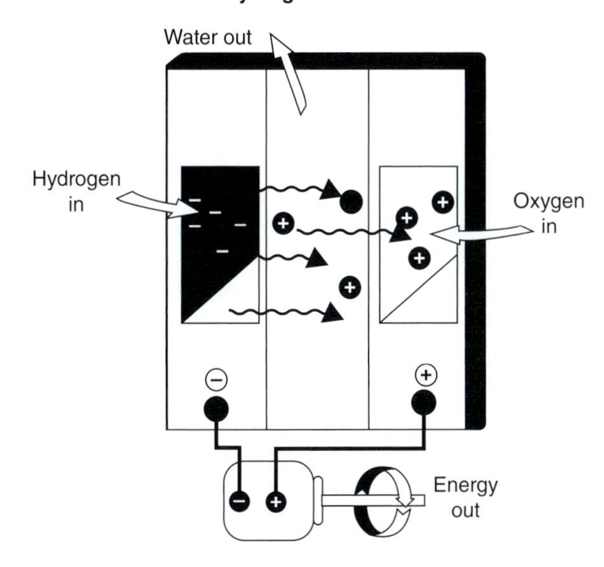

Figure 6.10 In a hydrogen fuel cell, hydrogen gas is split into protons which flow across from an anode to a cathode where they combine with oxygen to form water, generating current in an external circuit. (Alternative Energy News, http://www.alternative-energy-news.info/technology/hydrogen-fuel/)

produced in this manner. However, a new technology referred to as "solar hydrogen" uses the excess electricity from solar PV panels to power electrolyzers that produce hydrogen. As this technology expands, hydrogen fuel cells could play a major role in our sustainable energy future, particularly because they are a mobile source of fuel generated by renewable energy that could power automobiles.

Fuel cells that directly convert biomass to electricity with assistance from a catalyst activated by solar or thermal energy are also under development. This hybrid fuel cell can use a wide variety of biomass sources that are waste products of agriculture and forest industries. These include cellulose and lignin from crop residues, as well as powdered wood, algae, and waste from poultry processing. This technology could be used in scalable units to provide electricity locally or regionally depending on demand and available biomass. With these and other types of innovations, the potential for hydrogen and other types of fuel cells as part of the renewable production of clean energy is great.

Nuclear Energy

Nuclear energy is a major source of electricity in some countries and regions and is valued for its lack of carbon emissions. Globally, it contributes to approximately 6% of energy production, but in some countries and regions, it represents a much higher proportion. For example, it provides 75% of the energy for electrical generation in France, 30% in Europe, and up to 40% in the eastern United States. The question is whether or not it is renewable. Nuclear power plants use nuclear fission of uranium to generate heat to make steam to run electric generators. Uranium is mined where there are high concentrations in the Earth's crust. More than half of the world's uranium comes from mines in Canada, Australia, and Kazakhstan. Other major producers are the United States, South Africa, Namibia, Brazil, Niger, and Russia.

The question of whether uranium used in nuclear energy is a renewable resource is often debated. Most supporters of nuclear energy point out that its low carbon emission should be the basis for considering it to be a renewable energy. However, we do not define renewables on the basis of their emissions but, rather, on their rate of depletion in relation to their rate of renewal. One of the major arguments against including nuclear energy in the list of renewable energies is the fact that uranium deposit on Earth is finite. However, others have argued that through the use of breeder reactors, we could produce uranium and essentially have an infinite supply to fuel the Earth for another 5 billion years.

Perhaps the argument should not be over whether nuclear power is renewable but, rather, whether it is sustainable. In terms of its zero carbon emissions and the opportunities it creates for energy independence, nuclear power is certainly a more sustainable form of energy than fossil fuels. However, other aspects contradict our criteria for sustainability, including the potential for nuclear catastrophes such as those at Chernobyl and Fukushima that have had devastating long-term effects on local and global scales. For example, the 1986 accident in Chernobyl, Ukraine, resulted in 28 initial deaths from radiation followed by 7,000 cases of thyroid cancer among individuals younger than age 18 years at the time of the accident. Also, the end products of nuclear reactions are harmful radioactive wastes for which we do not have a way to treat other than to store them. Finally, the conflation of nuclear power with nuclear arms proliferation has had a number of political implications, and the destruction of the environment through mining of uranium suggests that other forms of energy could better meet our criteria for sustainability.

As described previously, our current nuclear power is based on the **fission,** or the splitting of the heavy unstable nucleus of uranium molecules into two lighter nuclei to release energy and fission products, which are what constitute nuclear waste. There is much interest in developing nuclear fusion as perhaps a more sustainable form of energy. In fusion reactions, two light nuclei such as hydrogen combine, releasing vast amounts of energy and inert helium as the end product. This is the same process that powers the sun and other stars. Fusion-based nuclear power has the potential to produce endless amounts of clean, carbon-free energy while eliminating the problems of radioactive waste and the risks of nuclear proliferation and catastrophic accidents. The challenge is that the initiation of a fusion reaction requires temperatures of approximately 120 million °C in order to convert fuel atoms into plasma so that they can fuse. Generating this amount of heat, and confining the plasma and the fusion reaction, has proved to be extremely challenging. Magnetic fields generated by superconductors are used to contain the plasma inside of fusion reactors. Researchers have been able do to this on a very small scale for a short period of time. The challenge is to scale up the process to commercial, economically viable proportions for what may be the ultimate energy solution. Progress is being made, but unfortunately, we are a long way off from accomplishing this goal.

New Technologies to Manage and Distribute Energy

A fully renewable, carbon-free system based on a variety of sources, some of which are temporally variable in their production of energy, requires smart grid or supergrid technology to effectively manage and distribute energy. The traditional grid is relatively local and fragmented, and most energy sources tend to be within 50 meters from the point of use. A smart grid is transcontinental in nature, links a diversity of energy sources with users, and is multidirectional (Figure 6.11). It uses integrated sensing technology to gather and act on up-to-the-minute information such as the availability of electricity from specific suppliers in relation to consumer demand.

This regional approach can connect very different sources of energy distributed over wide geographical ranges with each other and consumers. It is more reliable, flexible, and efficient. Its multidirectionality integrates the diverse renewable energy system with a wide range of consumers. It also allows smaller producers to effectively sell or return energy back to the grid, as well as peer-to-peer energy sales. Another advantage of the interconnectedness and wider geographical range is that fewer power sources can serve more people, thereby reducing capital costs, and there are greater opportunities for consumers to choose the source of their electricity.

Because these new grids will span and transport large energy loads over greater geographical areas, new and more effective ways to transport energy, such as superconductor cables, will be required. In the traditional grid, electricity is transmitted as high-voltage AC, which has large transmission losses and is inefficient for transport across long distances. The large interconnected supergrid will consist of high-voltage DC technology that will facilitate rapid long-distance transport. This integrates well with solar and wind because they produce DC current. These high-speed

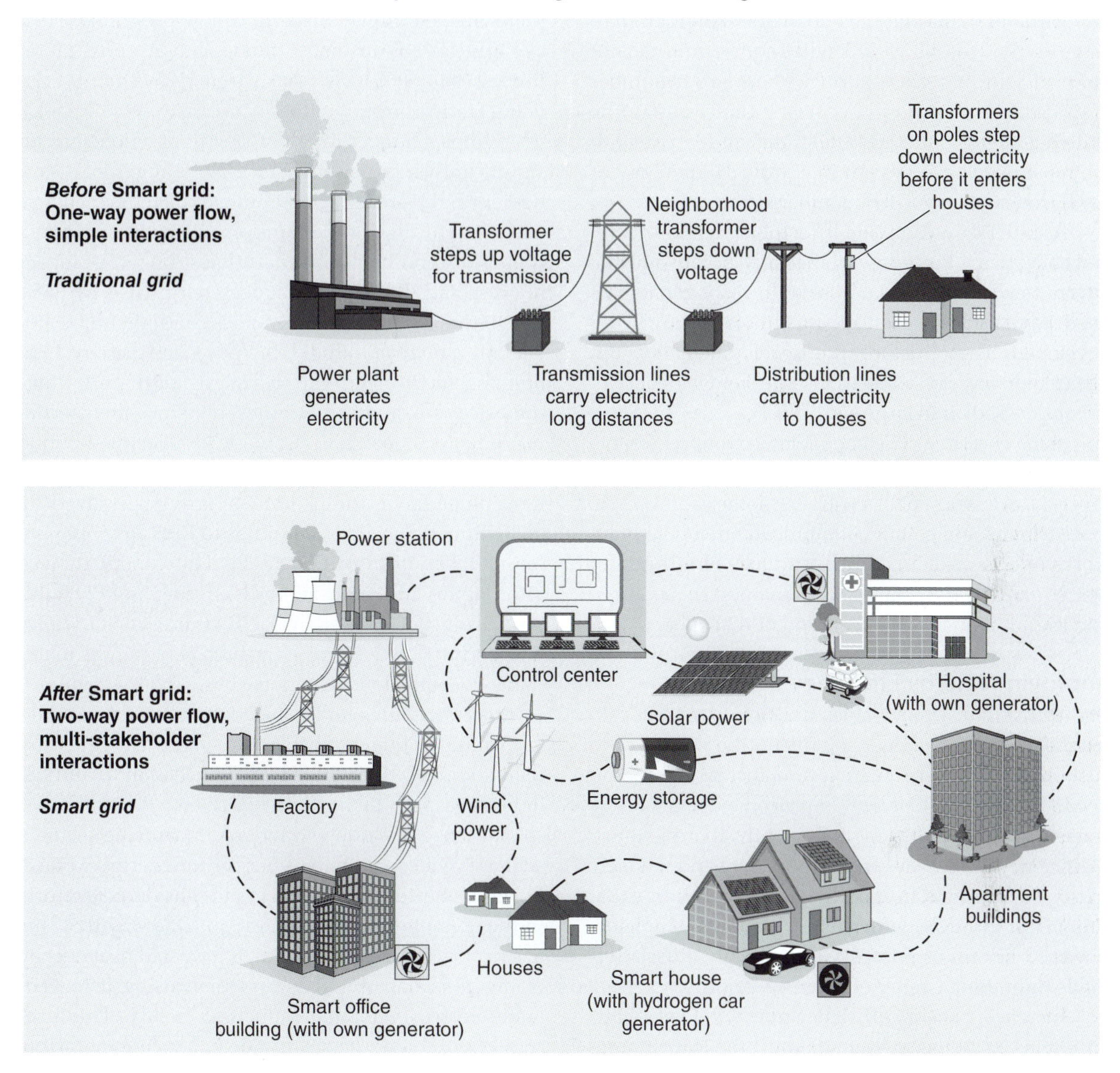

Figure 6.11 In contrast to the local and fragmented traditional grid, a smart grid is transcontinental in nature, links a diversity of energy sources with users, and is multidirectional. (https://goo.gl/images/JE52X9)

cables will be interconnected at strategically placed supernodes that can convert the current in either direction (AC–DC or DC–AC) and collect, integrate, and route renewable energy. The supernodes interconnect different regional and national grids for multilateral energy trading. The further development of low-temperature superconductor wire will reduce the amount of cable required to deliver the same amount of current and can be deployed underground, eliminating the massive deployment of aboveground high-tension wires, which are a common complaint of grid deployment.

There are a number of challenges in the further development of transcontinental smart grids. An obvious one is cost, but all indications are that capital investment in the smart grid is economically sustainable. One US Department of Energy study calculated that internal modernization of US grids with smart grid capabilities would save between $46 billion and $117 billion during the next 20 years.[8] Another challenge is that the new grid will require multilateral political will and cost sharing. There are a number of grid modernization projects well underway in the United States and Canada, Australia, and Japan, with the most developed system integrating the energy of the European Union (EU) member states.

Another key component to the transition to renewable energy is storage, especially because of the intermittency of wind and solar. In response to this need, advanced energy storage technologies are being developed. These include the development of giant lithium-ion battery installations, the storage of energy as compressed air in large geological caverns or vaults, and using excess electricity generated from renewable sources to cheaply produce hydrogen, which is in essence a stored form of electricity. Another approach is to distribute storage to geographically dispersed networks of energy-dense batteries in tens of millions of homes and use the smart grid to manage their charge and the subsequent redistribution of that energy.

Improving Energy Efficiency

The transition to renewables and new ways to distribute and manage energy will not be sufficient to sustainably meet our current and increasing energy needs without concurrent improvements in energy conservation and efficiency. Currently, there is much inefficiency in the ways that we generate and use electricity. For example, in 2008, the amount of energy that Russia lost through inefficiency could have completely powered Britain or France. Consequently, Russia has made improving energy efficiency a priority.[9]

Increasing energy efficiency now will lower GHG emissions and other pollutants until the transition to renewables is complete. It will also decrease the rate of our increasing energy demand, allowing for reduced rates of capital investment in renewable energy infrastructure. Increased energy efficiency provides cost savings for producers and consumers as long as the ROI of the improvement is within the lifetime of the item or entity being improved.

Energy efficiency can be increased through better energy management and distribution, as discussed previously. Another approach is to increase building and institutional energy efficiency, which is further discussed in the final section of the book when we explore sustainability at those levels. Energy-consuming products can be made more energy efficient, and there are a number of certification programs, such as ENERGY STAR, that use independent certification processes for products that save energy without sacrificing features or functionality.

Simple changes in the type of light bulb used can significantly increase efficiency. Switching from incandescent light bulbs to LED bulbs will reduce energy use by up to 75% for the same amount of light production. These bulbs also last 40 times longer. Compact fluorescent (CFL) bulbs reduce energy use by 10%, last 10 times longer than incandescent light bulbs, but contain mercury. Initially, the purchase price of LED bulbs seemed cost prohibitive at greater than 30 times the price of incandescent bulbs, but since their initial introduction, cost has dramatically declined while performance has improved. Since 2008, prices for LED light bulbs have dropped 90%. The money saved by switching 25 incandescent bulbs to LEDs over the life of the LED bulbs is nearly $7,000. This combination of affordability and cost savings is driving an LED lighting revolution. Goldman Sachs forecasts that LEDs will account for 69% of light bulbs sold and greater than 60% of the installed global base by 2020.[10] Numerous municipalities and institutions that have a long-term stake in their infrastructure and thus will easily realize the ROI have begun the conversion to all LED lighting, especially in street lights.

Because the industry and manufacturing sector is such a large consumer of energy and energy expense affects its bottom line, it has made considerable efforts in the area of energy conservation. One way that this has been achieved is by capturing wasted heat energy from industrial processes and then using it to generate electricity for the industrial facility. This process of combined heat and power, or **co-generation**, is becoming more popular, especially among high heat–producing industries such as steel and smelting facilities, but it is starting to be incorporated into industrial and non-industrial facilities that generate heat on smaller scales. Another large consumer of energy is transportation, and greater efficiency has been and can further be achieved in this sector, as discussed later in this chapter.

Increasing Access to Energy

Access to energy is an essential part of development and poverty reduction. In 2011, the UN General Assembly launched the Sustainable Energy for All initiative and in 2016 included increased access to clean energy as one of its Sustainable Development Goals (SDGs). The goals of these initiatives are to ensure universal access to modern energy services, double the global rate of improvement in energy efficiency, and double the share of renewable energy in the global energy mix, all by 2030.[11] This initiative embodies our definition of sustainability as best stated in the UN report by former UN Secretary-General Ban Ki-moon: "Energy is the golden thread that connects economic growth, increased social equity, and an environment that allows the world to thrive."

One large-scale project underway in response to the UN's agenda to increase global access to energy is the result of collaboration between the World Bank Group and Gigawatt Global Coöperatief U.A., a multinational renewable energy company focused on the development and management of utility-scale solar fields in emerging markets. They have constructed the first large-scale solar field in Rwanda, which will bring power to 12 million people and raise the nation's generation capacity by 6%. The 28,360-panel solar field was built on land owned by the Agahozo-Shalom Youth Village, a kibbutz-style orphanage for victims of the 1994 Rwandan genocide. This project is good for the environment, generates local employment and education, and empowers the country with access to electricity. Similar solar projects are underway in South Africa, Argentina, Nigeria, Jericho, and Turkey.

Solar alone cannot meet the objective of doubling the share of renewable energy in the global energy mix by 2030. Also, many state that merely doubling the share, which only brings the percentage of global energy consumption of renewables to 40–50%, is insufficient to slow climate change while meeting growing energy demands. To increase this share and access to the developing world, further investment in a diverse global portfolio of renewables will be needed.

According to the International Energy Agency, hydropower must be a significant part of the global energy portfolio in order to meet growing demand and reach those without access to energy. As previously discussed, we are currently extracting less than 20% of the technical potential of hydropower, and this number is much lower in the developing world. In energy-limited sub-Saharan Africa, there are large potential hydropower resources, and their proper development could make hydropower the African continent's most cost-effective, renewable source of energy. However, only 10% of this hydropower potential has been mobilized. This is beginning to change. For example, in the Democratic Republic of Congo, where only 2.5% of its hydropower potential is being tapped, a number of mid-size and large hydroelectric plants, including one of the world's largest hydropower sites, are being developed.

The expansion of hydropower is a cost-effective way to bring electricity to homes, power businesses and industry, light clinics and schools, spur economic activity, and create jobs in order to improve human well-being in the developing world, where access to energy is lacking. However, the environmental and social risks of hydropower need to be carefully managed. This will require support for environmental, institutional, social, and technical studies, as well as a full life cycle assessment of both impacts and benefits to maximize the sustainability of these projects. Much of this technical support is being provided by the World Bank, which recently provided a $73.1 million grant to the Democratic Republic of Congo to support these types of technical efforts in relation to their new hydropower projects.

In addition to large-scale, grid-based electrification with renewable energy, another way to more quickly and affordably increase access to energy is through smaller, decentralized community grids or mini-utilities that can meet the energy needs of households and small businesses. These mini-grids are often powered by biodiesel, biomass, hydro, wind, or solar, but on a small scale. Perhaps even more immediately effective and affordable are off-grid solutions that provide clean energy for light and cooking. Examples include cook stoves with higher efficiency, such as the BioLite clean-burning stove, which can be used both for cooking and for generating up to 10 watts of power to charge phones, lights, and other technological tools. Another innovation is the gravity light, which is driven by a weight pulley system that can provide 25 minutes of free, instant light each time the weight descends. Even simple rechargeable LED flashlights have been shown to improve the education, health, and security of those who have them. Other off-grid solutions include portable solar systems that can run lighting, charge batteries, and power small appliances, such as radios, mobile phones, and small computers, and larger fixed solar installations that can power refrigerators and larger appliances.

As with solutions presented in prior chapters, those that are the most sustainable involve community buy-in and responsibility, including individual payment of initial and ongoing services or products including those offered by for-profit companies. This is working with the rapid proliferation of the household off-grid solar industry in Bangladesh (Box 6.3). Assistance should be in the form of affordable financing or cost that allows for ownership and payment for energy-related products or the delivery of the energy. These costs will be offset by the elimination of other costs, such as reduced time spent searching for firewood and reduced money spent on kerosene and other fuels. Ultimately, access to energy will increase health, education, and economic opportunity, thus further developing the market for these products and the delivery of clean energy.

According to a report by the International Finance Corporation (IFC) of the World Bank, there is a significant market—$37 billion annually—that is currently being spent on kerosene used for lighting and biomass for cooking that could be used to invest in more sustainable forms of energy.[12] It also found that there are a number of emerging manufacturers, distributors, and service providers that sustain themselves economically and create jobs by offering this market clean energy and technological solutions with greater value, affordability, and fewer health and environmental impacts.

A program that embodies this idea is the Lighting Africa program, a joint initiative of IFC and the World Bank. The program fosters partnerships among local and global manufacturers and local distribution companies to meet the massive need for affordable off-grid

BOX 6.3

Stakeholders and Collaborators: Bringing Electricity to Rural Bangladesh

In Bangladesh, where nearly 50% households are without energy and the impacts of climate change are steadily becoming extreme, the need for renewable energy is an obvious priority. So how can renewable electricity be successfully brought to large numbers of households widely dispersed across remote, rural areas? The answer is through partnerships and ownership. The partnerships are a network of non-governmental organizations (NGOs) working with the World Bank and a government-owned financial intermediary as part of the Rural Electrification and Renewable Energy Development Project (RERED). Through an extensive network of NGOs, low-interest microcredit is offered to each household, which can then purchase its own off-grid solar system. These NGOs are known and trusted by rural households because they have historically provided microfinance for other income-generating activities. These systems are affordable, and the money spent on them is easily made up by eliminating the purchase of kerosene and other fuels and also by the elimination of time spent searching for combustible biomass to burn in stoves.

This program has been very successful. Since 2009, more than 50,000 solar home systems have been installed every month, creating more than 15 million solar households. As such, Bangladesh has the fastest growing solar home system program in the world. Access to electricity has greatly transformed the lives of the people in these new solar households. Indoor air pollution has essentially been eliminated. With electricity, small businesses such as tea stalls and repair shops are now possible in these homes. Home security has improved, especially for women. Children can now study and read in lighted rooms at night, and technologies such as computers and mobile phones are now an available option.

In addition to improving the lives of these people by providing access to renewable electricity, this project has resulted in a burgeoning solar industry, creating jobs and additional economic opportunities with hundreds of new companies such that Bangladesh now has the seventh largest renewable workforce in the world. These new local companies have a variety of missions, such as the production of high-quality monocrystalline silicon solar photovoltaic modules, installation and maintenance services, and provision of batteries and accessories. One company, Solaric, offers affordable, low-tech electronics so that off-grid populations can live high-tech lifestyles. This company alone has provided solar electricity to 35,000 households in 5,000 rural villages. It has replaced 130,000 kerosene lamps and reduced CO_2 emissions by 21,000 tons per year. The success of the RERED shows that sustainability is best achieved through partnerships, networks, and individual responsibility and ownership.

lighting products. For example, the Lighting Africa–Nigeria program was launched in 2015 with the objective to help 6 million people gain access to clean, modern, affordable lighting products while avoiding 120,000 metric tons of GHG emissions associated with current fuel-based lighting technologies.

Transportation

The transportation sector accounts for 27% of energy use worldwide. More than 90% of that transportation energy comes from petroleum, primarily for road transport of cars and trucks. In burning these fuels, transportation is responsible for more than 20% of global GHG emissions, with road transport responsible for more than 70% of this and air, rail, and shipping accounting for the rest. Clearly, we must change the way we transport ourselves and our goods in order to stop climate change in connection to the transition to renewable energy. Technical innovations that provide increased efficiency and renewable sources of fuel at the level of the vehicle are one way to accomplish this. Another is through a complete rethinking of our transportation systems.

Increased Efficiency

In the past 50 years, fuel efficiency has dramatically improved through the advent of fuel injection technology, the use of lighter materials and streamlined designs, and other mechanical innovations. However, this improvement has slowed, and the average fuel economy for automobiles has remained steady at 27.5 mpg, where it has been since 1990. A major obstacle to this has been the increased popularity of larger, inefficient SUV-style vehicles in the United States stimulated by artificially low gasoline prices due to government subsidies. The US government does mandate corporate average fuel economy for automobile manufacturers, referred to as CAFE standards. The current mandate is an average of 35.5 mpg across all models for each manufacturer, and this will rise to 54.5 mpg by 2025. The 2025 mandate is purported to cut GHG emissions in half. However, there has been recent political pressure to reduce the 2025 standard. Technologies that will improve vehicular efficiency include more efficient engines and designs, hybrid gas–electric vehicles, and completely electric vehicles. Electric car technology emerged in the mid-1800s and was popular until the introduction of Ford's mass-produced, gasoline-powered automobile. The demand and potential for mass-production electric cars re-emerged in the mid-1900s as concerns regarding fossil fuel use began to rise. However, many believe its further development and adoption was suppressed by automobile manufacturers, the oil industry, the hydrogen industry, and the US and state governments, as documented in the 2006 film *Who Killed the Electric Car?*

In recent years, the entry into the market of hybrids such as the Toyota Prius and fully electric cars such as the Tesla and the Nissan Leaf has stimulated interest and market demand for such vehicles. Electric cars can be recharged at home and through networks of publicly accessible charge points located on streets, at shopping centers, and at other specific charging locations. Numerous charging network initiatives are underway in countries and regions throughout the world. Some cities integrate car share programs with electric vehicles and offer priority parking for hybrid and electric vehicles.

The availability and sales of both hybrid and electric cars have steadily risen, but low subsidized fuel prices have limited their potential ROI and further expansion, especially in the United States. However, it appears that the automobile industry is anticipating large market growth for hybrid and electric cars. For example, in 2017, Ford announced that it will design and build seven new electrified vehicles in the next five years, including an F-150 hybrid pickup truck and a Mustang hybrid sports car. Volvo and Jaguar Land Rover have also pledged electrification of their future models.

Other efficiency-related technologies under development include waste heat recovery and regenerative breaking as ways to capture energy that can be used immediately or stored as electrical energy. Another technology is flexible fuel vehicles (FFVs), which run on gasoline or gasoline–ethanol blends of up to 85% ethanol; these have also increased in availability and use. There are also vehicles in development that run on clean-burning hydrogen fuel cells, and there are plans to develop hydrogen filling stations as well.

Although these new technologies improve efficiency and reduce fossil fuel use and GHG emissions, there is some question regarding the extent to which these new types of vehicles are sustainable. When the electricity used to charge these vehicles or make the clean-burning hydrogen fuel is generated using fossil fuels, the issues of renewability, fuel independence, and GHG emissions are not fully eliminated. This can be alleviated by linking electrical generation and hydrogen production directly to renewable sources of energy. This is being achieved by the development and expansion of solar electric and solar hydrogen filling stations. The problem with FFVs relates to those associated with agricultural ethanol production already discussed, but this can be eliminated through cellulosic production of ethanol using crop residuals and other wastes rather than food crops. Finally, for each vehicle type, a full life cycle analysis may shed light on other types of environmental social impacts, including the heavy reliance on rare-earth metals in their construction, which have been shown to have serious environmental and public health consequences on local communities where they are extracted.

Renewable Aviation and Shipping

Renewable fuels for other forms of transportation, such as flight and shipping, also need to be developed. With aviation, the goal is to develop so-called drop-in jet fuels that will mimic the chemistry of petroleum jet fuel and can be used in today's aircraft and engines without modification and without compromising performance. The most promising alternative is sustainable aviation biofuel that will allow airlines to reduce their carbon footprint and decrease their dependence on fossil fuels. Another motivating factor is to offset the risks associated with the high volatility of oil and fuel prices. A similar approach is being developed for marine shipping, which is central to the global economy. The focus for both aviation and shipping biofuels is to have them made from sustainable, non-food biomass sources.

Rethinking Transportation Systems

The increase in efficiency of individual vehicles can contribute to our sustainable energy future if it is directly linked to renewables, does not use biofuels that compete with our food and agricultural system, and if the new technologies have minimal environmental and social impacts during their complete life cycle. However, rethinking our entire transportation systems in connection with increased efficiency may more effectively move us toward sustainability. This can include enhanced use of individual vehicles, development of more effective mass transportation systems, and enhanced methods of shipping as the global economy expands.

Car share programs such as Zipcar, many of which now exclusively use electric vehicles, and separate highway lanes for high-occupancy or renewable-fueled vehicles will encourage more efficient use. New cab-sharing projects in dense urban areas that employ GPS and smartphone technologies, such as HubCab in New York City, could reduce the number of trips by 40%. Ride sharing services such as Uber and Lyft allow people to avoid purchasing a car, which results in a decrease in the overall number of cars on the road, reducing car emissions between 34% and 41% per year per each household using these services.

Integrating electric and hybrid-electric vehicles with the developing smart grid in what is referred to as **vehicle-to-grid** will enhance the overall efficiency of our renewable energy system. Recall that energy storage is a key component to the smart grid, and one way to achieve this is to distribute storage to geographically dispersed networks of energy-dense batteries. As the fleet of electric cars expands, it can provide storage capacity. The average car is used for less than five hours per day, and at any time of the day, 95% of them will be parked and connected to the grid fully charged. In times of high energy demand or shortage, a small portion of energy could flow from the batteries of these cars back to the grid. With two-way metering, the vehicle owners can and should be compensated or credited for their contribution to the grid.

Further development of passenger and freight rail service will be an essential component of sustainable transportation systems. Since the 1980s, light rail systems have been affordable ways to efficiently transport large numbers of people in urban cores. They also contribute to urban redevelopment, which, as discussed later in this book, is an essential component of sustainability. High-speed long-distance rail, which is well developed in Europe, needs to be expanded there and elsewhere in the world. As fuel prices for plane and automobile travel rise with decreasing fuel availability and subsidies, usage of these rail systems will increase. One way to mitigate the environmental impact of new rail construction is through road–rail parallel layout, which locates new railway tracks alongside existing highways, as was recently done in France and Germany.

Integrated development of urban transportation systems with the objective of replacing individual automobiles with mass transit systems can drastically reduce energy use and GHG emissions. Integrated transportation planning is an example of systems analysis or thinking that is best achieved by assessing current and future human behavior in conjunction with structural constraints of a city or region. This is examined further in Chapter 11 as we explore community, urban, and regional sustainability.

One example of a successful integrated transit system is in Curitiba, Brazil, which is home to and serves 3.5 million people. Rather than light rail, its transportation plan is based on bus rapid transport (BRT). BRT features exclusive bus lanes, traffic signal priority for buses, preboarding fare collection, and large-capacity bus boarding from raised platforms in tube stations with free transfers throughout overlapping bus services. Up to 80% of travelers in Curitiba use the BRT as their primary source of transportation, with an 89% user satisfaction rate. The BRT is saving 27 million liters of fuel annually, and as a result, Curitiba uses 30% less fuel per capita than comparable cities. Curitiba now serves as a model of integrated mass transit that is being adapted in other cities such as Cape Town, South Africa.

Another component of a sustainable energy system is to increase and improve opportunities for cycling, walking, and new forms of low-impact transportation. New urbanism is an urban design movement that promotes walkable and bikeable communities. Efforts to improve urban cycling include the development of a network of safe cycling lanes and increased bike parking opportunities. Bike share programs that are currently present in many cities can contribute to these efforts. Among the most bike-friendly cities in the world are Amsterdam, Copenhagen, Seville, Tokyo, and Rio de Janeiro. In Berlin and Vienna, there are a number self-governed residential projects in which residents work directly with architects to create amenities that support urban cycling and make the use of cars less convenient.

New forms of low-impact transportation include electric-assisted and fully electric bicycles or e-bikes that have ignited multiple entrepreneurship opportunities. Examples include the Copenhagen Wheel, which transforms any bicycle into a smart electric hybrid. The affordable fat-tire e-bike developed by Storm Sonders and the new Unimoke two-person utility e-bike designed and built by Urban DriveStyle (Figure 6.12) are new innovations that have been supported through crowdfund sites such as Indiegogo and Kickstarter. Another new start-up, Boosted Boards, led by visionary Sanjay Dastoor, is promoting electric power long boards as part of the future of sustainable transportation. These boards are portable,

Figure 6.12 The Unimoke utility e-bike designed by UrbanDrivestyle that can be adapted for personal, emergency, maintenance, delivery, and other forms of transportation is powered by a pedal assisted electric motor. Its development was supported through the on-line crowdfunding page Indiegogo. Pictured here is the author riding his own Unimoke on the of Campus of Muhlenberg College. *Photo by T. Amico, courtesy of Muhlenberg College.*

can travel a range of seven miles uphill, and can travel at speeds up to 22 miles per hour with breaking capacity. These are just a few examples of how the transition to sustainable energy and transportation systems linked with social entrepreneurship can spur innovation, create job growth, and stimulate the economy.

Barriers to a Renewable Future

Despite the obvious and impending negative consequences of fossil fuels, and the significant benefits and potential opportunities of renewables, the transition has had a slow start. According to the Renewable Energies Policy Network,[13] as of 2017, renewable energy production had remained steady at approximately 19% of global energy consumption for the prior 10–15 years and at less than 10% in the United States. This is the result of a number of political, economic, and social barriers to the market penetration of renewable energies. In this section, we identify these barriers. The following section explores ways to overcome them.

Government Subsidies Favor Fossil Fuels

An energy subsidy is a payment from the government that keeps the price of energy lower than the market rate for consumers or higher than the market rate for producers. Examples include direct payments to companies, tax exemptions and rebates, price controls, government funding for research and development, and government-provided insurance. Subsidies have favored fossil fuels, keeping their prices down, limiting investment in renewables, and adversely affecting the competitiveness of renewables in the marketplace.

According to the International Energy Agency, worldwide fossil fuel subsidies have been nearly six times higher than subsidies for the renewable energy industry at approximately $523 billion compared to $88 billion, respectively.[14] Government spending to reduce retail prices of gasoline, coal, and natural gas has steadily increased. This creates market distortions that encourage wasteful consumption of fossil fuels and discourage the transition to renewables.

There are two recent trends with regard to fossil fuel subsidies. First, direct subsidies on fossil fuels are being reduced while indirect subsidies in the form of tax breaks for exploration and extraction of fossil fuels and reduced lease rates on federal lands are increasing. Second, although there has been a recent increase in renewable subsidies, much of this has been for the production of biofuels as an indirect way to subsidize agriculture, but as discussed previously, this is quite problematic for agriculture, energy production, and the environment.

Those who are against eliminating or changing energy subsidies argue that subsidies keep fuel prices low and that without them the economy will suffer. There is also pressure on politicians to maintain subsidies, both from the corporations that finance their campaigns and from the public that benefits from lower fuel prices. Others make the case that for fossil fuels the margin of profit is low and risks are so high that without subsidies, further exploration might be inhibited. Still others consider the dollar amount of subsidy per unit of energy produced. Of course, because fossil fuels represent more than 80% of current energy consumption, this number is approximately 25 times higher for renewables. However, these arguments completely ignore the external costs of fossil fuels, including climate change, their rapid depletion, and increasing global energy demand. Subsidizing the dominant market player reduces competition, innovation, and the potential societal benefits of renewables. It contradicts the notion of free market capitalism and reflects the power that fossil fuel–related businesses have over government spending.

Economic Barriers

In addition to subsidies, there are a number of other economic and market barriers to the transition to renewables. Investment in renewables is limited by a highly controlled energy sector, as well as high investment requirements and capital for entrepreneurs and larger companies. Uncertainty due to the lack of information and awareness of renewables, in addition to the perceived high payback period or ROI, also limits financing possibilities. Higher costs because of economies of scale are also serious impediments to the growth of the renewable industry. Start-up costs have been a considerable barrier to bringing renewable energy to developing countries, many of which are currently without any form of power generation. Also, complex geopolitical forces in combination with new fossil extraction efforts have kept the prices of oil and natural gas at record lows, making it more challenging for renewables to compete and expand in the marketplace.

Infrastructure, Institutional, and Cultural Barriers

Another barrier to renewables is that the current energy infrastructure is based on the tradition of fossil fuels, in which large-scale power plants transmit energy in one direction to the point of consumption. Renewable energy systems will require a multidirectional grid infrastructure and other technological innovations in energy transmission. Institutional barriers exist because many industrial or physical plant managers are trained only to find low-cost solutions and may be unfamiliar with renewables. Also, there have been federal tax credits as high as 30% for investing in renewables, but tax-exempt entities such as universities, communities, and cities that are often scale-appropriate for renewable investment cannot take advantage of incentives paid through tax codes.

There are a number of cultural barriers to renewables. Even as the public increasingly recognizes the importance of transitioning to renewals and in spirit would support most renewable projects, it often opposes them. The bulk of this opposition comes from supporters who do not want the project near them. This "not in my backyard" or NIMBY argument crosses political lines and has come from ardent supporters of renewables. It is also reminiscent of the legacy of environmental injustice where energy projects were placed only in the poorest neighborhoods.

As previously discussed, the public rejection of renewable projects, whether they be small or large, is often due to aesthetics or environmental reasons. Large solar or wind farms are often deemed unsightly, destructive to the beauty of the environment, and could potentially lower real estate values. An offshore wind farm off of Cape Cod, Massachusetts—a location where residents would generally support renewables—was delayed for years partly because of aesthetic concerns. We have also explored many of the environmental concerns with specific types of renewable energy, led by protests against potential bird mortality from wind farms. However, we have recognized that none of these negative impacts come close to the devastating global effects of fossil fuels and climate change.

Similarly, there are a number of barriers to individuals interested in adding renewable energy systems to their homes. These systems include solar PV panels, solar hot water systems, wind turbines, and geothermal installations. Other individuals are interested in going completely off the grid. Barriers to these types of individual projects include cost, time of payback, obstruction from local utility companies, and zoning restrictions. The latter is particularly an issue in historic districts and communities with homeowners associations, and it is typically driven by aesthetic concerns.

Overcoming Barriers to a Sustainable Energy Future

We have seen that the transition away from fossil fuels toward renewable energy is vital for a sustainable future. Yet, there have been and still are a number of barriers to this transition. In this section, we review policy changes that will foster the move to renewables. Currently, renewable energy supplies between 10% and 15% of the global energy demand (see Figure 6.1). The International Energy Agency projects that due to rapidly growing global energy demand, coupled with the lack of new policy initiatives, fossil fuels will account for more than 90% of total primary energy consumption in 2020.[15] Continued dependence on fossil fuels will further catastrophic climate change and other environmental damage. There will also be more conflict over oil, less opportunity for energy independence, and continued limited access to energy in developing countries. Local, national, and international policies are needed that will resolve or prevent these problems from occurring.

Change Subsidy Policy

The most obvious policy recommendation would be for governments to stop subsidizing fossil fuels. Cutting those subsidies would be economically efficient, reduce overall energy consumption, and level the playing field with renewable power. The International Energy Agency suggests that removing fossil fuel subsidies would reduce CO_2 emissions by as much as 2.6 gigatons per year by 2035. That is half of what is required to prevent the planet's average temperature from increasing by 2°C or more per year. There will be short-term political and economic consequences for removing these subsidies; therefore, enacting new policy that eliminates them will be difficult.

Politically, it may be easier to create subsidies that encourage renewable development than to eliminate fossil fuel subsidies. Several governments provide capital subsidies for installation of renewable energy systems. Tax exemptions and credits and third-party, low-interest financing mechanisms are ways to provide incentive for residential and industrial renewable energy programs. The US federal residential renewable energy tax credit allows for a personal income tax credit for 30% of the total system cost for the installation of renewable energy systems with no upper limit. This was due to expire at the end of 2016 when it was extended with a gradual reduction in rate through 2022. However, in 2017, Scott Pruitt, the newly appointed head of the EPA, proposed eliminating this extension arguing that renewables should stand on their own and compete against coal and natural gas and other sources as opposed to being propped up by tax incentives and other types of credit while ignoring fossil fuel subsidies. As of 2018, the extension for this tax credit is still in place, but its future remains uncertain. Pruitt has since resigned from the EPA, but his replacement shares similar pro–fossil fuel/anti-renewable viewpoints.

Some have argued that creating subsidies for renewables as a way to balance out fossil fuels subsidies is equivalent to creating a monster to slay a monster. Perhaps the best political and economic approach is to first create subsidies that encourage investment in renewables, but with the ultimate goal of reducing or eliminating all government subsidies for energy. Also, international subsidies from aid agencies should link rural development where energy infrastructure is poor or lacking to the deployment of renewable energy systems. As the market for renewables grows and fossil fuels become more expensive, the transition to a renewable energy economy could occur in a less drastic way with reduced political and economic ramifications.

Internalize Externalities

Another type of subsidy that keeps the price of fossil fuel down for consumers and profits up for producers is lack of accountability of externalities. Producers are typically not held financially responsible for the environmental damage and related impacts on health that their activities cause. These costs are passed along either directly or indirectly to the public. One striking example of this is the complete exemption of liability for environmental and health damages caused by fracking companies in some states (see Box 6.1). By not accounting for externalities, the true costs of fossil fuels are kept artificially low, thereby creating an unlevel playing field that makes renewables less competitive. One way to solve this problem is to develop ways to calculate externalities so that corporations can be held financially responsible for the damages that they cause. This would help level the playing field between fossil fuels and renewable energy, and it would make renewables more economically attractive and profitable.

Support the Development and Deployment of Renewable Energy

Equally, if not more, important as changing or eliminating subsidies are policies that encourage research and development and investment in renewable energy. This support could come from redirecting subsidies and funding of the misguided policy focus on "clean coal" toward the development of renewables. As previously discussed, spending billions of dollars on new coal-fire energy plants is a bad long-term investment as well as bad for the environment.

Governments can support research and development of new technologies through competitive grants, the creation of technology incubators and public research centers, and private–public partnerships. The ultimate goal of these government projects should be to transfer that technology to the private sector for commercial development. Research efforts should be focused on new renewable technologies. This should include development of the infrastructure to manage the production and distribution of electricity from a variety of renewable sources based on demand and production information. Because renewable energy systems will span regions, nations, and continents, national and international funding support and cooperation will be needed for their development.

In addition to research and development, renewable energy projects require greater amounts of initial financing than comparable conventional energy projects already in place, and there is a perception of greater up-front risk. Thus, capital markets may demand higher lending rates for renewable projects. Policies that provide incentives for investment and reduced lending rates will ease the transition to renewables. They will also spur economic development and create jobs.

Creating policy that promotes renewable energy has been a priority for many EU countries in an effort to create more jobs, especially in Germany and Greece. India has used disincentives for using fossil fuels to then promote renewables by doubling the tariff on coal. It has then used these revenues to provide low-cost financing for renewable energy projects and to support clean energy research. This has resulted in the growth of a vibrant renewable energy economy in India. Domestic policy in China, including limits on GHG emissions, has stimulated investment in the development, deployment, and export of renewable energy technologies, making it a global leader in this newly emerging industry. It is also understood that investment in renewables will create large domestic supplies of affordable energy, which in turn will attract international industries to those countries.

A number of countries are developing policies for banning fossil fuel–based vehicles. For example, in 2017, the United Kingdom and France promised to ban sales of these vehicles by 2040. China, which has been the world's largest market for passenger vehicles since 2009 and is set to represent 30% of the global car market by 2025, is promoting a ban on non-electric cars, but it has not yet set a timeline for the ban. India, the Netherlands, Norway, Scotland, and Germany have also made such pledges to phase out gas vehicles before 2040. Even the state of California is exploring policy options for eliminating fossil fuel vehicles.

Critics of renewable energy development and policies that promote it have claimed that it will slow economic growth and increase the cost of energy. However, investments in renewable energy projects create significantly more jobs per unit energy produced than do investments in fossil fuels, and these projects reduce energy costs.[16] We have also seen that policies that promote electric and hybrid vehicles and the development of infrastructure to support them have spurred innovation and the expansion of that market. Despite the knowledge that such policies stimulate the economy, create jobs, allow for energy independence, and can slow climate change, the United States has failed to enact such policies. The American Clean Energy and Security Act of 2009 contained numerous carbon reduction measures, among which were important incentives for capital investment in renewables. However, this bill failed to pass in the US Senate. This has been a missed opportunity by the United States, where the political climate makes it unlikely that legislation to promote both the development of renewable energy and control of GHGs and climate change will pass in the near future.

Despite the lack of policy at the federal level, the US state-level commitment and investment have resulted in local economic gains. For example, although it is a smaller state, Massachusetts has created twice as many jobs in clean energy as Pennsylvania has created in the natural gas/fracking industry.[17] As this economic and environmental advantage becomes more apparent, more states may pass policies that will further promote clean energy investment and deployment. Even in the face of current national policy emphasizing coal and fossil fuels over renewables, US clean energy jobs have surpassed those in the fossil fuel industry by five to one.[18] More rational federal energy policies and increased resources to promote them could speed up the transition to a renewables-based economy as the United States plays catch-up in this rapidly emerging marketplace.

In the absence of strong national policy that promotes renewable energy, the development of public–private partnerships similar to those in other sectors may work better in the United States and other countries. The creation of strong public–private partnerships that provide income-generating solutions coupled with needed education in renewable technology will be necessary to make investment in renewables more attractive in the United States and other reluctant nations. These partnerships could potentially provide funds needed for the development, construction, and maintenance of capital-intensive renewable projects that are essential in the pursuit of a vibrant and sustainable clean-energy industry.

Guaranteed Markets

Because of existing barriers such as fossil fuel subsidies, renewable energy has been less able to compete in the energy market. A law that required energy suppliers to include renewable energy as part of the energy they provide could help solve this problem. Laws such as this exist in numerous EU countries and in the United States, in which the Renewable Portfolio Standard requires each retail supplier of electricity to provide a specified percentage of renewable energy in its electricity supply portfolio.

Carbon emissions trading policy, also known as cap and trade, is a market-based approach that could provide economic incentive to transition to renewables and further guarantee markets (see Box 3.2). A similar device is the trade of **renewable energy certificates (RECs)**. RECs can be traded like emissions credits but are based on the amount of energy produced renewably rather than on emissions. A green energy provider is credited with one REC for a specific amount of electricity (e.g., one for each 1,000 kilowatt-hours) that it produces from renewable sources such as a wind farm. The provider can then sell these certificates to states or companies that are not meeting a minimum for the proportion of their energy portfolio that is renewable. The minimum is determined by policy in the 29 US states that require some level of compliance. For example, a California law mandates that 33% of electricity produced must utilize renewable resources by 2020. The purchase of the REC is used to offset the buyer's non-renewable production of electricity as a way to meet state standards. Through RECs, these state policies create a guaranteed market for renewable energy even if it is produced in a different area of the country. Proposed carbon taxes (see Chapter 3) will also promote renewable energy without a direct mandate.

Green Investments

Government budgets alone will never be enough to initiate the transition to renewable energy or to mitigate and adapt to climate change. Funds from the private sector and institutional investors will also be required. One innovative way to fund clean energy, transportation, and other low-carbon projects is through green bonds. Green bonds provide investors high-quality, fixed-income investments that make a difference. They appeal to a certain class of investors focused on ethical choices such as social choice funds offered by many mutual fund companies.

Since 2008, the World Bank has issued nearly $8 billion worth of green bonds in 18 currencies, and the International Finance Corporation has issued $3.7 billion in green bonds. Much of these funds are invested in clean energy projects that will both provide an ROI and foster the transition away from fossil fuels to renewables. The growth of the green bond market from the private sector as well as local and national governments will be an important source of capital and investment for energy projects with environmental benefits. Institutions with large investments, such as endowed colleges and universities, are starting to divest their funds from fossil fuel companies, much like the large-scale divestment from South African industries during apartheid in the 1980s. This is an opportunity for these institutions to model and educate about and also influence the transition from fossil fuels to mitigate global climate change. The use of investment and divestment as a tool in effecting change is further discussed in Chapters 9 and 10.

Energy Sector Liberalization

Energy liberalization is a type of policy measure that restructures and deregulates the energy sector to introduce competition. One way that this is done is to decouple energy generation from energy distribution and offer the consumer choices among different generators. This is occurring in a number of US states and in the EU primarily as anti-trust policy to encourage competition among suppliers and choice by consumers.

By decoupling generation from distribution, a number of electricity providers are able to offer consumers the option to purchase renewable or green energy either directly from producers or indirectly through RECs, as described previously. One example of such a supplier in the United States is Green Mountain Energy, which through consumer choice and the use of RECs has prevented the emission of 24.5 billion pounds of CO_2 and has helped develop more than 65 solar and wind power facilities while providing renewable energy at relatively competitive rates. Currently, only 17 of the 50 US states offer retail choice programs to commercial and industrial consumers.

Energy Buy-Back

There should be policies that require or encourage electric utility companies to participate in energy buybacks from household producers of renewable energy. This can occur through two-way movement and metering of electricity either from or to the grid. In this way, households or other small producers of renewable energy can feed surplus electricity back to the grid. They can be credited for this through metering that shows the net consumption of power (i.e., the difference between electricity brought in from the grid and that produced by the household and pushed out to the grid). In this way, they can then be compensated for their net production of electricity, typically at specific rates referred to as feed-in tariffs or as renewable energy credits. Currently, feed-in tariffs for renewable energy exist in more than 40 countries, and they are widely considered one of the most effective ways to increase household solar energy. The rate is typically determined at the state or regional level. It should be 100% of the utility price, but it is often considerably less. Policy should be put in place to guarantee a fair and reasonable price to homeowners who sell renewable energy back to the grid.

In addition to selling energy back to the electric utility, policy should allow for direct **peer-to-peer energy sales**. This bypasses the power company and allows home producers of excess energy to earn a better rate while providing peers or neighbors with energy at prices lower than those generated and transmitted by power companies. This is something that Power companies tend to oppose because it cuts into their profits. As a result, peer-to-peer energy selling has been very slow to emerge. A notable exception is in the Netherlands, where a company called Vanderbron provides a platform for making peer-to-peer energy sales. This platform and policy have allowed excess energy from household producers to be sold and power approximately 20,000 households. A first step toward peer-to-peer energy sales is deregulation of the energy sector, as described previously. However, although this has occurred in a number of US states, we have yet to see peer-to-peer energy sales as a possibility. Policy and platforms such as those in the Netherlands are needed to encourage this novel way to distribute energy.

Local Policy

In many US municipalities, the placement of solar panels and other types of renewable energy infrastructure is restricted primarily for aesthetic reasons. These types of policies need to be changed or blocked. For example, in 2014 in Vermont, a bill to give towns the ability to restrict the placement of solar panels did not pass for fear that it would slow renewable energy growth. Vermont has established a goal to source 90% of its energy from renewable sources by 2050. Other states fight back against the restrictive nature of homeowner associations (HOAs) such as those in historic districts. For example, in April 2014, Minnesota's state legislature passed an energy bill that included the prevention of HOAs from prohibiting the installation of solar panels.

Another approach is for municipalities or regions to locally source their energy from only renewable sources. For example, the city of Boulder, Colorado, adopted an energy localization framework that requires energy to be generated locally or within the region using renewable resources rather than external fuel sources (Box 6.4). As more states and municipalities begin to understand that the need for renewable energy far outweighs aesthetic concerns and that it can give their citizens more control over their energy, these sorts of policies will begin to proliferate.

Microfinancing

An additional way to help support the expansion of renewable energy, particularly in developing countries, is through low-cost financing that is provided to homeowners so that they can purchase affordable solar systems. In Bangladesh, where nearly 50% of households or 78 million people are without electricity, there has been much success through the Rural Electrification and Renewable Energy Development Program. This program is supported by a government-owned financial intermediary, the World Bank, and other development partners that channel funds to non-governmental organizations, which in turn provide microcredit to households for buying the solar home systems. Through this program, they reach 50,000 households a month and have already improved the lives of more than 15 million people who previously were cooking and heating their homes by burning scavenged wood, using kerosene for lighting, and using diesel to run small generators. This approach has also stimulated local business development that builds, installs, and services these panels, resulting in Bangladesh now having the seventh largest renewable workforce in the world (see Box 6.3).

International Policy and Energy Conflict

International energy markets are undergoing a transformation. This is due both to new discoveries and sources of fossil fuels, such as that in the United States through unconventional extraction, and to changing energy demands such as the growth of consumption in China, which will soon surpass the United States in its scale of oil imports. Also, geopolitical change such as the breakup of the Soviet Union and new alliances like the EU change the dynamics of international resource supply and demand, as does the development of energy independence with increases in renewable energies. Finally, as fossil fuel supplies become depleted, prices will rise and incentives for acquiring

BOX 6.4

Collaborators and Stakeholders: Localized Power in Boulder, Colorado

The people of Boulder, Colorado, have been on a collaborative journey to make their city one that runs on sustainable energy. The journey began in 2002 when Boulder became one of the first cities in the nation to pass a resolution in support of the Kyoto Protocol, establishing the goal of reducing greenhouse gas emissions to 7% below 1990 levels by 2012. To do this, the city established a carbon tax in 2006. Then, after a 20-year contract with Xcel, a major regional energy company, expired in 2010, the City of Boulder passed two measures in an effort to localize its power supply for the sole purpose of reducing its impact on the planet. The first authorized the city to increase the utility occupation tax by up to $1.9 million a year to acquire the existing electric distribution system from Xcel. The second authorized the city to create its own locally owned municipal light and power utility.

The goal of creating its own local, municipal utility was so that Boulder could make its own energy decisions—decisions that would transform its energy system from one that was fueled mostly by carbon-intensive fossil fuels to a grid based on carbon-free renewables. The people were motivated by the environmental impacts caused by carbon-based energy production and increasing energy costs. They wanted to support local firms and innovators that are responding to a rapidly changing energy market and also to make investments in a resilient infrastructure for the city's future generations.

The people of Boulder have largely succeeded, but not without a battle. Originally, the city worked directly with Xcel to get it to develop a plan to transition to renewable energy. When that failed, the city started the process of democratizing its energy. This was done through grass-roots efforts that brought together stakeholders from throughout the community, including non-profit organizations, youth advocates, and religious institutions. But Xcel fought back by first spending more than $1 million in a campaign of fear to kill the measures. When democracy won, Xcel did not stop. It was motivated not just by the potential loss of $35 million in profits but also by fear that this would serve as a model for other communities. So Xcel funded a ballot initiative to undo the community's work. Then to promote it, Xcel spent millions more than the advocates for local energy—an obscene example of corporate control over democracy.

Although they were outspent, the people of Boulder won and defeated the ballot initiative, and on September 14, 2017, the Colorado Public Utilities Commission delivered a ruling saying that Boulder could move forward with its negotiations with Xcel to acquire the utility. The City of Boulder is prepared to reliababy take control of its electric utility, and to provide electicity to its people at lower rates than those of Xcel, following a specific framework to fully transition to renewable energy sources. However, as of July 2018, they were still working on finalizing the municipalization process, and a number of financial road blocks have further slowed the process. In the meantime, Xcel continues to fight to maintain its control by offering to generate 55% of its electricity with renewables by 2026 if Boulder was to kill its municipalization effort. These offers are much less ambitious than Boulder's goals and are claimed to be laced with fine print. Because of this, Boulder is holding fast to its vision for a community-owned, renewable electric utility to both fight global climate change and to keep the model of democratizing energy alive for hundreds of other communities.

This is a story about democracy. It is about different stakeholders coming together to make a change. They empowered their community to say "no" to fossil fuels and have created a model for other communities to do the same. It has also been a lesson on how slow and cumbersome the democratic process can be, particularly when the goal of that process is for the people to legitimately seize control from a powerful corporation in order to achieve a sustainable future for their community and the planet.

territory with those resources through conflict will increase. All of these factors will further impact the geopolitical and international security consequences of energy.

In a 2014 publication *Fueling a New Order?*,[19] it is argued that both global institutions and leadership of current and emerging powers must develop policy that will reduce conflict over energy. Despite a recent proliferation of global institutions with this focus, they have not yet developed an effective system for global energy governance. This must be a priority if we are to resolve and prevent international security issues related to energy, but tense geopolitical issues are currently a major obstacle.

A Global Energy Agenda

Compared to other sectors, there has not been a clear and consistent global energy agenda or policy. Given that so many people in the developing world lack access to energy, and the strong connection between energy access and poverty, this needs to change. Although improving access to clean, renewable energy was not one of the Millennium Development Goals, the UN now recognizes that universal access to clean energy is a foundation for meeting all of the goals, and making it a key priority on the global development agenda is one of the 2016 SDGs. There will be devastating consequences if we do not address the increasing global energy demand among the poor by providing access to affordable clean energy as part of development. This can be achieved through the implementation of policies previously discussed, such as microfinancing and the different ways to make renewable energies affordable.

Within the next 5–10 years, the renewables market is expected to become one of the fastest growing in the world. Developing countries need the opportunity to participate in that market as one of many ways to help bring their people out of poverty. Moreover, if we do not make this transition to renewables as quickly as possible, climate change will continue to have serious consequences for the world, but it will disproportionately affect developing countries. The development of a global strategy for renewable industry that stops pollution and climate change, creates equitable economic opportunity, and brings social justice through poverty reduction is a key component to the sustainability of our people and planet.

Assesing Our Sustainable Energy Future

It is easy to develop a number of indicators to assess the transition to renewables. An obvious one is the increase in the percentage of energy used that comes from renewables by a specific year. These goals can be made and assessed at community, national, and international levels. For example, many have argued that the goal to power the United States with 100% renewable carbon-free energy by 2030 is reasonable and achievable. Similar indicators can be developed for increasing access to energy in areas of the developing world without electricity. Goals and indicators for reducing GHGs and mitigating climate change, energy independence, and reduced conflict over energy must also be central to our assessment process.

There are also formal ways to assess integrated transportation systems. One example is the Low Emission Transport Toolkit offered for free by the Low Emissions Development Strategy Global Partnership (http://en.openei.org/wiki/LEDSGP/Transportation_Toolkit). The toolkit supports development planners, technical experts, and decision-makers at national and local levels to plan and implement low-emission transportation systems that support economic growth. It provides formal methods of assessment, including baseline data, the development of alternatives, and monitoring improvement from those alternatives.

For each form of renewable energy or transportation, we must perform a full life cycle analysis that includes both environmental and social impacts. None of the solutions presented here are perfect. For example, wind power has been criticized for its impact on bird populations. However, this must be considered in light of the fact that if we do not quickly transition to renewables, climate change may have a greater impact on birds than that of all wind turbines combined. There will always be compromise, but the goal is to minimize the negatives. At this point, solar PV seems to have the least environmental impact, is the most easily deployed, and is currently the most affordable form of renewable energy. Clearly, it will play a major role in the portfolio of energy sources that sustains our energy future and mitigates climate change.

Another approach to assess the sustainability of a particular policy or technology is through systems thinking. Through the development of causal models, systems thinking offers a holistic view that helps us better understand how what is being examined is affected either directly or indirectly by other factors in the system. Such models consist of feedback loops that reveal how a factor might indirectly influence itself over time (Figure 6.13).

A systems approach has been used to analyze the sustainability of wind energy.[20] Figure 6.13 shows an example of a feedback loop and a causal model generated in the analysis. The feedback loop reveals that an increase in the number of wind projects will ultimately result in further increases through the generation of profit and further investor interest (see Figure 6.13).

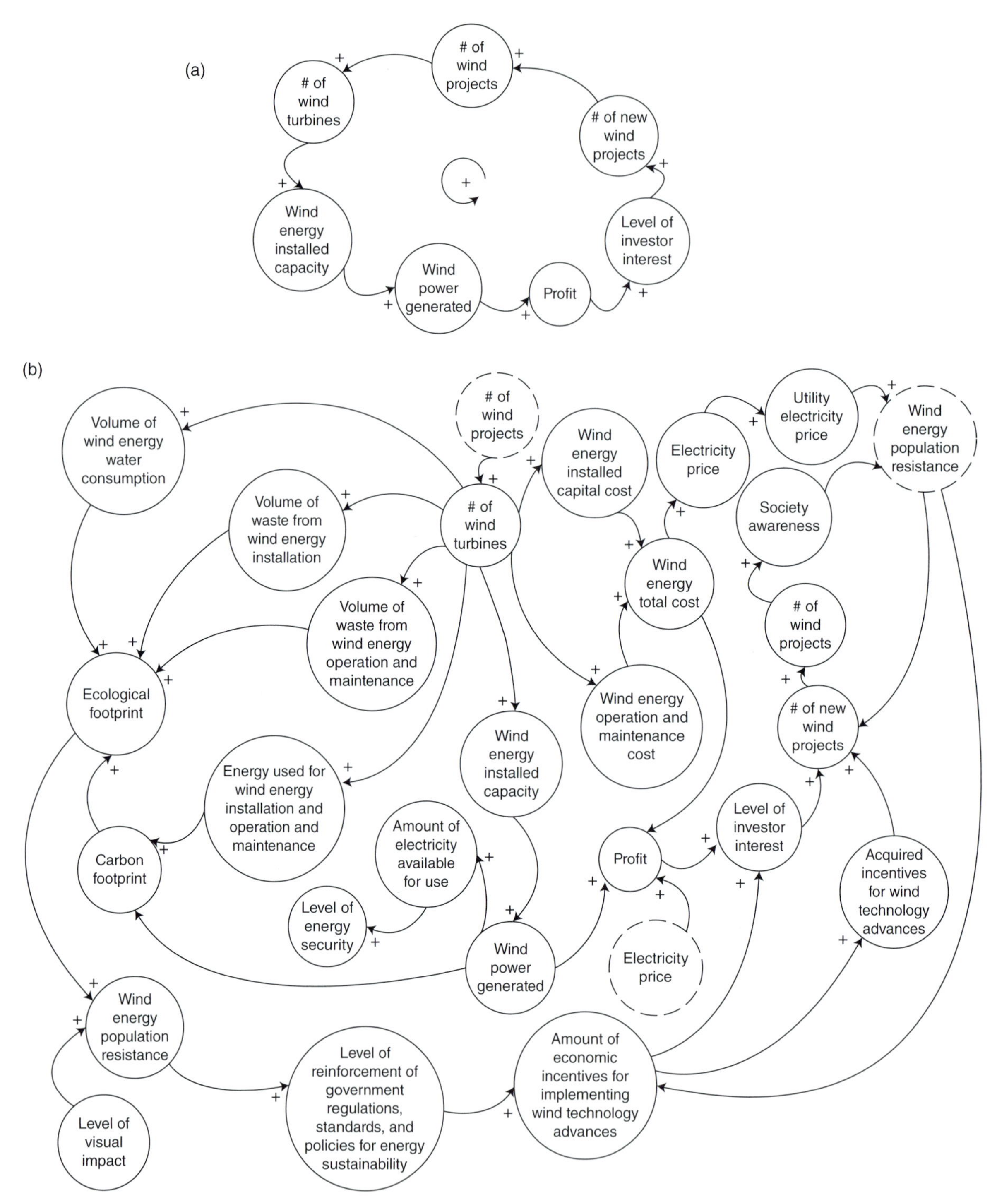

Figure 6.13 Systems thinking to analyze wind energy sustainability. A. An example of a feedback loop as part of a systems model for analyzing wind energy sustainability. Circles represent elements or factors, and arrows represent relationships in which factors can influence other factors either positively or negatively. B. A subset of a wind energy sustainability causal model integrating a number of feedback loops. (Adapted from Tejeda, J., & Ferreira, S. (2014). Applying systems thinking to analyze wind energy sustainability. *Procedia Computer Science* 28: 213–220.)

The more complete causal model shows how a multitude of factors in all three realms of sustainability either negatively or positively influence each other. For example, economic incentives increase the number of wind energy projects through increased investor interest. Also, societal awareness increases the amount of economic incentives offered by decreasing population resistance to wind energy (see Figure 6.13).

The Transition is Underway

There are many obstacles to achieving the transition from a fossil fuel economy to one based on renewables. These obstacles are political, institutional, cultural, and economically driven. However, these barriers are beginning to be broken. The economic barriers that once limited solar and wind energy are now gone. As market demand and innovation in other forms increase, this will happen with other forms of energy. Barriers are also broken when solutions involve community buy-in and responsibility, including individual payment for initial and ongoing services or products including those offered by for-profit companies focused on benefiting society.

In the United States, many view political barriers to renewable energy as insurmountable, primarily because of corporate influence over politics leading to the promotion of a climate change denial movement. This occurs because campaign finance allows for well-funded super political action committees (PACs) such as Citizens United to spend unlimited amounts of money furthering political causes. For example, the Koch family, which has annual revenues well over $100 billion from the fossil fuel industry, is funding numerous PACs to preserve its fossil fuel interests, including trying to halt walking, biking, and mass transit projects. One way to deal with this is to link the climate change and renewable energy movements to campaign finance reform.

There is hope. Voters from both parties are beginning to see through the well-funded climate change denial movement. A recent survey of Republican and Republican-leaning independent voters showed that the majority believe that climate change is happening and that something needs to be done about it. An even larger majority of this group (77%) believe the United States should use more renewable energy sources in the future.[21] Yet, 72% of elected Republicans are not responsive to this view in their own constituency, continue to ignore climate science and deny climate change, and argue that it is a conspiracy that threatens our future.

This disconnect between voters and politicians is dismaying, but it also represents a potential opportunity. As increasingly more voters across all political lines see that climate change is real, that the fossil fuel economy is not sustainable, and that renewables can solve these problems while creating economic opportunity, there is potential to eliminate this political barrier through their actions and their votes.

The newly emerging pro-coal, deregulation, and anti-renewable political agenda, including attempts to roll back the Clean Climate Plan, is misguided for a variety of reasons. The plan focuses on electrical generation, which in the United States contributes 30% of GHG emissions. The Clean Climate Plan will not only help limit GHG emissions but also result in major reductions in premature deaths, heart attacks, and days of missed work due to air quality–related illness. The new policy direction makes no economic sense. Coal plants are no longer economically viable. Renewables create between three and six times more jobs for the same amount of energy produced as fossil fuels create. The solar industry alone supports better paying jobs than the oil, coal, and gas industries.

The transition to renewables is well underway. It is becoming central to the global economy and accepted as part of the global development agenda. Investors, corporations, and even oil-producing nations such as Saudi Arabia and the United Arab Emirates are investing billions in this new, sustainable economy, driving a recovery in global clean energy investment to $64.8 billion in early 2107. It is an economy that has the potential to mitigate climate change, reduce impacts on health and environment, and increase the quality of lives of the billions of global citizens who participate in it. Breaking down barriers, creating new technologies and integrative systems, and empowering people and communities to make sound energy choices will speed this vital transition to a sustainable energy future.

Chapter Summary

- Coal, oil, and natural gas are fossil fuels that are the basis of our economy and fuel everyday life. They are not sustainable. These resources are finite and rapidly becoming depleted. Fossil fuels have been a source of international conflict and limit the possibility for security and energy independence. They negatively impact people and the environment and are a major source of GHGs, which are driving global climate change.

- Energy is crucial to human well-being and human development. However, 1.2 billion people in developing countries currently lack access to electricity, and that number is rising with increasing population sizes. Without universal access to energy, sustainable development and poverty reduction cannot occur.
- Transitioning from fossil fuels to renewable energy is the only sustainable way that energy needs can be met, global climate change can be mitigated, and the global economy can continue to grow in an equitable way that will create opportunity without harm to health and environment.
- Despite the negative consequences of fossil fuel use and the significant benefits and potential opportunities of renewable energy, there have been a number of economic, political, and social barriers to the transition from fossil fuels to renewables. These include subsidies that favor fossil fuels as well as economic, infrastructure, institutional, and cultural barriers.
- A number of policy solutions could foster the transition to renewable energy, including changing subsidies to level the playing field and a variety of economic policies that will encourage businesses to develop and deploy renewable energy solutions. International and global policy should focus on reducing conflict over energy and on the development of a global strategy for renewable energy.
- Renewable energy technologies include solar, wind, hydro, geothermal, biological fuels, ocean energy, and fuel cells. There is some disagreement regarding whether nuclear energy is sustainable even though it is carbon neutral. Each of these forms of renewable energy has advantages and disadvantages in terms of sustainability, but none of the disadvantages come close to those associated with fossil fuels. As we develop a portfolio of renewable energies, every effort should be made to reduce environmental and human impact during the life cycle of each technology.
- The transition to renewable energy will require the development of smart grid technology and infrastructure to effectively manage and distribute energy. These grids must be transcontinental, information based, multidirectional, and have storage components.
- In addition to renewable sources of energy and the development of the smart grid, improvement in energy efficiency will be essential to sustain our energy needs.
- Increasing access to renewable energy to those who currently do not have it can be achieved through large-scale, grid-based projects or small, mini-utility projects and off-grid technologies. These types of projects are most sustainable when there is community buy-in and responsibility, stakeholder support, and when they create economic opportunity. One of the best ways to do this is through affordable financing that allows for ownership or payment for products and delivery of energy.
- In connection with the transition to renewable energy and in order to mitigate climate change, we must also rethink transportation. This must include increasing fuel efficiency, the use of renewable-based electric and hybrid electric vehicles, and other renewably generated fuels for vehicles, aviation, and shipping. In addition, we need to redesign our transportation systems and cities so that they favor renewably fueled vehicles, mass transit, and new forms of low-impact transportation.
- We should be optimistic about our energy future. Indicators are already showing that barriers to renewables are coming down, that access to clean energy is increasing, and that this transition favors an equitable economy that in turn could mitigate climate change.

Digging Deeper: Questions and Issues for Further Thought

1. In what ways will the transition to 100% renewable energy rely on systems thinking? Draw a diagram of a model renewable energy system showing all of it elements and how they interact.
2. What are the reasons behind the climate denier movement in the United States? How can they be addressed? What kinds of evidence would you use to counter this movement?
3. What are the possible connections between gender equity and access to energy?

4. What are the barriers to transitioning from fossil fuels to renewables, and what are some examples of successful ways to overcome them?
5. France derives approximately 75% of its electricity from nuclear energy with its 58 nuclear reactors. What are the advantages and disadvantages of using nuclear energy as a primary energy source?

Reaping What You Sow: Opportunities for Action

1. In the United States, 30% of GHG emissions come from electrical generation. Where does the other 70% come from? What personal changes can you make to reduce GHG emissions in those areas?
2. Using the EPA Power Profiler (https://www.epa.gov/energy/power-profiler), compare the fuel mix and air emissions rates of the electricity in your region to the national average and determine the air emissions impacts of electricity use in your home, school, or business. Based on this, develop a clean energy plan for your community, school, or business.
3. Visit the Low Emissions Development Strategy website at http://en.openei.org/wiki/LEDSGP/Transportation_Toolkit and access the Transportation Toolkit. Use this toolkit to generate some recommendations that would make the transportation system in your community more sustainable.
4. Research what it would cost to install a geothermal system for an average home in your area. Based on current costs of heating and cooling, determine the length of time for return on investment (ROI). Explore other solutions that might be more cost-effective and offer a lower ROI. Is ROI the only consideration you would use in making an energy improvement for your home? Why or why not?

References and Further Reading

1. Environmental Protection Agency. n.d. "Power Profiler." http://oaspub.epa.gov/powpro/ept_pack.charts
2. Strahan, David. 2008. "The Great Coal Hole." *New Scientist* 197 (2639): 38–41.
3. Southall, B. L., T. Rowles, F. Gulland, R. W. Baird, and P. D. Jepson. 2013. "Final Report of the Independent Scientific Review Panel Investigating Potential Contributing Factors to a 2008 Mass Stranding of Melonheaded Whales (*Peponocephala electra*) in Antsohihy, Madagascar." Independent Scientific Review Panel.
4. San Sebastián, M. D., and D. J. Córdoba. 1999. "The 'Yana Curi' Report: The Impact of Oil Development on the Health of the People of the Ecuadorian Amazon." Departamento de Pastoral Social del Vicariato de Aguarico, London School of Hygiene and Tropical Medicine Medicus Mundi.
5. Panwar, N. L., S. C. Kaushik, and Surendra Kothari. 2011. "Role of Renewable Energy Sources in Environmental Protection: A Review." *Renewable and Sustainable Energy Reviews* 15 (3): 1513–24.
6. Intergovernmental Panel on Climate Change. 2014. "Climate Change 2014: Impacts, Adaptation, and Vulnerability. Part A: Global and Sectoral Aspects." In *Contribution of Working Group II to the Fifth Assessment Report of the Intergovernmental Panel on Climate Change*, edited by C. B. Field, V. R. Barros, D. J. Dokken, et al. Cambridge, UK: Cambridge University Press.
7. Richter, Daniel deB, Dylan H. Jenkins, John T. Karakash, Josiah Knight, Lew R. McCreery, and Kasimir P. Nemestothy. 2009. "Wood Energy in America." *Science* 323 (5920): 1432–33. doi:10.1126/science.1166214
8. Kannberg, L. D., M. C. Kintner-Meyer, D. P. Chassin, R. G. Pratt, J. G. DeSteese, L. A. Schienbein, S. G. Hauser, and W. M. Warwick. 2003. "GridWise: The Benefits of a Transformed Energy System." Pacific Northwest National Laboratory under contract with the United States Department of Energy. p. 25. arXiv:nlin/0409035
9. Bashmakov, I. 2009. "Resource of Energy Efficiency in Russia: Scale, Costs, and Benefits." *Energy Efficiency* 2 (4): 369.
10. Goldman Sachs. 2015. "The Low Carbon Economy: GS SUSTAIN Equity Investors Guide to a Low Carbon World, 2015–25." *Equity Research*, November 30.
11. The United Nations Secretary General's High Level Group on Sustainable Energy for All. 2012. "Sustainable Energy for All: A Global Action Agenda." The United Nations. Accessed March 16, 2015. https://www.seforall.org/sites/default/files/l/2013/09/9-2012-SE4ALL-ReportoftheCo-Chairs.pdf

12. International Finance Corporation, World Bank Group. 2012. "From Gap to Opportunity: Business Models for Scaling Up Energy Access." https://www.ifc.org/wps/wcm/connect/topics_ext_content/ifc_external_corporate_site/sustainability-at-ifc/publications/publications_report_gap-opportunity
13. REN21. 2017. "Renewables 2017: Global Status Report." http://www.ren21.net/gsr-2017
14. International Energy Agency. 1998. *World Energy Outlook*. Paris: International Energy Agency.
15. Panwar et al., "Role of Renewable Energy Sources in Environmental Protection."
16. Jackson, T., and P. Senker. 2011. "Prosperity Without Growth: Economics for a Finite Planet." *Energy & Environment* 22 (7): 1013–16.
17. Massachusetts Clean Energy Center. 2016. "2016 Massachusetts Clean Energy Industry Report." Accessed October 24, 2017. http://www.masscec.com.
18. US Department of Energy. 2017. "2017 U.S. Energy and Employment Report." Accessed October 24. https://www.energy.gov/downloads/2017-us-energy-and-employment-report
19. Jones, Bruce, David Steven, and Emily O'Brien. 2014. "Fueling a New Order? The New Geopolitical and Security Consequences of Energy." The Brookings Institution. Accessed February 22, 2015. http://www.brookings.edu/research/papers/2014/04/14-geopolitical-security-energy-jones-steven
20. Tejeda, J., and S. Ferreira. 2014. "Applying Systems Thinking to Analyze Wind Energy Sustainability." *Procedia Computer Science* 28: 213–20.
21. Maibach, E., C. Roser-Renouf, E. Vraga, B. Bloodhart, A. Anderson, N. Stenhouse, and A. Leiserowitz. 2013. "A National Survey of Republicans and Republican-Leaning Independents on Energy and Climate Change." Fairfax, VA/New Haven, CT: George Mason University/Yale University, Yale Program on Climate Change Communication. http://climatechangecommunication.org/sites/default/files/reports/Republicans%27_Views_on_Climate_Change_2013.pdf

CHAPTER 7

FOREST AND MINERAL RESOURCES: MATERIALS TO MAKE STUFF

In addition to the resources that we need to survive, such as air, water, those needed to grow our food, and the energy that powers our daily lives, there are those that are required to make the "stuff" that we use. These include the raw materials used in building, construction, and the manufacturing of the products we use every day. Some of this—mainly plastics—comes from fossil fuels. Other materials are also extracted from the natural environment, such as forest and mineral resources. The sustainability of these resources depends on the extent to which they are renewable and the impact their extraction, manufacturing, use, and disposal has on the environment and human health.

In this chapter, we consider both forest and mineral resources, which make up the ingredients of many of the material things we use. We consider the extent to which they may be renewable and meet our criteria for sustainability. We also contrast the use of these resources, and the consequence of that use, between developed and developing countries. As we did in previous chapters, we examine policy, including ways to manage these resources, and innovations that can enhance our use of materials within the context of sustainability. Finally, we use these two types of resource extraction to further develop our use of systems thinking as a way to develop sustainable solutions. The use of plastic as a primary material pervades daily life in developed countries and has been rapidly expanding in the developing world. We explore the sustainability issues related to the production, use, and disposal of plastics in Chapter 8.

PLANTING A SEED: THINKING BEFORE YOU READ

1. What everyday products and materials are made from forest resources?
2. What is the connection between food security and forest management?
3. What about forests' resources make them inherently renewable? What disrupts this renewability?
4. In contrast to forest resources, what makes minerals inherently non-renewable?
5. List five mined minerals or metals that are in the consumer goods you use on a daily basis. Where are they mined?
6. What are some of the social justice issues related to the mining of precious metals and gems?

Conservation is the application of common sense to the common problems for the common good.
—Gifford Pinchot

Forest Resources

Forests provide timber, wood, and paper and other pulp products. They also provide a number of non-timber products that include raw materials for medicines, foods, resins and chemicals, ornamental plants, and other materials such as cork and fibers. Many of these resources are of great economic importance. Forests also provide vital ecosystem services. **Deforestation** is the destruction of forests in order to extract their resources or make the land available for other uses such as agriculture. Historically, developed countries experienced high rates of deforestation as part of that development. For example, the United States went through a period of intense deforestation between 1600 and 1900 so that by 1920 more than two-thirds of existing forests had been leveled at least once, including the vast majority of eastern forests. Currently, the total amount of forested lands in the United States is approximately 70% of the original cover. However, much of this cover is in the form of tree plantations that do not contain the ecological wealth and diversity of the original forest.

Today, more than 1.5 billion people, approximately 25% of the world's population, depend on forests for their livelihoods in both developed and developing countries. For the global poor, forests are the primary source of fuel, medicine, and materials for shelter and clothing. They also contribute to national development, poverty reduction, and provide opportunities for income generation. Forests can enhance food security for vulnerable populations, especially in rural areas, where they provide a supplemental source of food or income generation that can complement that from subsistence agriculture.

Forests are the habitat for much of the Earth's biodiversity, especially in the tropics, and a major source of the planet's primary productivity. They provide a variety of ecosystem services, such as soil conservation, the preservation of biodiversity, and the buffering of aquatic systems from potential pollution. Forests also provide a considerable amount of carbon sequestration and storage, playing an important role in climate change mitigation and resilience. In addition, forests provide economic, cultural, social, spiritual, and recreational functions.

Despite the value of forests, or perhaps because of it, they continue to disappear at alarming rates. According to the United Nations Food and Agriculture Organization (FAO), an estimated 18 million acres (7.3 million hectares) of forest, which is roughly the size of the country of Panama, are lost each year. That amounts to approximately 36 football fields' worth of forest lost every minute. Nearly 50% of the world's tropical forests have been cleared. This continued loss of forest is responsible for as much as 12% of annual global carbon dioxide emissions.

In this section, we survey the materials and services that forests provide and discuss whether or not this provision is sustainable. We examine the root causes of deforestation and the connections among forests, deforestation, and global political economics. We then explore how policy, management, and innovation—including possible alternative materials—could improve the overall sustainability of forest resources and the planet. Later in the chapter, we do the same for non-forest, mineral resources.

Forest Materials

Forest materials certainly have the potential to be renewable. After all, they are composed of plants that can grow and replace themselves within a relatively short period of time. Forests, trees, and other plants might be considered non-renewable resources if they are cut down faster than they grow back. This in turn will transform the environment and could alter it in such a way that it may be impossible for that original ecosystem to recover, resulting in the loss of both habitat and biodiversity. Also, soils tend to replenish themselves at much slower rates than plants, so harvest practice that does not conserve soil will severely limit the potential for forest renewal. Here, we examine some of the forest materials that are harvested and the extent to which the extraction of forest resources can be sustainable. Later, we examine a variety of solutions that may make these resources more sustainable.

WOOD PRODUCTS

Wood is a porous and fibrous structural tissue found in the stems and roots of trees and other woody plants. **Timber** is the collective term for cut tree trunks, but it can include branches and bark. Uses of wood include timber that can be cut into **lumber** or boards, or used to make plywood or veneers. Wood products are also used for posts, poles, and railway ties. Wood is also commonly used as fuel wood or to make charcoal. Cut timber is also used to produce paper.

Timber can be classified as either softwood or hardwood. **Softwood** comes from cone-producing gymnosperms or conifers that usually remain evergreen. Because they tend to grow faster, their wood is frequently softer. In general, **hardwood** comes from trees that are angiosperms and often lose their leaves each year. They tend to be slower growing, so their xylem or wood cells are more densely packed, resulting in harder wood.

The bulk of the timber produced is from softwoods and most often pine. Other commonly used softwoods include cedar, fir, juniper, redwood, spruce, and yew. Softwoods have a wide range of applications and are found in building components such as fiberboard, framing windows, doors, and flooring. They are also used in the production of paper. Paper is made by first extracting the fibers from the wood in a process referred to as **pulping**. The fibers are then pressed, rolled, and heated and dried to make paper. Paper production represents a significant proportion of timber harvest. For example, even with recycling and the trend toward electronic readership, each *New York Times* Sunday edition requires the harvest of 150 acres (approximately 60 hectares) of forest.

Hardwoods are more likely to be found in high-quality furniture, decks, flooring, and construction that are made to last. Commonly used hardwood trees from temperate forests include oak, cherry, beech, hickory, and maple. A number of fine or high-quality tropical hardwoods, including mahogany, teak, Spanish cedar, and rosewood, are used for more expensive furniture and woodwork. These trees are often highly valued. In Brazil, a single mature mahogany tree generates approximately $6,000 in sawn timber after costs. However, the bulk of that income, approximately 90%, goes to international buyers, whereas a Brazilian logger earns approximately 10% of that income. Indigenous tree farmers who sell mahogany trees to loggers rarely receive more than $20 per tree.

The largest individual producer of timber products is the United States, with Canada and Brazil as the two next largest producers. Beyond these countries, tree production is more evenly spread among nations. The management of forests to maximize timber production, referred to as **silviculture**, is done in a variety of ways. The traditional management approach, particularly with softwood timber (which is the dominant product), is the agronomic approach to forestry in which trees are viewed as crops. This is reflected by the placement of the US Forest Service within the US Department of Agriculture. This is basically a plantation approach to forestry in which even-age stands are planted as monocultures and clear-cut or harvested in one operation either entirely or in patches or strips (Figure 7.1). Sometimes seed trees and shelterwood are left behind to help regenerate a new stand on the harvested plot. Hardwood can be produced in the same manner. Fine tropical hardwoods are often extracted from intact tropical forest, but to gain access to these valuable trees, parts of forest are often removed to create space. Both approaches require the construction of logging roads.

Figure 7.1 Aerial view of monoculture eucalyptus plantations showing logging roads in Matubatuba, KwaZulu Natal Province, South Africa. *Richard du Toit/Getty Images.*

The agronomic approach to forest production is in many ways renewable, especially when harvested portions are managed for regeneration. However, they are anything but sustainable. Cutting native forest and replacing it with monoculture stands disrupts the function of the ecosystem and essentially eliminates local biodiversity. Other problems with this type of forest management include erosion and landslides, leeching of soil nutrients, disruption of normal surface drainage, and increased opportunities for the establishment of invasive species. Clear-cutting also directly impacts humans by eliminating a number of small-scale economic opportunities, such as fruit-picking, sap extraction, hunting, and harvest of non-timber forest products. It also ruins the aesthetic and recreational value of the land. However, there are sustainable alternatives to these approaches, which are discussed later in this section.

NON-TIMBER FOREST PRODUCTS

Useful substances that are obtained from forests that do not require the harvest or logging of trees are referred to as **non-timber forest products (NTFPs)**. These include a wide variety of edible products, such as nuts, seeds, berries, maple syrup, and mushrooms. There are also a number of spices, flavorings, and fragrances, such as allspice, cinnamon, and winter green. Other products include oils, gums, resins, rubber, cork, bark, bamboo, grasses, and non-timber fuel wood. Floral greens and decorative plants such as small palm fronds often are harvested for the floral industry. Throughout the world, NTFPs can be a significant contributor to rural livelihoods. For example, in communities in Cross River National Park, Nigeria, income from NTFPs is approximately 13% of the total annual income.[1]

Medicinal plants are also important NTFPs. Usually thought of as traditional or alternative medicine, the reality is that globally, nearly 90% of rural populations still rely on plants as their only source of medicine. The local harvest and sale of medicinal plants, especially by healers and shamans who possess the knowledge on how to use them, can be a significant part of the economy in rural and forest communities.

Many of the medicinal herbs based on tradition or folkloric medicine and originally found in forests and other natural areas are now mass cultivated and sold as nutritional supplements. This comprises nearly a $100 billion global market. In addition, 25% of prescription medicines are based on compounds discovered in forests, accounting for more than 40% of revenues in the $300 billion per year global pharmaceutical industry. Often, the potential healing power of these plants is based on indigenous knowledge. Despite this, the people from the areas where these plants grow and those who possess the original knowledge of their healing potential rarely receive any portion of the hundreds of billions of dollars in revenue from the massive nutritional supplement and pharmaceutical industries. For example, the pharmaceutical industry learned from indigenous peasant farmers in Oaxaca, Mexico, about a steroid compound found in a locally cultivated yam that is now a main ingredient in female oral contraceptives. However, once the compound could be made synthetically, the farmers received no share of the large profits even though their traditional knowledge was crucial in the development of this drug.

Are NTFPs sustainable? Because most are plant parts and their harvest does not kill the entire plant, most NTFPs are renewable. For example, the harvest of maple syrup has been shown to have no negative consequences for healthy trees. The same is true for most other plant parts or products, although overextraction can damage populations. Medicinal plants such as ginseng, where the whole root is harvested, have been overdepleted in many regions. However, if properly managed, the harvest of NTFPs can be ecologically sustainable.

The economic importance of NTFPs and the equitable income generation from them significantly contribute to their sustainability. For example, maple syrup production generates well over $100 million annually in North America, supporting mostly family businesses. A comprehensive study of the sustainability of NTFPs in the less developed world indicates that their harvest is both ecologically sustainable and represents a significant income-generating opportunity for the rural poor that can reduce poverty at least in the short term (Figure 7.2).[2] However, further research on management practice that can maximize the long-term ecological and economic sustainability of forest gathering is needed. Compensating the people from the sites where medicinal plants are found and for their knowledge of these plants would be the just thing to do and would increase the earning potential from forests.

Figure 7.2 Boy from Petén, Guatemala, showing medicinal plant leaf and souvenir made from non-timber forest products, both of which can be used for income generation. *Photo by R. Niesenbaum.*

Forests and Food Security

Much of the current discourse on food security is focused on increasing and expanding agricultural production. The provision of food is often thought of as antithetical to forest protection in that agriculture has been a major cause of deforestation. However, intact forest has a major role to play in providing food security and nutrition that can supplement small-scale farming. Forests provide a wide range of nutritious food to more than 1 billion people globally.[3] In rural areas of developing countries, forest products such as leaves, seeds, nuts, honey, fruits, mushrooms, edible insects, and other forest animals including bushmeat are consumed directly by people living in and around forests. These products are also sold, generating income for rural populations.

The expansion of large-scale industrial production systems in tropical regions, often for exported products, threatens forest ecosystems and their potential contribution to food security, diets, and nutrition. Also, growing populations are putting increased pressure on forest resources. The role of forests in supporting human food security and nutrition and how best to manage these activities are not well studied, and they are often ignored in national and international policymaking related to food security and forest management.

Forest Ecosystem Services

In addition to the goods they provide, forests perform a number of vital ecosystem or environmental services. Natural forests are important reservoirs of biological diversity and habitats for wildlife. Many of the world's most threatened and endangered species live in forests. Forests play a major role in sequestering carbon from the atmosphere, water and air purification, and nutrient and water cycling. They protect watersheds and the rivers and streams that run through them, and they play a major role in soil protection and formation. When deforested, soils are rapidly eroded and nutrients are leached from the soils, which can cause eutrophication. Forests also provide social and cultural benefits, such as recreation, a chance to connect with nature, and spiritual opportunities. Because forests are valued for their timber and products, these other benefits which are often lost are not adequately considered either economically or for their intrinsic value.

One particular kind of forest, consisting of mangrove trees, provides numerous ecosystem services in coastal tropical habitats. Mangrove forests are composed primarily of tropical maritime trees or shrubs of the genus *Rhizophora* and are habitat for many other plant and animal species. These forests grow along coasts that come in contact with tidal brackish waters throughout the tropics, covering 15.9 million hectares globally. They stabilize coastlines and reduce flooding and erosion from storm surges, currents, waves, and tides that will increase with climate change. Their intricate prop root systems provide habitat and food for fish and other organisms.

Mangrove forests are one of the world's most threatened tropical ecosystems, with more than 35% of them already lost. Mangrove forests are cleared to accommodate agriculture, aquaculture, human settlements and infrastructure, industrial areas, and coastal tourism and recreation. In addition, mangrove trees are overharvested for firewood, construction wood, wood chip and pulp production, charcoal production, and animal fodder.

Forests and Climate Change

Forests have a great capacity to capture and store carbon. An obvious way to help mitigate global climate change is to not only stop deforestation but also reforest areas that have already been cut and lost. Younger forests do have higher rates of primary productivity and sequester nearly 30% more carbon compared to older forests. As such, some have argued that converting old-growth forests into more productive tree plantations will help mitigate global climate change and at the same time provide a necessary resource.

However, this argument makes an assumption about how much carbon is stored in a forest and how much will be lost in its conversion into final product. If 100% of the old-growth carbon that is harvested remains stored in the wood product, then this argument might be correct. But in a classic study of this question, it was revealed that when timber is harvested from a forest, 15% of wood is left behind to rot. Also, the bark is removed and approximately 40% of the wood is lost as sawdust or scrap during the conversion of timber to lumber. In paper production, only 46–58% of the tree is recoverable fiber. As all of these materials decompose, the stored carbon is lost to the atmosphere. This can be more than 50% of the stored carbon in a cut forest. This loss is not made up by a 30% increase in productivity of the younger plantation.[4] Not surprisingly, those making this argument typically have support from the timber industry.

Deforestation

As previously discussed, historical deforestation in what are now developed countries has been extensive, and current and projected rates of deforestation in the developing tropics are alarmingly high. In addition to the loss and damage to forests through the harvest of timber and NTFPs and the conversion of forest to tree plantations, there are many other pressures that have been accelerating the rate of deforestation. These pressures are determined by the interaction of environmental, social, economic, cultural, and political forces for a particular country or region. In this section, we examine the causes of deforestation as well as forest recovery. Then we discuss ways to manage forests more sustainably so that they can continue to provide their valuable resources and services for future generations.

POPULATION GROWTH

It was shown in Chapter 1 (see Figure 1.5) that the rate of deforestation has been directly related to population growth. Today, most of this population growth is in developing countries, where there is an even stronger relationship between it and deforestation, and this trend is projected well into the future (Figure 7.3). With increases in population size come increases in a variety of pressures that lead to deforestation. One of the primary pressures is the increased demand for food. We need to feed 200,000 additional people every day, and with this comes the conversion of forest into field for crop production and cattle grazing. This is made worse by shifting dietary habits that have increased the demand for beef in developing countries and by the increasing amounts of agricultural imports from developing to developed countries. Commercial agriculture and clearing forest for cattle ranching have been responsible for 70% of tropical deforestation (Figure 7.4).

As populations grow, they remove forest through a variety of methods, including burning and clear-cutting for agriculture. One common practice in tropical rainforests, where pressure for cropland is intense, is subsistence farming. This often entails cutting portions of forest and then burning the wood so that crops can be grown on the land. In many tropical environments, most of the nutrients are tied up in the plant material; the soils tend to be nutrient-poor. The ash from the burned trees provides some nourishment for the initial planting of crops, but depending on the intensity of cultivation and soil type, those nutrients can become rapidly depleted. When this occurs, the farmers must move on to a new patch of land and begin the process again. Because the abandoned patches are so nutrient-depleted when left idle, forests often never recover and the land remains in a degraded state. This and other forms of subsistence agriculture are responsible for 25% of tropical deforestation (see Figure 7.4).

POVERTY

Poverty plays a major role in deforestation. The world's rainforests, primarily located in the Global

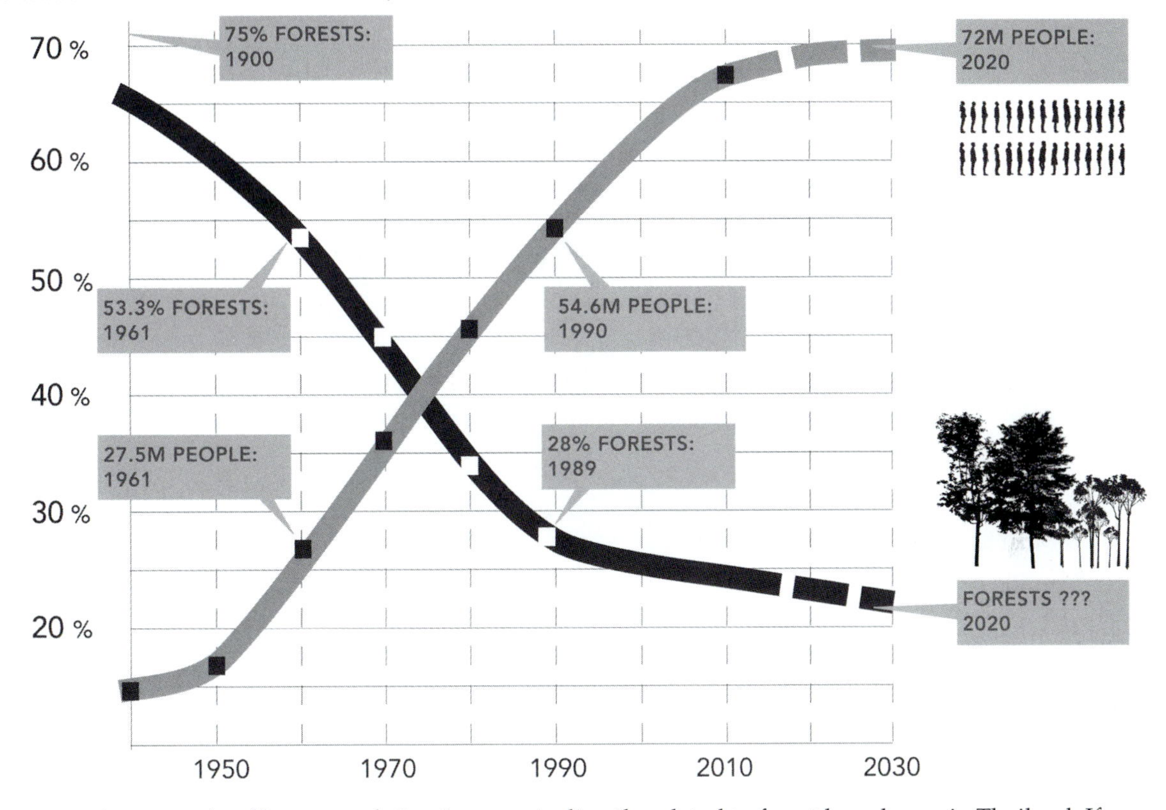

Figure 7.3 An example of how population increase is directly related to forest loss shown in Thailand. If no action to protect forests is taken, it is projected that continued population growth will result in the loss of nearly 75% of original forest by 2030. *Luke Yeung & Architectkidd.*

South, are found in the poorest areas of the world. Up to 40% of tropical deforestation is caused by subsistence activities on a local level by impoverished people who have no other choice but to use the rainforest's resources for their survival. As discussed in other chapters, because of the lack of access to clean energy throughout rural developing countries, wood is harvested and used for indoor cooking and heating. This creates an extreme amount of pressure on forests, and it is a major source of indoor air pollution. Also, the search for wood is time-consuming and impacts earning and education potential, especially among girls and women. In addition, the combination of population growth and poverty results in increased urbanization, which in turn results in further deforestation to make more land available for expanding cities (see Figure 7.4).

ACCESS TO LAND

In most developing countries, access to land that is suitable for farming is limited by a number of factors, resulting in the expansion of agriculture through deforestation. One cause of this limitation is land tenure, or the rules that define how property rights are allocated. In many countries, owned land passes on from one generation to the next, where offspring either share or divide that land. Thus, the number of people a family farm plot supports becomes larger with each generation, rendering it insufficient to meet needs and economically infeasible. Access to land by individuals

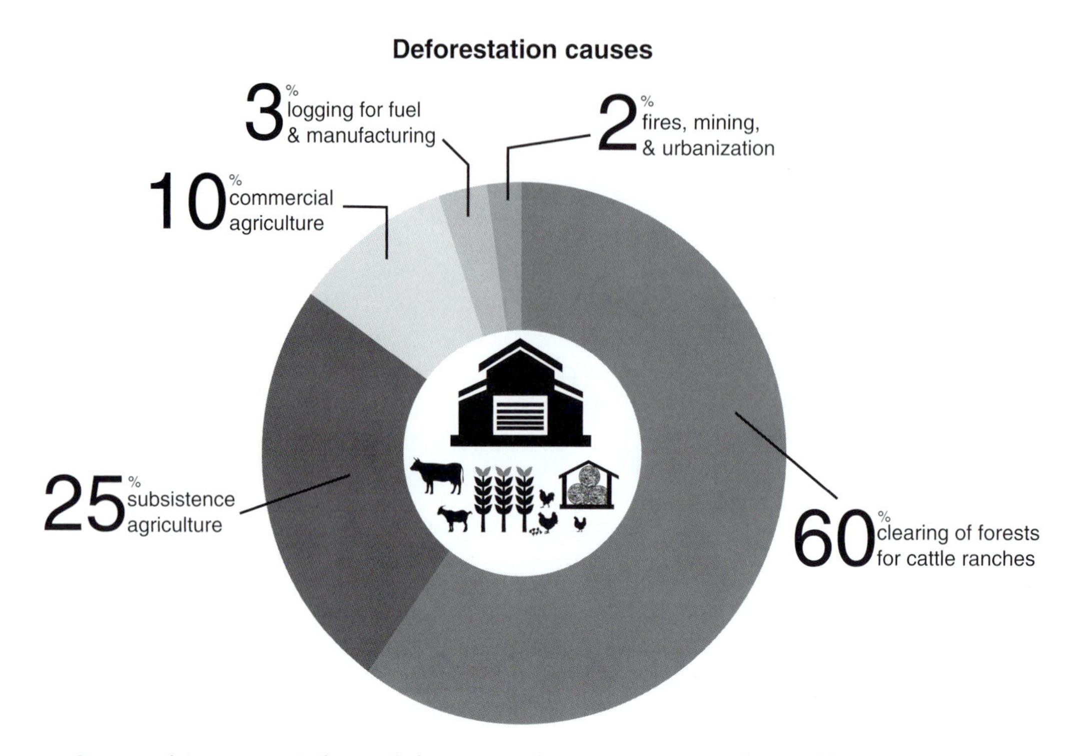

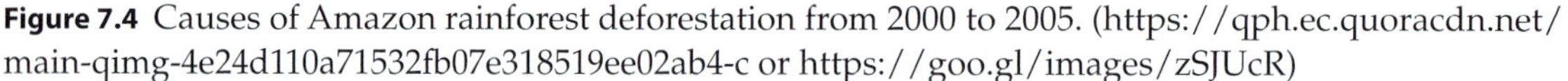
Figure 7.4 Causes of Amazon rainforest deforestation from 2000 to 2005. (https://qph.ec.quoracdn.net/main-qimg-4e24d110a71532fb07e318519ee02ab4-c or https://goo.gl/images/zSJUcR)

is also limited because much of the best arable land is held by large landowners and corporations. This has been particularly true in Latin America, where, for example, in Chile the largest 7% of the farms occupy 81% of the land.

POLITICAL FORCES

The policies and institutional weakness of governments have significantly contributed to deforestation. Increased market demand has motivated governments to incentivize large-scale forestry and agriculture with an emphasis on commodities or cash crops. Examples of such policy include subsidies and low-interest loans for the expansion of forestry and large-scale agriculture. The resultant increase in deforestation is made worse as small farmers are displaced and relocate to forests, where they cut forest for subsistence agriculture. Also, many countries have instituted resettlement programs that relocate peasant farmers to what were once forested border areas as a means of exercising sovereignty. One well-known example of this has been along the shared borders of Colombia, Ecuador, and Peru in the Rio Putumayo watershed. In each of these countries, farmers have been relocated to forested border areas, where they have no choice but to clear them for subsistence farming in order to feed their families.

Pervasive corruption is symptomatic of weak government institutions, which in turn contribute to deforestation. Payoff or campaign support is often provided to government officials at all levels in exchange for timber concessions, approval for clearing forest for large-scale agriculture, and turning a blind eye to illegal logging.

Fiscal policy resulting in large foreign debt has further increased deforestation in developing countries. According to the World Bank, in 2010 the total external debt in developing countries was $2.1 trillion, and that number has continued to rise.[5] This massive debt service requires that many countries liquidate their natural resources, especially forests. It also diverts funds from programs that would discourage deforestation. As such, there is a direct relationship between the amount of foreign debt and deforestation.

MARKET FORCES

Globalization, the expansion of international trade, and the rising cost of land and labor in developed and rapidly developing countries have led to increasing demand for the import of forest and agricultural products from poorer tropical countries. The expansion of industrial agriculture into the tropics to feed both the developing and the developed world is a major cause of deforestation. Figures 7.5 shows where this expansion is occurring, and Figure 7.6 shows that the threat to tropical forest is in those same locations. For example, large amounts of deforestation in Latin America have been attributed to crop production for export to China and the United States. Many consumers of exported agricultural products, such as bananas, grown on larger scale farms or plantations are not aware that their demand for these products is a major cause of rainforest destruction.

FOREST RECOVERY

In many cases, forests can recover. After a historical period of forest decline due to agricultural and timber extraction, eventually a large proportion of deforested land can revert back to forest. The US and many developed countries in Europe experienced such **forest transitions** during the 19th and 20th centuries. For example, after the large-scale deforestation of the eastern United States that occurred prior to the 1900s, the majority of that land recovered and is now covered by forest. Such forest transitions have also occurred in developing countries such as Bangladesh, China, Costa Rica, Cuba, the Dominican Republic, El Salvador, Rwanda, and Vietnam.

Forest transition theory identifies a number of socioeconomic factors responsible for the eventual increase in forest cover with development, including the concentration of agriculture on more productive land, the development of industrial economies, increased imports of agricultural and timber products as national economies become integrated into global markets, and the implementation of forest conservation policy. Once deforestation is stopped and these lands are abandoned, they often revert back to forest both through spontaneous regeneration and through

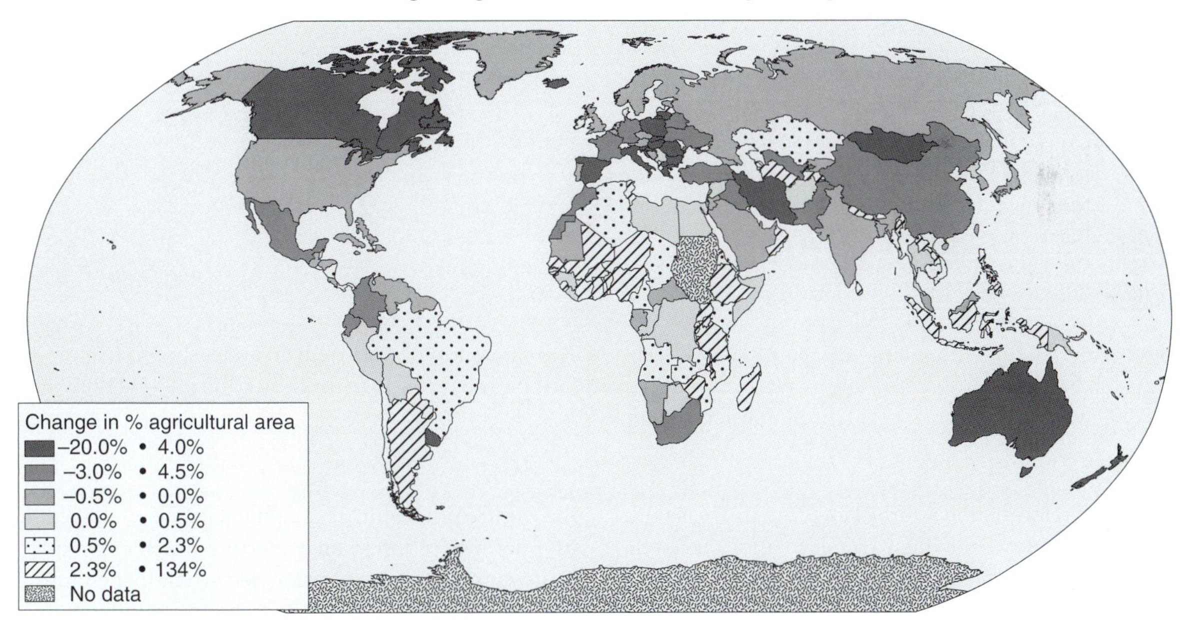

Figure 7.5 Change in agricultural areas by country from 1998 to 2011. The expansion has primarily occurred in developing countries to meet the growing demand for food in China and the United States and is related to the risk of deforestation shown in Figure 7.6. (The Nature Conservancy, https://thebreakthrough.org/images/elements/change_in_ag_land.png Attibution: The Nature Conservancy 2014, Map by J Fisher, Data Sources ESRI, FAO.)

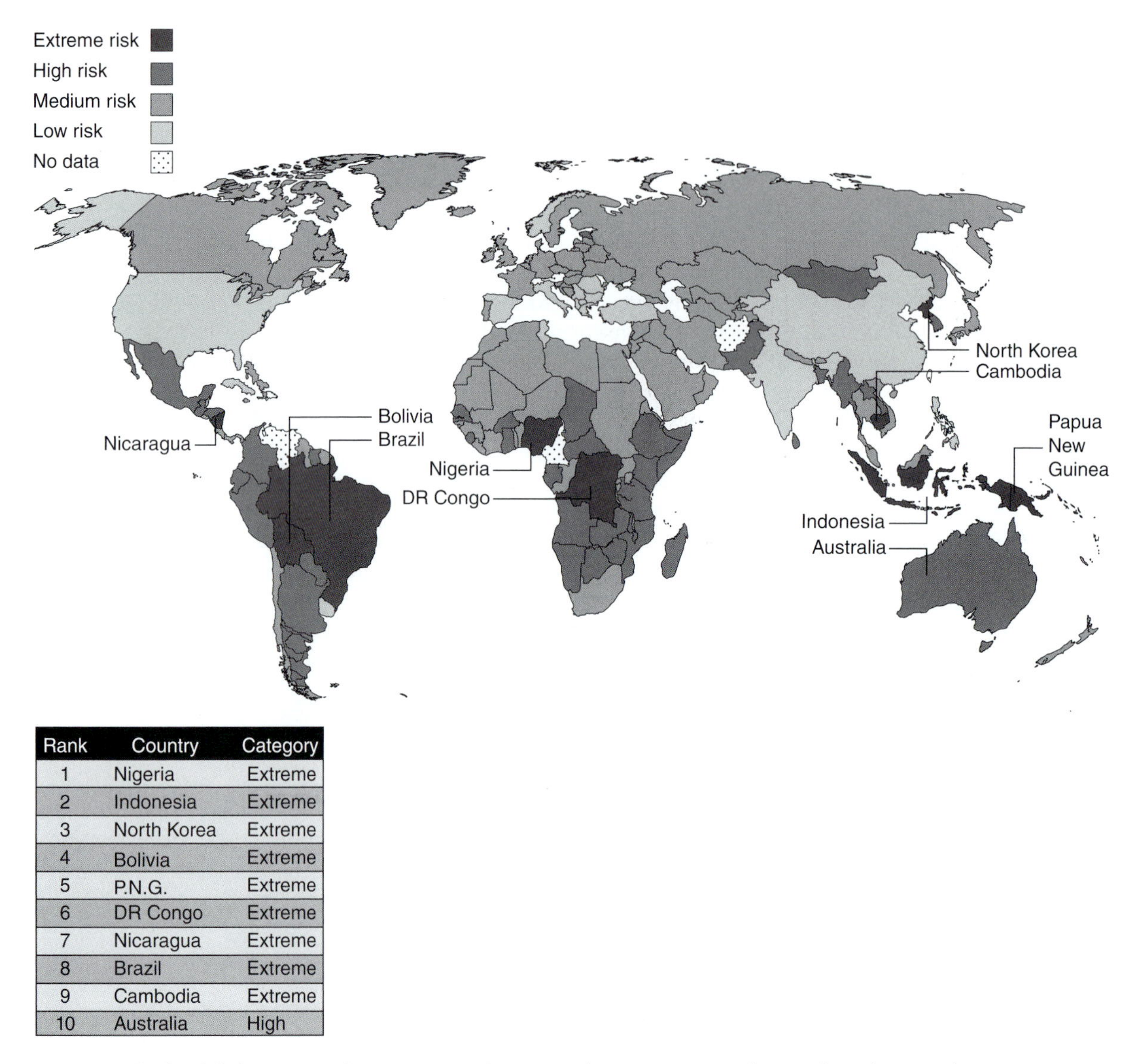

Rank	Country	Category
1	Nigeria	Extreme
2	Indonesia	Extreme
3	North Korea	Extreme
4	Bolivia	Extreme
5	P.N.G.	Extreme
6	DR Congo	Extreme
7	Nicaragua	Extreme
8	Brazil	Extreme
9	Cambodia	Extreme
10	Australia	High

Figure 7.6 Risk of deforestation by country in 2012 matches expansion of agricultural areas shown in Figure 7.5. (Maplecroft 2011, The Towers, St Stephens Road, Bath BAI 5JZ, United Kingdom, +44 0 1225 420 000, www.maplecroft.com, infoAmaplecroft.com)

promoted reforestation efforts. However, in areas with poor soil quality and intensive slash-and-burn-style agriculture, the potential for forest recovery may be more limited.

The net forest area gain that occurs as countries become more developed seems like good news. However, because forest transitions are facilitated by imports of timber and crops, local increases in forest cover often shift the pressure on natural forests elsewhere in the world. As a result, development may actually result in a net loss of forest on a global scale. For example, post-war economic expansion and strengthened forest conservation policy in Vietnam led to a rapid net gain of forest in the early 1990s. But this could only occur with increased timber imports from poorer neighboring countries, where deforestation rates resulted in a net decrease in forest cover over the broader region. This type of displaced demand creates the illusion of increased gains in forest cover with development but actually shifts resource extraction and deforestation to less developed countries, resulting in an overall net loss of forests.

Forest Management

Forest management is not new. Traditional forest communities have managed their resources sustainably in the past. In his book *Sacred Ecology*, Fikret Berkes explores a variety of examples of how local and indigenous knowledge has been used to sustainably manage forests such as the subtropical dry forest of the Dominican Republic.[6] However, over time, as the forces responsible for deforestation described previously have come into play, such traditional practice and knowledge have been lost. Early modern forest policy was aimed at maximizing the short-term yield of forest resources, but not on sustaining them for future generations. However, the development of a variety of new approaches to forest policy and management—and in some cases a return to traditional approaches—may be very effective at achieving sustainability.

EARLY MANAGEMENT

Early forest use was managed by households and communities at very local levels, mostly to increase the yield of whatever resource was being extracted. During the early development of monarchies, forests were viewed as an endless resource. Forests were clear-cut to open up the land for farming and to supply timber for ships, houses, and fuel. As early as the 1100s, Europe was rapidly becoming deforested. In England, as much as one-third of the forest was lost early in its history, and this continued through the Industrial Revolution to the degree that less that 10% of it remains today.

Much of early forest policy was rooted in colonialism. For example, in the late 1700s through the 1800s, the British commercialized the forests in the countries they had colonized, seizing control of forests and forest dwellers in India and Burma (now Myanmar). British policy not only restricted forest dwellers from using their indigenous resources but also restricted their rights to conserve the forests on which they depended. Policies were focused on putting the wider national interests of the colonizers ahead of the needs and claims of local communities located in or near forests. In North America, European settlers quickly harvested much of the available timber for housing, industry, the creation of railroads, and to clear land for farming, especially in the eastern portion of the nation, which in many areas it was completely deforested.

In the post-colonial era, the newly independent nations developed centralized national forest policies. Forests were state-controlled. For example, in 1961 in Thailand, the government decided that 50% of the country should be forest land and as such started evicting encroachers to reach that target. In India, a Conservation Act required central permission to change the legal status of any forest, and permits were required for any new use. This created contention between tribal forest dwellers and national forest departments.

THE US FOREST SERVICE

US federal forest management dates back to 1876 when Congress created an office in the US Department of Agriculture to assess the nation's forests, which were designated as national forest reserves in 1891. In 1905, President Theodore Roosevelt formed the US Forest Service, which was led by Gifford Pinchot, within the Department of Agriculture, and the forest reserves were renamed national forests. Much of the focus of early Forest Service policy was on watershed protection, forest restoration, and wildfire management and suppression.

After World War II, there was a post-war housing boom in the United States. The associated increased demand for timber led to more widespread use of commodity-oriented harvesting techniques such as clear-cutting. Public and scientific concern about this led to several laws that were enacted to protect forests. These laws formed the basis of the conservation, or managed use, movement. Central to this was the **multiple-use approach** to forest management. This approach maximized potential for outdoor recreation, watershed protection, and the production of timber, wildlife, and fish. Strict environmental preservationists were and remain opposed to this multiple-use approach.

Sustaining Our Forests

Globally, we depend on forest resources and forest ecosystem services. Because forest resources are composed mainly of plants with relatively high growth rates, they should be renewable. Unfortunately, the rate of extraction of forest resources and deforestation for other reasons far exceeds the time frame for regeneration. Many other aspects of the way we interact with forests are unsustainable, including management techniques that clear-cut forests and essentially wipe out the ecosystem and its biodiversity.

Many of the global poor depend on the forests economically, for food, medicines, and other resources. Forest communities have rich knowledge of the uses of forest plants, but this knowledge has essentially been stolen and used to reap great profits for large corporations in the developed North. The destruction of forest is also a major contributor to global climate change as carbon sinks are lost. Nothing about this story seems consistent with our model of sustainability. However, the forests and the people who depend on them can have a sustainable future.

An important first step in sustaining forests is to slow the rates of deforestation. To do this, we must directly address the social, economic, and geopolitical forces responsible for it. Slowing population growth and reducing poverty through education, empowerment, and sustainable income generation are central to this. So are the enhancement land tenure policies, the elimination of subsidies for large-scale agriculture and forestry, the strengthening of governmental institutions, and external debt relief (Box 7.1). Market forces

BOX 7.1

Policy Solutions: Swapping Debt for Nature

One major contributing factor to deforestation is the large external foreign debt owed by many developing countries, particularly in Latin America. This is because the need to repay that debt requires that many countries liquidate their natural resources, especially forests. Thus, one way to slow deforestation is through debt relief. One approach to this developed by Thomas Lovejoy of the World Wildlife Fund is the debt-for-nature swap, in which a portion of a developing nation's foreign debt is forgiven in exchange for local investments in environmental conservation measures.

There are two types of debt-for-nature swaps. In a commercial debt-for-nature swap, an NGO such as Conservation International, World Wildlife Fund, or The Nature Conservancy purchases the debt of the indebted country from commercial banks holding that debt. The NGO then cancels the debt in exchange for the debtor country committing to conservation programs (Figure 7.1.1). In the other type of swap, the debt is held by a foreign government that forgives a portion of it in exchange for environmental commitments by the debtor country (see Figure 7.1.1).

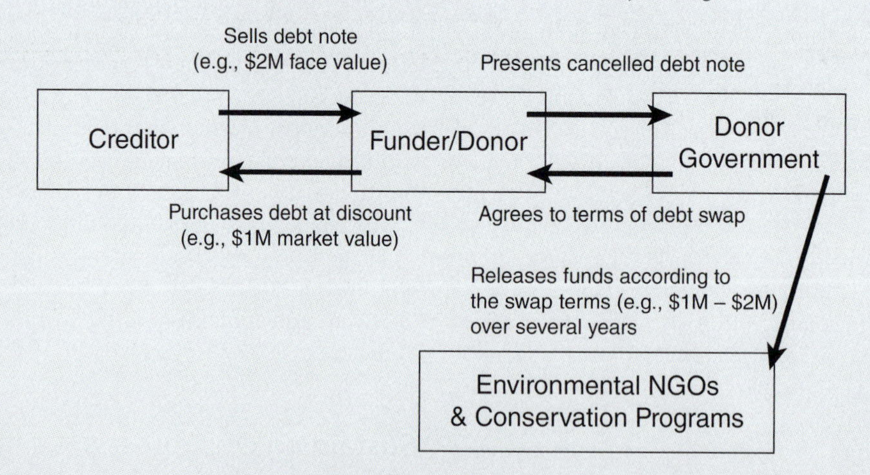

Figure 7.1.1 Debt-for-nature swaps are financial transactions in which a portion of a developing nation's foreign debt is forgiven in exchange for local investments in environmental conservation measures. This is accomplished when a funder or donor purchases the debt owed by a debtor government at a discount and then cancels that debt in exchange for their investment in conservation programs. *DexDeerstalker/Wikipedia.*

Debt-for-nature swaps benefit creditors by allowing them to get full or partial payment on risky loans that they can then place in higher yielding, more secure investments. Debtor countries benefit through a reduction in their total external debt and the opportunity to pay for conservation initiatives. This debt reduction also allows them to invest funds in their countries that would otherwise be sent out as debt payment. They also reap the environmental benefits such as improved ecosystem services and potential revenue from ecotourism. The NGOs benefit by being able to use this tool as a means to achieve their conservation objectives.

From 1987 to 2010, the total number of conservation funds generated through debt-for-nature swaps exceeded $1 billion in approximately 40 different countries. The top three beneficiaries were Peru ($123 million), Costa Rica ($112 million), and the Philippines ($59 million). As an example, in the Philippines in 1988, the World Wildlife Fund purchased $390,000 of Philippine debt at a discounted cost of $200,000. The Philippine central bank then invested over two years the full face value of the loan ($390,000) in designated conservation projects.

A major benefit of debt-for-nature swaps has been their ability to influence conservation over the long term. This is because the central bank of the debtor country usually pays out the amount of swapped debt for conservation projects over numerous years, and in some cases, it endows conservation funding that will exist in perpetuity. For example, in Ecuador, the central bank is paying $10 million in swap proceeds facilitated by the World Wildlife Fund and the Nature Conservancy to the Ecuadorian conservation organization, Fundación Natura, over nine years, with a percentage each year being placed into an endowment fund.

There are some criticisms of debt-for-nature swaps. Some argue that these swaps, which amount to approximately 1% of total external debt, are insignificant and that debt forgiveness would be a more substantial and just approach. Also, the bulk of debt-for-nature swaps have occurred in just a few countries and not necessarily where the needs are the greatest. In some cases, these swaps overlooked the predominantly poor or indigenous communities residing on the land set aside for conservation. However, when considering the needs of these communities and employing equitable application of debt-for-nature swaps, they can represent a significant contribution to sustaining forest resources.

that promote deforestation need to be considered when developing international trade policies. Finally, the rate of consumption and import of forest and agricultural products by industrialized and developing countries needs to be addressed. This can be done both by considering alternative products and through sustainable practice in both forestry and agriculture. In this section, we explore ways to improve the management of forest resources, innovative approaches to using forest resources, new and more sustainable alternatives, and more effective recycling of materials that are extracted from forests as ways to sustain them.

Management Solutions

In the inception of post-colonial centralized forest management and the development of the multiple-use approach in US national forests, rates of global deforestation soared—as did a major concern for the environment and the rights of people who live near or in these forests. From this came new approaches to forest management that included both the outright protection of forest ecosystems and their biodiversity and more sustainable management of forest resources. These sustainable approaches were focused on increasing the renewability of the resources, protecting the environment, as well as considering the economic and social needs of people who depend on the forest for their survival. It is now widely accepted that sound forest management is essential to sustainable economic development and can meet the growing demands for food, fiber, biofuel, shelter, and other bioproducts as the world population increases to 9 billion people by 2050.

SUSTAINABLE FORESTRY

The concept of sustainable forestry grew out of the international sustainability and sustainable development movements that emerged in the 1980s. At the 1992 Earth Summit in Rio de Janeiro, forest principles were adopted that captured the general international understanding of sustainable forest management at that time. This led to a definition of

sustainable forest management that emphasized the maintenance of biodiversity while maximizing forest health, productivity, and regeneration capacity so that current and future ecological, economic, and social functions could be met. This would provide a way to meet the increasing demand for forest products and benefits while preserving forest health and diversity. In recognition of the fact that this balance is critical to the survival of forests and to the prosperity of forest-dependent communities, FAO and many international non-governmental organizations (NGOs) adopted this definition and developed specific approaches, criteria, and indicators for implementing and assessing sustainable forestry practices.

DE-CENTRALIZATION

As previously mentioned, post-colonial forest policy emphasized centralization, putting management and enforcement in the hands of newly emancipated national governments. This was somewhat effective in achieving broad conservation objectives, but it was ineffective in meeting the needs of communities that relied on those forests. Since the 1980s, a number of countries have diminished the government's role in forest management and have let market forces and private companies drive it. This has been a catastrophic trend, particularly in developing countries, where poor extractive management has led to both the depletion of the resources and environmental destruction.

A policy solution to increase sustainable practice would be to both de-centralize and de-privatize forest management responsibilities and shift that responsibility to state and local governments. As previously discussed, top-down rural development projects are likely to fail, and projects that empower local communities to manage their own resources are more sustainable. This is true for sustainable forestry projects as well. Community-based forestry projects, especially when partnered with an NGO such as the Rainforest Alliance, have been able to more systematically manage their forest resources.

These kinds of NGO–community partnerships not only keep forests standing but also do so while improving the livelihoods of members of those communities. This is done through the development of small enterprises within those communities that sustainably harvest forest products, including timber. But it is done in ways that protect and regenerate the resources for future generations and preserve ecosystem function and biodiversity.[7] Other community-level policies typically encourage value-added activities that use and further develop the forest products rather than exporting the raw product. This keeps more of the income-generation potential within the community. For example, in the Petén region of Guatemala, forest communities are milling their own lumber and manufacturing furniture. This creates jobs and economic opportunity within the community (Box 7.2).

BOX 7.2

Stakeholders and Collaborators: Cutting the Trees to Save the Forest in the Mayan Biosphere Reserve

The Man and the Biosphere Programme (MAB) emphasizes the human partnership with nature by integrating human activity, scientific investigation, and preservation. MAB reserves typically consist of a core area designated strictly for preservation, a surrounding buffer zone in which research and carefully monitored human activity such as low-impact ecotourism occur, and a transition zone designated for sustainable development (Figure 7.2.1). In MAB, groups of stakeholders are empowered to collaborate in the formulation and management of their own sustainable development plan. This is what has happened in the Bethel Cooperative situated in the Mayan Biosphere Reserve (MBR) in the Petén region of northern Guatemala.

The Maya Forest, located in Mexico, Belize, and Guatemala, is the largest contiguous area of tropical forest remaining in North and Central America. It has incredibly high biodiversity, a large degree of endemism (species that only occur there), and large numbers of rare and endangered species. It serves as an important migratory corridor and provides local and global ecosystem services. It is thus an obvious priority for conservation. However, it encompasses regions of escalating population growth and intense poverty. These pressures have led to the recognition that managed resources use and income generation are an essential part of the conservation agenda and that without them, typical forms of

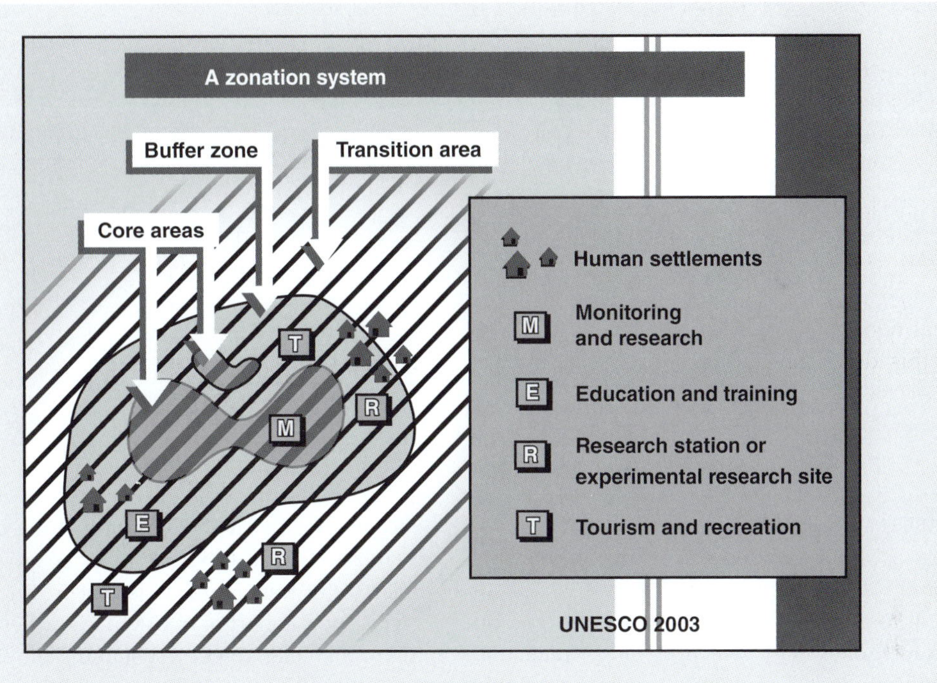

Figure 7.2.1 MAB zonation system. (UNESCO 2003.)

deforestation, such as clear-cutting and slash-and-burn agriculture, will eventually destroy this forest. It seems that the only way to save this forest is to cut some trees.

Bethel is one of six cooperatives that was created in the 2.1 million ha Guatemalan portion of the MBR. Bethel is unique in that it is populated by a number of different post-war refugee groups that have returned to Guatemala from Mexico, bringing together displaced communities from different regions of Guatemala. With the help of a number of international NGOs and broad stakeholder participation, the cooperative was formed and a conservation plan was developed that included opportunities for income generation and poverty reduction. The plan included the development of low-impact ecotourism in the buffer zone and sustainable timber and non-timber product harvest, sustainable agriculture, and some human settlement in the transition zone. Scientific research programs were established to assess the sustainability of these activities.

The forestry program relies on the extraction of the high-value timber such as Spanish cedar and mahogany using the polycyclic felling system, in which only mature trees are harvested from 1 of 20 harvest plots annually. Because harvest is limited to 1 plot each year, a minimum portion of forest is impacted that has 20 years to recover until the next age-specific harvest. Extraction of trees was done in a way to minimize forest impact, maintain mixed-aged tree structure, and allow for regeneration of new seedlings. A preliminary scientific study demonstrated that this technique could generate significant annual yield and income as younger trees grew into the harvestable age group during the subsequent 20 years and that post-harvest seedling regeneration was significant and could potentially perpetuate this resource.[1] This suggests that the forest could remain essentially intact while providing a sustainable source of income from cut trees on an annual basis.

1 Niesenbaum, R. A., M. E. Salazar, and A. M. Diop. 2004. "Community Forestry in the Mayan Biosphere Reserve in Guatemala." *Journal of Sustainable Forestry* 19: 11–28.

continued

continued

The problem is that as the economic needs of the community grow, there may be greater pressure to harvest more trees than the plan allows. One man said, pointing to a Mahogany tree in a plot not slated for harvest, "We could buy a very nice truck with the proceeds from that tree, but we won't allow it to be cut, because our children will depend on income from that tree to feed, clothe, and educate their families." To reduce the pressure to cut more high-value trees, additional income generation and value-added activities are needed. The Bethel Cooperative addressed this with the help of some NGOs by developing the capability to mill its own lumber and produce furniture within the community. This allows more of the value of each tree to be realized within the cooperative as opposed to selling raw timber directly to foreign buyers that would then reap those economic benefits. Coupled with the sustainable harvest of non-timber forest products, increased product marketing, the development of ecotourism activities within the forest, and sustainable agriculture practice in the transition zone, this may be the only way that the valuable forest can be protected from more destructive forms of resource harvesting.

SUSTAINABLE MANAGEMENT

There are a variety of techniques for harvesting timber and NTFPs that increase forest sustainability. Central to them is the ecosystem approach. This approach integrates management of land, water, and living resources that promote conservation and sustainable use in an equitable way. Obviously, this approach eliminates clear-cutting and the establishment of plantations. Rather, individual trees are removed from forests in ways that minimize impact through environmentally sound forest harvesting and transport operations. This is achieved through better planning, roads, felling, extraction, long-distance transport, and post-harvest assessment. For example, felling should be done to accommodate extraction and avoid damage to trees that are not being removed. This is accomplished by employing directional felling techniques.

Trees can be harvested in a variety of ways that are alternatives to clear-cutting and that can maximize regeneration. One method, **age-specific selective harvest**, extracts only older specimens of economically valuable trees. This leaves younger trees to be harvested at a later date, and these can also serve as seed trees that establish seedlings in areas where older trees were removed. Selective harvest also requires policy and enforcement that prevent illegal poaching of trees that would undermine the sustainable management program.

Another technique that incorporates selective harvest, careful extraction, and maintaining seed trees is the **polycyclic felling system**. Under this system, 20 harvest plots are established for a 20-year cutting cycle, in which trees are selectively harvested from each plot once every 20 years. Minimum cutting diameters are established based on species-specific growth rates. The objective of this approach is that within each 20-year cycle, a new cohort of trees will grow beyond the minimum diameter, thereby sustaining the supply of new trees indefinitely. Seed trees of each commercial species are protected to allow for re-establishment of the youngest cohort. Incorporating this into a community-based forestry program has been shown to be sustainable in the Mayan Biosphere Reserve of Guatemala and other tropical forests (see Box 7.2).

FAO provides the international community open access to tools and guidelines for sustainable forest management. This includes details on how to manage natural forest production and forest and landscape restoration. There are also modules on climate change mitigation and adaptation, development of forest-based enterprises and global forest products, investment in value-added on-site processing, and ways to get involved with the carbon market. This toolkit can be accessed at http://www.fao.org/sustainable-forest-management/toolbox/en.

INDEPENDENT CERTIFICATION

A key aspect of sustainable forestry is third-party certification. Certification provides forest owners with independent recognition of their sustainable management practices that protect biodiversity and improve livelihoods. It allows consumers to know the origin of forest products and to take responsibility, through their informed purchases, for how the world's forests are used. This is similar to consumer choice based on certified organic or fair trade products. A number of organizations perform these certification, verification, and validation services, including the Programme for the Endorsement of Forest Certification and the Rainforest Alliance for international certification and the Forest Stewardship Council in the United States. Producers of wood products such as building materials, furniture, and even guitars made with certified materials can then promote their products as sustainable.

A criticism of forest certification is the expanding number of certification initiatives—currently, more than 90 different types worldwide. The variety of schemes has the potential to confuse both consumers and producers of forest raw materials and products. This could undermine the value of forest certification as a tool to both communicate good environmental practice and promote sustainable forestry on the ground. To overcome these problems, one policy effort should be to establish global standards and make available comparative information on the world's forest certification schemes.

MANAGE FOR FOOD SECURITY

Through proper management and policy, improving food security does not necessarily have to result in deforestation. In fact, there are a number of examples in which the two go hand in hand. In Vietnam, for example, economic and agricultural reform have changed the way local communities view and use the forest. Through de-centralized, community-based farm and forest management, as well as loans and technical assistance for managing small farm and forestry enterprise, forest cover is actually increasing as food security for the growing population improves.

Similar approaches have been implemented with the same type of success in the Republic of Gambia in West Africa and in Costa Rica. In the Gambia, the focus again is on local management because the government has transferred ownership of 200,000 hectares of national forest to villages, actually increasing forest cover.[8] In Costa Rica, the emphasis has been on promoting sustainable agriculture while enforcing stronger legal controls over land use and providing funding for forest protection. It has also instituted a program for paying landowners for ecosystem services. Under this program, farmers receive payment for keeping forest on their land intact, but they are permitted to sell limited amounts of timber for increased income. This has been an incredible success resulting in a dramatic increase in forest cover.[9]

COMPENSATE FOR INDIGENOUS KNOWLEDGE

Many of our prescription medicines are based on indigenous forest knowledge and chemicals from plants found in these communities. However, these communities typically receive no portion of the massive profits that the medicines yield. Only a small fraction of rainforest species have been examined for their potential benefits. So now a major part of the search for new cures involves bioprospecting, which is the search for the huge diversity of physiologically active plant chemicals found in tropical forests. One way to narrow this search is to work with indigenous healers who possess much knowledge of plants and their potential for curing disease.

Our principles of sustainability dictate that these communities should benefit from these discoveries, and the profits should also contribute to conserving biodiversity, given that it is a repository for potential future cures. The best example of how this is being implemented is through an agreement between the Costa Rican National Biodiversity Institute (INBio) and the pharmaceutical giant Merck & Co. INBio provides Merck's drug-screening program with chemical extracts from wild plants, insects, and microorganisms, and in return Merck provides INBio funds for research and sampling, royalties on any commercial products that result, and technical assistance and training to help establish drug research in Costa Rica. INBio agreed to contribute 10% of the up-front payment from Merck and 50% of any royalties to Costa Rica's National Park Fund to help conserve national parks and forest ecosystems.

A major problem is that the tribes that hold this indigenous knowledge are becoming isolated and extinct as habitats are destroyed though deforestation. For example, in the Amazon, there are hundreds of indigenous groups that possess thousands of years of ancestral knowledge about their environment—a knowledge that could provide new medicinal compounds and cures. But since these tribes depend on the ecosystem to maintain their way of life, their existence is threatened by the high rates of deforestation in the Amazon. The Amazon Conservation Team, led by ethnobotanist Mark Plotkin, partners with indigenous peoples of the Amazon to protect both their cultures and their ancestral lands (Figure 7.7). This is accomplished through ethnographic mapping, strengthening indigenous culture, indigenous land protection, developing integrated health care that incorporates indigenous knowledge, and sustainable development.

Figure 7.7 Ethnobotanist and conservationist Mark Plotkin of the Amazon Conservation Team discusses local plant and animal names with Trio shaman and apprentices in Kwamalasamutu village in Surinam. *Mark Plotkin and Amazon Conservation Team.*

MAN AND THE BIOSPHERE

A more comprehensive approach to sustaining forest environments is the United Nations Educational, Scientific and Cultural Organization's (UNESCO) Man and the Biosphere Programme (MAB). This intergovernmental scientific program aims to establish a scientific basis for the improvement of relationships between people and their environments by creating biosphere reserves. MAB has established 651 biosphere reserves in 120 countries throughout the world.

MAB biosphere reserves protect biodiversity by establishing a core conservation zone that is legally constituted for long-term conservation and that is essentially untouched and accessed only by scientists who monitor it. The core is surrounded by a buffer zone that is identified for activities compatible with conservation, such as research and ecotourism. The buffer zone, in turn, is surrounded by a transition area designated for sustainable development that can include a variety of income-generating activities, including sustainable forestry and agriculture (see Box 7.2).

BIOCULTURAL CONSERVATION

This chapter has been making the case that the conservation and preservation of forests works best when the social and economic needs of the people who live in or near those forests are also preserved. This approach is consistent with what is often referred to as **biocultural conservation**. It is not too different from the sustainable approaches already discussed, such as the MAB biosphere reserves; however, it places even greater emphasis on the conservation of culture and heritage as an essential aspect of habitat conservation. Biocultural conservationists recognize that indigenous cultures have developed more sustainable lifestyles and they act as wise stewards of biodiversity. Thus, biocultural conservation strategies strongly consider indigenous people as potential allies in the protection of biodiversity and serve to protect their heritage and culture, including language, folklore, and indigenous knowledge. The biocultural conservation approach has been central to the MAB biosphere reserve projects and others, such as the Guanacaste National Park Project in Costa Rica, that empower and enable people and communities to actively protect their environment.

FOREST PRESERVATION

When setting policy for establishing forest reserves, the priorities are typically to protect those with the highest biodiversity. The most diverse forests occur in the tropics, which are often the poorest regions of the world, making sustainable approaches that improve the livelihoods of individuals an essential component of their preservation. Other priorities include ecosystem function, watershed protection, and the establishment of **ecological corridors**. Because deforestation results in forest fragmentation, many efforts have been made to protect corridors that connect forest fragments to allow movement and dispersal of plants and animals across a fragmented landscape (Figure 7.8). Recently, Colombia announced plans to create one of the world's largest ecological corridors across northern South America. The reserve would cover approximately 135 million hectares (1.35 million km^2), linking the Andes Mountains to the Atlantic Ocean via the northern Amazon rainforest and making it the largest protected area in the world.

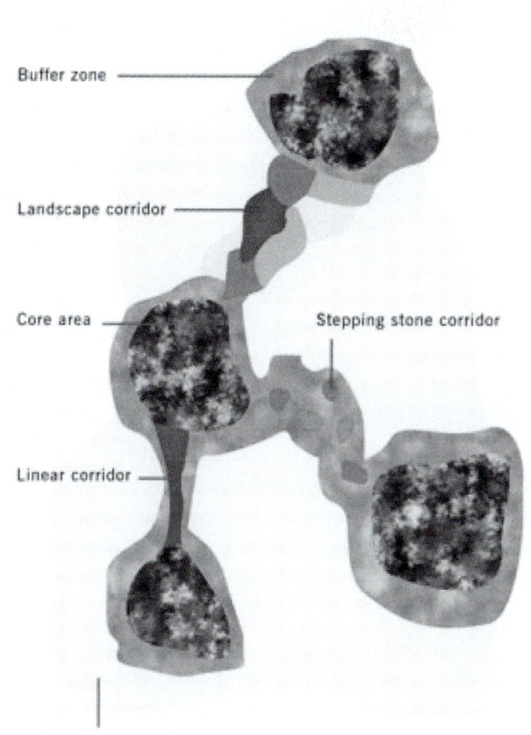

Figure 7.8 Ecological corridors link preserved areas and allow movement and dispersal of plants and animals across fragmented landscapes. Three types of corridors are shown: linear, landscape, and stepping stone. (Lawton, J.H., Brotherton, P.N.M., Brown, V.K., Elphick, C., Fitter, A.H., Forshaw, J., Haddow, R.W., Hilborne, S., Leafe, R.N., Mace, G.M., Southgate, M.P., Sutherland, W.J., Tew, T.E., Varley, J., & Wynne, G.R. 2010. Making Space for Nature: a review of England's wildlife sites and ecological network. Report to Defra.)

Forest preservation can occur either through national designation—for example, the creation of national parks or forests—or through international efforts such as the Biosphere Reserve Program. Forest preservation occurs more locally either by directly purchasing and protecting forest ecosystems or through the creation of conservation easements. A **conservation easement** is a voluntary, legally binding agreement that limits certain types of uses or prevents development from taking place on a piece of property now and in the future, thus permanently conserving the land. The creation of the easement often provides tax benefits for the property owner, and a good policy solution to protect forests would be to increase the size and opportunity for these tax benefits. Preservation by purchase or through conservation easements usually occurs through partnerships between NGOs such as The Nature Conservancy or Land Trust Alliance and private landowners.

REFORESTATION

Policy should also promote and encourage active reforestation of deforested areas. This should be done to restore habitat and biodiversity, potential resource use, and for mitigating climate change. There have been many efforts to restore forests throughout the world. Wangari Maathai, 2004 Nobel Peace Prize recipient, founded the Green Belt Movement that uses a watershed-based approach to help communities conserve biodiversity, restore ecosystems, and reduce the impact of climate change. The movement has planted more than 47 million trees to restore the Kenyan environment. Another organization, Trees for the Future, is dedicated to planting trees within rural communities in the developing world to enable them to restore their environment, grow more food, and build a sustainable future.

CARBON CREDITS

Because deforestation accounts for approximately 20% of greenhouse gas emissions, policies and programs that provide carbon credits or offset funds to pay for forest preservation, reforestation, and to provide substitutes for forest-based products should be expanded. This will serve both to mitigate climate change and to enhance forests globally. The program REDD+ (Reducing Emissions from Deforestation and Forest Degradation) is designed to achieve this goal. The basic goal of REDD+ is to preserve trees that take up and store carbon and would otherwise be cut down by increasing their value. It accomplishes this by enabling companies, conservation groups, and countries to invest in forests as offsets for carbon emissions. This is formally linked with existing carbon trading schemes in which carbon emitters pay to protect forest carbon sinks to offset their emissions. Non-profits such as Stand for Trees provide opportunities for individuals to purchase carbon credit certificates that directly go to preserve forests under the REDD+ model. As discussed in Chapter 3, expansion and global acceptance of carbon trading schemes should be a policy priority. This can occur through their inclusion in international climate change protocols.

The sustainability of such programs will depend on whether or not it is worthwhile either for landowners or for governments to participate. This means the payment or value for an intact tract of forest must be higher than what could be made through cutting down the trees. In countries such as Costa Rica, where nature is the primary source of revenue through ecotourism and the impact of climate change is already being felt, the use of offsets to protect forest has been quite successful. In fact, Costa Rica intends to become the first carbon-neutral nation by nurturing its forests, promoting carbon-neutral technologies, and charging tourists and businesses a voluntary tax to offset their carbon emissions, which will then be used to fund conservation, reforestation, and research in protected areas.

The application of the REDD+ model for preserving ecologically important, diverse mangrove forests has proven to be a challenge. This is problematic because mangrove forests may contribute up to 10% of total global deforestation emissions, despite covering just 0.7% of tropical forest area.[10] The barriers to inclusion of mangroves in REDD+ include the lack of carbon accounting methodologies specific to the unique nature of mangrove forests, thus making certification difficult. A pilot project applying REDD+ to mangrove forest preservation is underway in Kenya. Its success is largely due to the people's understanding of mangrove deforestation. The program also focuses on the social benefits of mangrove preservation by providing financial rewards for avoiding deforestation for community projects such as school and village infrastructure improvement and demonstration of the linkage between mangrove deforestation and fish stocks on which communities depend.

PROMOTE RECYCLING

Policies that promote or require recycling of forest products such as wood and paper will help sustain our forests. Many communities in the United States now have mandatory recycling programs, mostly as a result of soaring landfill costs (see Chapter 8). The use of reclaimed wood from demolished buildings, old wine barrels, and wood pallets is another form of recycling that reduces pressures on forest resources. Often, the residues of wood and paper production, such as sawdust, wood shavings, and shredded paper, are used to make construction materials. Recycling is less common in developing countries; however, reuse is more common there. Recycling can only be successful at sustaining forest resources if there is a market for the material to be recycled and if consumers purchase recycled materials.

Alternatives to Forest Resources

In addition to sustaining forest resources through effective policy management and increased recycling, another effective way to sustain these vital resources is the development of non-forest-based replacements for forest products. Although wood remains the predominant material in house construction, the green building movement has led to the development and use of a variety of newer materials. Examples include autoclaved, aerated concrete blocks that take the place of a number of components used in a standard construction, including wood, insulation, house wrap, and drywall, all in a single product. Another product, Rastra, is a type of insulated concrete form made mostly of recycled polystyrene with some cement. Once stacked into walls, the blocks are reinforced with steel and filled with concrete to form finished walls. Another option is Durisol, which is a hollow-core block made from mineralized wood shavings and cement that can be stacked into walls and then finished with reinforcing steel and concrete. In addition to replacing wood as the primary construction element, these products offer increased insulation, fireproofing, mold and insect resistance, and greater strength than conventional materials. These materials are becoming cost competitive with wood-framed construction, especially when accounting for these advantages.

A number of wood replacements are being constructed from recycled plastic materials. Some materials, such as fencing, are now made completely of recycled plastics, which are durable and easily maintained but are limited in that they lack the aesthetic quality of wood. Wood composites that combine recycled plastic with wood residues look and feel like wood. They are also more durable than wood. An example is the eco-friendly composite decking by Trex, which uses recycled wood and plastics and environmentally sound manufacturing to make attractive, low-maintenance decking, framing, and railings. This company has worked with members of the US Green Building Council to transform the design and construction of buildings to create environmentally and socially responsible spaces.

There are other biological materials that serve as sustainable replacements for wood. One option is to replace trees that are being overharvested and in some cases driven to extinction with species that are less often used. There are hundreds of tree species in most forests, but the majority of our wood products are made from just a handful of highly valued species. Many of these species are now threatened with extinction due to overharvesting. These can be replaced with less frequently used species that have similar characteristics and aesthetic qualities to those that are being overexploited. According to the World Wildlife Fund Global and Forest Network,[11] there are a number of lesser known tropical timber species that could serve for sustainable replacements for highly valued species such as mahogany, teak, rosewood, ebony, oak, and maple. This assumes, of course, that they are harvested following sustainable forestry practices.

Another sustainable alternative is to use plants other than trees that can be used for their fiber and pulp and also to make construction materials such as flooring and fiberboard. These include hemp, kanaff, cork, straw, and bamboo. Many of these can be grown in areas where deforestation has occurred and the soils are so degraded that the forest itself cannot be regenerated. A good example of this is bamboo, which is viewed as one of the most sustainable sources of fiber and wood in terms of environment, social benefit, and entrepreneurial activities (Box 7.3).

BOX 7.3

Innovators and Entrepreneurs: EcoPlanet Bamboo and Bamboo Bikes

The rise in demand for wood and fiber-based products has resulted in increased rates of deforestation of temperate and tropical forests. Much of the deforested land, particularly in the tropics, is so degraded that it cannot be reforested or planted with crops. This severely limits the income generation and agricultural capability of local communities. One solution is to plant native bamboos (Figure 7.3.1), which are capable of growing on this degraded land. Bamboo is highly productive and can produce the same yield as hardwood and softwood timber on one-fifth of the land. It can initially be harvested five years after planting and then sustainably harvested year after year. It can be grown by individual farmers or on industrial scales. It offers great potential for carbon sequestration, helps in rural

Figure 7.3.1 Bamboo, a grass, can be grown for fiber on degraded land as a sustainable alternative that can reduce poverty. *Camille Rebelo, EcoPlanet Bamboo.*

continued

continued

development through job creation, promotes soil and water conservation, and just may be the sustainable way to decrease pressure on remaining forests.

Recognizing the benefits of bamboo cultivation, social entrepreneur Troy Wiseman founded EcoPlanet Bamboo. His goal as a *conscious capitalist* is to provide sustainable alternatives that can reduce poverty and stop and repair the degradation of the planet. To achieve this goal, Wiseman created EcoPlanet Bamboo, which provides a source of sustainable, certified raw fiber through innovative closed-loop and chemical-free manufacturing technologies for products that can meet current and future market demand.

EcoPlanet has established three plantations in West and South Africa and in Central America that can provide more than 3 million tons of raw fiber annually to the pulping industry. These plantations were started on land that was so degraded that food or trees could not be planted, so they do not compete against uses that ensure food security and they do not limit reforestation potential. Currently, there are plans to expand plantations to other areas where land degradation and poverty are worsening. Through research and development, the company has developed clean, closed-looped, integrated manufacturing processes that shorten supply chains and reduce environmental impact. The goal is to create bamboo-based solutions for the four industries that are responsible for using the majority of the wood and timber supply globally: textiles, energy, paper, and engineered wood products timber. Other products under development include activated carbon that can purify water and absorb mercury. Bamboo can also be used in construction, bioplastics, medicines, and even vehicles.

Is EcoPlanet Bamboo a sustainable business? Wiseman believes that entitlement is the biggest killer of opportunity. So his focus is to create a viable business that creates jobs where impoverished people can work for fair wages. With the three plantations, the company has delivered profits while providing thousands of jobs in extremely impoverished areas. It also empowers women. EcoPlanet employs 25–30% women in leadership positions in countries in which opportunities for women have been extremely limited. Environmentally, these plantations will sequester 1.5 million tons of CO_2 and will restore thousands of acres of degraded land. The company has been certified by the Forest Stewardship Council and Verified Carbon Standard (VCS), and it has received gold-level Climate, Community & Biodiversity Alliance (CCBA) certification. EcoPlanet Bamboo is a model of social entrepreneurship and conscious capitalism, which are an essential part of a sustainable future.

Another program is the Ghana Bamboo Bikes Initiative, which was started by teenage social entrepreneurs Winnifred Selby, Bernice Dapaah, and Kwame Kye. The aim of this endeavor was to address both the transportation needs and the unemployment problems in their community while simultaneously addressing climate change, poverty, and rural–urban migration. The initiative does so by creating jobs for young people, especially women, through the building of high-quality bamboo bicycles (Figure 7.3.2). Compared to the production of traditional metal bicycles, bamboo bikes require less electricity and no hazardous chemicals. Not only are the bikes light and stable, they can handle rough terrain and can carry large farm loads and passengers. The company employs approximately a dozen people full-time, produces between 60 and 100 bicycles per month, and provides hundreds with a form of transport—truly a sustainable solution.

Figure 7.3.2 The Ghana Bamboo Bikes Initiative addresses transportation needs, unemployment, and poverty. It creates jobs, especially for women, through the building of high-quality bamboo bicycles. *Camille Rebelo, EcoPlanet Bamboo.*

Can We Sustain Our Forest Resources?

Given that forests have the biological potential to regenerate, the opportunity to use forest resources sustainably exists. This, of course, will rely on many of the approaches and solutions presented previously that will preserve forests, relieve pressure on their resources, and provide for their sustainable management. There are a number of common themes that have emerged as we explored these solutions that suggest certain recipes for success. One of the common elements of successful sustainable forest management includes a shift from national to locally focused projects. These projects emphasize community-based management, decision-making, and control of resources that involves all stakeholders.

Another important element is integrating management with rural livelihoods, poverty reduction, and improving food security, including funding for the development of sustainable enterprises. Funding for stronger and enforceable policies and incentive for forest protection, and reforestation through private–public partnerships, is also needed for success. Finally, sustainable forest projects are most successful when forest conservation is linked to the conservation of heritage and culture, including the preservation and valuation of indigenous knowledge.

Measures of success with regard to our sustainability criteria can be and have been well documented. Standard indicators include positive or negative changes in forest area, biodiversity, ecosystem function and services, and potential for climate change mitigation for a particular area over time. These should be considered with indicators of social and economic improvement, community impact, income generation, and other elements that relate to our sustainability criteria and objectives. This monitoring and assessment should occur on a regular basis. FAO offers a sustainable forest management toolbox that provides resources for implementation and assessment (http://www.fao.org). With best practices, forest resources can be used to the benefit of individuals and communities while maintaining biodiversity, providing ecosystem services, and mitigating climate change long into the future.

Mineral Resources

In addition to forest resources, another set of primary ingredients of the things we manufacture or possess are mineral resources. Minerals are inorganic, solid materials that naturally occur in the Earth's crust or can be found on the ocean floor. They are typically formed by physical or geological processes rather than biological processes. Minerals are different than rock, which is usually an aggregate of many different minerals. Minerals occur as single elements such as gold and silver, but more often they occur as chemical compounds. All have an ordered atomic structure that determines their properties.

There are nearly 5,000 different minerals, but only approximately 20 are considered economically essential and are abundant enough for industrial use. Ores refer to minerals that occur in large accumulations that are easily extracted and have economic value as a commodity. Those minerals that we use the most and are in the form of ore are iron, aluminum, copper, manganese, zinc, chromium, lead, titanium, and nickel. Approximately another 20 are highly valued as gems, such as diamonds, rubies, and sapphires, because of their aesthetic value, durability, and rarity. Rare-earth minerals contain one or more rare-earth metals. These metals are a collection of 17 chemical elements, all of which exhibit similar chemical properties. They are gray to silver or grayish metals that are typically soft, malleable, and reactive. Despite their name, they are found commonly in the Earth's crust but are not concentrated enough to be economically viable to recover like ores; however, more recent technologies have begun to create a large demand for rare-earth metals.

Minerals have a multitude of uses—perhaps too many to mention except for a few examples. Some, such as potassium, phosphate, and potash, are mined primarily for their use as fertilizers, which are in demand to meet the needs of food production. Others, such as clay, sand, gravel, rock, and gypsum, are used heavily in construction of buildings and roads, including cement. Still others are used to produce our primary metals, such as copper, iron, nickel, tin, aluminum, and of course, steel. Chromium and manganese are used in the production of steel alloys. Zinc is used with copper in the production of brass and other metal alloys, and it is also used to produce batteries (as is lithium). There are also precious minerals such as silver, gold, and platinum, and gems such as diamonds, rubies, and sapphires.

Our ever-expanding technology is critically dependent on minerals, and both the amount and the diversity of those being used are rapidly increasing. Smartphones, computers, and the other technology that underpins modern life would be impossible without rare-earth minerals. Today, the average automobile contains more than one ton of iron and steel, aluminum, copper, silicon, zinc, and more than 30 other minerals, including titanium, platinum, and gold. In the 1980s, computer chips contained 11 mineral-derived elements; they contain more than 60 today. Most sustainable technologies are also dependent on minerals. For example, platinum and palladium are metals used in air pollution control technologies. Other metals that are mostly rare-earth elements are used in magnets for wind turbines, compact fluorescent and LED light bulbs, and in batteries and other components of hybrid and electric vehicles. In addition to the mineral silica, solar photovoltaic cells also contain the rare-earth minerals cadmium, indium, and tellurium.

Some minerals are also mined from the ocean floor, including sand, gravel, manganese nodules, and cobalt crusts. Manganese nodules are lumps composed mainly of manganese, iron, silicates, and hydroxides that grow around a crystalline nucleus at a rate of only approximately 1–3 mm per million years. Economic interests lie in the small concentrations of cobalt, copper, and nickel, which make up a total of approximately 3% of the nodules by weight. In addition, there are traces of other significant elements, such as platinum and tellurium, that are important in industry for various high-tech products. Cobalt crusts form along deep ocean volcanoes or vents and contain high concentrations of cobalt, manganese, iron, and other metals that accumulate as they are formed.

In this section, we examine whether minerals that are an essential component of the global economy and the growing technology sector are sustainable. We do this in terms of renewability and future availability and also how their extraction, use, and disposal impact the environment and human well-being. We then offer solutions and alternatives that could improve the overall sustainability of mineral resources and the planet. We do not do this for every mineral but, rather, focus on some specific cases.

So Are Mineral Resources Sustainable?

Because minerals are created through physical, geological processes, their formation is essentially on a geological timescale. Thus, from a human use perspective, mineral resources are finite limited resources that are non-renewable. The global use of most mineral resources has steadily risen from the 1970s to the present, with the United States consuming the most—approximately 20% of all minerals. It is difficult to predict how long specific mineral resources will last because this depends both on the amount that is used and the rate of discovery of new deposits of those minerals. Minerals that occur in extensive, widespread seams, such as potash, phosphate, copper-containing minerals, and iron ores, are predicted to last on the order of hundreds of years. Minerals that are in high demand but occur in more localized deposits, such as lead and rare-earth minerals, will have shorter lifetimes. It is believed that the abundance of rare-earth minerals that are critical for our technologies, including renewable energy systems, cannot be sustained given the increasing demand for them.

A number of economically essential minerals are considered endangered and will be up to 80% depleted by 2040 if extraction does not decrease. These include gold, silver, mercury, sulfur, tin, tungsten, and zinc, all of which are experiencing increased rates of extraction—some exponential—as demand and value rise. Marine minerals remain in great abundance because of the expense of mining them from ocean depths as deep as 4,000 meters. However, as the values of these minerals rise and terrestrial sources become depleted, there will be more pressure to extract them.

Like oil, mineral resources are not equally distributed among nations throughout the world and are concentrated in certain areas. For example, 85% of the world's supply of rare-earth minerals comes from China. This means that consumers are dependent on other countries for them, and access is dependent on international export. This can lead to tension among nations and the formation of mineral cartels, or groups of countries with mineral resources that band together to control supplies and prices. Many valuable mineral resources come from areas of conflict such as African nations that are embroiled in civil war. Thus, the purchase of these resources may support violence and human rights violations, or these resources may simply be unavailable.

The Impacts of Mining

Both surface and underground mining can have severe environmental and human health impacts. Most minerals are extracted from surface mines such as open pit mines or through strip mining or mountaintop removal. As discussed with regard to surface mining of coal (see Chapter 6), this devastates the landscape and has other severe ecological impacts by removing the vegetation and topsoil. It also contaminates the air with dust and toxic substances and causes toxic compounds to percolate into the groundwater, both of which impact the health of mine employees and neighboring communities.

Both surface and underground mines produce an incredible amount of waste material, also known as mine tailings, which must be disposed of properly. They have the potential to clog aquatic systems with sediments. Tailings often have toxic compounds in them such as uranium, lead, zinc, and arsenic, which can further pollute the soil and aquatic systems as they are leached from the waste rock material. These metals also impact human health as they enter sources of drinking water. Other tailings contain sulfides that when dissolved in water produce sulfuric acid in what is called acid mine drainage, which acidifies nearby aquatic systems. Many mining processes require the use of excessive amounts of water, which is a limited resource. Forests and wilderness areas are destroyed in an effort to reach mines (see Figure 7.4). Marine mining for manganese nodules and cobalt crust can have a major impact on the ocean floor, especially on benthic or bottom-dwelling communities, as the ocean bottom is scraped and the waters are clouded with sediments. As much as 120 km^2 of deep sea habitat are destroyed through ocean mining each year.

PROCESSING MINERALS

The processing of minerals includes the grinding down of the mined material and then either using heat or chemical processes to extract the base metals from their ore. These processes can have severe environmental and health impacts. Mineral processing is one of the largest users of energy worldwide and therefore contributes heavily to air pollution and global climate change. Smelting is the process of separating the metal from impurities by heating the concentrate to a high temperature to cause the metal to melt. Smelting the concentrate produces a metal or a high-grade metallic mixture along with a solid waste product called slag. The primary environmental impact of smelting is air pollution. Some ores are rich in sulfides and so when smelted release sulfur oxides, which form acid rain (see Chapter 3). Other air pollutants include particulates and toxic metals such as arsenic, cadmium, and mercury. These pollutants can then contaminate the soil or impact forest ecosystems surrounding the smelters, as well as the health of communities living near them.

One example of extreme environmental impact from smelting occurred at the site of the New Jersey Zinc Company Smelter in Palmerton, Pennsylvania. During the nearly 70 years of operation of this smelter, airborne emissions averaged approximately 47 tons of cadmium per year, 95 tons of lead per year, 3,575 tons of zinc per year, and between 1,400 and 3,600 pounds of sulfur dioxide per hour. The acid precipitation from the sulfur dioxide contributed to the complete deforestation of surrounding mountains. The zinc, cadmium, and lead accumulated in the soil reached concentrations that are toxic to plants and also contributed to deforestation and inhibited the re-establishment of vegetation, resulting in a nearly 1,000-hectare barren wasteland. A larger copper–nickel smelter in Sudbury, Ontario, Canada, had much greater impacts, including the loss of vegetation on more than 46,000 hectares and acidification of more than 7,000 lakes.

These smelters also impacted the health of the communities near these sites. In Palmerton, elevated levels of lead and cadmium were detected in the blood and hair of children. Lead is a heavy metal that is toxic at very low exposure levels and is a multiorgan system toxicant that can cause neurological, cardiovascular, renal, gastrointestinal, and reproductive damage. Cadmium is also toxic to humans, mainly affecting kidneys, and is a carcinogen by inhalation. Cadmium is accumulated in bone, which may serve as a source of exposure later in life. Horses and cattle from farms close to these smelters developed illness and fatigue from high concentrations of lead and cadmium.

RARE-EARTH MINERALS

There are environmental, public health, and social issues associated with the increase in demand, extraction, and process of rare-earth metals. Due to their high value, they are often mined illegally with little or no environmental or health precautions. In China, from which 85% of these minerals are obtained, the government provides no financial support to encourage or enforce environmental standards. When mined improperly, large amounts of dust containing fluorine, hydrofluoric acid, and sulfur oxides are generated. When processed, the remaining tailings often contain large amounts of radioactive thorium. Processing 1 ton of rare-earth metals produces 2,000 tons of toxic waste; one of the largest mining sites in China produces 10 million tons of waste water per year.

Those who work and live near the sites where rare-earth minerals are mined suffer high cancer rates and lung disease. One of many tragic examples is what happened in Dalahai Village in the Inner Mongolia region. Once a place of agrarian beauty, rare-earth mineral mining has transformed it into a barren landscape void of life, with high cancer rates among the remaining local population.

There are social and environmental justice issues surrounding the extraction and use of rare-earth minerals. As the wealthier, industrialized world's demand increases for these resources, which are located predominantly in the poorest areas of Asia, environmental and health impacts go unchecked. In places such as Malaysia, this has resulted in social unrest. The land and the people are exploited as these minerals are sold in ways that minimize cost and maximize profits for a few. This is a classic example of how resources are extracted from poorer regions without concern for environmental or health impacts and then sold cheaply to wealthier nations while the costs of those impacts are externalized to local communities.

GOLD

The mining of gold also has severe environmental and human health consequences. The primary impact is through the use of mercury and cyanide in the processing of the ore to purify the gold. Both mercury and cyanide are environmental and human health toxins. These impacts can be seen at large industrial gold mining processing operations or where smaller artisanal gold mining occurs, which is becoming more common as the value of gold increases.

One such example is the Abangares region of Costa Rica. Until the mid-1930s, this location was dominated by large-scale mining operations held by North American interests. Local waterways where the people obtained their drinking water were heavily polluted by mercury and cyanide such that an aqueduct system that drew water from above the mining areas and brought it to the communities below had to be constructed. The Great Depression brought an end to large-scale industrial mining in this region. However, local groups of artisanal minors called *coligalleros* continued to mine the tunnels that were left behind by the industrialists, hammering stone by hand under very dangerous conditions. These miners bring their extracted stone to local sites that have mechanical grinders called *rastras* that pulverize the stone to free the tiny flecks of gold that were embedded in the rock. They use mercury to form an amalgamation with the gold flecks, making them easier to collect. They then gather the gold–mercury amalgam using their bare hands and heat it to separate the mercury from the gold. No effort is used to protect those working at the *rastras* from mercury or mercury fumes. There are no respirators, protective gloves or clothing, or ventilation hoods or equipment in use. These *rastras*, or small-scale processing sites (Figure 7.9), are located in the backs of homes, near rivers and farmers' fields, and even close to schools. Prior to 2000, when the price of gold was less than $300 per ounce, there were less than 50 of these *rastras* dotted throughout the community. However, since then, the price has spiked as high as $1,800 per ounce, remaining on average well above $1,200 per ounce. As a result, there are now more than 300 *rastras* with close to 400 families making a living from mining or processing gold.[12] Not only are miners exposed to considerable amounts of mercury, but also much of it is released into the environment, where it can be converted into more toxic methylmercury. Although they attempt to recapture the mercury, much of it is lost in the process. It has been calculated that as much as 6,000 kg of mercury is released into the environment each month. It easily enters the rivers, from which it ultimately runs into the Pacific Ocean, where it can accumulate in the large global fishery along the Central American coast. Mercury exposure and release into the environment is not limited to this site in Costa Rica; it has become a global problem. Small-scale artisanal gold mining has become the world's leading source of mercury emissions, contributing nearly 40% of the 2,000 metric tons released into the global mercury cycle each year.

CONFLICT MINERALS

Conflict minerals are those mined in conditions of armed conflict and sold to support that conflict. They typically originate from war-torn African nations. For

Figure 7.9 Backyard *rastra* or small-scale gold processing site where rocks containing gold flecks are ground and mercury is added to form an amalgamation with the gold. Later, the mercury is burned off to separate it from the gold. *Joseph E. B. Elliott, Muhlenberg College.*

example, in the eastern provinces of the Democratic Republic of Congo, where tin, tungsten, gold, and other valuable minerals are extracted, various armies and rebel groups have profited and funded their violent actions during wars in the region through mining. This mineral-funded civil conflict in the Congo has resulted in more than 5.4 million deaths. That is greater than the number of deaths from the US revolutionary and civil wars and the Vietnam and Korean wars combined. Often, those working to extract the minerals are essentially enslaved, forced at gunpoint or otherwise threatened. The atrocities associated with this industry are immense and disheartening, and it is nothing short of genocide.

Valuable gems such as diamonds and jade are also often mined in areas of conflict or social injustice. One example of a conflict mineral is so-called "blood diamonds" mined in countries embroiled in civil war, such as Angola, Ivory Coast, Sierra Leone, and Zimbabwe. In these cases, the diamond industry funds rebel groups, resulting in atrocities, mass murder campaigns, enslavement, and the systematic relocation of individuals. Mining of other gems in Burma (now Myanmar), Madagascar, Cambodia, and Brazil has resulted in land confiscation, extortion, child labor, and unsafe working conditions. Large profits are made because workers have been paid slave wages.

SAND

Even the extraction of sand, which appears to be universally abundant, no longer meets our criteria for sustainability. A $70 billion industry, sand has become the most widely consumed natural resource on the planet after fresh water.[13] As a major ingredient of concrete, the primary use of sand is in the construction industry, which has been drastically expanding with development and increased urbanization. Sand is also the source of silicon used in the manufacture of photovoltaic solar panels and computer chips and microprocessors. It is also a source of titanium and zirconium used in industry, and it is an ingredient in some detergents and cosmetics.

There are both environmental and social justice issues related to the extraction of sand. Sand mining is damaging forest, reef, beach, river delta, marine, and mangrove ecosystems and also the animals that live within those systems, on which many people depend. In Indonesia, approximately a dozen small islands are believed to have completely disappeared because of sand mining. The demand for sand has resulted in the expansion of illegal sand mining, regional conflict, and the creation of so-called "sand mafias" that illegally mine sand. In many countries, such mafias control large parts of the construction industry through bribery and violence.

Making Mineral Resources More Sustainable

We rely heavily on mineral resources, and our dependence on them is steadily increasing. Yet, these resources are not sustainable: Their abundance is finite, and they are not renewable. They are geographically concentrated, which has consequences for mineral independence, international trade, and security. Mining and processing them have serious environmental, human health, and social consequences. The economic demand for certain resources has the potential to fund violent conflicts and for enslavement, child labor, and in some cases, genocide. As bad as this seems, there are a variety of solutions that can make this situation more consistent with our criteria for sustainability. These are presented next.

Policy Solutions for Mineral Resources

For the most part, state ownership of mineral resources results in domestic legislation that regulates mining operations. As such, there are no international, binding laws regarding the extraction of mineral ores. One exception is the deep sea mining of manganese nodules and deep ocean crusts that often occurs outside of territorial waters and into international waters. Thus, deep sea mining falls under the jurisdiction of the UN Convention of the Law of the Sea, which monitors and imposes fees for the mining of a globally shared resource. However, the United States has argued that its citizens and corporations have the right to mine the deep seabed and may do so without following the Law of the Sea. The United States affirmed that deep seabed mining is a high seas freedom open to all nations; however, only the wealthiest nations can afford to harvest this globally shared resource. To make deep ocean mining more sustainable, all nations should be subject to the UN Law of the Sea's provisions for deep sea mining.

In most developed nations, existing environmental policies govern mining. For example, in the United States, the General Mining Law of 1872 has been the long-standing basis for mining public lands. It allows foreign and domestic companies to take valuable minerals from public lands without paying any royalties, and it still permits public land to be purchased at the 1872 price of less than $5.00 an acre. The 1872 mining law contains no environmental provisions, allowing hardrock mines to wreak havoc on western water supplies, wildlife, and landscapes. To address this, the Hardrock Mining and Reclamation Acts of 2010 and 2014 were introduced to replace the antiquated 1872 law, which mandated mining as the best use for public land. These new laws would encourage environmental protection, clean-up of abandoned mines, and preservation of critical habitats and national parks and monuments. There would also be a tax to fund reclamation or repair of sites mined on public land and provisions for enforcement and oversight. The 2010 bill died in committee, and the 2014 bill has yet to be passed. Positive action on this type of bill would improve conditions from mining on US public land.

For mining on private land in the United States, there are more than three dozen federal environmental laws and regulations, and at least five separate mining acts that cover all aspects of mining. Among these are the Clean Air Act; the Clean Water Act; the National Environmental Policy Act, which requires an interdisciplinary approach to environmental decision-making; and the Federal Land Policy and Management Act, which prevents unnecessary degradation of federal lands. The US Bureau of Land Management regulates mining claims, patents, and surface use and management, and it serves to improve abandoned mining sites. The US Geological Survey also serves to support mineral resources stewardship. The problem is that mining falls under the jurisdiction of so many bureaucratic entities. As such, there is often conflict between federal policy and either states' rights or individual landowners' rights, which has led to numerous disputes. A strong federal policy that centralizes mining regulation and related impacts is important to ensure access to mineral resources and environmental protection and reclamation.

Historically in Europe, each country had its independent civil laws governing mining. Today, however, under the European Union (EU), there is now a regional focus on the sustainability of mining through targeted initiatives and research funding. These include the development of new technologies for finding further deposits in established mines, economically and environmentally viable exploration and processing of ores with low levels of mineral or metal concentration, and reducing mining's surface footprint.

The type of mining regulation in the United States and the EU is absent or weak in much of the rest of the world. Globally, mining has become increasingly deregulated over time, which has resulted in increased environmental damage, threats to human health and miner safety, and conflicts over the international trade in strategic minerals. One recent exception to this is El Salvador, which in 2017 made history as the first country in the world to impose a nationwide ban on metal mining.

This was motivated by the country's dwindling supply of clean water—finally, a recognition that water is more valuable than gold. El Salvador's action could serve as a model for other nations or communities that oppose mining, especially in environmentally sensitive areas.

Because of the general lack of national mining policy, international regulation is of vital importance. This has been initiated through the establishment of the Extractive Industries Transparency Initiative (EITI). The EITI is a global standard for the governance of mineral resources as well as oil and gas. It recognizes that wise use of natural resource wealth should play an important role in sustainable development and poverty reduction, but that if not managed properly, it can create negative economic, environmental, and social impacts. The standard is implemented and reported on by stakeholders from member countries, including governments, mining companies, community members, institutional investors, and partner organizations such as development banks and intergovernmental councils.

Currently, there are 48 implementing member countries, 31 of which are compliant with the EITI standard. Member countries are located primarily in Africa and Asia. Notable countries going through the process for membership, but not yet fully compliant with the EITI standard, are the United States, the United Kingdom, and Ukraine. The benefits to countries that join and their stakeholders include improved investment climate, stabilization of political unrest through increased governmental transparency, and increasing the amount of information available to the public, which can be viewed online by anyone.[14]

Other than the EITI, there is little other international policy or guidance on mining. The UN recognizes the importance of strong and effective regulatory policy to ensure that the mining sector delivers economic and social benefits while minimizing social, health, and environmental impacts. Given the increase in demand for mined minerals, their potential depletion, and the role they can play in sustainable development and poverty reduction, the UN and other international entities should develop policy that improves the sustainability of mineral extraction and processing, and they should help governments to do so as well.

RARE-EARTH MINERALS

As previously discussed, the increase in demand for rare-earth minerals has been devastating for the environment and the health of people who either mine them or live near mining sites. China essentially has a monopoly on these resources, providing 95% of global rare mineral output. Interestingly, in order to improve environmental conditions, the Chinese government has begun intensifying control over the exploitation, processing, and export of rare-earth minerals. These restrictions have limited the amount that can be exported and have resulted in increased prices. In turn, the United States, EU, and Japan, which are the primary importers, have opposed these restrictions as violations of World Trade Organization (WTO) rules on exports and have questioned their environmental motives. Transparency in China's environmental regulation on rare-earth mineral mining and processing and also the recognition and willingness of importing countries and the WTO that the costs of environmental and human health protection should be the burden of the purchaser are vital.

ARTISANAL MINING

As the prices of precious metals have sharply increased, so has small-scale artisanal mining and all of the environmental and health impacts that go with it. Banning such activity—for example, making the use of mercury in artisanal gold mining illegal—would not be a sustainable policy. It would intensify the tension among miners, their community, the government, and environmentalists. Because of the economic potential of such activities, banning them would force these activities to be done in a clandestine manner. This would likely result in greater risks for miners and processors who have no other economic opportunities. It would also lead to increased impacts on the environment and human health as workers try to hide their activity and are denied access to training and safer methods.

A better approach would be to enact policies that support safe mining and processing through training and incentives. These policies must be developed through a participatory approach in which all stakeholders are involved. One possibility is to develop centralized locations for processing that have better environmental and safety standards. Working at these sites could provide alternative employment for reasonable wages and offer a better price for services and finished products. Any costs associated with improving the environment and health conditions should be passed on to the consumer. Finally, certification programs for sustainable precious metals, such as "green gold," that monitor and ensure better environmental, economic, health, and social conditions by offering the consumer the opportunity to pay a premium for this exist and should be expanded (Box 7.4).

BOX 7.4

Individual Action: Choose Ethical Sources of Diamonds and Gold

Ironically, both diamonds and gold are used as expressions of love and commitment, yet their extraction and processing can support violence, result in worker exploitation, and severely impact the environment and human health. So-called "blood diamonds" have funded rebel groups, particularly in many African nations, which are then able to conduct acts of violence in these regions, creating unrest, turmoil, and despair. The atrocities involved in this industry are immense and disheartening. They include bloody civil wars; the enslavement and abuse of workers, including women and children; the systematic relocation of individuals and communities; and mass-murder campaigns.

Nearly 90% of those who work in gold mining are artisanal or small-scale gold miners. Comprising 15 million people in Africa, Asia, and Latin America, these workers represent the impoverished and the vulnerable who have little other options for work and income. They work in harsh and dangerous conditions. Their use of mercury or cyanide in the extraction process is both a public health hazard and a global environmental risk. Both artisanal gold miners and those who work for large-scale industrial mines are rarely paid a fair wage or a significant proportion of the market value of the gold they mine.

So how can individuals make ethical choices when they purchase diamonds or gold? The answer is to boycott the suppliers that violate human rights and damage the environment. This can be achieved by purchasing only those products that have been certified to meet a specific set of standards. For diamonds, in 2002 the Kimberley Process Certification Scheme (KPCS) was implemented to combat human rights atrocities throughout the world that are often linked to the diamond trade. The KPCS was developed in Kimberley, South Africa, as a practical way to prevent illicit, conflict diamonds from entering the legitimate diamond trade, and it certifies diamonds as conflict-free. The certification process is managed and enforced by a coalition of states, NGOs, and business bodies interested in the diamond trade. Today, as a result of this process, 99% of diamonds in the marketplace are conflict free as defined by KPCS. However, KPCS has been criticized because of its narrow definition of conflict diamonds, which is restricted to those that support rebel movements, does not address the broader risks to human rights posed by the trade in diamonds, and only applies to rough-cut diamonds. Thus, many are calling for a revision and expansion of this process.

The process of certification to ensure that the gold individuals purchase is mined under improved conditions for workers and reduced environmental impact is managed by Fairtrade International, much the same way that coffee and cocoa are certified (see Chapter 5). Fairtrade gold miners are empowered to organize in cooperatives or associations that are owned and democratically governed by the miners. This gives them better bargaining power with traders to get a fairer return for their gold and greater control over the jewelry supply chain. Fairtrade miners can also establish and join trade unions, and they can collectively negotiate their working conditions. Fairtrade certification requires responsible use of chemicals and mandatory use of protective gear, which are achieved through required health and environmental and safety training for all miners.

By purchasing diamonds or gold through these types of certification programs, individuals are making sustainable choices. They are allowing for more equitable income, not contributing to and thus supporting violence and human rights violations, and are helping protect the environment and human health. These certification systems are limited and need to be revised and expanded for all precious minerals and metals that are valued purchases for individuals. Perhaps the most effective way to reduce the negative impact of these highly valued resources is to better educate the consumer about the deplorable circumstances under which they are extracted. This may cause them to think about choosing to express their love in ways that do not require the purchase of materials that, because of their extreme value, will always have the potential to result in the exploitation of humans and the environment. Perhaps the ultimate sustainable solution would be for individuals and couples not to make expressions of love and commitment that are symbolized by objects but, rather, more gallantly choose to make a more sustainable home and life together.

CONFLICT MINERALS

The purchase of minerals that support violence and human rights violations must be stopped. Policies and provisions that monitor and identify the source of conflict minerals are essential. In addition, the supply chain by which these minerals may travel through other countries and could potentially hide the conflict source must be identified. Then sanctions on sources of conflict minerals and prohibitions of their purchase must be put in place and enforced. Certification programs that allow consumers to avoid or boycott the purchase of conflict minerals, such as "conflict-free" diamonds, must be further enhanced (see Box 7.4).

The UN Security Council has used sanctions to restrict nations that are known sources of conflict minerals, such as the Democratic Republic of the Congo, Sierra Leone, and Liberia. These sanctions need to be expanded and enforced. In the United States, the 2010 Dodd–Frank Wall Street Reform and Consumer Protection Act requires manufacturers to audit their supply chains and report conflict mineral usage. The US Securities and Exchange Commission now requires an annual conflict minerals report including audits for supply chain analysis for any company that uses minerals that potentially have their origin in regions of conflict. From a sustainability and social justice perspective, tracing and banning the use of conflict minerals must be a global priority, as should addressing and alleviating the root causes of such conflicts.

Technical and Innovative Solutions

Given the essential but unsustainable nature of mineral resources and relatively little policy to improve their sustainability, technical innovations related to mineral resource use will be essential. First, we need to find ways to decrease our demand for mineral resources and increase opportunities for mineral recycling. We need innovations that offer cleaner ways to mine and process minerals. We must also find new sources of existing minerals and alternatives to those that are the most problematic or limiting. New techniques in reclamation of mined sites must also be developed. A number of these types of innovations are described next.

DECREASE DEMAND

A first solution to sustaining our mineral resources is to decrease our dependency on them. This can be achieved through reductions in population growth but, more important, through decreases in consumption. We have become a throwaway society, in which the things we purchase have planned obsolescence. Redesigning mineral-containing products so that they are more durable and not disposable will decrease our demand for minerals. For example, the average longevity of an automobile is 10–12 years. By doubling that longevity, mineral use by car manufacturers would be cut in half. Car companies that have a reputation for longevity, such as Volvo, are able to charge a higher initial price, but the cost over the lifetime of their vehicles is often less than that for cheaper, less durable vehicles. Even better than extending the longevity of cars is to build less of them and rely more on public transit. These solutions would in turn significantly reduce the demand for minerals, as vehicle manufacturing is one of largest consumers of them.

Innovations in design could also reduce our dependency on minerals. For example, structural beams might be designed to be equally strong while using less steel. Better design and use of rechargeable batteries so that their life is extended will also significantly decrease use of toxic minerals such as lead, zinc, and cadmium. Engineers are developing new materials from non-metallic materials that when produced could have less environmental and human health impact. One example of this is glass ceramics, which can be substituted for various metals in industrial products, including engines.

RECYCLING

One obvious way to decrease mineral demand is through recycling. Most commonly used metals, such as steel and aluminum, are recyclable. In the case of copper and steel, recycling is already more energy efficient and is nearly as cost-effective as processing ore. However, recycling minerals embedded in technology is more challenging, and most consumer electronics become obsolete very quickly. The average life of a cell phone is only 18 months, the average life of a personal computer is two or three years, and the average life of a flat-screen television is three or four years. In the United States, we dispose of more than 400 million electronic devices each year, the majority of which (75% or 3.4 million tons) end up in landfills. Almost all electronic devices today contain toxic yet useful metals, including mercury, lead, and rare-earth elements. Placing them in landfills impacts the environment but also increases the need to mine and process minerals for the next generation of electronic goods.

Referred to as e-waste, electronics are difficult to recycle. The majority of recyclers in the United States do not recycle the e-waste they collect. They can make more money by exporting it to waste traders rather than by processing it in the United States. These traders send it to poverty-stricken countries, mostly in Asia and Africa, where the e-waste ends up in backyard recycling operations. Workers and their immediate environment are exposed to toxic fumes as they break down and "cook" the e-waste to extract the recyclable minerals. E-waste recycling operations have also been established in US prisons, where worker health and safety protection standards are not enforced.

A solution to this problem is to develop an environmentally safe and just recycling infrastructure. As the price of minerals used in electronics rises, there will be market pressure to increase recycling, and there should be an expansion of this industry. This recycling infrastructure should be adapted to include decommissioned renewable energy generators and hybrid and electric vehicles, which contain large amounts of recoverable rare-earth elements. This expansion must, however, be coupled with enforced sustainability standards and include bans on shipping e-waste to countries without those standards.

As issues arise over the mining of sand, new approaches to recycling concrete from demolition projects need to be developed and expanded. This is occurring in the United States, where 140 million tons of concrete are recycled each year. Concrete can be recycled and turned into aggregates that can be combined with virgin aggregate to make new concrete. Used concrete can also be reclaimed and reused in a variety of applications, such as making paving blocks and serving as a base for asphalt pavement, and in some cases, it has been used to generate new marine reef habitats. The extent to which this reduces the demand for sand, however, is unclear because there is no way to completely recycle used concrete into new concrete for construction.

EXTRACTION AND PROCESSING

In addition to decreasing use and increasing recycling, more sustainable mining processes need to be developed. One of the foci of this research has been on water conservation and reclamation in mining practice. The process of mining minerals is typically water intensive, and run-off can impact downstream water quality and aquatic systems. A number of innovative practices have been put in place to conserve water, especially ways to capture and reuse water rather than continuously draining the local water supply. New technologies that allow water to be collected and treated before being released into the environment have helped reduce the impact on aquatic systems. Emissions control techniques described in Chapter 3 can be used to limit the impact of smelters, thereby keeping them in compliance with clean air policy. Renewable energy sources should be used for the extraction and processing of minerals in order to reduce greenhouse gas emissions and other forms of air pollution. Sound land use planning should also be employed with the objective of preserving habitat and biodiversity.

NEW MINERAL SOURCES

Another way to sustain minerals and our mineral-based economy is to identify new sources. One possible source that is ripe for expansion is deep sea mining, as long as it can be done using environmentally sound techniques. Currently, deep sea mining is limited by cost. This will change as terrestrial sources become depleted and if the externalities of current unsustainable mining practices are internalized in the price of the minerals. Deep sea mining would be an effective way to obtain a large amount of rare-earth minerals, and environmentally benign ways to do this are being developed.

Another method links the production of fresh water from desalination to mineral production. One of the end products from desalination is a concentrated brine solution that is usually dumped back into the ocean. Research engineer Damian Palin has developed a revolutionary process that uses bacteria to extract the minerals present in the brine. Bacteria are introduced to the brine, which then increase in number and through their metabolic activity produce a small electrical charge that attracts the minerals. Referred to as biological mining, this process can extract large amounts of calcium, potassium, sulfur, and magnesium from what would be a waste product from the desalination process. This can be done without the ecological and health impacts of conventional mining, and it will allow countries without mineral resources to generate their own, in turn offering further economic opportunity and security. If the desalination component is powered by renewables, this may be an ideal, sustainable solution that alleviates the problems of depleted drinking water and mineral resources.

Another example of biological mining includes a more sustainable way to extract copper from ore that has low concentrations of the mineral and would otherwise be commercially unviable to do so. Several companies are currently using microorganisms to liberate copper from rocks. Because this extraction is the most energy-intensive part of conventional mining practices, this approach would reduce cost and environmental impact. This process is not only less energy intensive but also can significantly increase yields from low-grade materials that are currently treated as waste. This can increase the total amount of copper extracted from a mine by 30%. Further research in biological mining should be a priority, as should the development of effective ways to extract minerals from tailings and waste and ways to extract minerals from low-grade ores, thereby increasing the total yield of a particular mine. Both of these have great potential to make mineral use more sustainable.

Space is another potential source of minerals. Asteroids and other planets are known to have high concentrations of precious metals such as gold, silver, and platinum, as well as more common elements such as iron and nickel. Because of the difficult nature and cost inherent in extraterrestrial mining, few companies or governments are currently considering it. One company, Planetary Resources, which refers to itself as the "Asteroid Mining Company," is actively developing the technologies and strategies necessary to make asteroid mining economical.

RECLAMATION

There have been many advances in the science of mine reclamation. Reclamation is the action taken to return mined sites to their former state or as safe and healthy repurposed sites. Mine reclamation involves the physical stabilization of the landscape, restoring topsoil, re-vegetation, and the return of the land to a useful purpose. The first step is the removal of the toxic contaminants from the area. This often consists of isolating and treating contaminated material from the soil and water near the mine. Once this is achieved, steps are taken to restore the site to the ecosystem that existed at the mine site before it was disturbed. Mine reclamation often involves using native plants to gradually restore natural ecosystem processes.

Relatively recently, community stakeholders have been brought together to identify the most beneficial use of mine sites. This has led to a variety of creative and successful reuses of a number of decommissioned mine sites throughout the world. Repurposing of mine sites makes the areas safe, restores the environment, and contributes to the local economy after the mines have closed. A number of mines have been repurposed as mining museums or other visitor attractions. In Poland, the Wieliczka Salt Mine was converted into a museum that features works of art carved in the salt. It has been designated a UNESCO World Heritage Site. Some sites have been reclaimed for agriculture, forestry, aquaculture, and natural and recreational areas.

PHYTOREMEDIATION

Phytoremediation is a process of decontaminating soil or water by using plants. The plants can sequester or stabilize toxins near their roots, extract the toxins and incorporate them into their biomass, or actually chemically transform the toxins into less toxic substances. At the same time, these plants can stabilize disrupted soil conditions. This technique has been used to remediate a number of sites that were heavily contaminated by mining and smelter operations. The possibility of recovering and reusing the valuable metals extracted from the soils by the plants in a process referred to as "phytomining" is currently being explored by a number of companies. Limitations of phytoremediation include the fact that only shallow depths occupied by the roots can be remediated and that it is a time-intensive process.

Is Sustainable Mining Achievable?

The mining and minerals industry supports roughly 45% of the world's economic activities. Recently, it has come under tremendous pressure to improve its social, developmental, and environmental performance, but it has been relatively slow to respond. Whether our large global dependency on such a finite resource that negatively impacts the environment and humans in such significant ways is sustainable is a reasonable question. The current situation appears to be bleak, but the policy and technical solutions presented here do offer the potential to make mineral extraction, processing, and use more sustainable.

Mining has the clear potential to benefit developing countries, reduce poverty, and stimulate economic opportunity. But this development will be sustainable only if society meets its mining and mineral needs more responsibly and by protecting the environment, workers, and surrounding communities. Our use of mineral resources is a clear example in which the cost of externalities—whether they damage the environment or have negative impacts on human health and well-being—should be internalized and incorporated into the price of those minerals. This will raise the price of most products, but that will stimulate innovation, decrease use, and overall move us toward sustainability.

As demonstrated in the discussion of e-waste, we are in dire need of responsible life cycle management of products from creation to disposal. As discussed here and also in Chapter 8, which more generally covers waste, a sustainable future will require a transition from a linear economy to one that is more circular. A linear economy follows products from cradle to grave, where they are manufactured from raw materials, sold, used, and ultimately discarded. A circular economy is one that is waste-free or cradle-to-cradle, where products have greater longevity and are repaired, reused, and recycled. This approach will be essential to sustaining our use of minerals as a central component of our economy.

Systems Thinking and Sustainable Resource Management

Analyzing the sustainable management of resource extraction is often limited in its scope. For example, best practices might be developed for a particular forest, mining community, or conservation site without considering how local activities affect surrounding areas, the region, or even global processes. Another problem is that not all of the relevant stakeholders impacted by such practices and management decisions are always identified and involved in the decision-making process. As a result, certain issues or problems may not be addressed. Finally, the management of a particular resource may not consider how competing functions or interests might interact. For example, the extraction of timber might increase economic opportunity for some, but it could negatively impact those who depend on NTFPs or ecotourism projects in the same area.

A way to avoid these limitations is through systems thinking. Examples of systems thinking in the analysis of land use management (see Chapter 2) and wind energy (see Chapter 6) have already been discussed. Here, we explore how it can be applied to increase our understanding of the broader ecological and economic role of mangrove forests and potential impacts from management decisions. Mangrove forests cover 15.9 million hectares of brackish coastline globally. They provide numerous ecosystem services, such as reducing erosion, offering flood protection, providing habitat and food for marine organisms, and sequestering carbon dioxide. They also offer economic opportunity through fisheries, provision of wood, and through ecotourism projects that offer tourists opportunities to explore the wildlife in and around these habitats.

Systems thinking can help us better understand the value of the mangrove forests and the reasons for the rapid rates of destruction. A first step when considering this approach is to identify the factors or elements that either influence or are influenced by mangrove forests. Examples of such factors include stakeholders such as local community members, tourists, ecotourism development, and those who fish or harvest shrimp. Other factors could be ecosystem services and habitat and biodiversity loss. Once the list is generated, causal relationships among factors can be established. For example, ecotourism could provide economic opportunity for the local community in a way that motivates local conservation of mangrove forests. A negative relationship could be the impacts that ecotourism development has on mangrove forests and wildlife.

Applying the systems approach more broadly—perhaps beyond the scope of a single forest—reveals even more about sustainability and mangrove forests, and it could guide both management and individual decision-making. For example, many people avoid meat for environmental reasons, specifically for the contribution of industrial meat production to greenhouse gas emissions and climate change. Seafood is often viewed as an environmentally friendly alternative. Mangrove forests are rapidly being destroyed for the creation of both inland shrimp ponds and beef pastures. Researchers have compared carbon storage and release in various components of intact mangrove forests to those of converted shrimp farms and pastureland in order to better understand the true environmental costs of each of these food types. They found that both shrimp and cattle production resulted in drastic reductions in carbon storage compared to intact mangrove forest. The amount of carbon released per kilogram of production showed that this type of shrimp farming actually releases more carbon into the atmosphere compared to typical means of beef production.[15]

When integrating carbon storage and emission data with other factors, such as social and economic data, it becomes evident that the potential economic gain from these production systems hardly makes up for the local and global environmental costs associated with exploiting them for that gain. This is compounded by the fact that as sea level rises with climate change, intact mangrove forest could serve as a nature-based solution for flood prevention. Such systems thinking can guide management and development decisions on use of ecosystems that include both income potential and environmental costs. It can also inform consumer choices, such as avoiding shrimp cocktail and steak dinners.

Chapter Summary

- Many of the raw materials that are the ingredients of the things we manufacture and use in our daily lives come from forest and mineral resources. Both forest and mineral resources represent a significant part of the global economy. However, the extraction and processing of these resources often have environmental, health, and social impacts.
- Forest resources consist of timber and non-timber products. Because these resources are plants, they should be renewable. Forests also provide a number of ecosystem services, including carbon sequestration, air and water purification, watershed protection, and soil stabilization. They are the habitat for most of our biodiversity and wildlife. Forests also provide important recreational, cultural, and spiritual opportunities.
- There are a number of social, geopolitical, and economic reasons why deforestation and clear-cutting occur at alarming rates. This destruction of forest resources impairs ecosystem function, significantly contributes to global climate change, and is directly related to global poverty.
- Forest transition theory offers a number of explanations for why, after a long period of forest decline, a large proportion of deforested land eventually reverts back to forest. These include the concentration of agriculture on more productive land, the development of industrial economies, increased imports of agricultural and timber products as national economies become integrated into global markets, and the implementation of forest conservation policy. However, such increases in forest cover with development often shift the pressure on natural forests elsewhere in the world. Thus, development may actually result in a net loss of forest on a global scale.
- Early forest management focused on maximizing short-term yield, but management has since transitioned to more conservation-related approaches such as multiple-use, which maximizes the potential for protection and is consistent with conservation goals.
- Policy solutions to make our use of forest resources more sustainable include the de-centralization and de-privatization of forest management responsibilities and shifting that responsibility to state and local governments with a focus on community-based forestry projects.
- Forests can be sustainably managed using ecosystem approaches, by eliminating clear-cutting and plantation approaches, using selective harvest techniques, and developing sustainability certification programs. Community-level programs that provide sustainable use and income generation for those living near forests are a key component of sustainable forestry practice.
- Many prescription medicines are based on indigenous forest knowledge and chemicals from plants found in these communities. However, these communities typically receive no portion of the massive profits that these medicines yield. Mechanisms to compensate indigenous people for the knowledge that promotes conservation of both the forest and their culture are an essential component of sustainable forestry practice.
- Because poverty is directly linked to deforestation, policies that promote sustainable development and poverty reduction should help protect forests. UNESCO's Man and the Biosphere Programme is focused on meeting the needs of the people while conserving natural resources and biodiversity in more than 600 reserves in 119 countries. Central to this idea is that biocultural conservationists recognize that indigenous cultures have developed more sustainable lifestyles and that they act as wise stewards of biodiversity. Thus, biocultural conservation strategies strongly consider indigenous people as

potential allies in the protection of biodiversity and serve to protect their heritage and culture, including language, folklore, and indigenous knowledge.

- In addition to improving the livelihoods of those who depend on forest reserves, other essential components of forest preservation include ecosystem function, watershed protection, the establishment of ecological corridors, and acquiring and protecting land from development.
- Policy should also promote and encourage reforestation of deforested areas. Carbon credits should be used to promote both forest preservation and reforestation.
- Promotion of recycling and the development of forest alternatives such as concrete products, and the use of bamboo and other fiber-producing plants that grow in adverse conditions, will help sustain forest resources.
- In addition to forest resources, another primary ingredient of the things we manufacture or possess is mineral resources, which are inorganic, solid materials that naturally occur in the Earth's crust or can be found on the ocean floor.
- The demand for minerals is rapidly rising, representing greater than 45% of the world's economic opportunity. This is especially true of minerals, such as rare-earth metals, that are used in the technology sector.
- Minerals have a finite abundance, and they are not renewable. Mining and processing them have serious environmental, human health, and social consequences. The economic demand for certain resources has the potential to fund violent conflicts, enslavement, child labor, and, in some cases, genocide.
- There have been few policies to increase the sustainability of minerals except those that prevent the purchase of and analyze the supply chain for those minerals that support conflict and human rights violations.
- Decreasing demand, recycling, development of cleaner extraction processing methods, finding alternatives to current minerals, and improvements in the reclamation of mined sites will make their use more sustainable.
- There are many ways that mineral resources could be more sustainable. However, as with most resources, a transition from a linear use economy to a circular economy that focuses on reuse and recycling will be necessary.

Digging Deeper: Questions and Issues for Further Thought

1. What are some of the social, economic, and geopolitical forces responsible for deforestation, and how can they be addressed to better sustain forest resources?
2. For a particular mining or forestry operation or approach, develop and draw a causal model that could guide systems thinking, like that in Figure 6.13. What are the major factors or elements that impact those activities? What are the causal relationships among these factors? Who are the stakeholders?
3. What is the connection between food security and forest management?
4. What personal choices can you make to improve the sustainability of forest or mineral resources? How will these choices support sustainability criteria?

Reaping What You Sow: Opportunities for Action

1. Find out whether your school or workplace has a purchasing policy. If it does, request a copy. Either suggest modifications of the existing policy or start from scratch and develop a sustainable purchasing policy with regard to resource use and extraction. Consider environmental, social, and economic aspects of sustainability.
2. Quantify the average amount of paper you use in a given week and how much of that is from recycled or raw materials. Use the Paper Calculator at http://c.environmentalpaper.org to quantify the impacts of your paper use. Consider ways to reduce that impact.

References and Further Reading

1. Ezebilo, E. E., and L. Mattson. 2010. "Contribution of Non-Timber Forest Products to Livelihoods of Communities in Southeast Nigeria." *International Journal of Sustainable Development & World Ecology* 17: 231–35.

2. Stanley, D., R. Voeks, and L. Short. 2012. "Is Non-Timber Forest Product Harvest Sustainable in the Less Developed World? A Systematic Review of the Recent Economic and Ecological Literature." *Ethnobiology and Conservation* 1.
3. Sunderland, T. C. H., B. Powell, A. Ickowitz, S. Foli, M. Pinedo-Vasquez, R. Nasi, and C. Padoch. 2013. "Food Security and Nutrition: The Role of Forests." No. CIFOR Discussion Paper. Bogor, Indonesia: Center for International Forestry Research (CIFOR).
4. Harmon, M. E., W. K. Ferrell, and J. F. Franklin. 1990. "Effects of Carbon Storage on Old-Growth Forests to Young Growth Forests." *Science* 247: 699–701.
5. International Bank for Reconstruction and Development/The World Bank. 2012. *The World Bank Global Financial Report.* Washington, DC: The World Bank.
6. Berkes, F. 2008. *Sacred Ecology.* 2nd ed. New York: Routledge.
7. Samii, C., L. Paler, L. Chavis, P. Kulkarni, and M. Lisiecki. 2014. "Effects of Decentralized Forest Management (DFM) on Deforestation and Poverty in Low and Middle Income Countries: A Systematic Review." *Campbell Systematic Reviews* 10(10).
8. Thoma, W., and K. Camara. 2005. *Community Forestry Enterprises: A Case Study of The Gambia.* Rome: Food and Agriculture Organization of the United Nations.
9. Pagiola, S. 2008. "Payments for Environmental Services in Costa Rica." *Ecological Economics* 65(4): 712–24.
10. Donato, D. C., J. B. Kauffman, D. Murdiyarso, S. Kurnianto, M. Stidham, and M. Kanninen. 2011. "Mangroves Among the Most Carbon-Rich Forests in the Tropics." *Nature Geoscience* 4: 293–97.
11. World Wild Life Fund Global and Forest Network. 2013. *Guide to Lesser Known Tropical Timber Species.* Washington, DC: World Wildlife Fund. Accessed April 13, 2015. http://www.worldwildlife.org/publications/guide-to-lesser-known-tropical-timber-species
12. Niesenbaum, R., and J. Elliott. 2019. *In Exchange for Gold: The Legacy and Sustainability of Artisanal Gold Mining in Las Juntas, Costa Rica.* Champaign, IL: Common Ground Research Networks.
13. UNEP Global Environmental Alert Service. 2014. "Sand, Rarer Than One Thinks." https://wedocs.unep.org/bitstream/handle/20.500.11822/8665/GEAS_Mar2014_Sand_Mining.pdf
14. EITI International Secretariat. https://eiti.org
15. Kauffman, J. B., V. B. Arifanti, H. H. Trejo, M. García, J. Norfolk, M. Cifuentes, D. Hadriyanto, and D. Murdiyarso. 2017. "The Jumbo Carbon Footprint of a Shrimp: Carbon Losses from Mangrove Deforestation." *Frontiers in Ecology and the Environment* 15 (4): 183–88. doi:10.1002/fee.1482

CHAPTER 8

SOLVING OUR GARBAGE PROBLEM

So far, we have discussed the sustainability of the environmental resources that humans require to survive, to maintain certain lifestyles, and that are essential for economic development. However, our human footprint does not end after we produce, buy, and use or consume things; the final impact occurs when we discard items. The things we create and use with our planet's resources that eventually no longer have any apparent utility are ultimately discarded as garbage or waste—and we produce a lot of it. Globally, we generate more than 3.5 million tons of solid waste per day. A World Bank study projects a 70% increase in global solid waste production to more than 6 million tons per day by 2025, and that is expected to more than triple by 2100.[1] These increases will primarily come from urban areas in developing countries facing the greatest economic, environmental, and social challenges.

PLANTING A SEED: THINKING BEFORE YOU READ

1. In your community, what is the fate of your garbage? What percentage is recycled? Where do the non-recyclables go?
2. List all the items you use once and then dispose of either in the trash or by recycling in a single day.
3. What are the barriers to participation in recycling where you work or go to school or in your community?
4. What is meant by planned obsolescence, and what is its connection to sustainability?
5. What are the connections between garbage and climate change?

Use it up, wear it out, make it do or do without.

—Women of the Great Depression

This rise and production of garbage will have both drastic economic and environmental costs. For example, in the United States, waste disposal costs currently exceed $100 billion annually. The global cost of dealing with garbage is expected to rise to $375 billion by 2025. Solid waste management is often the largest cost to municipalities and local governments, which spend more on it than on parks, recreation, public safety, or libraries. The steepest cost increases will be in developing countries. Environmentally, our garbage is filling landfills, polluting our water and air, clogging our rivers and oceans, and emitting climate-changing greenhouse gasses (GHGs).

Unless we radically change the way we use resources and generate waste, the production of garbage will undermine any of the other sustainability efforts we have previously discussed. As we make decisions to discard rather than reuse or recycle our limited resources, their rate of depletion will continue to accelerate. But it does not have to be this way. In this chapter, we examine the garbage we produce and how we currently manage it. We then explore policy and innovative technologies that will allow us to reduce waste production and actually use it as a resource. Ultimately, we reframe our thinking about linear resource use from that of extraction-to-production-to-consumption-to-disposal, a model that is often referred to as **cradle-to-grave**. Rather, we consider ways to decrease our use of disposable resources and to make our resource use more circular, where wastes become the raw materials for other products and processes; in other words, **cradle-to-cradle**. We will find that shifts in behaviors, new manufacturing models, and other innovative solutions will offer enormous economic, environmental, and social benefits that can contribute to a sustainable future.

Figure 8.1 A California family photographed with the garbage they produce in one week. *Gregg Segal, 7 Days of Garbage, www.greggsegal.com*

The Garbage We Produce

We have a global garbage problem. The wealthiest developed countries produce almost half of the world's trash. The United States leads the way, where the average person produces four to seven pounds of solid waste per day, or the person's own body weight in trash approximately every three months (Figure 8.1). That is more than double the amount produced in 1960, and it is 50% more than the amount currently produced by Western Europeans.[2] As the developing world becomes urbanized more rapidly than it can develop strategies for managing and reducing waste, the global garbage heap will increasingly pile up.

There are generally three types of solid waste. **Municipal solid waste (MSW)** refers to the stream of garbage collected through community sanitation services. It is the stuff we throw away from our homes and institutions. It mostly consists of packaging, discarded products, and organic materials including food waste (Figure 8.2). **Industrial solid waste** is that which is produced by industrial activity. It includes any material that is rendered useless during a manufacturing process and includes waste from factories, mills and mining operations, construction, or excavation. It also includes sludge from waste water treatment plants, ash from energy plants and factories, and nuclear waste. Finally, medical or **biohazardous waste** includes solid materials that come from health care facilities or laboratories that have the potential to transmit disease. However, all

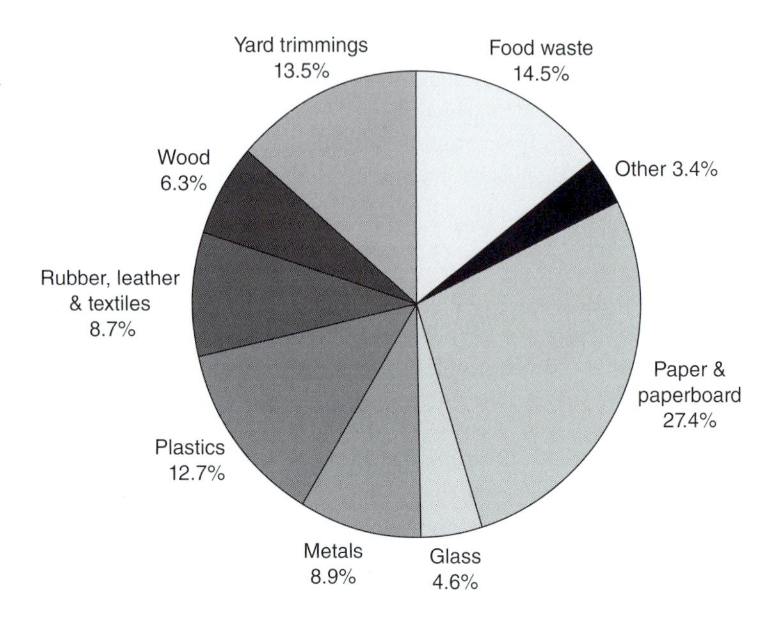

Figure 8.2 Proportion of different types of waste produced in the United States. (US EPA, 2012, MSW Facts and Figures.)

waste has the potential to be hazardous to human and environmental health.

It is difficult to conceive of how much MSW we actually produce, but here are some disturbing numbers. Each day in the United States, we discard enough garbage to fill 63,000 garbage trucks. In a given day, these trucks are filled with more than 80 million plastic shopping bags, 50 million pounds of polystyrene packing peanuts, and 2,000 tons of disposable diapers, all of which can take up to 1,000 years to degrade. Each day, we throw away tons of obsolete personal computers, cell phones, and a multitude of other appliances that no longer work; these are items that are designed to become obsolete. These trucks will be filled with paper, bottles, and cans. Even in the age of electronic and social media, each person still receives nearly 700 pieces of junk mail per year, the majority of which is never opened nor recycled. We will use and toss approximately 2.5 million non-returnable bottles every hour. Each year in the United States, we dispose of more than 290 million used automobile tires, 5.7 million pounds of carpet, and approximately 28 billion pounds of food. Even as the demand for food and nutrition continues to rise, more than one-third of the food produced globally is lost, wasted, or discarded, and in the United States, this represents approximately 15% of the total garbage produced (see Figure 8.2). These are just a sampling of the kinds and amounts of things we put in the garbage. Even more disturbing is the rapid rise in global trash production and where it ultimately ends up.

Managing Our Waste

So where does all of this garbage go? The majority of it goes to dumps or landfills. The rest is incinerated, dumped in the oceans, or accumulated as trash or litter. Very little of it is recycled, composted, or put to use in other ways. Almost 40% of the world's waste ends up in huge rubbish dumps located mostly near urban populations in poor countries in Africa, Latin America, the Caribbean, and Northern Asia, posing a serious threat to human health and the environment. These dumps are situated near or within populated urban areas. Nearly 65 million people—all of them poor—live within 10 km of the world's largest dump sites. Most dumps are also near environmentally sensitive natural resources such as rivers and lakes, and they can easily contaminate aquifers. The annual waste taken to the world's 50 largest dump sites is 21.5 million tons, and they have accumulated nearly 400 million tons spread over 2,175 hectares (ha).

DUMPS

Many dumps are open trash pits that tend to be unregulated and are often illegal. They contaminate our air and water, release GHGs, often burn uncontrollably, and attract vermin and insects that are vectors of disease. Open dumps are a major contributor to the global burden of disease. These include diarrheal and respiratory diseases, adverse birth outcomes, and cancer. In poorer countries, they attract thousands of human scavengers who have no other economic options but to risk injury and disease as they sift through garbage searching for materials of potential value.

LANDFILLS

The development of the sanitary landfill was an important step in beginning to minimize the adverse effects of open waste disposal. They are regulated and monitored for potential pollution of air and water. In contrast to dumps, landfills are engineered and controlled dumping sites. They are constructed by first excavating a large area for garbage disposal, which is then lined along the bottom with either clay or synthetic material to isolate the trash from the environment. This lining also prevents the toxic **leachate**—water that has percolated through the waste—from contaminating the water supply. Each day, the garbage is covered with a layer of soil to minimize odor, pests, and wind disturbances. Many landfills have liquid collection systems at the bottom that recover any leachate, which can then be treated.

Unlike open dumps, because the garbage is covered, it decomposes **anaerobically** or without oxygen. This results in the production of the potent GHG methane, which has roughly 30 times the global warming potential of carbon dioxide (CO_2). At many landfills, this gas is vented and burned or collected and used as a fuel source. The garbage and soil layers are piled hundreds of feet in the air, and they are eventually capped with a thick layer of soil that is then revegetated. The transition from open dumps to technology-advanced landfills was fairly extensive in the United States after they were mandated by the Resource Conservation and Recovery Act of 1976. The transition has been slower elsewhere, particularly among poorer nations.

Landfills are among the largest human-made structures in the world. At one time, the Great Wall of China was the largest constructed object, and it is visible from space. The second largest was the Fresh Kills Landfill in New York City, also visible from space. When it was closed in 2000, it covered 2,200 acres (890 ha) filled with more than 150 million tons of waste, and it actually surpassed the Great Wall in total size. Since the closure of Fresh Kills, the Apex Landfill in Nevada has become the largest, receiving 200–300 tons of garbage every hour and covering 12,000 acres (4,850 ha).

The number of landfills available to receive garbage has significantly declined as the rate of garbage production has increased. In 1979, approximately 18,500 landfills were available in the United States. By 1990, just 11 years later, this number had decreased by nearly 60%, and by 1998, there were 7,924 landfills available. In 2006, there were only 1,754 left. Landfills are similarly in decline in Europe and other developed countries. Although the number of landfills has declined, it appears that overall capacity has not, because landfills are now larger and can receive more garbage. However, if our rate of garbage production does not decline, capacity will become an issue. Whether or not this is the case, fewer landfills mean that on average garbage has to travel farther from its source. Twenty years ago, garbage from Manhattan would have traveled just a few miles by barge to the Fresh Kills Landfill on Staten Island. Today, it is exported to Ohio, Pennsylvania, or West Virginia. These longer trips mean significantly more fuel use and GHG emissions.

INCINERATION

Burning household garbage and trash in open piles was one of the earliest forms of waste disposal, and it is still done in many rural areas, even in the United States. Large-scale incineration was developed in the 1890s, and by the 1920s it was a common method of waste disposal. Initially, incinerators were simply used to reduce the solid mass of the waste and did so by up to 80%. The remaining ash still needed to be placed in dumps or landfills. Prior to the development of flue gas cleaning systems or stack scrubbers, particulate matter, heavy metals, dioxins, furans, sulfur dioxide, hydrochloric acid, and mercury were a significant pollution component to incinerator stack emissions. Other pollution issues include odor, dust, and the need to dispose of toxin-containing ash.

Modern garbage incinerators are governed by clean air policy described in Chapter 3 and effectively limit emissions as required by law. They emit less than modern coal- and oil-fired power plants but typically not less than plants powered by natural gas. Today, most incineration is linked to energy generation in **waste-to-energy (WTE)** plants, especially in Europe. There is growing global interest in WTE plants because they seem to solve two problems simultaneously: waste disposal and energy production. Because garbage can and is being used to generate energy in WTE plants, many consider it to be a renewable energy source. However, this is not without controversy and debate. We perform a more thorough analysis of the potential for trash to energy as a sustainable solution later in this chapter.

OCEAN DUMPING

Prior to 1972, the world's oceans served as a major dumping ground. Until 1931, New York City dumped most of its garbage in the Atlantic Ocean, and major open dumping of millions of pounds of waste each year continued well into the 1960s. These materials included polluted dredged spoils, industrial and military wastes, sewage sludge, garbage, and construction materials. Between 1946 and 1970, nearly 90,000 containers of radioactive wastes were dumped at sites in both the Pacific and Atlantic Oceans. In addition, a significant amount of wastes enter the ocean through rivers, atmospheric and pipeline discharge, construction, offshore mining, oil and gas exploration, and shipboard waste disposal. Unfortunately, the oceans had become the ultimate dumping ground for the by-products of society's consumption.

Recognizing the environmental catastrophe that was taking place, the US Congress passed the Ocean Dumping or Marine Protection, Research, and Sanctuaries Act in 1972, giving the US Environmental Protection Agency (EPA) power to monitor and regulate ocean dumping. Also, the 1972 United Nations (UN) Conference on the Human Environment in Stockholm urged all nations to control the dumping of waste in "their oceans" by implementing new laws, and the UN later developed and implemented the Convention on the Prevention of Marine Pollution by Dumping of Wastes in 1975. The International Maritime Organization was given responsibility for this convention, and what is referred to as the London Protocol was finally adopted in 1996, a major step in the regulation of ocean dumping. The 1982 UN Convention on the Law of the Sea also directs nations to adopt laws and regulations on ocean dumping.

The previously mentioned US and international reforms did slow overt ocean dumping, but they have not eliminated the entry of massive amounts of garbage into the oceans. Illegal ocean dumping and shipboard waste disposal still occur. Also, untreated sewage and storm run-off still enter streams, rivers, and estuaries, not only polluting them but also eventually carrying plastic waste materials to the ocean. The United States passed the Ocean Dumping Ban Act of 1988, which prohibits all municipal sewage sludge and industrial waste dumping into the ocean, yet because of poor infrastructure, it still occurs. The growth of urban slums and garbage heaps along coastal areas in developing countries also contribute garbage to the oceans, especially with rising sea levels, storm surges, and major events such as tsunamis. For example, the 2011 earthquake and subsequent tsunami that occurred in Japan swept approximately 5 million tons of debris into the Pacific Ocean.

The amount of garbage that still enters the oceans is staggering. Despite the laws and conventions against ocean dumping, a number of countries, mostly in Asia, dump up to 12 million tons of garbage into the ocean each year, with China as the major contributor. As of 2004, the United States was still releasing more than 850 billion gallons of untreated sewage and storm run-off every year, and as sewage treatment infrastructure becomes more antiquated and overburdened, this number will increase. This sewage and run-off carries cups, plastic bags and bottles, and items such as cotton swabs, condoms, tampon applicators, and dental floss that are flushed into the waste stream. All of these items end up at sea and along our beaches.

In 1988, medical waste, including syringes feared to be contaminated with HIV, washed up on New York and New Jersey beaches. It was believed that this waste had been dumped at sea illegally by hospitals and disposal companies. But after investigation, it was found that most of the debris was garbage that had washed into New York Harbor from the Fresh Kills Landfill or waste from New York City's sewage system that included syringes and vials that had been thrown away by individual drug users and diabetics. The resultant closing of New York and New Jersey beaches cost the states approximately $1.5 billion in tourist revenue.

THE GREAT PACIFIC GARBAGE PATCH

Much of the garbage that enters the ocean is floatable and non-biodegradable. As such, natural currents concentrate this debris in the North Pacific Ocean in what is referred to as the Great Pacific Garbage Patch. Really two patches combined, it spans waters from the West Coast of North America to Japan and is bounded by the North Pacific and North Equatorial circular currents (Figure 8.3). Twice the size of Texas at approximately 1.4 million km^2, the Great Pacific Garbage Patch stretches for hundreds of miles across the North Pacific Ocean, making it the world's biggest dump. Even more shocking is that the seafloor beneath the Great Pacific Garbage Patch may be a more massive underwater trash heap. Oceanographers recently discovered that approximately

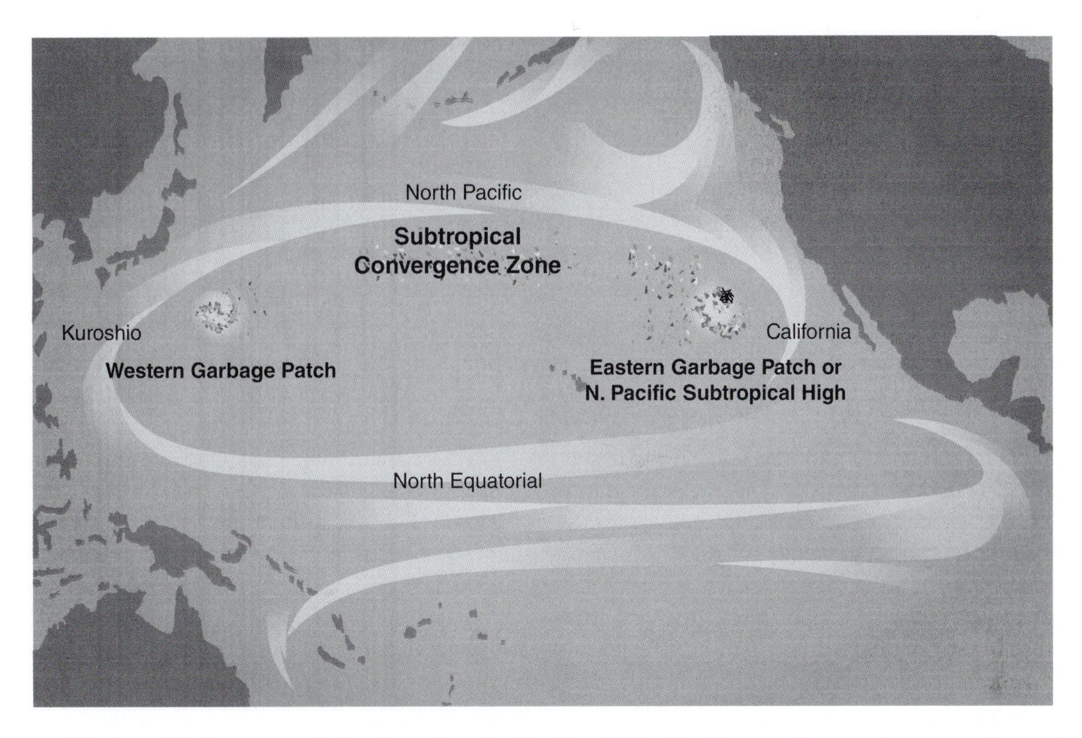

Figure 8.3 Marine debris accumulation locations in the North Pacific Ocean where major currents create gyres forming The Great Pacific Garbage Patch. (NOAA Marine Debris Program.)

70% of marine debris eventually sinks to the bottom of the ocean. Similar floating garbage patches and underwater trash heaps occur in other ocean gyres.

Most of the debris in the floating garbage patch is non-biodegradable plastic. The world produces more than 100 billion pounds of plastic each year. Approximately 10% of it ends up in the ocean. Larger pieces can entangle marine organisms and birds, and smaller pieces can clog the intestines or gills of organisms that consume them. This plastic debris results in an estimated $13 billion a year in losses from damage to marine ecosystems, including financial losses to fisheries and tourism as well as time spent cleaning beaches. Over time, sunlight eventually photodegrades the bonds in the plastic polymers, reducing it to increasingly smaller pieces of plastic less than the size of a grain of salt. These so-called **microplastics** are the most abundant type of plastic debris in the ocean. Another major source of microplastics is the microbeads used as abrasives in cosmetics, facial scrubs, and toothpastes that eventually make their way into the ocean through sewage discharge. A single tube of toothpaste can contain as many as 300,000 of these plastic microbeads.

Plastic is inherently toxic, containing chemicals that have been linked to various environmental and health problems. Human exposure to plastic is so common that our bodies and urine contain significant concentrations of the plastic-containing compound bisphenol A (BPA), a known endocrine disrupter. Plastic floating in the oceans has also been shown to absorb pre-existing organic pollutants such as polychlorinated biphenyls (PCBs) from the surrounding environment. The toxic compounds associated with plastic enter the marine food chain as it is consumed by marine organisms. These compounds become concentrated as they move up the food chain through **biomagnification**, a process in which they impact larger marine animals and enter marine fisheries, offering further opportunities for human exposure.

Another perhaps disturbing phenomenon is the discovery of microplastics floating on the surface of the ocean serving as a novel ecological habitat colonized by thousands of species of bacteria and microalgae; researchers refer to this as the **plastisphere**. The environmental impacts of the interaction among microplastics, the microbes that inhabit them, marine food chains, and global currents are poorly understood and are the focus of current research by a number of interdisciplinary research teams. Included among them are student and faculty researchers with the Sea Education

Association, an internationally recognized leader in undergraduate ocean and sustainability education.

Recycling and Composting

Obvious alternatives to landfills and incineration are recycling and composting. Recycling is the process of separating, collecting, and remanufacturing or converting used or waste products into new materials. The potential environmental and economic benefits of recycling seem obvious. When products made with natural resources are recycled, less of that resource must be harvested. Recycling not only conserves those resources but also reduces the environmental impact of their extraction. For example, recycling a single run of the Sunday *New York Times* would save 75,000 trees. Recycling also lessens the amount of material placed in landfills, where rates of decomposition are extremely low. It also takes less energy to process recycled materials than to process virgin materials, thus reducing the environmental impact of energy, including reduction in climate-changing GHG emissions, and energy dependence.

According to the EPA, in 2014 the United States recycled 89 million tons or 34% of its MSW. This reduced net GHG emissions by more than 168 million metric tons of CO_2 equivalents. This is comparable to the annual GHG emissions from more than 33 million passenger vehicles. Producing aluminum from recycled materials uses 95% less energy than that which is needed to make the same amount of aluminum from a virgin source. Recycling glass produces 50% less carbon than producing it from raw materials. A 2011 report by the Tellus Institute found that if we increase the US national recycling rate to 75% by 2030, we would reduce GHG emissions by 515 million metric ton CO_2 equivalents. This would be equivalent to shutting down 72 coal-fired power plants or taking 50 million cars off the road.[3]

Recycling creates economic growth and tax revenue through the creation of new businesses and jobs. In the United States, recycling is a $100 billion business, and the recycling industry employs more than 1.1 million people typically at wages higher than the national average. Many of these jobs are located in poorer urban centers where job creation is vital. Recycling offers cost savings to manufacturers. For example, it costs 30% less to produce glass bottles from recycled glass than it does to make the bottles from raw materials. The recycling industry also generates billions of dollars in federal, state, and local tax revenues. The Tellus report found that moving to a 75% recycling rate would create 1.5 million new jobs. The state of California has committed to doing this by 2020 and anticipates that this will create more than 100,000 new jobs.

Despite the obvious benefits of recycling, the proportion of trash that is sent to landfills or incinerated rather than recycled varies greatly by country (Table 8.1). Many countries send very little of their garbage to landfills, and most of it is sent either to trash-to-energy incinerators or for recycling and composting. This is not the case for poorer European countries such as Hungry, Poland, and Lithuania, which landfill more than 70% of their garbage. Nor is it the case for the United States, which is the largest producer of garbage.

Although the United States makes up only 5% of the world's population, it produces 30% of the world's waste, including 251 million tons in 2012. Unfortunately, 80% of all products manufactured in the United States are used only once and then discarded, and less than 35% of what is discarded is recycled, with the majority of it going to landfills (see Table 8.1). Among the things we throw away, paper and paper products have the highest rates of recycling—approximately 65%—but represent the largest single category of waste generated: 68.2 million tons or approximately one-third of all waste produced (see Figure 8.2). In 2012, less than 9% of plastic, 34% of glass, and 20% of aluminum wastes were recovered through recycling.[4] The good news is that recycling has been increasing in the United States. From 1995 to 2015, the number of communities with recycling increased from 500 to 10,000, making recycling available to 90% of the total population. However, not everybody takes advantage of that opportunity.

Rates of recycling are governed by municipal and institutional priorities and by individual behavior. Many of the materials that are collected for recycling are of high value, as is made evident by the rate of theft of recycled materials from curbside recycling containers. However, the market price for these commodities varies greatly by the specific material and over time. For example, the price for used plastics—a petroleum product—is driven by the price of oil, which exhibits regular periods of decline. When this occurs, it makes more economic sense for companies to buy freshly made plastic than recycled plastic. Recycling other materials, such as metals and glass, offers greater economic benefit.

TABLE 8.1 Fate of Garbage by Country

COUNTRY	LANDFILLS (%)	RECYCLING/COMPOSTING (%)	INCINERATION (WASTE-TO-ENERGY; %)
Germany	0	66	34
Netherlands	1	60	39
Austria	1	70	29
Sweden	2	49	49
Belgium	4	48	49
Denmark	4	60	36
France	32	34	34
Italy	45	43	12
Finland	46	36	18
UK	48	40	11
Spain	52	39	9
Portugal	62	20	18
USA	54	34	12
Hungry	72	18	10
Poland	78	21	1
Lithuania	78	21	1
Bulgaria	100	0	0

Source: European data from Eurostat 2012; US data from EPA. 2014. "MSW Facts and Figures for 2012."

For most communities, recycling viewed in isolation is not profitable—that is, it costs more to collect, process, and sell the materials than the income that is generated from those sales. However, when considered within the context of landfill costs, the economic benefits become clearer. In New York City, even though recycling itself loses money, it still costs 20% less than disposing of these materials in landfills. The rise in landfill **tipping fees**, the charge levied upon a given quantity of waste received at a landfill or other waste processing facility, has motivated the expansion of municipal recycling programs. In addition, these market-driven motives for recycling initiatives do not take into account the external costs of not recycling, such as increased energy use, resource depletion, and environmental and health impacts.

Why do individuals and organizations often not participate in existing recycling programs? Noncompliance is typically due to perceived inconvenience, increased cost due to the need to sort and store recyclables, and a general lack of knowledge about the benefits of recycling. Misperceptions often limit individual compliance in recycling programs. For a variety of reasons, individuals often have the sense that even though they place their materials in recycling containers, these are not recycled and end up either in a landfill or in an incinerator. In many cases, these perceptions are accurate because the contamination of recycled items with food and other materials seriously limits rates of recycling. Currently, recycling processors are limited in their ability to deal with that contaminated material. On average, approximately 20% of materials collected for recycling end up being sent to landfills and incinerators because of contamination. In general, the market for recyclables declines with fuel prices, which occurred rather drastically in 2015 and 2016. This is particularly true for plastics. As the market for recyclables drops, the proportion of materials collected for recycling by waste haulers declines, and much of these materials do end up being sent to landfills and incinerators.

Recycling is often limited because the end products differ from the original products and are of lesser quality. This reduction in quality with recycling is referred to as **downcycling**. Downcycling reduces reusability and has a limited number of iterations, with the material eventually ending up as waste. With each iteration of downcycling, material quality and value are diminished. Most materials—with exception of glass, which can be recycled indefinitely—are actually downcycled rather than recycled. Although most recycling is actually downcycling, it still diverts material from waste streams at least temporarily. The downcycled products are often less expensive than those made from raw materials, so it can help conserve those resources. Because it takes more energy to create new products from raw materials than from downcycling, it reduces energy consumption and related impacts on water, air, and climate change. Later in this chapter, we consider more efficient ways to reuse and recycle material, including **upcycling**, in which the end product has more value than the original material.

COMPOSTING

Globally, the largest proportion of the solid waste that we generate is organic garbage, such as food, forestry, yard, and landscaping wastes. A natural process called **composting** can be used to recycle or repurpose these materials. Composting is the controlled decomposition of organic materials by naturally occurring microorganisms and results in the production of partially or completely decomposed organic material. Effective composting requires the aeration of the decomposing organic material. **Vermicomposting** uses red worms (*Eisenia fetida*) to break down food waste into high-value compost called castings.

In addition to diverting organic wastes from landfills, composting offers a number of other benefits. The end product is a marketable commodity. Finished compost is valued by organic farmers, gardeners, and landscapers because it can be used to improve soil quality, block weeds, and works as a natural fertilizer, thereby reducing the need for water, fertilizers, and pesticides. It can also be used to replace excavated soil for landfill cover layers and caps and to control water run-off and erosion. Because composting more typically employs aerobic decomposition, CO_2 is produced as opposed to methane, which would be produced if the material were placed in a landfill, where it would decompose in the absence of oxygen. Anaerobic systems that generate and capture methane for use as an energy source add value to composting without further increases in GHG emissions, especially when substituting for fossil fuels.

Most composting efforts are either small-scale, household composting projects or part of municipal yard waste collection and management programs. Large-scale, urban food composting programs have been slower to catch on, but given the volume of food waste, there is growing interest in this area. San Francisco and Seattle have taken the lead and now require food scrap recycling. New York City, which sends 1.2 million tons of food waste to landfills each year at a cost of nearly $80 million per ton, is now developing a pilot food waste composting project with an initial goal of composting 10% of it. Private food composting businesses are also beginning to emerge to serve institutions and smaller communities. There is significant entrepreneurial opportunity related to composting in the areas of collection, production, and marketing and sale of final product both in developed and less developed countries. Thus, composting has the potential to offer economic opportunities, environmental benefits, and potential for improving soil quality and agriculture.

Challenges associated with establishing large-scale composting projects include a lack of infrastructure for collecting, storing, composting, and distributing composted products. Household collection, compliance, sorting and handling of the materials, and potential for odor are concerns. Another issue is the composting of post-consumer food waste. **Pre-consumer food waste** is generated during processing and food preparation, or food that has not been served but has gone bad. Post-consumer food waste is food that has been handled by the consumer and then discarded. As a result, there is often concern about the risks of contamination or infection from human pathogens when handling post-consumer food waste, but this risk can be eliminated through proper handling and composting techniques. Another limit to expanding composting projects, particularly in developing countries, is a lack of understanding of the biological processes involved in composting and a lack of vision or marketing plan for the final product.

DIVERSION

A concept related to recycling and composting is diversion. The diversion rate expresses how much waste is diverted or stopped from entering the landfill. In most cases, the waste is either recycled or composted as an alternative to being deposited in a landfill. Sending garbage to WTE plants is also considered diversion. Another form of diversion is when trash or garbage is repurposed to make sellable products or art such as bags, belts, and other gift items. This is common in developing countries, where these products are often sold through cooperatives or networks to consumers in the developed world.

Hazardous Waste

Waste that poses a substantial or potential threat to public health or the environment is referred to as hazardous waste. It can be produced commercially as a by-product in the manufacture of useful materials and goods. In the past, because these hazardous wastes were not useful to the industry producing them, they were typically disposed of in the easiest, least expensive way. After World War II, as industrial processes expanded, there was a sharp rise in the production of these wastes, and millions of tons were produced per year. It was also estimated that prior to the regulation of hazardous waste, only 10% of it was disposed of properly.

The treatment, storage, and disposal of hazardous wastes are now heavily regulated in the United States and other developed countries. However, the safe disposal of toxic waste has become a global challenge. Each year, 440 million tons of toxic waste are produced globally, and many countries lack the regulation and technical ability to dispose of it in ways that do not impact human and environmental health.

Another source of hazardous waste is generated at the household level. One example is electronic waste and the impact it has on the environment and human health, which we discussed in Chapter 7. Other household hazardous wastes include toxic chemicals and other substances for which the homeowner no longer has a use, including cleaning solutions, pesticides, spent motor oil, paint, drain cleaners, batteries, and fluorescent bulbs. Currently, there is very little regulation regarding the disposal of household hazardous wastes, and much of it enters the municipal trash stream unchecked. In response to this, local municipalities have been developing collection programs that collect and then properly dispose of these materials. However, participation is optional, and the level of compliance is unknown.

Historically, there has been little or no regulation of hazardous waste disposal in developing nations. In addition, more than 90% of the millions of metric tons of hazardous waste generated globally each year originates in the major industrialized countries, where disposal facilities have become scarcer and environmental laws have become stricter. This, coupled with the proliferation of free trade agreements and economic globalization, has resulted in the rapid expansion of the export of toxic waste to poorer, less developed countries. This transboundary export of hazardous waste has become a multibillion dollar industry that has provided opportunities for corporations to dispose of their waste much more cheaply. But this of course comes at a price. The nations that receive this waste typically lack the means to dispose of it properly, and so their people and environment are put at great risk. Often referred to as **toxic colonialism,** the export of toxic waste and its disposal have become a major issue of environmental injustice on a global scale.

There are a number of well-known cases of the export of hazardous waste from developed to less developed countries. In 1986, a ship carrying toxic incinerator ash from Philadelphia because the ash could no longer be legally disposed of in the United States surreptitiously dumped 4,000 tons of this waste on a beach in Haiti and the rest of its load at sea. In 1988, five ships transported 8,000 barrels of hazardous waste from Italy to the small town of Koko, Nigeria, in exchange for $100 monthly rent that was paid to a Nigerian for the use of his farmland as a dump site. Several thousand tons of highly toxic and radioactive wastes in leaking drums, including 150 tons of PCBs, which are both carcinogenic and toxic, ended up at this site.

Garbage and Poverty

Like most environmental hazards, deficiencies in waste management also disproportionately and unjustly affect the poor. The people who live closest to open dumps, landfills, incinerators, and sites polluted with toxic wastes are most often poor and members of minority and immigrant communities. It has been shown that in the United States, race is the strongest determinant of who is exposed to such environmental hazards. The largest hazardous waste landfills in the United States are located in predominantly Black or Hispanic communities. Toxic waste clean-up has also been shown to occur in White, wealthier communities long before it does in poorer, minority communities.[5]

In addition, the poor of the developing world are subject to even worse injustice, especially in urban areas. Garbage is often dumped in land adjacent to slums, and in many poorer, urban environments, it is never collected. Many people in developing countries, especially those with limited education, lack economic opportunities, and waste picking may be their primary source of income. The World Bank estimates that 2% of the world's urban population consists of waste pickers.[6] A significant number of workers in this informal waste sector are women and children, who are exposed to hazardous substances in wastes such as lead, mercury dioxins, and infectious agents. Recycling of e-waste, mostly from the United States, occurs in small backyard operations primarily in Asia where workers, including children, are exposed to toxic chemicals. Poorer nations are increasingly becoming the dumping grounds for our hazardous wastes. Any sustainable solution regarding waste must take into account these inequities as they relate to poverty and race.

Policy Solutions for Managing and Reducing Waste

The current way that we deal with waste is not sustainable. Not only are we running out of space to place our garbage but also we are polluting our bodies and our environment as we dispose of the garbage that we produce. Consumption, resource use, and waste production are complexly intertwined. Our cradle-to-grave mentality implies that the only direction of flow of resources is from point of origin to manufacturing, consumption and use, and ultimately being laid to rest as waste. This model ensures the continued depletion of resources and flow into the waste stream. Moreover, the manufacture, distribution, and disposal of the things we produce all result in the release of GHGs and climate change.

We teach our children the "three R's": reduce, reuse, and recycle. Yet we are not effectively or sufficiently doing these. We also need ways to undo the damage already done by our reckless management of waste, and we need to ensure justice as we do so. In this section, we consider new or strengthened policies that will help us better manage our waste and our resource use. Then we consider technological and innovative approaches to reducing, reusing, and recycling as well as cleaning and repairing current environmental damage from poor waste management. Finally, we consider new industrial and economic models that consider waste production and resource depletion in concert with each other.

National Policies and a Global Issue

Policy for waste management is typically developed and enforced at municipal, state, and national levels. However, we have seen that solid waste management—or the lack of it—has global impacts. These impacts include increased GHG emissions, ocean pollution (especially with plastics), and the export of garbage (particularly e-waste) to developing countries that lack the infrastructure and regulation to safely and effectively manage it. Obviously, the best solution for dealing with waste will vary by locality, and thus local, regional, and national policies are important. However, international policy, enforcement of that policy, and funding for the development of infrastructure to manage waste must become a global priority.

As we examine policy and other sustainable solutions to our waste problems, we should not only consider how best to manage it but also consider ways to prevent its production, which is directly related to consumption. Thus, we need to begin to explore ways to reduce consumption both in the developed areas of the world and as an integral aspect of sustainable development. In addition, rather than simply improving ways to manage waste, we must consider creative ways to reuse discarded goods or the resources that compose them. We should continue to promote recycling and composting, and we must develop ways to recover value from waste, particularly in the form of clean, useable energy.

SANITARY LANDFILLS

Open dumps are extremely problematic from both environmental and public health perspectives. In the United States, this has been dealt with first through the 1965 Solid Waste Disposal Act and then through the Resource Conservation and Recovery Act of 1976. Both of these acts encourage states to develop comprehensive plans to manage solid waste by prohibiting open dumping and setting criteria for municipal solid waste landfills and other solid waste disposal facilities. Landfill standards include location restrictions prohibiting them from being sited near geological faults, flood plains, wetlands, and communities. Liners, leachate collection, and compaction and covering with soil are also required. Hazardous materials are banned from landfills, groundwater must be regularly monitored, and landfill companies must have funds set aside for environmental clean-up, fines for violations, and post-closure care.

When landfills are closed, the law requires that they be cared for so that no subsequent environmental contamination can occur. The first step of landfill closure is to install a final cover system to minimize infiltration of liquids and soil erosion. By law, this cover layer must consist of an infiltration layer of at least 18 inches of soil covered by an erosion layer of at least 6 inches that is capable of sustaining native plant growth. Once closed, a leachate collection system, groundwater monitoring, and methane monitoring are required. The land occupied by closed landfills is often reclaimed for other uses, such as wildlife reserves, parks, and golf courses, although these may require more specialized capping materials. Examples include Mount Trashmore Park in Virginia Beach, Virginia, and the design and engineering of Freshkills Park from what was once the world's largest landfill. Freshkills Park will occupy 2,200 acres, making it nearly three times the size of Central Park and the largest park developed in New York City in over 100 years. It will offer a variety of public spaces and facilities, including playgrounds, athletic fields, kayak launches, horseback riding trails, large-scale art installations, and much more. Several parts of the park are now open to the public. The remainder is scheduled to be opened in phases, through 2036. (Figure 8.4).

The use of landfills as a primary means of waste management and their regulation are essentially limited to the United States. Other countries either lack the space or lack the resources to effectively develop, manage, and properly close sanitary landfills. In other developed countries, incineration—primarily for energy—or recycling are more common (see Table 8.1). Middle-income countries tend to have a combination of poorly managed landfills and open dumps, whereas the poorest countries almost exclusively use open dumps and burning as the primary ways to deal with garbage. Poorly managed waste has an enormous impact on health, local and global environments, and the economy. Thus, international policy, program development, and funding for more effective waste management and waste reduction must be a priority for developing countries.

INCINERATION

With the rise of waste incineration, mostly in the form of WTE, policies ensuring stringent environmental control are essential. In the United States, state and federal governments have enacted legislation and regulations under the Clean Air Act (see Chapter 3) to do just that. Incinerators must comply with a combination of federal, state, and local regulations. For example, a municipal waste incinerator site in California must comply with federal laws, as well as get a permit

Figure 8.4 The transformation of Fresh Kills Landfill on Staten Island, New York, into a world-class park with miles of biking and running trails, serpentine water features, a wind farm, and a 9/11 memorial is underway. *Cryptome; Rendering courtesy of Field Operations/The City of New York, https://goo.gl/images/wUuDme*

from the California Air Resources Board and one from the local air pollution control district.

The extent to which incinerators effectively limit emissions is under intense debate, particularly in the United States. However, in Sweden and Denmark—which burn as much as 80% of their waste, and 20% of heat production and 5% of electricity are generated through trash incineration—there is little or no debate. The application and enforcement of modern emission standards and the use of modern pollution control technologies have reduced emissions to levels at which pollution risks from waste incinerators are now generally considered to be very low. Another environmental advantage to controlled incineration is that burning trash releases CO_2, whereas if this waste were landfilled, methane (the more potent GHG) would be emitted as the garbage decomposed.

There is no conclusive evidence of non-occupational health effects from incinerators, but it is argued that small, chronic effects might be virtually impossible to detect. Others argue that even the cleanest incinerators release a wide range of toxic air pollutants, many of which are not monitored. Also, end products such as captured pollutants and the highly toxic fly ash must be safely disposed of at sites approved for toxic wastes.

There are environmental justice issues associated with trash incineration and WTE. These plants are most often placed in poorer neighborhoods. For example, there were plans to build a WTE incinerator in the low-income Curtis Bay neighborhood of Baltimore, Maryland. State environmental officials had approved a permit that would allow the plant to burn 1.4 million tons of tires, plastic, and construction waste each year, potentially releasing tons of fine particulates, mercury, heavy metals, and other toxins. This neighborhood is already impacted by the presence of a 200-acre coal plant, a fertilizer plant, one of the nation's largest medical waste incinerators, chemical plants, fuel depots, and an open-air composting site. In 2009, Curtis Bay was ranked among the top areas in the nation for the release of toxic emissions. Higher rates of asthma and cancer have been well documented in the area.

The fight to stop this south Baltimore plant was led by 16-year-old student activist Destiny Watford and fellow members of a student-led group called Free Our Voice. Their endeavor was ultimately successful after four years of concerted effort. Watford was recognized for her leadership in this successful effort with the prestigious Goldman Environmental Prize that also provided $175,000 to allow her to continue her work in protecting her community. This is an excellent example of how the action of an individual can truly make a difference. However, more than individual action is needed. Local governments must enact and enforce more effective policies with regard to emission standards, control, and enforcement, and the just placement of these facilities must be a priority. Even if such stringent policies were enacted, many would still argue that WTE is neither renewable nor sustainable, as discussed later in this chapter.

RECYCLING

Given the potential environmental and economic benefits of recycling or even downcycling, policies should be passed that promote, reward, or require recycling. The United States has no national recycling policy. The Solid Waste Disposal and Resource Conservation and Recovery Acts focus on the elimination of open dumps and the creation and management of landfills. But they barely mention recycling beyond recommending that the federal government should increase its purchase of products made with recycled content.

The EPA publishes manuals and offers workshops on implementing curbside recycling programs, but there is no federal policy or funding to support this. However, because of the decreasing accessibility of affordable landfill space, a number of municipalities, counties, and states have enacted recycling policies. These policies require and implement mandatory curbside recycling programs, and currently there are nearly 10,000 of them nationwide. Similar polices in other developed nations have likewise led to increases in recycling.

Mandatory recycling imposes fines for placing recyclable items into the garbage stream rather than into recycling containers. These laws are imposed on households, businesses, and most often on construction and demolition projects. Also, licensed haulers that collect MSW may have their licenses revoked for non-compliance in recycling programs when they discard rather than recycle required materials, even when the market for these materials declines. It has been shown that enforcement programs that focus on compliance rather than punishment are more effective. This includes provision of recycling containers, educational programming, and clearly communicated policies and procedures. Compliance increased when those who placed recyclables in their trash were first warned and then fined.[7]

There are other types of policies that promote recycling. For example, pay-as-you-throw (PAYT) systems have been shown to increase recycling when there is a charge for disposal of garbage but not for recycling. They also remove some of the cost burden of waste management from property tax bills. There are numerous PAYT programs in the United States, Europe, and Asia, and they have been shown to be quite effective. Where PAYT programs have been implemented, they decreased residential waste by 30% and have similarly increased rates of recycling.

Other programs encourage recycling through a mix of education, rewards, and impact metrics. Through companies such as the RecycleBank in New York City, household recycling can be tracked using radio frequency identification or global positioning system technology installed on recycling containers and in recycling trucks. They are thus able to track which households are actually recycling and offer financial incentives for households and communities that are the most compliant. Financial incentives such as these are quite effective; however, in most cases, the rate of compliance returns to the pre-incentive level if and when the incentive program ends.

Returnable glass bottles that enabled reuse or recycling and prevented entry into the trash stream used to be common in the beverage industry. But because of the associated expense placed on the beverage companies, returnable bottles are essentially extinct. One of the major players in the elimination of returnable bottles was Coca-Cola. It achieved this by supporting and promoting the expansion of municipal recycling programs that would replace returnable bottles. In so doing, it was able to promote itself as pro-environment. However, what the company really did was convert the internal cost of bottle collection, cleaning, and reuse to an externality ultimately funded with public money. The eventual conversion from glass to plastic bottles was the final nail in the coffin of the returnable bottle.

In response to the elimination of returnable bottles, a number of states have enacted container deposit laws or "bottle bills." These bills are a proven, sustainable method of capturing beverage bottles and cans for recycling. A **container deposit law** requires a minimum refundable deposit on beer, soft drink, and other beverage containers in order to ensure a high rate of recycling or reuse. The refund value of the container provides a monetary incentive to return the container for recycling. The presence of a bottle bill in a state generally results in much higher rates of material recovery and reduced litter. The collection and recycling of containers in bottle bill states has created tens of thousands of new jobs in retail, distribution, and recycling. In some states, unclaimed deposits are deemed abandoned by the public and are therefore property of the state, which then uses these monies to fund other environmental programs or to cover the administrative costs of the deposit program.

A container deposit law is an example of a sustainable policy solution that promotes more efficient resource use, stimulates the economy, and creates income opportunities for the poor. Despite this, only 10 US states have such laws. There is strong opposition against such bills from the beverage and retail industries that use their political power to prevent the laws from being enacted. They oppose them because the bills make beverage producers financially responsible for the recycling of beverage container waste and make retailers responsible for their collection. They argue that such bills duplicate curbside recycling, are a public health threat, are inefficient, are outdated, are a regressive tax, and will damage local businesses and lead to closures or layoffs. However, their effectiveness has been demonstrated in the states that have them. A policy priority should be the expansion of container deposit laws to other states that include aluminum cans and plastic bottles. This will require overcoming strong corporate opposition through public support, consumer and voter education, and working with legislators and public interest and environmental groups.

The trend in local and organic foods (see Chapter 5) is another phenomenon that is beginning to reinvigorate the returnable glass model. People are choosing sustainable milk producers that provide milk in returnable/reusable bottles similar to those used in the days of the once common milkman. They do this for health and environmental reasons as well as the increased convenience of delivery services. Similarly, a number of purveyors of locally produced natural juices offer their products in a returnable glass jar for which the customer leaves a deposit. These examples illustrate ways that sustainable food choices can also promote efforts to reduce waste.

COMPOSTING

The largest proportion of the solid waste that we generate is organic garbage, including food and yard waste. All of this could be diverted from landfills through composting programs. In addition to diversion from the primary waste stream and associated cost savings, composting produces a highly useable end product that can be used to amend and add organic material to soil. As a soil amendment, compost sequesters carbon, improves plant growth, conserves water, reduces reliance on chemical pesticides and fertilizers, and helps prevent nutrient run-off and erosion. Organic waste can also be used to generate biogas in anaerobic composters. The development of composting policy and programs is a key sustainability strategy that will protect the environment, create jobs, and build resilient local economies. For example, composting in the state of Maryland employs two times more workers than landfills and four times more than the state's trash incinerators on a per ton basis. Composting supports twice as many jobs on a per dollar capital investment basis compared to landfills.[8]

Despite the clear benefits of composting, municipal policies and programs that promote or require it have lagged behind other types of recycling. Yard waste collection and composting have become relatively common, unlike opportunities for composting food waste (which makes up the bulk of the organic fraction of garbage). Thus, many have dubbed the composting of food waste as "recycling's final frontier." Limitations include the need to sort kitchen wastes from recyclables and non-recyclable trash. There are concerns that food wastes attract pests or create odor, and there could be health risks associated with post-consumer food wastes. However, these problems can be dealt with through proper handling. Because of the clear benefits of composting food waste, many US states and European nations have enacted or are in the process of enacting policies that will promote or require composting of all organic wastes. Such composting policies are also consistent with climate action plans and soil preservation efforts.

Another type of policy that promotes both composting and recycling is based on diversion goals. In Europe, the Landfill Directive required European Union (EU) nations to divert biodegradable waste away from direct landfilling into alternative forms of treatment with targets of 50% diversion (relative to 1995 levels) by 2013, rising to 65% diversion by 2020. In California, existing law requires local municipalities to divert through recycling and composting 50% of solid waste disposed by their jurisdictions, and it establishes a statewide diversion goal of 75% by 2020. Other policy creates mandatory recycling and composting ordinances requiring all households and businesses to participate. In 2009, the City of San Francisco passed the first local municipal ordinance in the United States to universally require separation of all organic material, including food residuals. Other policy efforts to promote composting include implementation of per ton surcharges on disposal facilities, moratoriums on building new trash incinerators, state composting infrastructure development policies, state compost usage encouragement policies, and statewide economic incentive policies.

The growth of urban food composting may necessitate changes in zoning policy at the local level. At the state level, composting is regulated as a solid waste disposal activity. The current standard for most US municipalities is that households, community gardens, or small urban farms that compost in an area less than 25 m^2 (approximately 300 ft^2) and that compost only yard wastes, animal wastes, food scraps, and other appropriate additives are not subject to zoning requirements or regulation. This assumes that composting at this scale is conducted in a manner that does not constitute a health or environmental hazard via odors or noise. Larger facilities for composting collected food scraps or larger urban farms and community gardens will need to comply with more stringent requirements that include regular testing. Zoning policy must be modified to allow for such activity. The state of Ohio has been a model for developing zoning codes that promote composting and organics waste diversion.[9]

Policies that promote food scrap composting are gradually spreading through Europe and the United States, but there are a number of barriers. One of these is the power of the US waste management lobby and industry that wants to prevent revenue loss from the diversion of food waste. From 2009 to 2016, waste management companies spent between $4.5 and $7.1 million per year on federal lobbying. These companies also support political action committees that spend hundreds of thousands of dollars on federal candidates to stop pro-composting legislation. Such lobbying efforts have been even greater at the state level. However, economics are now beginning to favor the creation of composting policy. As the cost of landfills rises, composting is viewed as a sustainable solution. New York City has been placing 1.2 million tons of food waste in landfills every year at a cost of nearly $80 per ton. Huge economic benefits come from the diversion from landfills. Other cities are catching on and are establishing curbside composting programs. These efforts are enhanced through private–public partnerships that take advantage of existing for-profit compost efforts and infrastructure. School composting has also been viewed as an important part of municipal composting programs. By starting in the school districts, municipalities see significant savings in disposal costs, which can then be put back into the school system. The schools can serve as pilot programs that will promote broader acceptance by the residents, enhance sustainability curricula, and allow children to become the strongest advocates for expanding local programs and adopting composting policy.

OCEAN DUMPING

There are already both national and international policies in place to restrict ocean dumping, which is a serious environmental issue. Despite this, millions of tons of garbage are still entering our oceans each year. Much of this is because of ongoing illegal dumping, sewage and storm run-off carrying the waste materials to the ocean, and coastal dumps and garbage heaps in urban slums mostly in developing countries. The main reason for this is that the existing policies that restrict ocean dumping and sewage run-off lack compliance and enforcement mechanisms. Policy that develops, implements, and funds these mechanisms must be a priority. It is essential that sewage and storm water infrastructure be improved to prevent ocean contamination. Alternatives to coastal slums and dumps must also be found. Finally, because plastics are such a major and persistent component of ocean garbage, policies related to plastic use and disposal should be developed.

PLASTICS

Each year, nearly 300 million tons of plastics are produced globally, and the number has been steadily rising at approximately 5% per year. Approximately 4% of the petroleum consumed worldwide each year is used to make plastic, and another 4% is used to power plastic manufacturing processes. However, recovery and recycling of plastics remain insufficient, and millions of tons of plastics are sent to landfills, incinerated, or end up in the oceans. In the United States, the rate of plastic recycling is less than 10%. The rates are higher in Europe, but they are much lower in developing nations, where plastics are increasingly being marketed but the infrastructure for recycling them is lacking. Recycling rates fluctuate with the price of oil. As the price declines, it becomes more cost-effective to make plastic from raw materials than from recycled plastics.

The volume of plastics that enter the waste stream, their direct linkage to fossil fuels, and their negative environmental and health impacts are extremely problematic, yet there is essentially no policy regulating the specific production, use, and disposal of plastic. One recent policy success has been the passing of the Microbead-Free Waters Act of 2015 by the US Congress and signed into law by the president. This law prohibits the manufacture and sale of rinse-off cosmetics containing intentionally added plastic microbeads. Other policy regulating production and disposal of plastic will be vital for a sustainable future.

Recycling policies, including container deposit laws described previously, must include plastics. It should be illegal to market plastic bottles in countries that lack infrastructure for recycling them, and corporations that produce and sell beverages in plastic bottles should be required to contribute to the development of such infrastructure. Subsidies described in Chapter 6 that artificially keep oil prices low and thus economically dis-incentivize plastic recycling should be eliminated. There is growing legislation that ranges from restricting the provision of plastic shopping bags to requiring them to be collected and recycled by retailers that provide them and even outright bans that should be expanded further (Box 8.1).

BOX 8.1

Policy Solutions: Plastic Shopping Bag Legislation

Each year, an estimated 1 trillion plastic shopping bags are used worldwide, which amounts to more than 2 million plastic bags used per minute. Less than 1% of them are recycled. These plastic bags have an extremely low decomposition rate, persisting for up to 1,000 years. Plastic bags litter our cities, fill our dumps and landfills, and often end up in our oceans, where they impact marine organisms (Figure 8.1.1). Littered bags are unsightly, can hurt aquatic and terrestrial animals by suffocation or entanglement, and often block drains and cause flooding. Their clean-up has cost local communities more

continued

continued

Figure 8.1.1 Plastic shopping bags are a major source of litter. *AP Photo/Rich Pedroncelli.*

than $400 million per year. According to data from the Ocean Conservancy's annual International Coastal Cleanups, plastic shopping bags are consistently among the top 10 pieces of trash collected on beaches throughout the world. Our addiction to plastic bags also fuels our dependency on oil. In the United States, where more than 100 billion plastic bags are used each year, approximately 12 million barrels are consumed in their production.

The environmental concerns over plastic shopping bags have led to the development of much legislation regarding their use. The first plastic bag laws were passed in 2002 in Ireland and Bangladesh. Currently, there are plastic bag ordinances in 20 countries and in many municipalities in 22 US states. These policies have ranged from a required per bag charge or offering a discount for using reusable bags to outright bans on manufacturing and prohibiting stores from providing single-use plastic bags. Others allow for bags to be provided by retailers but require those that do provide bags to offer opportunities for recycling. Many view fees as more effective than outright bans. Because customers will continue to require a means to carry their purchases, a ban that does not address other types of carryout bags and does not successfully encourage reusable bags will typically result in customers switching from one bag type to another, such as from thin plastic to paper or thicker plastic bags. Fee or reward policies that allow customers to make a conscious choice about whether they require a bag in the first place or to bring their own reusable bag have been shown to be effective. Charges of 5–10¢ cause a significant reduction in bag consumption. For example, Los Angeles County's 10¢ per bag fee has reduced single-use plastic bag use by 95% and paper bag use by 30%. Another solution is to outlaw non-degradable plastic bags and replace them with those that are compostable or biodegradable, but opportunities and infrastructure for composting them must be a part of such a policy.

Opposition to plastic bag legislation comes mostly from the plastic bag manufacturers and conservative groups that oppose any restrictions on retailers or customers. They argue that environmental concerns are exaggerated and that any bag legislation will cause the elimination of manufacturing jobs, negatively affect retail businesses, and take personal choice away from the consumer. Others complain that per bag fees go to the grocery stores, which profit from the fees, and some have suggested the state government should receive the fees and use them for environmental clean-up programs. Others claim that bag legislation is not effective because it is difficult to enforce, so there must be consequences for failure to comply. For example, the state of Hawaii's bag legislation includes $100 to $1,000 fines for those retailers that do not comply.

Shoppers in most locations are still confronted with the question of paper or plastic as they check out. Environmentally, paper bags are not much better than plastic, and their production causes pollution and consumes energy and water. Paper bags are inefficiently recycled and are not biodegradable in landfills. So when asked to choose between paper or plastic, the answer should be neither, and consumers should choose to instead bring their own reusable bags. Bag legislation has been shown to make shoppers much more aware of the problems that surround disposable bags and encourage them to use their own bags. The makers of the award-winning film *Bag It: Is Your Life Too Plastic?* offer the Bag It Town Tool Kit with step-by-step instructions and resources to initiate a campaign for plastic bag legislation (http://www.bagitmovie.com/downloads/bagittown_toolkit.pdf).

Companies and organizations could be required to disclose their production, use, and handling of plastic and plastic waste. The Plastic Disclosure Project sponsored by the United Nations Environment Programme and other foundations promotes this type of transparency.[10] By measuring the quantity of plastic that flows through an organization, efficiencies can be gained in cost reduction, waste reduction, new design, new materials, and better recycling. In essence, it calculates a "plastic footprint" for corporations and institutions such as hospitals, universities, government offices, and other large organizations. This type of program has the potential to reduce plastic use because better plastic management will reduce costs and spur innovation in material use and design. Ultimately, as we move away from a fossil fuel economy, there should be policy that promotes the development of alternatives to plastic that are more sustainable.

HAZARDOUS WASTE

Currently, the management and disposal of hazardous waste are well regulated within the United States by the Resource Conservation and Recovery Act and the Comprehensive Environmental Response, Compensation, and Liability Act (CERCLA); in Europe by the Hazardous Waste Directive; and in Canada by the Canadian Environmental Protection Act. These policies established permitting standards for disposal, developed tracking systems for toxic wastes, and require clean-up and redevelopment of contaminated sites. In the United States, CERCLA, commonly known as Superfund, created a tax on the chemical and petroleum industries and provided broad federal authority to respond directly to hazardous waste sites that could endanger public health or the environment. Through this program, billions of US tax revenue went into a trust fund to pay for cleaning up abandoned or uncontrolled hazardous waste sites when no responsible party could be identified.

Less regulated household hazardous waste often ends up in landfills, incinerators, or enters the environment through waste water disposal. These wastes represent a significant contribution of toxic chemicals in the environment. Because of this, stricter regulation, better education, and improved collection programs will be a vital part of managing these wastes. Reducing the broad availability of products that contain hazardous ingredients and educating the public about non-toxic alternatives could go a long way toward limiting human and environmental exposure to these toxic chemicals. In the United States, California is an example of a state that has taken the lead in providing the public with convenient collection locations for household hazardous waste, encouraging the use of alternative products, and requiring producers to assume responsibility for stewardship of their products and materials from production to disposal.

Developing countries have lacked policy for the management of hazardous waste. The initial expansion of the hazardous waste trade from developed countries to developing countries, as previously described, led to the adoption of the Basel Convention on the Control of Transboundary Movements of Hazardous Wastes and Their Disposal in 1992. This global treaty prohibits exporting hazardous waste to countries that lack the technical, administrative, and legal capability to manage the waste in an environmentally safe manner. It also establishes procedures for notifying the importing countries about the elements of the hazardous waste and the risk involved, and it allows signatory nations to ban imports or require exporters to gain consent before sending toxic materials to them.

After the initial adoption of the Convention, some countries and environmental organizations claimed that it did not go far enough and argued for a total ban on shipment of all hazardous waste to less developed countries. This was formally entered into the Convention in 1995. This international ban would put an end to the export of toxic wastes to poorer countries from the 24 wealthiest nations that generate 98% of the 400 million tons of toxic waste produced each year, most of which is then exported to Africa, Asia, Latin America, and the Caribbean.

By 2015, the Convention with the ban was ratified by 182 nations and the EU. The United States, which is the world's top exporter of toxic and electronic waste, is the only developed country that has not yet ratified the Convention due to the refusal of Congress. United States industries that lobby and financially support members of Congress protested the ban. Lobbyists advanced the argument that poorer countries should be given an opportunity to import, process, and repackage hazardous waste produced by "First World" corporations. The US Chamber of Commerce has urged the US government to meet with developing countries to convince them that it would be in their best economic interest to support free trade of hazardous wastes including toxic e-waste. Of course, this argument neglects the true environmental and health costs of importing those wastes.

The Basel Convention has been considered a failed policy by some who believe that its initiatives are not sufficiently funded and that widespread corruption has allowed the continued export of wastes to poorer nations. Critics also argue that the absence of the United States as a participant has undermined the Convention, resulting in the exposure of millions of poor people to very toxic chemicals. The remedy to this continued injustice is to better fund the initiatives of the Convention, which would include enforcement of the ban, reduction in corruption, and the transfer of toxic waste management and remediation technologies to poorer nations. There should be investment in new technologies that would result in a reduction in the production of hazardous wastes. Finally, the United States and its corporations should recognize the true external costs of exporting toxic wastes to poorer countries by ratifying and complying with the Convention.

Poverty, Environmental Justice, and International Waste Management

Landfills, dumps, and incinerators are most often located in the poorest areas. Free trade and tighter restrictions in developed countries have led to the export of waste from the largest producers to poorer nations that lack policy and technology to deal with those wastes. Finally, garbage production is rising most rapidly in the poorest nations of the world in direct connection to urbanization, and that garbage is primarily disposed of in unsanitary open dumps or is openly burned without any emissions control in those urban communities. The global garbage problem is one of our most serious environmental justice issues.

We need policy that will protect the poor from the placement of landfills, incinerators, and dumps in their communities. Funding is needed to clean up polluted waste management sites already located in those communities. Tougher policy that protects poorer nations by banning transboundary export of garbage must be enforced. Developed countries must be required to manage their own waste within their own boundaries, and this must include waste reduction.

Poorer nations typically lack infrastructure to properly manage their garbage. For example, in Peru, where there is a serious shortage of official waste disposal facilities, illicit dumping of trash has become routine, and it is difficult to track and control through enforcement. Of the 6,000 metric tons of garbage Lima produces every day, 14% or 840 metric tons is disposed of improperly. Innovative approaches, funding, and technology transfer to help poorer nations manage their wastes in ways that protect the environment and human health are needed. In the case of Peru, one very innovative approach to deal with the problem of illegal dumping is the Gallinazo Avisa or "Vulture Warns" Initiative. Through this creative initiative, black vultures have been fitted with GoPro cameras and global positioning satellite units to track and map illegal dump sites (Figure 8.5). This initiative has had much success in controlling and cleaning up illegal dumping in Lima, and it has also raised public

Figure 8.5 Vultures which seek out garbage can be fitted with GoPro cameras and GPS to track illegal dumping. This is being implemented in Peru (equipped vultures can be tracked at http://www.gallinazoavisa.pe/). *istockphoto/Igor Alecsander.*

awareness about the amount and the negative effects of illegal dumping.

One organization that is contributing to improving waste management globally is WasteAid International (https://wasteaid.org), an initiative of the solid waste management industry whose mission is to connect poor and vulnerable communities to waste management resources and experts. Improving waste management in poorer nations will require a stakeholder approach such as that taken by the Global Partnership on Waste Management (Box 8.2).

At the same time as open dumps are eliminated in poorer nations, it should be recognized that millions of people worldwide make a living picking, sorting, and selling materials from these dumps. Waste is often a primary and critical source of income for households and waste pickers, and it contributes to local economies. Pickers also provide a critical service by supplying useable materials to others who are better able to profit from them, and they perform an environmental service by diverting material from the waste stream. One amazing example of creative reuse through trash picking is in Cateura, Paraguay, a slum built on a landfill where the 2,500 families who live there survive by separating garbage for recycling. They have created a children's orchestra that plays instruments constructed out of picked trash. Referred to as the *Orquesta de Reciclados de Catuera*, they have been the subject of a documentary called the *Landfillharmonic* and now tour worldwide (Figure 8.6). The protection of this economic and now cultural segment and its workers must be an integral part of the improvement of global waste management.[11]

As we consider sustainable development, it is clear that rates of consumption and production are at the core of waste production. Until recently, waste production has not been considered in development plans. The concept of sustainable consumption and production (SCP) is taking hold and is now a priority for the UN. SCP is consumption that responds to basic needs and improved quality of life in developing countries. It aims to do this by minimizing the use of

BOX 8.2

Stakeholders and Collaborators: Global Partnership on Waste Management

We have a global garbage problem. In developing countries, population growth, economic development, and urbanization have changed patterns of resource consumption and have caused a rapid increase in both the volume and types of waste that is being produced. As urbanization, industrialization, and consumption-based economic development in developing countries increase, both municipal and industrial solid waste generation rates will continue to rise and approach those of developed countries. However, most developing countries lack both solid waste management plans and infrastructure and technologies for safe disposal and recycling. In newly urbanized areas of developing countries, 30–60% of all the urban solid waste remains uncollected despite the fact that these municipalities spend 20–50% of their available budget on solid waste management.

To address this issue, the United Nations Environment Programme established the Global Partnership on Waste Management (GPWM) in 2010. The GPWM is a voluntary and collaborative relationship among a diversity of international stakeholders that include international organizations, governments, non-governmental organizations, municipalities, private companies, and academic institutions. The GPWM facilitates the development of synergistic relationships and international cooperation among stakeholders in order to identify and fill information gaps, share that information, and avoid duplication so as to complement existing work. Stakeholders engage in dialogues in an effort to develop coherent international policy required to address the challenging issue of waste management. They also provide technical support and work to enhance funding to facilitate the development and implementation of integrated waste management, including infrastructure at regional, national, and local levels. The goal of the GPWM is to use this holistic, stakeholder approach to overcome environmental, public health, social, and economic issues that result from insufficient waste management in poorer areas of the world and to develop sustainable solutions to help solve the global garbage crisis.

To learn more about the GPWM, visit its website (http://web.unep.org/ietc/what-we-do/global-partnership-waste-management-gpwm).

natural resources and toxic materials as well as the emissions of waste and pollutants over the life cycle of the service or product to protect the needs of future generations. However, the problem with SCP is that it is aimed at developing countries and ignores the tremendously disproportionately higher rates of consumption and waste production in wealthier nations. The objective of reducing resource use, consumption, and waste production must be applied not just in those developing countries but also in all countries in order to achieve global sustainability.

Applied Solutions for the Global Garbage Problem

The global garbage crisis cannot be solved by policy alone. Legislation must include funding for enforcement and infrastructure, and public compliance—whether it be economically or ethically incentivized—will be necessary to confront corporate opposition to such policies. Increased education about the waste problem and its solutions, especially through supporting curricula, and in-school composting and recycling programs will be vital to this effort. In this section, we examine applied solutions to dealing with waste that include new technologies and innovative business concepts. Later, we consider consumption and waste production more broadly within the context of current and potentially new economic and business models.

Waste Treatment Processes

There have been a number of advances in waste treatment processes, and research continues in this area. For landfills, new alternative cover designs in landfill caps are being developed to replace clay cap designs currently employed. These new cover designs are constructed from various materials, including specialized plastic layers, clay geotextiles, and stone. The purpose of these covers is to better isolate the waste from the environment and prevent water from infiltrating the landfill, which results in the production of toxic leachate. Also under development are aerobic bioreactor landfills that will be designed to have improved leachate management and greater rates of methane capture for energy production. In these new landfills, oxygen and moisture levels can be manipulated to increase the rate of microbial decomposition from decades to years. This will result in a decline in the mass of landfills, increasing capacity by at least 30%.

Single-stream waste technologies will take the burden of sorting materials from households and institutions and integrate it as part of the waste management system. In what is referred to as mechanical biological treatment, recyclables and biodegradable materials are separated from the waste stream using a variety of mechanical processes, including size-sorting vibrating screens, drum separators, magnets, and near-infrared optical sorters (Figure 8.7). Eddy current technology is used to sort aluminum by moving it through a field of electrons, which attach themselves to the cans and give them a negative electrical charge. Developing

Figure 8.6 Children from the Recycled Orchestra of Cateura, Paraguay, a village essentially built on top of a landfill where garbage collectors browse the trash for sellable goods. Now referred to as the Landfill Harmonic, this orchestra, which uses instruments built from materials acquired from the landfill, has offered these youth an improved quality of life. *The Landfillharmonic, http://www.landfillharmonicmovie.com/*

Figure 8.7 Single-stream recycling sorting equipment employing magnetic, eddy current, air, and optical sorters integrates sorting as part of the waste management system. *Machinex Industries, Inc.*

technology to clean lower valued contaminated recyclables in energy- and water-efficient ways should be a priority for future research and development.

New approaches to food composting are also becoming available. Aerobic in-vessel rotary drum technology can convert food scraps into high-quality, nutrient-dense compost in five days. Other technologies for large-scale mixing and the use of geotextiles for odor control are being developed. Large and scalable worm bins are now available for institutional, commercial, and municipality scaled vermicomposting. A number of entrepreneurial activities have grown from the increased interest in food scrap composting, including services that provide weekly food scrap pickup from homes and businesses for a very reasonable fee such as Compost Cab in Washington, DC, and Farmer Pirates in Buffalo, New York. They then compost those materials for use by local farmers. In addition to selling or providing compost for soil enrichment, new marketable uses are being developed, such as compost-filled geotextile mesh tubes or "compost socks" that are quickly becoming the standard for run-off and erosion control at construction and other disturbed sites as required by most municipalities.

Waste-to-Energy

Another way to approach the global garbage problem is to not think of it as waste but, rather, as a potential resource. As we are confronted with limits to landfills and recycling, climate change, and diminishing fossil fuels, using our garbage as a resource to produce energy is now being widely considered as a sustainable solution. However, this view is not without controversy. Some say garbage is a renewable source of energy. Others argue against this because it is a resource that is not regenerated through natural processes and 90% of garbage could potentially be reused, recycled, or simply not produced. There are serious environmental concerns regarding WTE, and there is fear that as we invest in the needed infrastructure, we will create a demand for more garbage that will distract from efforts to reduce it and that funds will be diverted from the development of more sustainable forms of energy. The main ways that waste can be converted to energy are through incineration, methane capture, and biofuel production. Here, we examine each of these approaches, discuss both the advantages and the disadvantages of each, and consider ways to make them more sustainable.

INCINERATION

The most common way to generate energy from garbage is by burning it to generate heat, which can then boil water to create steam that is then fed into a turbine to generate electricity or can be used for heating. Three tons of typical garbage contain as much energy as one ton of fuel oil. In addition to generating electricity, incineration can reduce the volume of waste by

90%. Worldwide, there are more than 600 WTE plants, most of them in the EU, Japan, the United States, and more recently, China. The most efficient WTE facilities recover as much as 1,000 kilowatt-hours (kWh) of electricity and heating per ton of solid waste. In Europe, WTE was supplying 38 billion kWh in 2006, and this is expected to increase to 98 billion kWh by 2020. This could supply 22.9 million households with electricity and 12.1 million with heat. In the United States, 86 WTE plants currently process 30 million tons of trash each year or approximately 12% of the solid waste that the country produces. This generates enough electricity to power approximately 2 million homes.[12]

WTE has the potential to meet our energy needs, making garbage a commodity rather than a problem. Sweden has taken the lead in doing this with its 32 WTE plants. Of the 4.4 million tons of household waste the nation produces each year, half is converted into energy through WTE, and less than 1% of the waste goes to landfills. Given its large WTE capacity, Sweden now also generates income and energy by importing 800,000 tons of waste per year from other EU countries, which pay for this service. If the United States were to employ Sweden's approach, it could meet 12% of its total electricity demand while diverting 3.8 million tons of garbage from landfills.

There are a number of new advanced thermal treatment technologies under development that will make WTE more sustainable. One example, **pyrolysis**, involves the anaerobic decomposition of organic material at elevated temperatures. This process leads to the production of pyrolysis oil that can be used in much the same way as a crude oil but with much lower sulfur concentration. It also produces **syngas**, which is a mixture of carbon monoxide, carbon dioxide, and hydrogen. Syngas can be used like natural gas to generate electricity and also as a precursor in the manufacture of ammonia, methanol, synthetic hydrocarbon lubricants, and fuels.

Another technology, referred to as **gasification**, uses very high temperatures to convert organic wastes in the presence of oxygen to produce hydrogen and methane, which are then burned to generate electricity. Still relatively unproven but with much potential, **plasma gasification** does not burn the waste but, rather, transforms carbon-based materials in an oxygen-starved environment using ionized, superheated air or plasma up to 14,000°C. This results in the production of syngas, and it transforms the remaining materials into their constituent atomic elements, such as metals and silica, that can be recovered for reuse.

Is using garbage as a fuel to generate energy green? Most people associate incineration with pollution-belching smoke stacks—and that was certainly true when garbage incineration was in its heyday in the late 1800s and early 1900s. However, regulations from the Clean Air Act of 1970 that were further strengthened by the enactments of Maximum Achievable Control Technology regulations in the 1990s have forced WTE plants to meet stringent clean air standards. Many facilities have since installed new equipment or shut down in response to these stringent emission standards. Such equipment includes filters to capture particulate matter and carbon injections to absorb heavy metals such as mercury and also dioxins and furans. Computer systems closely monitor pollutant levels to ensure they remain as low as possible. Test data show that emissions of toxins such as dioxins and mercury are now found at levels barely detectable by the most sophisticated instrumentation.[13] However, some communities and activists express concern that even undetectable emissions may not be benign and that emission control and standards are insufficient.

Most WTE plants have approximately the same or less environmental impact as energy produced from natural gas, and they have much less impact than oil or coal plants. Concerns regarding GHG emission are to a certain extent misplaced. It is true that burning trash releases CO_2, but if it were landfilled, it would actually produce a more potent GHG, methane, and the energy that would have been generated through WTE currently would most likely otherwise be generated with fossil fuels. Hydrogen produced from gasification and syngas is a clean, carbon-free fuel that has much potential to power vehicles. There are also concerns about the ash produced by WTE plants, but in the United States, all ash is tested in accordance with EPA rules before leaving the WTE facility to ensure it is safe for disposal or beneficial reuse (including metals recovery and recycling).

Another criticism of WTE is that it has the potential to directly compete with recycling; however, there has actually been a positive correlation between WTE use and recycling. This is the case in Sweden, which despite its large WTE capacity has a 48% recycling rate—one of the highest in the world (Figure 8.8). One way this is achieved is through the removal of recyclables prior to combustion, which is often an integral part of WTE. Also, because sorting occurs at the plant, neither individual behavior nor trash collection

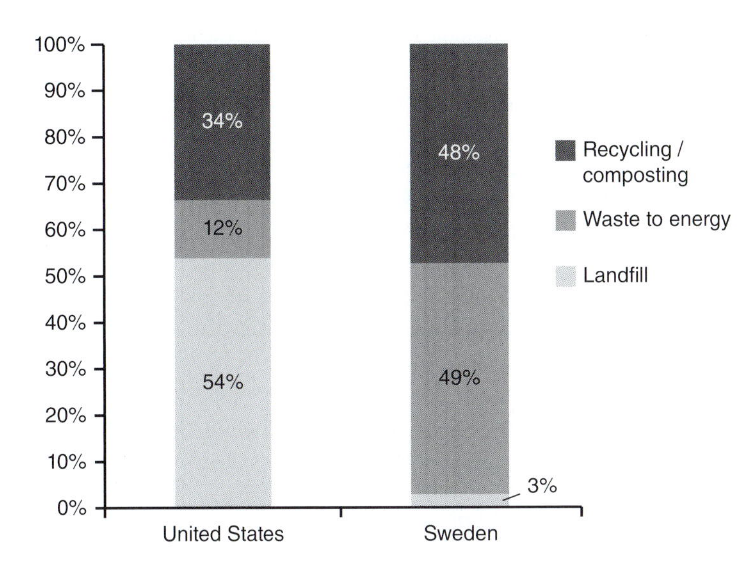

Figure 8.8 A comparison of waste management methods between the United States and Sweden shows that recycling and composting rates are actually greater with greater use of WTE. (Data from Matt Williams, ACORE, https://www.scribd.com/document/122133655/Waste-to-energy-success-factors-in-Sweden-and-the-United-States, page 13.)

procedures need to be modified for WTE to be effective at increasing rates of recycling.

WTE produces emissions that are typically 50% lower than emission standards, and it diverts trash from landfills that are space limited, have rising costs, and are a source of atmospheric methane. It reduces our dependency on fossil fuels, cuts GHG emissions, and actually increases rates of recycling. WTE seems like the ultimate sustainable solution. Unfortunately, there is one problem that is often neglected when considering the sustainability of WTE, and that is related to the cost of the infrastructure and operation. The upfront capital needed to build a WTE plant that meets environmental standards is between $100 and $300 million depending on size, and to operate at maximum efficiency, the incinerators must burn trash continuously. Not only does this cost make WTE essentially unavailable for poorer nations but also the economic benefits of WTE may take several years to be fully realized. A new WTE plant must operate for 20- to 30-years to be economically viable. Thus, a common pre-condition to the construction of a WTE plant is a long term contract that requires a consistent, specified volume of trash from the municipality, which must pay if it cannot provide it during the contract period. Thus, WTE creates a long-term need for garbage and may undermine efforts to reduce waste. Within the next 30 years, new technologies that improve recycling and decrease waste production will likely be developed, but we will have made long-term commitments to provide garbage. Investment in WTE will divert funding from more renewable, cleaner forms of energy and will perpetuate our linear, cradle-to-grave economy, thereby hampering our transition to a more sustainable circular economic model.

BIOFUELS FROM WASTE

In Chapter 5, it was shown that the anaerobic digestion of animal wastes to generate methane can be used to run electrical generators, gas-powered lights, stoves, and heaters. Also, as discussed in Chapter 6, ethanol-based biofuels are being produced from crops and crop residues. Similar processes are now being developed to convert garbage into biofuels. For example, a project in Oregon called Food for Watts collects and creates biogas from post-consumer food waste to make biomethane, which is then burned to generate electricity. A Scottish energy startup has developed a way to use the large amounts of residue left over from Scotch whisky production to produce millions of gallons of biobutanol that contains 25% more energy than ethanol and can be used to fuel automobiles. A London-based company, bio-bean, is turning waste coffee grounds into biofuels and biomass pellets that can create enough power to heat 15,000 homes. Because all of these fuels come from waste materials that would have released CO_2 or methane as they decompose, they are considered carbon neutral.

INCREASED METHANE CAPTURE

We know that garbage in sanitary landfills decomposes anaerobically, producing the potent GHG methane. New technologies to capture these gases and use them as fuel are under development and should be a priority as long as we manage our solid waste landfills. One example of this is an energy-generation facility being constructed at Nevada's Apex Landfill, the largest in the United States. The plant will extract the methane from the landfill and prevent it from entering the atmosphere. It will then be used to generate electricity and, in so doing, will convert the methane into the less potent GHG CO_2.

Plastics

Plastics come from fossil fuels, increase our carbon footprint, do not biodegrade within reasonable time frames, and are a major source of environmental pollution that has consequences for wildlife and human health. Thus, there is a real need for alternatives. One obvious alternative is to return to glass bottles. Unlike plastic, which is made from fossil fuels, glass is made from sand. Glass does not contain chemicals that can leach into one's food or body, and it is easily reused or recycled. One limitation of returning to glass is that it is heavier than plastic, so there will be fuel consumption and emissions consequences during transportation.

There are a number of non-fossil fuel–derived alternatives that are also biodegradable. Referred to as bioplastics, their production lowers the carbon footprint of fossil fuel–based plastics by 42%. Bioplastics have been made from casein, the principal protein in milk that can be converted into biodegradable material that matches the rigidity and compressibility of plastic. Its use in the production of clothing, furniture cushions, packaging, and other products is being explored. Other sources of biodegradable plastics include starches from corn and potatoes. Although these are advances over fossil fuel–derived plastics, the use of the already overburdened agricultural system to make plastic substitutes has many of the same disadvantages as crop-based biofuels. This is especially true if they are produced using conventional agriculture, which is both water and chemical intensive and soil depleting. It also diverts cropland away from food production, and most of these products are genetically modified organism dependent, which can deter from sustainability if not used in a fair and safe way.

As with biofuels, a better alternative to crops is to use agricultural residual material to make bioplastics. There have been a number of recent developments along these lines. Bagasse is the pulp left over when juice is extracted from sugar cane or beets. It can be pressed into a cardboard-like material used to make waterproof food containers, and it is easily composted. A new material being developed by Nobel Environmental Technologies is Ecor, which is made from waste cellulose. It can be used to make large fiberboard-like panels that can be manufactured into a variety of products from furniture to book bindings, and it is completely and readily recyclable. Other wastes or residues that can be used to make plastic-like materials include chicken feathers and lignin, a by-product of paper mills. Feathers are composed almost entirely of keratin, which along with lignin, is an extremely durable compound that can be used to make biopolymer materials that have the feel and appearance of plastic but are 100% biodegradable.

There are also potential bacterial sources of plastic. The use of bacterial-produced cellulose to make biocomposite materials and biodegradable plastic-like sheets for use in biomedical, packaging, and agricultural industries is being developed. In other research, *Escherichia coli* are being manipulated to directly produce butanediol, a chemical compound used to efficiently make products ranging from spandex to car bumpers without fossil fuels. Inserting these same plastic-producing genes into algae holds even greater promise because they have the capacity to produce larger amounts of the plastic product and are fueled by sunlight. This has the potential to produce the same plastics currently on the market with a significantly reduced carbon footprint; however, these products would not be biodegradable.

In addition to the development of new types of non-fossil fuel, biodegradable replacements for plastics, innovations and improvements in plastic recycling and clean-up are important. Plastics are at best downcycled and diminish in quality with each cycle more than any other material. There is an effort to increase the end product value of recycled plastics. For example, plastic waste has been successfully converted into a high-performance aggregate for use in making lightweight concrete, mortar, and block used in construction that reduces water consumption to 10% from the 15% required with clay-based materials. The use of polyethylene terephthalate (PET) to make plastic bottles allows for increased value in recycled end products that can be used to make clothing, bedding, and carpet. New technical innovations can enhance the use of recycled plastic bag end products such as in enhanced road construction and concrete brick production, composite decking, and nanotube membranes used in electronics, wind turbines, and sensing devices.

There is so much plastic in our environment, including the 460,000 pieces of plastic per square mile on the ocean surface. Because it is so persistent in the environment, even if we completely stopped using and discarding plastic, it would still remain a major pollutant requiring remediation or clean-up. One way is through citizen clean-up programs such as those organized by the Ocean Conservancy. One of the largest volunteer events to benefit the world's oceans involved 560,000 participants in 91 countries, including Canada, Chile, Ghana, Australia, China, and Greece. They removed more than 16 million pounds of trash strewn along 13,000 miles of shoreline. Mapping marine debris found on shorelines to identify where waste accumulates has been shown to improve community clean-up efforts such as those sponsored by the non-governmental organization Living Oceans around northern Vancouver Island shorelines in Canada.

A new invention developed by entrepreneur Boyan Slat when he was 19 years old will allow the removal of plastic from large areas of ocean surface. Slat, who is now CEO of the Ocean Cleanup Foundation, predicts that his system, which consists of a 2,000-m floating boom that passively collects plastics as wind and ocean currents push debris through it without trapping ocean life, will remove 44,000 tons of plastic, or half of the Pacific Garbage Patch, over a five-year period (Figure 8.9). These systems will be autonomous, powered by solar energy, scalable, and possibly even revenue neutral as collected plastic could be sold on the recycling market.[14] The massive plastic clean-up is scheduled to start in the Pacific in 2018 and globally in 2020.

Figure 8.9 A 100-km-long floating structure, reported to be the largest in history, will be dispatched to garbage patches around the world to begin removing the 88,000 tons of floating plastic. The structure consists of an anchored network of floating booms and processing platforms that can span the radius of a garbage patch acting as a giant funnel as the ocean moves through it. The angle of the booms forces plastic in the direction of the platforms, where it is collected and stored for recycling. The use of booms rather than nets eliminates danger to marine animals. *Ocean Cleanup Foundation, http://www.theoceancleanup.com*

Reducing Packaging and Disposable Products

In the United States, packaging makes up approximately 40% of all MSW. Annually, we throw away nearly 39 million tons of paper/paperboard, 13.7 million tons of plastics, and 10.9 million tons of glass that are only used in packaging. The total amount of packaging waste increases approximately 1.8% annually. An obvious waste solution is to reverse this trend. At least 28 countries now have laws that encourage packaging reduction. The United States is not one of them. Ways to reduce packaging include personal choice—for example, buying products that use less packaging such as bulk foods. Alternative materials can also be used to reduce the impact of packaging, including the use of biodegradable and even dissolvable starch-based products such as those frequently used to make packing peanuts. Another is the use of synthetic materials such as Durabook paper that when recycled can be infinitely broken down and remade into new products without any downcycling.

We have become too dependent on products that are disposable. One way to reverse the trend in garbage production is to focus our buying power on reusable items such as non-plastic water bottles and bags (see Boxes 8.1 and 8.3). We can choose not to use or frequent establishments that use disposable food service containers and utensils, especially those made of Styrofoam and plastic. We can stop using products that individually package each serving. One example is the Keurig coffee makers, which results in the contribution of 9 billion plastic K-cups to landfills each year. Even the inventor of the K-cup, John Sylvan, now expresses regret about the environmental impact of these plastic cups, which have widely expanded since he sold the company for $50,000 in 1997. Reusable cups that also provide savings for the consumer are now available and are slowing the deluge of landfills.

Last, moving away from or making choices not to purchase products with planned obsolescence will reduce our garbage production. One way to do this will require companies to inform consumers how long their products will last and require them to repair or replace the products if they do not last. A new French law does exactly this for appliances. Inventing longer lasting materials or ways to extend the lives of products will also reduce the flow of trash. One example is the Keepod, invented by social entrepreneurs Nissan Bahar and Francesco Imbesi. This device, which costs only $7, contains an entire computer operating system that can be plugged into any USB port. This allows for easy upgrades of outdated computers, and it enables groups of people to share old computers that have had their hard drives removed. Each person can plug his or her own Keepod into the shared computer, giving each one access with individual password-protected settings, programs, and files.

BOX 8.3

Individual Action: Stop Using Single-Use Disposable Products

Most of the products that we consume are used once and then thrown into the garbage and persist for hundreds of years. One individual action that can reduce both resource consumption and waste production is to eliminate the use of such products. This can be done by using reusable rather than disposable plastic shopping bags (see Box 8.1). But there are many other single-use items that you can easily eliminate from your daily life. For example, you can stop using disposable plastic bottles by carrying reusable water bottles such as those made by Sigg or ECO Canteen, both of which are easily cleaned and lack the toxic chemicals such as dioxin and BPA that are often in reusable plastic bottles.

Each year, Starbucks provides more than 52 billion disposable cups, most of which are never recycled. This can be eliminated by carrying your own reusable, insulated coffee mug. Many coffee shops now offer incentives for doing this, including discounted prices or express lines; choose them. Other ways to reduce your use of single-use items is to carry your own silverware such as the innovative spork, reusable plates or bowls such as the collapsible Orikaso bowl, and a cloth napkin. A champion for this approach has been professional cameraman Dave Chameides, who is also known as Sustainable Dave. If you want to make a difference, follow Dave's lead in the effort to reduce the use of single-use, disposable items, and follow his advice that no one person can do everything, but everyone can do something.

Another approach to extending the life of consumer electronics is to eliminate the need for upgrades through modular electronics such as the Phoneblok, the brainchild of Dave Hakkens (Figure 8.10). Such modular electronics will allow the consumer to update specific components, such as the processor, battery, or screen, as needed rather than disposing of and replacing the entire item. It also offers a greater degree of product flexibility for consumer preferences and as those preferences change with time.

The problem of e-waste is also being addressed in the laboratory in the new field of "transient electronics." One advance in this area is the development of a new kind of electronic circuit board that completely disintegrates within a month when exposed to a mild acid. These new electronics are made of organic polymers rather than silicon, and the circuits are made of iron, which fully disintegrates and is non-toxic. The circuit board is made of a thin sheet of cellulose, a biodegradable plant-derived molecule. These electronics degrade into non-toxic components, disappearing entirely within 30 days when exposed to an acid with a pH of 4.6. They are also less than 1 μm thick, very lightweight, flexible, able to conform to irregular surfaces, and require less power to operate. The design of these transient electronic devices is based on the function of human skin, an excellent example of biomimicry.[15]

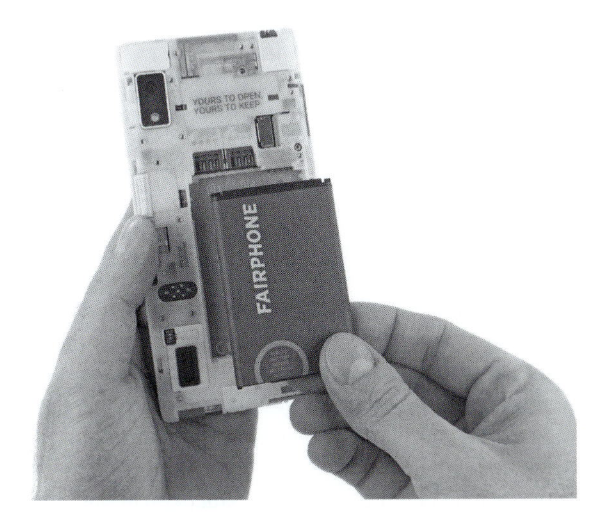

Figure 8.10 FairPhone has adopted the Phoneblok concept to produce a modular, mobile phone that one can easily upgrade, repair, and customize in order to create less e-waste. *Courtesy of Fairphone, https://www.fairphone.com/en/*

Garbage and the Linear Economy

There are a number of interesting and sustainable solutions to the garbage problem, including policies and innovations that allow us to better manage our waste through improved recycling and composting, reduced GHG emissions, and the use of waste as a supply of reusable resources and energy. They protect and clean the oceans, reduce the use of disposable materials, and ensure greater justice by protecting the poor and providing them with the technology and ability to manage their own waste. These solutions have the potential to meet many of our criteria for sustainability. They will protect our environment, maximize renewal, and reduce waste. Many of the improvements we are seeing and proposing will increase equitable economic opportunities and protect poorer communities from being the dumping grounds of the wealthy.

As hopeful as these solutions seem, most of them are just Band-Aids aimed at healing the inefficient pathway from resource extraction to production, consumption, and disposal. Even if these improvements are effectively implemented, our global garage problem will continue to plague the environment and people. Solutions that create a demand for more waste over the long term, such as WTE, are not sustainable solutions. Recycling that is essentially downcycling only slows the ultimate depletion of resources and flow of materials into the waste stream. A global economy and models for development that depend on increased consumption, single-use disposable products, and planned obsolescence cannot sustain itself. In the long term, we will need more than Band-Aids to heal this systemic resource use, consumption, and waste problem. What we need is radical, invasive surgery that will fundamentally alter the ways we do business and manage our economies.

Our linear economic model of extraction, manufacturing, distribution, consumption, and disposal is broken and unsustainable. We need to redesign our resources in a way that is focused on reduced consumption, increased reuse, infinite recyclability, more efficient design, and economic models of growth and development that do not depend on consumption and disposal. Ways to approach these are described here. Some are long-term and perhaps more radical recommendations. However, they are being contemplated by many. They are supported by new technology, and they have the potential to meet all our criteria for sustainability, including improved and just environmental and human conditions, while favoring business and equitable economic opportunity.

Reduce

Most, if not all, of our approaches to dealing with the garbage issue have focused on disposal, energy recovery, and recycling. Few have focused on reuse, and even fewer have focused on reducing or preventing waste. To be sustainable, we must flip this waste management paradigm to favor the elimination and reduction of waste over disposal, recovery, and recycling following what is referred to as the waste hierarchy (Figure 8.11). The goal of implementing the waste hierarchy is **dematerialization**, or the reduction of raw material inputs in conjunction with the reduction of waste outputs. The waste management options at the top of the hierarchy offer the greatest opportunity to reduce the need for virgin materials and environmental impacts associated with material processing and disposal.

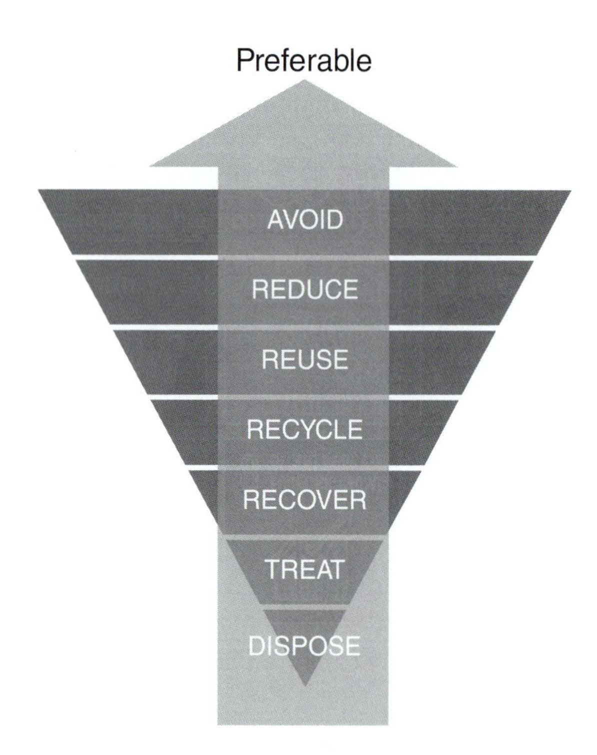

Figure 8.11 The waste hierarchy flips our current waste management paradigm to favor the elimination and reduction of waste over disposal, recovery and recycling. (Based on https://www.missouristate.edu/Sustainability/reduce-and-reuse.htm)

This paradigm shift has been central to the development of waste management policy, especially in the EU, and should guide the future development of community waste management plans. Focusing management options on the top of the hierarchy has been shown to reduce emissions of GHGs and other pollutants, save energy, conserve resources, create jobs, and stimulate the development of green technologies. The realized economic gains are even greater when the external, unpriced environmental and health costs of waste production and management are considered. However, implementing the hierarchy is often a challenge, and more complete life cycle analysis could be more revealing (e.g., when transportation costs and associated environmental impact for a specific management option outweigh other benefits).

Decrease Food Waste and Increase Composting

Globally, approximately 1.3 billion tons of food for human consumption is lost as waste each year. This is nearly one-third of the food that is produced. In the United States, this represents 14% of the solid waste produced, and this proportion is much higher in developing countries. In the United States, we throw out more food than plastic, paper, metal, and glass. Food waste contributes to food scarcity and hunger and also contributes to the environmental impacts of waste management. When food waste is sent to landfills or otherwise anaerobically decomposed, the potent GHG methane is produced. The global contribution and impact of GHGs from decomposing food is more than twice the emissions of US road transportation.

This problem must be addressed both through decreases in food waste and through increased composting of that waste. The decreases in food waste must occur at the level of production, distribution, and preparation and also at the points of sale and consumption. Solutions that focus on preserving food through non-biodegradable packaging such as air-tight plastics may solve one problem by creating another. One new innovative product can actually increase the life of fruits and vegetables by two to four times even in areas that lack refrigeration (Box 8.4). The design of new products and food systems such as this and those presented in Chapter 5 must be a priority in order to reduce or eliminate food waste.

BOX 8.4

Innovators and Entrepreneurs: Kavita Shukla and FreshPaper

Approximately 30% of the food produced globally is lost, wasted, or discarded food waste. Most of this waste is due to spoilage. An interesting discovery by Kavita Shukla has led to the development of a new product that can extend the life of produce two to four times longer. That invention is a simple piece of paper with a distinctive maple-like scent. It can be placed in a box, bag, or refrigerator drawer along with fruits and vegetables, keeping them fresher longer and thereby helping reduce food waste through spoilage. This low-tech, inexpensive paper can be used by large-scale growers and producers and also in the developing world, where 1.6 billion people lack refrigeration.

There is an interesting story behind this invention. When Shukla was visiting her grandmother in India, she accidentally drank untreated water. Her grandmother created a solution of spices and herbs and told her to drink it and that she would be fine. She was skeptical, but after drinking it, she did not get sick. This sparked her curiosity and prompted her to take the mixture home do some experiments, from which she found that the spices in the mixture stopped the growth of some of the common bacteria and fungi that she found in her kitchen. She then tried dipping fruits and vegetables in the mixture and found that they lasted longer. This eventually led to the development of the innovative product that is called FreshPaper.

FreshPaper is infused with edible, organic extracts and is entirely biodegradable and compostable. It is these extracts that give it its maple scent, which indicates that it is active. It can be used to help keep produce fresh from harvest to consumption. The company's name, Fenugreen, derives from one of the active ingredients, the spice fenugreek. The product has received rave reviews for its effectiveness and design. Shukla captures the true spirit of social entrepreneurship. Her simple product addresses the enormous global problem of food waste. By selling FreshPaper to consumers in the developed world, her company is able to make it available to the billions of people living without refrigeration in the developing world, as well as food banks and food pantries here at home, whose mission is to provide fresh, healthy food to the hungry.

In the United States, most of the food waste occurs at the points of sale and consumption. Supermarkets are a major source of food waste, with food products comprising 63% of their waste stream. Individual chains report disposing up to 60,000 tons of food waste per year, or the equivalent of 3,000 pounds of food per employee. In so doing, they are throwing away $165 billion each year.[16] There are a number of reasons for this. It is assumed that customers are more likely to buy produce if it is from a fully stocked display, so stores overstock them strictly for appearance. Customers expect cosmetically perfect food, so retailers dispose of items that do not meet that demand. This includes product packaging that gets damaged during shipping even when the food inside is fine. Expiration or "sell by" dates indicate peak freshness but do not mean the food is spoiled. They are not required by law, and it is in the producers' best interest to have their product discarded prematurely. Yet, most consumers will not purchase edible products that are near their "sell by" date, causing supermarkets to throw away perfectly edible food. Foods prepared by supermarkets, the food service industry, and restaurants are typically disposed of if they are not consumed the day they are prepared.

Policy and infrastructure that promote food waste collection, donation, and composting are of vital importance. Examples of policies for food waste include a French law that bans groceries from throwing away unsold food and requires them to donate it to charities or for use in animal feed or compost. Others are eliminating expiration labels for certain foods. In the United States, federal law is designed to encourage the donation of perishable prepared foods to charities that serve people in need by providing the donor protection from litigation and offering tax deductions. However, most often, there is no financial incentive to do so, and delivering those donations has costs associated with it. Finally, educating businesses and consumers in order to change marketing and display practices and about what "sell buy" dates really mean and what constitutes healthy food will both save money and decrease food waste.

Stop Downcycling and Start Recycling

When diversion of waste and reuse are not options, most of us turn to recycling. However, many products that we think are being recycled are at best downcycled, which results in the production of materials of lesser quality, and with each iteration of downcycling, the quality declines until it is no longer of use. Downcycling merely delays the movement of the material into the waste stream, and it serves only to slow the linear process of resource use and uses much energy to do so. Even when we are committed to recycling paper and plastic, we are actually only downcycling.

The simple solution to this problem is to use only those materials that can be fully recycled, such as glass, steel, aluminum, synthetic papers such as Durabook, and PET. Better yet, we should develop ways to upcycle materials. Upcycling is the opposite of downcycling; when a product is upcycled, its materials and parts are recycled into higher quality products. Examples of products that can be upcycled are PET bottles, which can be "upcycled" into higher quality products such as textiles. Shaw Carpet and Nike are examples of companies that have worked to create products that can be infinitely recycled. Other examples include the repurposing of discarded items such as the use of discarded wood pallets to make furniture or making jewelry from used computer chips and electronics parts. The development of innovative and efficient materials that can be infinitely recycled or upcycled should be a priority.

Eliminate Single-Use Plastic

Given the serious global environmental and health effects of plastic, single-use plastics should be banned. During his 2015 visit to the United States, the Pope recommended replacing single-use plastics with more sustainable reusable items. Only those that can be recycled, upcycled, are biodegradable, not made from fossil fuels, and are free of chemicals that threaten human health should be permitted. This is a reasonable solution because there are many alternatives, such as reusable bags and bottles. There have been successful bans of single-use plastic bags (see Box 8.1). The Plastic Pollution Coalition is an excellent resource for educating the public and providing resources with the aim of moving toward a plastic-free world.[17]

A number of communities and institutions are undertaking initiatives to ban the use of plastic bottles, polystyrene, and other plastic containers. The region around the historic site of Machu Picchu in Peru was becoming so littered with plastic water bottles that they were banned at that site in 2017. Costa Rica, a global environmental leader that has committed to being carbon neutral by 2021, also recognized that it has a plastic problem. It will be the first country in the world to initiate a comprehensive national strategy to eliminate single-use plastics, also by 2021. It will do this by working with diverse stakeholders through municipal incentives, policies, and institutional guidelines for suppliers, in addition to research and investment, in order to develop practical replacements for single-use plastic products. In sharp contrast, a policy that was put in place in 2011 to either ban or discourage the sale of bottled water in US national parks was eliminated in 2017 by the Trump administration.

Zero Waste

The practice of changing our behavior and redesigning our systems and resource use so that all products or their components are reused, resulting in no waste being sent to landfills or incinerators, is referred to as **zero waste**. Zero waste in essence captures the cradle-to-cradle or circular economic models that emulate sustainable natural cycles. Zero waste is a whole systems approach of restructuring production, distribution, and consumption to completely eliminate waste. Zero waste is achievable, but it requires commitments from individuals, industry, and government. Zero waste has been shown to be economically beneficial to both consumer and producer.

There is a movement among individuals frustrated by government inaction and a culture of waste that has taken on the zero waste challenge and has done so successfully. For example, blogger and author Bea Johnson has adopted a zero waste lifestyle in which her family of four produces only a single mason jar of waste each year. Her blog, *Zero Waste at Home* (https://zerowastehome.com), and best-selling book[18] have launched a global movement, inspiring thousands of people throughout the world to live simply and join the zero waste movement.

Zero waste has moved beyond the action of individuals. The zero waste movement that is rooted in sound ethical and economic thinking is effecting changes in every sector of society. It is shifting our culture from one of waste production to one of resource recovery. This has been best achieved through economic incentives for all stakeholders from manufacturers to consumers and also through changes in design and holistic system analysis. The Zero Waste Alliance has been a leader in this movement by partnering with industry, institutions, and communities to address barriers to and actionable work that can achieve zero waste. They work to establish zero waste business networks that allow for the movement of waste from one industry to another, where that waste can serve as a resource. By linking these business networks with local governments and non-profit organizations, zero waste districts and communities can be developed.

Institutions such as universities are rising to the challenge with the support of the Zero Waste Toolkit published by the College and University Recycling Coalition and the University of Oregon's Campus Zero Waste Program.[19] Municipalities throughout the country motivated by both economics and ethics have passed zero waste resolutions as guiding principles for government operations and for outreach activities and waste reduction strategies within their communities. Boulder County, Colorado, and many of its communities have been leaders in this endeavor and have seen it result in job growth. The zero waste community movement is becoming global, and an international alliance of zero waste communities has been formed to further this objective.

Recognizing the economic benefits of zero waste by cutting out landfill costs and allowing for more efficient production, a number of manufacturers are moving in this direction. One example is General Motors, which plans to make approximately half of its 181 plants worldwide landfill-free by selling scrap materials and increasing reuse and recycling. Other companies like General Motors, such as Subaru, Toyota, and Xerox, are also designing landfill-free plants but have not completely eliminated WTE as part of their operations, which many argue should not be included in the zero waste paradigm. The EPA works with manufacturers to minimize and eventually eliminate waste through its WasteWise program.

A New Paradigm for Economic Growth and Development

In the eyes of many, the continued growth of the industrialized economy relies on market expansion and increased consumption at home and extending those markets globally. As a result, increased consumption has become intrinsic to international development. Given that consumption is inextricably linked to resource depletion and waste production, consumption-based development is not sustainable development. In response to this, the UN has begun to integrate ideas and guidelines for sustainable consumption and production in its development plans. It has developed a number of indicators that measure whether consumption and production patterns are becoming more sustainable through eco-efficiency and the development of green industries with the aim of both bringing about more equitable sustainable development and poverty reduction.

The UN's sustainable consumption program neglects an essential part of the problem. This problem lies in the consumption inequities that exist where the industrial world, particularly the United States, will need to rethink its own economic model if we are to have a sustainable future. We must adopt what has been coined "small is beautiful economics" or "economics as if people mattered,"[20] which redefines what is meant by well-being and will require us to completely rethink what we consider to be wealth and progress in a way that will ultimately reduce consumption rooted within the context of the linear cradle-to-grave economic model.

New models for economic growth must incorporate sustainable business concepts. Sustainability makes business sense, and increasingly companies are moving in this direction. Businesses are beginning to benefit by fully integrating sustainability into the products and services they provide through improved reputations and cost savings. This includes incorporating ethical, inclusive, and transparent treatment of labor and also sourcing of materials and waste flows such as the Plastic Disclosure Project. These have been shown to increase labor productivity, provide opportunities to meet evolving customer requirements, and minimize business disruptions.

Even ways to reduce consumption are generating new business models. One example is the Sharing Platform Model. In developed economies, up to 80% of the things stored in a typical home are used only once a month. This innovative model uses technology to create new relationships and business opportunities for existing companies, entrepreneurs, and consumers to rent, share, swap, or lend their idle goods. This will allow for decreased resource consumption and waste flow while creating opportunities for income generation and consumer savings. In the following chapters, we explore other examples, tools, and models that will help individuals, organizations and businesses, communities and regions, and national and international efforts to contribute to our sustainable future.

Chapter Summary

- Our ever-growing rates of consumption are not only depleting our resources but also increasing our rate of garbage production. Garbage production has economic, environmental, public health, and social implications and contributes to global climate change. Unless we change the way we use resources and generate waste, the production of garbage will undermine other sustainability efforts.
- Currently, waste is placed in dumps or landfills, incinerated, or not managed at all. Much of it ends up in the oceans. Recycling, composting, and other forms of diversion offer alternatives to landfills and incineration, but they are underutilized or ineffective. Hazardous waste requires special laws and handling to protect the environment and people.
- There are many justice issues related to waste including toxic colonialism, in which toxic waste is exported from developed to less developed countries. The poor are disproportionately impacted by deficiencies in waste management; placement of landfills, dumps, and incinerators; and the lack of environmental clean-up.
- A number of policy solutions can help improve waste management. These include laws that govern landfills, apply stringent emission standards for incineration, encourage recycling and composting, prevent ocean dumping, enhance recovery and recycling of plastics, deal with hazardous wastes in a way that protects the environment and human health, and ensure environmental justice.
- A number of new technologies and innovative business concepts can enhance waste management, including enhanced approaches to recycling and composting.
- There are also methods of capturing energy from garbage, such as waste-to-energy incineration, biofuel production, and methane capture. Although there are limits to some of these approaches, there is potential to meet some of our energy needs using wastes while reducing our impact on global climate change.
- Alternatives to plastics, reducing packaging and disposable products, and technologies that prevent food waste will also contribute to solving our global garbage problem.
- New economic and business models that move us from a linear or cradle-to-grave economy to one that is more circular will be vital to both reducing resource use and reducing garbage. Priorities should be placed on reducing consumption and waste production, decreasing food waste and increasing composting, improving the effectiveness of recycling, eliminating single-use plastic, and creating zero waste households, institutions, and communities.
- Future economic growth and development must incorporate concepts that reduce consumption and waste production in ways that will create opportunities for income generation and consumer savings and protection.

Digging Deeper: Questions and Issues for Further Thought

1. A waste-to-energy plant is being proposed to be built to serve your community. A number of environmentalists are protesting this proposal. What arguments would you use to make the case for this project, and what are some of the limitations or problems that might be associated with it?
2. What is the connection between a circular economy and life cycle analysis?
3. Rank the top three sustainable solutions in the area of waste in terms of the problems they solve and the practicality of implementation. Justify your prioritization.
4. How does food waste vary between developed and developing countries?

Reaping What You Sow: Opportunities for Action

1. Identify all of the stakeholders for the issue of food waste at your institution.
2. In conversation with those stakeholders, develop a method to estimate the amount of food waste your

institution produces. For example, many colleges and universities have employed "Weigh the Waste Campaigns" that quantify the amount of food waste produced per number of consumers over time. Others have developed surveys to ascertain perceptions of food waste.
3. In collaboration with stakeholders, implement your plan to estimate food waste on a regular basis.
4. Based on your findings and through stakeholder collaboration, implement a plan to reduce food waste at your institution. This could include an awareness campaign, trayless dining, portion control, changes in pre-consumer practice, or a composting program.
5. Conduct the previously described activities with respect to recycling or single-use plastic.

References and Further Reading

1. World Bank. 2013. *What a Waste: A Global Review of Solid Waste Management*. The World Bank Urban Development Series. Washington, DC: World Bank.
2. US Environmental Protection Agency. 2016. "Municipal Solid Waste Generation, Recycling and Disposal in the United States: Facts and Figures for 2014." Washington, DC: US EPA.
3. Tellus Institute. 2011. "More Jobs, Less Pollution: Growing the Recycling Economy in the US." Cambridge, MA: Tellus Institute.
4. US Environmental Protection Agency, "Municipal Solid Waste Generation."
5. Bullard, Robert D. 1990. *Dumping in Dixie: Race, Class, and Environmental Quality*. Boulder, CO: Westview.
6. World Bank, *What a Waste*.
7. Lord, K. R. 1994. "Motivating Recycling Behavior: A Quasiexperimental Investigation of Message and Source Strategies." *Psychology & Marketing* 11: 341–58. doi:10.1002/mar.4220110404
8. Platt, B., N. Goldstein, C. Coker, and S. Brown. 2014. *State of Composting in the US*. Washington, DC: Institute for Local Self-Reliance.
9. Rodríguez, A. A., and C. German. 2012. "Urban Agriculture, Composting and Zoning." GD No. 1011. Columbus: Ohio Environmental Protection Agency.
10. Plastic Disclosure Project. http://www.plasticdisclosure.org
11. Marello, M., and A Helwege. 2014. "Solid Waste Management and Social Inclusion of Waste Pickers: Opportunities and Challenges." Boston: Global Economic Governance Initiative, Boston University.
12. US Environmental Protection Agency, "Municipal Solid Waste Generation."
13. US Environmental Protection Agency. 2005. "Air Emission from Municipal Solid Waste Combustion Facilities." National Emissions Inventory Data. Washington, DC: US Environmental Protection Agency.
14. The Ocean Cleanup. http://www.theoceancleanup.com
15. Lei, T., M. Guan, J. Liu, et al. 2017. "Biocompatible and Totally Disintegrable Semiconducting Polymer for Ultrathin and Ultralightweight Transient Electronics." *Proceedings of the National Academy of Sciences of the USA* 14 (20): 5107–12.
16. Gunders, D. 2012. "Wasted: How America Is Losing Up to 40 Percent of Its Food from Farm to Fork to Landfill." Issue Paper 12-06-B. New York: Natural Resources Defense Council.
17. Plastic Pollution Coalition. http://www.plasticpollutioncoalition.org
18. Johnson, B. 2013. *Zero Waste Home: The Ultimate Guide to Simplifying Your Life by Reducing Your Waste*. New York: Scribner.
19. University of Oregon Campus Zero Waste Program. 2014. "University and College Zero Waste Campus Toolkit." Eugene, OR: University of Oregon and College and University Recycling Coalition.
20. Shumacher, F. 1973. *Small Is Beautiful: A Study of Economics as if People Mattered*. London: Blond & Briggs.

CHAPTER 9

SUSTAINABILITY AT THE MOST LOCAL LEVEL—THE INDIVIDUAL

In the previous chapters, we examined how the needs and wants of humans are met through the use of various natural resources. We did this within the context of our definition of sustainability. We found that these resources tend to be poorly managed, and their extraction, use, and disposal have serious environmental, health, and social consequences. Even resources that should theoretically be renewable are being depleted or degraded well beyond their comparatively short rates of renewal. These rates of depletion and degradation are accelerating with continued population growth and associated rise in resource use and consumption.

PLANTING A SEED: THINKING BEFORE YOU READ

1. What personal actions have you taken to contribute to a sustainable future? What else would you be willing to do?
2. Give examples of how the actions of one or a few individuals have resulted in significant change.
3. What are some barriers to personal behavioral change, and how can they be overcome?
4. What are the limitations of individual action in achieving significant change?
5. What are the benefits of individuals coming together in large-scale social movements and events such as the Climate March, March for Science, Women's March, and the Black Lives Matter movement, and what role does this play in effecting change?

I am only one, but I am one. I cannot do everything, but I can do something. And I will not let what I cannot do interfere with what I can do.

—Edward Everett Hale

Currently, the activities of industrialized nations and their people are responsible for the bulk of resource extraction, consumption, depletion, destruction, and waste production. The negative consequences of these actions disproportionately impact the poorest and most disenfranchised people on the planet. Although the current trajectory of the state of our planet, our use of it, and the resultant human and environmental condition currently seem unsustainable, there is much to be optimistic about. There are many opportunities for change and great possibility for creating a sustainable future for our planet and its people. The previous chapters offered specific examples of how policy, technology, and the actions of individuals, groups, institutions, and unique collaborations among stakeholders with very different interests can contribute to sustainability. Many of these solutions are already yielding measurable indicators of success and movement toward a more sustainable future. We are also starting to see opportunities for transformational, systemic change within the context of sustainability.

So far, our examination of sustainability has been through the lens of specific, resource-related issues. However, we have already seen that these issues are interconnected. For example, agriculture, energy, and water are all tightly linked with each other in a variety of ways. Another approach to getting at these interactions is to step back from the specific themes as they have been presented and to focus on sustainability at a variety of organizational levels.

In this chapter, we start by examining how the actions of individuals can make significant strides toward the goals of sustainability. Subsequently, we examine how sustainability is and can be approached by institutions; communities, cities, and regions; and on national and global scales. By working at different levels of organization, we will better be able to see the interconnectedness of multiple issues and how compromises among them are often made. Our goal is to use this organizational approach to broaden our quest for solutions to complex, interconnected problems that will move us toward sustainability in a significant way.

Individual Action

It is often argued that individual action is insufficient to move us toward sustainability objectives and that although the action of individuals can be meaningful, it is really only a Band-Aid on what amounts to a hemorrhaging wound. Many use the analogy of individual action as the mere rearrangement of the deck chairs on an already sinking ship, and they argue that what is required to achieve significant results are top-down policy approaches that make societal and global change happen. However, it will be shown that individual action can be an effective agent of change, and that is also a vital part of sustainable problem solving.

One of the sources for this disbelief in the power of the individual action is the perception that our personal contributions to environmental and social problems are small compared to those of industry and larger institutions. This makes our own initiatives seem inconsequential. However, this is a false perception primarily because the individual contribution to environmental degradation and associated negative social impacts are growing, while industrial contributions have declined with increased regulation and efficiency.

There is considerable evidence for the rise in individual contribution of pollutants.[1] We contribute to air pollution directly through our use of electricity, fuels, and transportation and indirectly through our purchase of goods and services that use energy in their production and delivery. Most of this air pollution we cause results from the burning of fossil fuels, such as coal, oil, natural gas, and gasoline to produce electricity and power our vehicles. For example in the United States, individuals directly emit one-third of all greenhouse gasses (GHGs) and smog-producing pollutants. We also release into the environment 50 times more benzene and 5 times more formaldehyde than all industry combined. Similar numbers exist for other pollutants as well. We also indirectly support worker abuse as we seek out cheap goods or crave items associated with social exploitation.

Given our significant and growing contribution to poor environmental and social conditions, there is actually great potential for each of us to make a real difference by changing our behaviors and patterns of consumption. Moreover, top-down structural changes imposed on people without the support of their ongoing actions and values are often met with resentment and opposition. Thus, the engagement and action of individuals play an important role in the success of top-down regulatory approaches.[2]

In this chapter, we explore how individual action and behavior can further sustainability objectives through seemingly small, incremental steps. We also discuss examples of how individuals can serve as catalysts for larger scale, transformational change. A variety of integrative and technical tools are presented that can support individuals as change makers. The role of individuals in influencing government is also considered, including laying out some priorities for policy changes that would promote sustainability objectives. In addition, we consider how to increase access to sustainable solutions to individuals from diverse populations and socioeconomic backgrounds, and changes in the way that we define success to be more in line with sustainability objectives are suggested. Finally, we explore how individuals can organize and work together, thereby amplifying their effect.

The Making of Effective Individuals

What makes an individual care about environmental and social causes? A number of factors have been attributed to this. Studies have revealed certain intrinsic qualities, such as openness to new experiences, an awareness of or empathy toward the suffering of other people, compassion, and knowledge and an understanding of the inevitable costs of destroying the environment.[3] People with these qualities tend to pursue their commitment to environment and justice even if they think that it will not make a difference; they simply follow a moral imperative. Those who do not have these qualities typically are against or lack such a commitment, and they will not support and often actively oppose any change. These intrinsic characteristics develop usually early in life, as individuals are influenced by people who are important to them, such as family, friends, teachers, and spiritual leaders who shape and model them. They also develop through transformative experiences such as travel or the death of someone close to them that allow them to directly witness environmental degradation, poverty, and environmental effects on health.

There are also a number of external forces that influence individuals to act in accordance with sustainability principles. One study revealed that the most important factor that influences sustainability-related behavior is monetary incentive. This includes financial incentives for such actions or financial penalties for non-compliance. Learning about the benefits of taking action—and the dangers of not taking action—also motivates people to try to make a difference. Another important external influence is the action of peers. For example, it was discussed in Chapter 8 that recycling compliance was less influenced by information about its importance and more so by the actions of others. Those who were given notice about not recycling and why they should recycle were less likely to change their behavior than those who were informed about the numbers of their neighbors who were participating.

Behavioral change or the decision to act or not act can be explained by the **theory of planned behavior**. This theory is based on the powerful way that an individual's beliefs (attitudes) interact with social influence (normative pressures) and perceptions of external controls (perceived behavioral control) to influence individual intent and behavior (Figure 9.1).[4] Individual action within the context of sustainability values could be influenced by each of these. For example, the decision to take public transportation instead of driving one's own car could be influenced by an individual's beliefs imparted through education, family tradition, or other sources of knowledge, such as messages

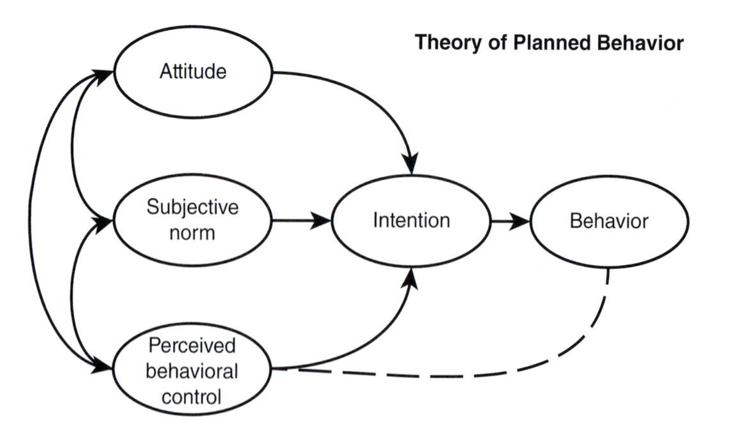

Figure 9.1 The theory of planned behavior predicts that that an individual's beliefs (attitudes) interact with social influence (normative pressures) and perceptions of external controls (perceived behavioral control) to determine intent and behavior. (Adapted from Aizen, 1991, https://www.researchgate.net/figure/5407720_fig1_Theory-of-Planned-Behaviour-Ajzen-1991).

from the media. It could also be influenced by how an individual perceives how others think about, value, or act on the outcome of this decision. Taking public transportation could be viewed negatively (low status) or positively (good citizen) by peers. Finally, this decision could also be influenced by outside control, such as economic status, the amount of traffic, rigid parking laws, or government subsidies for fuel or public transportation.

Messages from the media, government, and nonprofit organizations can also strongly influence attitudes and beliefs and thus behavior.[5] This can have both negative and positive influences within the context of sustainability. One positive influence on individual behavior is the role that celebrities who are genuinely committed to sustainability often play. There are many examples of this. One is Leonardo DiCaprio, who started a foundation that works on pressing environmental and humanitarian issues through grant-making, campaigning, and media projects. This included the production of the documentary film, *The 11th Hour*, which garnered attention about the environmental state of the Earth that included input from politicians, scientists, and activists. Actor Mark Ruffalo has been very outspoken against fracking and helped publicize the effort that led to its ban in New York state. He also joined protestors in North Dakota to support the Standing Rock Sioux Tribe's opposition to the Dakota Access Pipeline, bringing even more public attention to this issue.

Musicians and bands including Macklemore, Pearl Jam, Pete Seeger, Jack Johnson, Metallica, and Neil Young have been inspiring individuals to act on behalf of social and environmental causes for generations (Figure 9.2). Young, who has been singing about our negative impact on the Earth since the 1970s, was one of the very first to use and promote sustainability by directly incorporating it into his tours, fueling

Figure 9.2 Celebrity musician/activist Neil Young, whose song "Who's Gonna Stand Up?" is a call to action for environmentalists. The song appears on *Storytone,* his 35th studio recording, released on November 4, 2014, on Reprise Records. *Takahiro Kyono/flickr.*

all transportation solely with biofuels. To further promote sustainability, he has converted his 1959 Lincoln Continental convertible into a hybrid that runs on carbon-neutral cellulosic ethanol from crop residues and other wastes that he tours in to promote his ideas about sustainable fuels. This led to his creation of the Lincvolt project with the goal of inspiring new generations of automobile enthusiasts by integrating high-performance, zero-emissions propulsion technology with classic design elements (http://www.lincvolt.com).

Barriers to Change

Despite an increased understanding about what makes individuals act, strong barriers to changing behaviors and inspiring action remain. These barriers often stop people from caring about or even cause them to deny environmental or social problems. One such barrier is fear related to economic uncertainty. Increasingly, most people in the United States and elsewhere in the world are uncertain about their economic future. Those who argue against sustainability measures that might, for example, reduce climate change often try convincing people that the change will be bad for the economy and reduce wages and employment opportunities—although this is often not true. They create an atmosphere of fear. This, coupled with the continued loss of well-paying union jobs and the decline of the middle class, fuels these economic fears. This causes individuals to be less likely to support actions that could improve the environment and social condition. In addition, people from wealthy nations stand to benefit economically in the short term from inaction, so this is another cause of or reason for denial.

Another barrier that undermines individual contributions to sustainability efforts is the constant exposure to marketing and advertising that promote consumerism and link it to happiness. Most of this is aimed directly at children in their schools and through various forms of media to which they are regularly exposed. Corporations now sponsor public school curricula, and their logos appear in educational materials and as poster-sized displays on school walls and buses. The ongoing message of such media efforts falsely converts wants to needs. It promotes the opulent economy that causes hyperconsumption and favors single-use, disposable, upgradeable products. It also causes us to crave unhealthy foods and beverages that are responsible for a public health crisis for which we, not the companies that have created it, bear the costs.

The reality is that most people neither support nor oppose sustainability efforts but, rather, are apathetic or ambivalent toward them. What is the source of this apathy? It has been shown that apathy is not the result of being misinformed; actually, it seems to be the opposite. As individuals have increased knowledge about major sustainability issues such as climate change, ocean plastics, water depletion, and social injustice, they are less inclined to believe that they themselves can make a difference. This is in part because these problems are often presented in a negative manner instead of in a way that demonstrates that real solutions are possible and that individual action actually can make a difference.

It has been shown that individuals need more than just knowledge about a problem to motivate them to act. Rather, they need to be presented with practical ways to solve the problem before they reach the *tipping point* to change their behaviors or actions.[6] When individuals are made to feel fearful, guilty, and helpless without clear ways to make changes, they often become apathetic or even deny the need to respond to serious issues such as climate change. It has been the goal of this book, and a growing number of sustainability educators and curricula, to change this by increasing engagement in these issues through a more positive and empowering solutions-based approach.

People find it difficult to make day-to-day decisions when pressing matters seem remote from their everyday lives. For example, when people do not directly experience the effects of climate change, they are less likely to act on it than an issue that currently impacts their lives, such as a new factory or construction project in their neighborhood. Also, ambivalence or the tension between what people want to do and what they believe they actually can do is often due to perceived constraints of daily life. For example, one might believe in the use of public transportation as a way to live more sustainably but rarely use it because it is perceived as less convenient and having real or perceived added cost. At one point, the regional transportation system serving the Philadelphia area proposed shutting down one of its underutilized rail lines. There was a major objection to this by people who rarely used the line but wanted it to remain. There was an obvious disconnect between what they believed they wanted and their chosen behaviors. They liked the "idea" of having public transportation available more than actually using it.

Another cause of apathy is structural in nature. This is seen particularly in industrialized nations, where people are spatially separated from their sources of resources such as water, food, and energy and where their waste ultimately ends up. As a result, people are less connected or less protective of those sources or locations. For example, we are now very disconnected from where our food comes from. As such, we know less about how it was produced and are less likely to object to issues of environment, health, or social welfare that are consequences of that production. This was much less so the case when consumers had the opportunity to know and talk to those who produced their food. Structural changes that increase opportunity and convenience are often needed to cause otherwise unmotivated people to act. For example, an individual who is pro-environment may not participate in recycling until the municipality provides ways to make it convenient and cost-effective.

Overcoming Barriers to Change

So how do we overcome these barriers to individual change and increase engagement in the area of sustainability? Because the major motivator for individual action comes from intrinsic qualities that are typically imparted at a young age, we can create a new generation of problem solvers by modeling these qualities, teaching them about the world, and offering experiences that will shape their consciousness in ways that will cause them to effect the best change for the planet and its people.

Many of us who are engaged as adults were inspired by a teacher or other influential person or experience when we were children. It is difficult to understand how anyone could object to teaching compassion, empathy, and action to our young because these are traits a person from any background or belief system would value. Perhaps curricular and pedagogical changes in early education should emphasize these traits, and teachers should be rewarded for helping children become creative thinkers who want to apply their knowledge to make a difference. For many of us who are activists, our best teachers were those who "lit a fire" under us and then gave us the right tools to analyze situations and problem solve.

Because monetary incentives and disincentives are a likely way to influence one to act on behalf of sustainability, these should be considered as we establish new policies. Also, the role of personal and organizational influence and motivating individuals to act should not be underestimated, but these are more effective when done in a constructive rather than an antagonistic way. Most people do not respond well to being called out or criticized for not participating. Teaching environmental and social problems in ways that help people understand that short-term gains are just that—short term—and offering positive perspectives with pragmatic mechanisms to effect long-term change should also tip them toward action.

Reducing the constant exposure of our children to the marketing and advertising campaigns that promote conspicuous consumption, unhealthy foods and lifestyles, and extractive economies will also be effective in reducing participation in these activities in the future. One of the problems is that such marketing is tax deductible in the United States. One way to reduce the negative outcomes from this type of marketing would be to change this tax policy. Another way is to limit the times or venues for such marketing to limit exposure to children. Organizations such as Campaign for a Commercial-Free Childhood and Corporate Accountability International are playing a major role in advocating for these types of policy changes. There is evidence that limiting media exposure can reduce the negative consequences associated with it. One example was when tobacco advertising was limited both in television and printed media, directly resulting in reduced rates of smoking.

Another approach is to actually use media to promote sustainable behaviors. **Social marketing** uses commercial marketing approaches to influence behaviors that benefit individuals and communities for the greater good. Typically used by governmental or nonprofit organizations, social marketing differs from traditional marketing in that it is aimed at instigating positive changes in behavior or attitudes rather than selling a product or service. Social marketing is most successful when the marketer understands their audience and when a campaign offers a positive, optimistic message that inspires individuals to act. The online Community Tool Box of the University of Kansas Community Health and Development Group offers excellent resources for developing an effective social marketing campaign.[7]

One recent successful social marketing campaign was the "Choose How You Move" campaign piloted in Worcester, UK. It highlighted recent improvements in public transportation, bicycling routes, and pedestrian access, and it offered a variety of incentives and personalized travel advice. The campaign resulted in a 4% reduction in car trips, an 11% increase in walking trips, a 19% increase in bicycle trips, and a 20% increase in bus trips. Another success was the "Save The Crabs—Then Eat 'Em" campaign aimed at reducing individual lawn fertilizer use in the Chesapeake Bay area. The campaign significantly reduced lawn fertilizer use through stakeholder engagement, branding, and novel messaging that included TV spots, lawn signs, and restaurant coasters. The National Social Marketing Centre showcases these and a number of other successful campaigns.[8]

Policy changes that limit exposure to consumption-based marketing and promote socially beneficial marketing are beginning to achieve much success in the areas of health and sustainability by changing individual behavior. They are also beginning to alter the notion that equates consumption to happiness and success, which in turn may help place us on the path toward a more meaningful, sustainable form of joy and measure of success.

Individuals as Effective Agents of Change

There are two basic ways that individuals can be agents of change. First, individuals can change their daily behaviors in ways that will promote sustainability. These changes can be as simple as recycling or using reusable bags, or they can be more significant, such as becoming a vegan or purchasing an electric car. When an individual changes his or her behavior, this is viewed as a small step in what needs to be a transformative process. Many people are critical of the effectiveness of such small steps and suggest that they are actually distractions from more top-down approaches, but others have shown that coordinated action at the individual level can make a significant difference in, for example, reducing GHG emissions.[9]

The second type of change is when an individual's act is so significant that it leads to transformation. The act becomes a movement that can touch the lives of many and result in institutional and policy change. Next, examples of each are presented that demonstrate that both types of individual action can create change for a sustainable future.

The Importance of Small Steps

One can make personal choices like those already presented in this book, such as reducing one's carbon or water footprint, refusing to use single-use plastic or to purchase bottled water, eating less meat, or going "zero waste." These single choices alone will not likely make a significant difference in terms of the fate of the planet and its people. They do, however, have the potential to contribute to gradual societal change that can lead to rapid transformation. These small steps also provide opportunity for individual buy-in, which is essential for top-down or policy approaches to be successful.

Change by way of small steps can occur due to the power of virtue. Most people have a need to feel virtuous among their family and friends and in their community. This is supported by evidence that shows that when a particular behavior such as refusal to recycle is demonized, people often push back rather than comply. However, when such a behavior is presented as virtuous, people respond favorably. Children who learn that recycling and not polluting are acts of goodness are often able to convince their parents to accept that this is so because the parents want to be seen as virtuous in the eyes of their children.[10]

As we engage the people we know in this positive way, collective change can begin to occur as individuals become invested in causes at a personal level. This is a key step that allows people to care about and advocate for change in their communities and other levels of society.[11] These kinds of individual actions or personal changes that have the potential to contribute toward sustainability can be supported by a number of integrative tools that are publicly available on the internet (Box 9.1) or as smartphone applications (Box 9.2).

Although personal action is viewed as a vital part of launching a sustainability revolution, there is the potential for it to undermine such a movement. This may occur when individuals feel good about doing something minimal but then stop there. For example, consumers might feel good as they make "ethical purchases" to appear like an environmentalist or to relieve personal guilt but then do little else to further sustainability objectives in what is akin to individual "greenwashing." Also, a recent study showed that when individuals act on specific issues, they may be less likely to support larger scale solutions. For example, when individuals curbed their own energy use, they were less likely to support a tax on carbon emissions.[12]

BOX 9.1

Individual Action: Internet-Based Tools:

There are a number of internet tools that can help individuals reduce their environmental impact, lead a healthier lifestyle, and make behavioral and lifestyle changes that will allow them to contribute to sustainability objectives. Social media is an obvious tool for both receiving and sharing information about sustainability-related issues. Most of the sustainability-related organizations and leaders in the field regularly post to Twitter and Facebook. Examples include Waste & Recycling News, Circular Ecology, Energy Poverty, Food and Water Watch, and climate change activists such as Bill McKibben and Naomi Klein. In addition to the use of social media, there are a number of practical online resources that are more commonly recognized and used. Examples of these are listed here.

Footprint Calculators

The *ecological footprint* represents an essential accounting of our increasing demand on global ecosystems, and it has emerged as a leading sustainability indicator. Footprint calculators can be useful for individuals who wish to know the impact that their current lifestyle is having on the planet and concrete ways to reduce that impact. There are broad footprint calculators to estimate overall impact and others that are more specific.

The **Global Footprint Network** provides an ecological footprint calculator (http://www.footprintnetwork.org/en/index.php/GFN/page/calculators). This broad calculator allows the user to enter detailed information about their diet, trash production, purchases, housing, energy use, and transportation through an appealing interface. It then calculates the user's ecological footprint as the number of global acres of the Earth's productive area that are required to support that lifestyle. It also provides a carbon footprint and estimates the number of Earths it would take to sustain one's lifestyle if everyone lived that way. The calculations are location specific, and they allow individuals to account for their current demand on global ecosystems and make adjustments that will reduce that demand. Many other online ecological footprint calculators are available.

Carbon footprint calculators allow individuals to estimate their carbon footprint. The US Environmental Protection Agency (EPA) provides one (http://www3.epa.gov/carbon-footprint-calculator). Others are provided by the Nature Conservancy (http://www.nature.org/greenliving/carboncalculator), Terrapass (http://www.terrapass.com/carbon-footprint-calculator), the Center for Climate and Energy Solutions (http://carbonfootprint.c2es.org), and Carbon Footprint (http://www.carbonfootprint.com/calculator.aspx). Stanford University offers one that is geared toward students (http://web.stanford.edu/group/inquiry2insight/cgi-bin/i2sea-r1b/i2s.php?page=home#), and there is one specifically for youth (http://calc.zerofootprint.net). Another is focused on sustainable international travel (https://sustainabletravel.org/utilities/carbon-calculator), and one calculates a carbon footprint based on e-waste recycling (http://www.allgreenrecycling.com/ewaste-recycling-calculator).

Resource-specific footprint calculators include the water footprint calculator from the Water Footprint Network (http://waterfootprint.org/en/resources/interactive-tools/water-footprint-assessment-tool), which allows individuals throughout the world to calculate their personal water use and compare it to national averages. It also allows individuals to see where and how they are using their water and which foods require the most water in their production. The French organization Alimenterre offers a food footprint calculator (http://www.alimenterre.org/en/ressource/food-footprint-calculator).

Promoting Sustainability

A number of sites help promote sustainability beyond the calculation of footprints. These include tips for reducing power and water use and also food waste. Others connect people to ecosystems or offer general tips for living a more sustainable life.

The **Food Waste Alliance** provides a very useful toolkit for measuring and reducing food waste (http://www.foodwastealliance.org). This toolkit is extremely informative and provides an in-depth look at causes, sites, and solutions to the problem of food waste at both the individual level and the systematic level. The EPA also offers a guide for individuals and households to reduce food waste (http://www.epa.gov/recycle/reducing-wasted-food-basics).

continued

continued

There are useful tools for individuals to reduce **energy consumption**. Progress Energy, like many other energy distribution companies, offers an energy savings calculator (https://www.progress-energy.com/carolinas/home/save-energy-money/energy-saving-tips-calculators/100-tips.page). The Energy Use Calculator (http://energyusecalculator.com) allows individuals to determine the electricity consumption and power costs of most of their electrical devices and home appliances (a similar one can be found at http://energy.gov/energysaver/estimating-appliance-and-home-electronic-energy-use). Energy Star allows home owners to conduct do-it-yourself home energy audits and to compare their home's energy efficiency to that of similar homes throughout the country and get recommendations for energy-saving home improvements (https://www.energystar.gov).

A number of sites offer general tips for improving **household sustainability**, including easy ways to save water, avoid the use of toxic chemicals at home, improve energy efficiency, and reduce waste. These sites include http://eartheasy.com and the Context Institute's website (http://www.context.org/iclib/ic35/30ways), which offers 30 ways to become sustainable. MeterHero (https://www.meterhero.com) uses a dashboard approach to monitor one's actual water and energy use and track savings. The Dartmouth Sustainability Project (https://www.sustainability.dartmouth.edu) provides concrete ways for individuals who work and live on university and college campuses to live more sustainably.

There are sites that also support individuals who wish to be **vegetarians** or **vegans**. The Vegetarian Resource Group (http://www.vrg.org) provides a large variety of guides for vegetarians, including recipes, nutritional information, and guidance specific to teens and athletes. People for the Ethical Treatment of Animals (PETA) offers a vegan fact sheet (http://features.peta.org/how-to-go-vegan).

Climate Change

In addition to the resources described previously that can help individuals reduce their carbon emissions, there are some more general tools for reducing and documenting climate change. One, iSeeChange (https://www.iseechange.org), is a community climate and weather environmental reporting platform that combines citizen science, participatory public media, and cutting-edge satellite and sensor monitoring of environmental conditions. The organization works with media and scientific partners throughout the country to help audiences document environmental shifts in their own backyards and relate those changes to the bigger picture of how climate change is transforming all of our lives and livelihoods.

Changing industry and the use of fossil fuels are typically the focus of GHG reductions and climate change. But studies have also shown that individual and household action in the United States could provide a behavioral wedge to rapidly reduce US carbon emissions.[1] Through some simple behavioral changes, the amount of carbon that the United States releases could be reduced by an estimated 123 million metric tons per year. One resource to help in this effort is "The Lazy Person's Guide to Saving the World," available on the UN Sustainable Development homepage (http://www.un.org/sustainabledevelopment/takeaction). The guide offers actions that one can easily take and links to other resources that provide simple ways for all people to become part of the solution.

Activism

For those who are interested in moving beyond individual behavioral change to more activist approaches, there are a number of helpful online tools and guides for doing so. The "Info-Activism How-To Guide" (https://howto.informationactivism.org) provides a number of tactics and digital tools that can help activists be more effective. These include tools for producing effective audio, images, print, outreach, and video.

Increasingly, parents are becoming interested in raising their children with a strong environmental awareness, a deep ecocentric philosophy, and the skills to redirect society toward a sustainable path. Parents can find and share resources at a site called "Raising an Ecocwarrior: How Do You Raise a Child for Life on a Changing Planet?" (http://raisinganecowarrior.net). The site offers sustainability-related readings and activities for children of all ages and also opportunities to share ideas. Recognizing the importance of food choices in relation to sustainability, the site offers healthy recipes that will appeal to children and that will also teach and engrain in them sustainable eating habits.

---

1 Dietz, Thomas, Gerald T. Gardner, Jonathan Gilligan, Paul C. Stern, and Michael P. Vandenbergh. 2009. "Household Actions Can Provide a Behavioral Wedge to Rapidly Reduce US Carbon Emissions." *Proceedings of the National Academy of Sciences* 106 (44): 18452–56. doi:10.1073/pnas.0908738106

BOX 9.2

Individual Action: Smartphone Applications

In addition to online tools for individuals wishing to lead a more sustainable lifestyle, there are many readily available smartphone apps that can be practical tools to help in this effort. A sample of these are listed and described here:

- Seafood Watch (free): Provides the user with information on which seafood is preferable to eat based on location and species population dynamics, fish farming strategy, environmental impact, and health.
- Good Guide (free): Assesses home and living products in the areas of health, environment, and society. Products can easily be searched for by scanning the UPC code or using the application's search engine.
- Happy Cow ($3.99): Location-based guide to vegetarian and vegan restaurants, including reviews of food and pictures.
- Greening Your Family ($0.99 for premium): A practical sustainability shopping guide for everything from food to cosmetics and toiletries.
- iRecycle (free): The premier application for finding convenient recycling opportunities at home and on the go.
- Buycott Barcode Scanner (free): Allows users to scan barcodes for products and either see or enter reasons for boycotting the product.
- Carbon Foot Print Calc ($0.99): A simple app for calculating individual carbon output.
- EverEarth CO2 (free): A mobile carbon footprint calculator for automobiles.
- Home Sustainability Mobile Assessor (free): A mobile tool that can be used to complete home sustainability assessments.
- SocialEffort (free): A platform for individuals to find volunteering opportunities and log hours. Organizations can use it to manage and monitor volunteer activity.
- EcoChallenge (free): A fun method to reduce impact on the environment through challenges with friends and other users.
- Fragile Earth ($2.99): A photography app that gives bird's-eye views of different locations over time to give a cohesive view of climate change over time.
- Species on the Edge ($1.99): A guide to species on the endangered red list, and it provides information, pictures, and locations for them.
- iNaturalist (free): Allows users to record their personal experiences with nature and connect to others doing the same.
- Virtual Water ($1.99): Designed to raise consciousness of the amount of water polluted and used in daily life by calculating the amount of water used to produced our daily food and beverages.
- Farmstand (free): An area-specific guide to locally grown food and farmers markets.

In addition, individual action is frequently driven by ego, which can limit the achievement of sustainability objectives because it may prevent involvement or collaboration with others. This is particularly a problem at educational institutions, where individuals are present for just a few years. Although an individual's efforts may be extremely well intentioned, they may not be sustained because the individual is more focused on making the project their own. Instead, they could have collaborated with others in a way that might provide continuity even after the individual departs the institution. As we adopt and promote individual actions and personal change, it will be important to keep these points in mind.

Individual Action Can Radically Transform the World

There are many cases throughout history in which the actions of individuals led to radical change. Small, personal actions by just a few individuals had a major impact during the 1800s abolitionist movement in England, where women who did not even have the right to vote created momentum for the movement that ultimately led to the abolition of slavery in the British colonies. They achieved this through the simple action of refusing to serve sugar, which was largely slave-grown. This eventually led to a large-scale boycott. Their simple act also created a dialog about slavery and enticed and empowered the women to become activists, resulting in a major global transformation.

Individual action and sacrifice led to the women's right to vote, the desegregation of buses, and liberated the oppressed on multiple occasions. There are countless stories of people, often children, who initiated efforts that grew into movements that now make a significant difference in the world, such as supporting efforts to find cures for diseases or clean up the environment. As Gandhi stated, individuals can and need to be the change we wish to see in the world, and they have been successful at doing so throughout history.

Specifically with regard to sustainability, there are numerous cases in which individuals have made significant strides. One example is philosopher Andrew Briggle, who moved to Denton, Texas, not knowing much about fracking. But within five years of moving there, he led a citizens' initiative to ban it, making Denton the first Texas town to successfully challenge the oil and gas industry.[13] There is also small farmer David Oien, who, when told by corporate agribusiness to "get big or get out," decided to take a stand. By planting organic lentils that fix their own nitrogen, he was able to break the hold that the industrial agriculture machine and fertilizer companies had on him. He then created an underground network of organic farmers who grow food on biologically diverse farm systems without chemicals.[14] Then there is ecologist and cancer survivor Sandra Steingraber, who effectively linked science and activism to effect change. She provided data from the scientific and medical literature that demonstrated the relationship between environmental factors and cancer. Steingraber has been a vocal critic of the imbalance between funding devoted to studies of genetic causes of cancer and that devoted to studies of environmental contributions.[15] Adhering to the practice of civil disobedience, she led the fight against the industrialization of the pristine Finger Lakes region of New York state and the anti-fracking efforts that resulted in a statewide ban.

In many cases, these agents of change have been among the youngest in our communities. For example, at the age of 17 and 15 years, respectively, Simone and Jake Bernstein began to develop a national website, volunteennation.org, to help teens find volunteer opportunities, many of which are socially or environmentally oriented. Since its launch, they have empowered thousands of young individuals to make a difference in their local communities. When Jamie Romeo of Rochester, New York, was 8 years old, she started organizing volunteers to pick up and keep records about trash on a local beach on Lake Ontario as part of the first International Coastal Clean-up Day. The first year, her volunteer crew consisted of seven friends. Today, thousands of volunteers participate in this annual clean-up and collect more than 10,000 pounds of trash from 26 miles of local shoreline. There are countless other stories in which the actions of individuals of all ages have saved habitats, cleaned the environment, and have made other transformative changes in the area of sustainability.

Households

The actions of individuals often occur within the context of the household because this is typically the unit of resource use. Choices about food, the consumption of water, the use of energy, the purchase of other products, and how to manage waste are often made at the household level. Household decisions are often based on economics and return on investment (ROI), but they can be influenced by top-down approaches and incentives or disincentives. An example of the former is found in colder regions, where the ROI of increasing the insulation capacity of a home can reduce energy consumption by 20%. This in turn reduces the negative effects of fossil fuel consumption, as previously discussed. Because adding insulation with greater resistance to conductive heat flow—called R-value—is relatively cheap, the ROI can occur within a few years, so most homeowners are willing to make this investment; this is a common home improvement.

Other sustainability-related household improvements with relatively short ROI include installing reduced-flow plumbing, rain barrel capture and solar hot water systems, replacing outdated appliances with more efficient Energy Star appliances, and switching to more efficient lighting, such as light-emitting diode (LED) or compact fluorescent lamp (CFL). These types of changes are often referred to as the "low-hanging fruit" because the decision to make them is relatively easy. Other decisions, such as purchasing an electric or hybrid car, installing complete solar photovoltaic or geothermal systems, or installing a gray water irrigation system, have a much longer ROI. These actions are less common and often limited to those who have already made a commitment to sustainability objectives independent of cost or ROI.

The other type of influence on household sustainability behaviors and changes is governmental top-down incentives or disincentives. For example, federal, state, and local governments offer solar energy tax credits and rebates that can offset 30% of the cost of a solar system, thereby making it more affordable for homeowners. In states where the total available rebate is high, the rate of household conversion to solar energy is much higher. It has also been shown that incentives and disincentives both increase rates of recycling and decrease littering or dumping.

The Role of Entrepreneurs

In the previous chapters, we discussed cases in which individuals created sustainability-related solutions through entrepreneurship. An entrepreneur is an innovative individual who recognizes and seizes opportunities by first identifying a need and then converts those opportunities into workable, marketable ideas. They do this by obtaining funding to develop those ideas, eventually offering an innovative product, process, or service that generates profit, and while doing so assume the risks associated with the competitive marketplace. The qualities of successful entrepreneurs typically include a tremendous amount of personal initiative, creativity, the ability to communicate and pitch their ideas in order to attract resources, management skills, and a willingness to take risk in the face of uncertainty.

One way that individuals can significantly further sustainability objectives is through **social entrepreneurship**. Social entrepreneurs are individuals who create businesses that have social and/or environmental missions. Social entrepreneurship business models can be non-profit or for-profit. A leveraged non-profit is created when an entrepreneur sets up an organization that can deliver a product or service that improves environmental or social conditions. Rather than generating profit, the organization subsidizes funding from members of the organization or through donations and grants. Outside fundraising is often a vital part of such an organization's mission and economic strategy. An example of this is the Ocean Cleanup Foundation, the foundation led by entrepreneur Boyan Slat, who invented a passive system for cleaning plastic from the ocean surface and is using the foundation to support its implementation, as described in Chapter 8.

A social entrepreneur can also take a for-profit approach. Because the business is supported directly by income generation through the provision of the beneficial product or service, it is likely to be more sustainable than a business that relies on fundraising to deliver its solution. There are a variety of ways in which this can be achieved. One could be the development of an innovative product that is sold for profit that directly promotes the core values of sustainability. Entrepreneur Josh Goldman's business, Australis Aquaculture, is an example of this (see Box 5.2). The company makes healthy food accessible to a broad market in a way that improves the health of the consumer, protects the environment, and provides equitable and just income for workers. Profits generated from product sales sustain the company and its sustainability mission. Similarly, Troy Wiseman's for-profit company, EcoPlanet Bamboo, produces a sustainable product in a way that reduces poverty, improves social conditions, and protects and repairs the environment (see Box 7.3).

Another for-profit model of social entrepreneurship has two separate customer or beneficiary segments. One purchases the product at a reasonable price in developed countries, and the profits generated from those sales either partially or fully subsidize the provision of that product in poorer areas with even greater need. An example of such a product is LifeStraw (see Box 4.4). The Swiss company Vestergaard markets its extremely effective point-of-use water filtration products to backpackers and travelers who can afford to pay full price for them, and then it uses a portion of the proceeds from those sales to provide in-school water purifiers in developing countries. It also partners with aid organizations that raise money to provide individual- and community-sized filters for free as part of public health campaigns or in response to complex emergencies such as the 2010 earthquake in Haiti. The latter approach is often referred to as a hybrid non-profit enterprise because the profits sustain the operation of the company, which then collaborates with non-profit entities to deliver those products in cases of extreme need.

When individuals create value for customers or beneficiaries through social entrepreneurship, there is real potential to have lasting positive effects across economic spectra. However, to effectively do this, one has to first identify the problem to be solved and the needs of all customers or beneficiaries. This information can then be used to develop a marketable solution or value proposition within the context of costs and revenue streams to determine if the delivery of that solution will be economically sustainable. It is also important to consider any unintended social and environmental consequences. All of this is best achieved through what is referred to as the scientific method of entrepreneurship. This approach to business development maps out each of these elements in a structured business model canvas. This canvass then provides a framework to examine each element as a hypothesis to determine the feasibility of the business in a way that minimizes risk and maximizes social value (see Box 9.3).

Systems thinking is becoming a vital part of social entrepreneurship. This is because it provides tools for entrepreneurs who are trying to integrate innovations into larger, existing systems such as communities, school districts, government agencies, and corporate structures. One example of this is when Evan Marwell started the non-profit EducationSuperHighway to provide internet access to all US public school classrooms. To do this, he had to engage stakeholders at a number of large institutions, including the federal government, industry, communities, and 14,000 school districts. Marwell created an *influencer map* that revealed the complex interactions among these institutions. By expanding his thinking beyond a narrower entrepreneurial opportunity or technical challenge, Marwell and his team were able to visualize the complex system they were trying to change, and they have made much progress in doing so.[16]

Social entrepreneurship has come under some criticism. Just because they are successful by themselves, these enterprises are not necessarily going to solve complicated issues and fuel transformative changes in the redistribution of power, resources, and opportunities related to sustainability. However, we have seen, particularly in the case of EcoPlanet Bamboo, that this type of transformation is possible. Through its business, it has created real economic opportunity for the poor and has empowered women in the process (see Box 7.3). Social entrepreneurs may have a greater positive impact when they focus on a specific need, such as ocean clean-up, rather than attempting to tackle broader issues. Also, when these businesses partner with community stakeholders and local non-governmental organizations (NGOs), they have been better able to meet the needs of those communities.

Social entrepreneurs who create a product or service that empowers individuals to provide for themselves have been the most successful. This is best exemplified by the establishment of the Grameen Bank in Bangladesh by Muhamud Yunnus. Also known as the "Banker to the Poor," Yunnus's goal was to help individuals escape from poverty by providing loans with realistic terms and by teaching them a few sound financial principles so that they could help themselves. This approach has advanced to the forefront of poverty eradication that has now been replicated in more than 100 countries, earning Yunnus the Nobel Peace Prize in 2006, and it clearly reflects the power of the social entrepreneur.

In order to meet their social or environmental objectives, entrepreneurs must assess the need or problem that they are trying to solve, or what is referred to as the "value proposition" of their business model (Box 9.3). This must be done in a transparent, consultative manner. An often criticized example of a flawed value proposition is Tom's Shoes. For every pair of shoes Tom's sells in wealthier countries, another pair is provided to a child in need in a poorer country (see Box 9.3, Figure 9.3.2).

Some have questioned whether providing shoes is the most critical need in these communities because it does not effectively reduce poverty, which is the root cause of why children lack shoes. Increased access to vital resources such as clean water and hygiene in these communities may be a more vital need than shoes. Also, central to their model is to reduce "First World consumer guilt" among "socially conscious shoppers" through their shoe contribution program. This may contribute to "individual greenwashing." and resultant inaction, as previously described. In Tom's defense, the company is responding to some of these criticisms by contemplating the manufacture of the shoes in the communities where they are given away to generate economic opportunity and reduce poverty. Also, shoes are often viewed as a necessity in order to obtain educational or employment opportunities.

Working with Organizations

One effective way to increase individual power and effectiveness is through collaboration with others who share similar ideas or tendencies in social movements or organizations. People with similar viewpoints often come together in order to have their beliefs reaffirmed

BOX 9.3

Innovators and Entrepreneurs: A Scientific Method to Social Entrepreneurship

An entrepreneur identifies a problem or need and then develops a service or product that a customer will presumably want to purchase to solve that problem or meet that need. Once the problem and solution are identified, a business plan can be developed that is then pitched to potential funders and investors and that can serve as a guide for the execution of the business. However, this is not the wisest place to start. There are some crucial first steps in the entrepreneurial process that involve a scientific process of developing a set of hypotheses that can be tested through customer identification and validation. This early part of the process will provide knowledge about the potential for the proposed solution to have value, thus reducing risk for both the entrepreneur and investors.

The hypotheses about the business are organized in what is referred to as a business model canvas (Figure 9.3.1). The canvas is used to map out the key building blocks of a business model in a structured way. These building blocks include the customer segments or users of what you plan to offer. For each customer segment, there is a specific value proposition that includes the products or services that create value for the customer. The canvas also includes channels for how that value will be delivered and details about the kinds of relationships the business will have with customers.

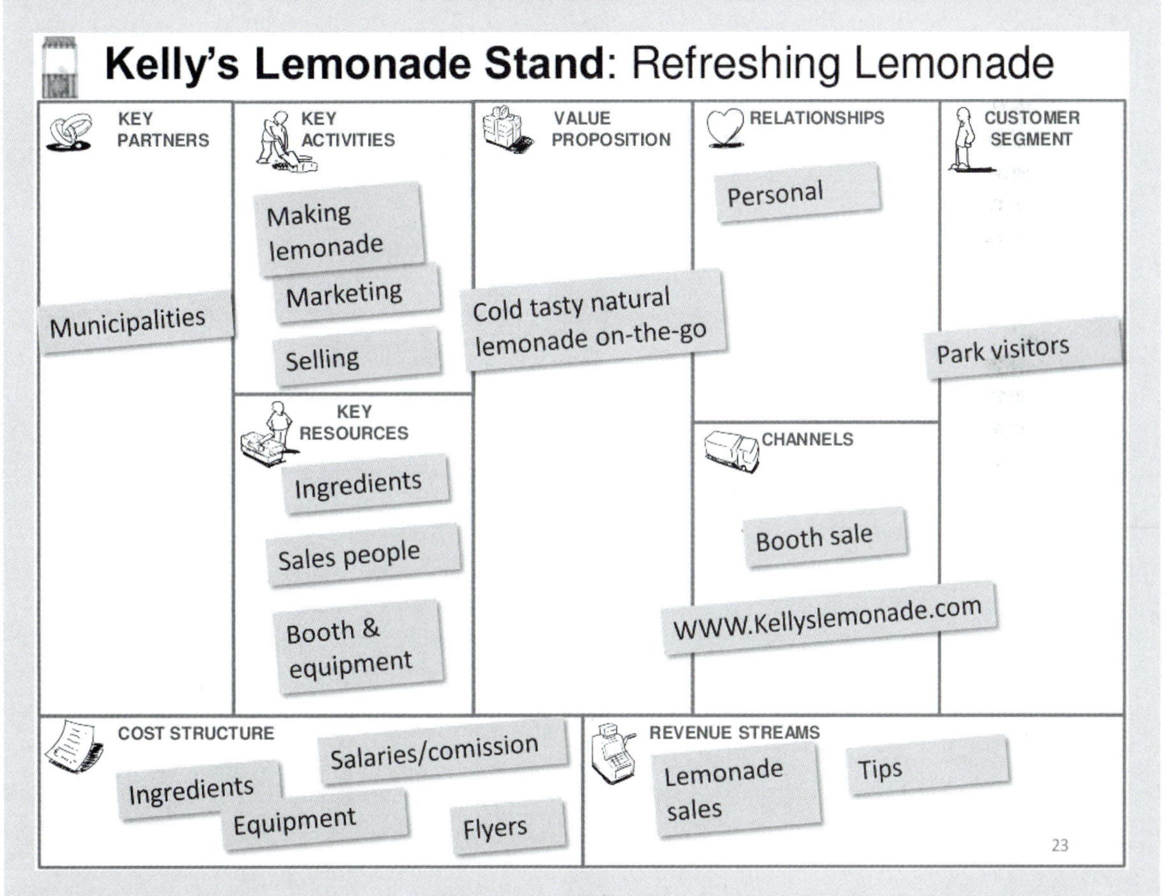

Figure 9.3.1 A business model canvas for a simple hypothetical business, Kelly's Lemonade Stand, that organizes the business model building blocks as hypotheses. *http://www.slideshare.net/featured/category/business, Creative Commons Attribution 4.0 International.*

continued

continued

The revenue streams make clear how the business model will capture value. The infrastructure and key resources required to create, deliver, and capture that value need to be described, and those resources that are indispensable to the business model are also added to the canvas. Key activities that need to be performed well and partnerships that can help leverage the business model are also part of the canvas. From all of this, a cost structure for delivering the solution can be established.

The canvas for social entrepreneurial endeavors adds additional categories such as stakeholders, social and environmental benefits, impact measures, unintended consequences, and financial sustainability. It can also be used to identify two types of customer or beneficiary segments—those that will pay full price and those that will pay a partially or fully subsidized price—showing the relationship between them. An example of this was presented previously with regard to LifeStraw, where customers in developed countries purchased water filters at full price, and revenues from those sales subsidized their provision in poorer countries. There are numerous variations on how the business model canvas can be adapted for social enterprises.[1] One example is shown in Figure 9.3.2, where a social lean canvas completed for Tom's Shoes.

Once the canvas is completed, these hypotheses are tested through the processes of customer discovery. The aim of this is to identify the customers or beneficiaries, whether the problem to be solved is important to them, and if the proposed solution is the right one to solve it. This is essentially a customer validation phase in which potential customers are interviewed to determine if there is a product–market fit. This should not be conducted as a sales or business pitch but, rather, as a way to gather information and impressions on the proposed solution. This is the extremely important

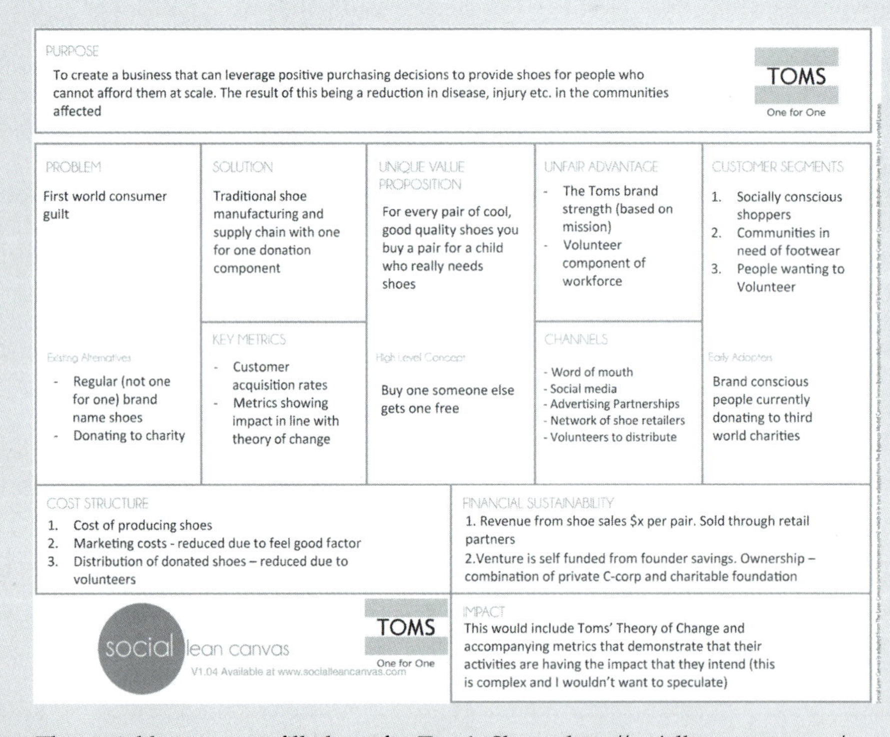

Figure 9.3.2 The social lean canvas filled out for Tom's Shoes. *http://socialleancanvas.com/, Creative Commons-Share Alike 3.0 Un-ported liscence.*

1 See http://www.socialbusinessmodelcanvas.com, http://socialleancanvas.com, http://www.businessmodelcompetition.com, https://www.blankcanvas.io/canvases/social-enterprise-canvas, and http://www.socialenterprisetrust.org

listening or transparent phase of social enterprise development. Once the customers are discovered and their interest in the value proposition is validated, then it is time to explore business execution, including the development of prototypes, customer creation, marketing research, and company building. This is when the business model can be completed and pitched to potential funders or investors.

Funding start-ups in general is challenging, but it is even more so for social enterprises because many investors are wary of values-based investing. Possible funding sources include venture capitalists who invest in exchange for equity in the company, angel networks or a collective group of investors, or crowdfunding, which is the process of raising money from many donors using an online platform such as Kickstarter, Indiegogo, and Crowdfunder. One can also fund the start-up using personal savings and those of friends and family, but then the entrepreneur is assuming most of the risk. All investors are looking for a clear strategy. When a social entrepreneur uses the the business model canvas to validate their idea, there is a greater chance that investors will be interested and that the enterprise will be financially sustainable and have a lasting, positive environmental and social impact.

Acknowledgment

I thank Rita Chesterton, Director of Innovation and Entrepreneurship at Muhlenberg College, for introducing me and my students to the scientific method, social entrepreneurship, and the business model canvass.

by being with others who share them and to amplify their individual effectiveness through power in numbers. Individuals can come together within the context of existing organizations that represent their views, or they can form new groups. Any such form of organized social or political act carried about by a group of people in order to address their needs is referred to as **collective action**.

These groups—whether they are informal social movements or established organizations such as NGOs or clubs—may take a variety of approaches. Some groups of individuals try to effect change through **direct action**. This is when a group of individuals commits an act that is intended to reveal an existing problem, highlight an alternative, or demonstrate possible ways to deal with a controversial issue. Direct action can include organized sit-ins, strikes, boycotts, workplace occupations, blockades, or refusal to obey certain laws, demands, and commands of a government in what is referred to as civil disobedience. Direct action has also been more aggressive or even violent and has included sabotage, property damage, computer hacking or hacktivism, political violence, and assaults and property destruction.

Recent examples of direct action include the Global Climate marches that occurred in 2014 in more than 175 countries and the Occupy movement that protests against global social and economic inequity. Another example of direct action occurred in 2014 when NGOs, farmers' groups, and indigenous organizations from throughout the world protested the World Bank's lending policies that promote the liberalization of developing country economies. Liberalization gives an advantage to large-scale land investment by Western corporations, and it disadvantages and actively dispossesses small farmers who currently feed 80% of the developing world.

ACTIVISM AND THE ARTS

Art is often seen as a form of activism, and artists are viewed as agents of change. An extremely creative form of direct action organized by the Brandalism Project challenged the corporate takeover of the UN COP 21 Paris climate talks in 2015. More than 80 artists from 19 countries created 600 fake ads that were posted in formal advertising space behind glass at bus stops throughout the city. The fake, satiric ads pointed out the irony that many of the sponsors have been significant contributors to climate change through their promotion of consumerism, fossil fuel consumption, and GHG production (Figure 9.3).

Jenny Kendler, the first artist in residence with the environmental law and policy non-profit Natural Resources Defense Council (NRDC), makes art that functions like activism. One example is her "Milkweed Dispersal Balloon" project, in which she handed out sustainable balloons filled with milkweed seeds for participants to pop to help spread the plant, which is essential for vulnerable monarch butterfly populations (Figure 9.4). The NRDC argues that her artwork allows viewers to initiate conversations about scientific-based issues that might otherwise seem intimidating.

Another excellent example of how art can reflect and fuel social movements is the creation of the image *Confronting Climate* by the artist by Rachel Schragis with the support of hundreds of other activists, artists, and organizers. This work serves as a flowchart of the ideas generated during and immediately following the first People's Climate March in New York City in 2014. It represents the complexity of climate change with the intention of igniting rich conversation about what it will take to confront this crisis. Because it is a tool to inspire action, effect change, and offer real solutions to the climate crisis, it has been adapted, with the permission of the artist, for the cover of this book.

Although direct action can be quite effective at mobilizing people and bringing issues to the forefront, it is often regarded as extreme and less effective in ultimately achieving the change the individuals seek. It may alienate stakeholders who could ultimately play a part in achieving the same objectives. Many individuals find other groups that better suit their sensibilities and may produce more demonstrable results. There are a great variety of groups, movements, and established organizations that bring individuals together to support and work on sustainability objectives. Among these are groups that take legal action, purchase land for conservation, participate in local environmental clean-ups, or run environmental education programs. Community gardening is a way for individuals to come together to create a more sustainable lifestyle. A way for individuals to assess the effectiveness of organizations and their ability to improve sustainability criteria is presented in Chapter 10.

CITIZEN SCIENCE

One interesting way that individuals can directly contribute to our knowledge about the environment is through citizen science. Also referred to as crowd or civic science, citizen science is a way for amateur scientists to contribute to large-scale research projects that often support sustainability objectives. Examples include the regular involvement of citizens in annual Christmas bird counts,[17] the tracking of cicadas that emerge once every 17 years by using homemade sensors to record and document emergence patterns,[18] and regular monitoring of the aquatic health of urban waterways (Figure 9.5).

A number of citizen science projects are playing an important role in documenting climate change. This includes the collection of large amounts of global data on phenology, which refers to the seasonal changes in plants and animals. The resultant data show that warming trends have significantly advanced leaf onset, serving both as another piece of strong evidence that climate change is well underway and as a tool for natural resource managers to better understand how plants and natural systems have and will respond to this change.[19] ISeeChange is a socially networked

Figure 9.3 Over 82 artists created 600 fake ads that were posted in formal advertising space behind glass at bus stops throughout the city to challenge the corporate takeover of the UN COP 21 Paris climate talks in 2015. *Artist: Jonathan Barnbrook Subvertising installed in Paris as part of Brandalism's COP 21 Paris Climate Talks intervention, 2015, Photo: www.brandalism.ch*

Figure 9.4 For her performance art piece "Milkweed Dispersal Balloon," Jenny Kendler hands out sustainable balloons filled with milkweed seeds for participants to pop to help spread the ecologically important plant. *Milkweed Dispersal Balloons by artist Jenny Kendler at the Pulitzer Arts Foundation, 2014 — Image courtesy of the artist.*

weather almanac for communities to collectively journal their climate experiences and observations and compare them to near-real-time climate information.[20] This groundbreaking environmental reporting project combines citizen science, participatory public media, and cutting-edge satellite monitoring of environmental conditions. The National Ecological Observatory Network (NEON) program, a continental-scale observation system for examining ecological change over time, not only engages citizens in their data collection but also offers a Citizen Science Academy that provides training to help increase the effectiveness of individuals and groups that contribute to its effort.

Working with Government

In addition to joining a group with common goals and interests, individuals have also been effective in helping make a difference within the realm of sustainability by working directly with local governments. This can be through membership on councils or commissions, by interacting with those groups, or by directly lobbying or supporting politicians. Many

Figure 9.5 Children from Milwaukee's 16th Street Community Health Center's summer science camp work with scientist Peter Levi as part of a citzen science program to monitor understudied urban waterways. *Adam Hinterthuer, Center for Limnology, University of Wisconsin-Madison.*

states have passed legislation that allows for the creation of sustainability and environmental advisory boards. Examples include environmental advisory councils (EACs) or conservation advisory commissions (CACs) that advise local municipal governments on sustainability-related matters. For example, both Pennsylvania and New York general assemblies have passed legislation that authorizes any municipality or group of municipalities to establish, by ordinance, an EAC and CAC, respectively.

EACs and CACs tend to focus on environmental conservation and improvement. Their activities include providing advice to local government officials and municipal agencies on environmental issues, performing natural resource inventories, and promoting community environmental programs. In essence, they provide governments with a pool of local talent to draw upon when making decisions that could affect environmental resources in their communities. Although decisions and recommendations are typically based on sound science, membership does not necessarily require professional training, and there are opportunities for individuals with a variety of backgrounds to influence environmental decision-making and programs where they live. Also, individuals not formally serving on advisory groups are able to attend and participate as per **sunshine laws**, which make their meetings and actions available for public observation and participation.

These types of advisory groups have been most successful when they develop sound proposals that will often provide cost savings to the municipality while meeting specific conditions of sustainability. Remaining positive and constructive, working and considering the needs of a broad array of stakeholders, and gaining the support of local officials early are also central to the effectiveness of such groups. When members of EACs or CACs situate themselves as antagonistic, environmental "watchdogs" of the municipality, they have alienated themselves from local officials and in many cases lost all opportunity to make significant contributions or have influence; in some cases, they have been rendered completely irrelevant. A variety of resources can be used to guide environmental advisory bodies, such as Pennsylvania's "EAC Handbook."[21]

Individuals can also become involved with government by getting appointed to local government authorities, boards, and commissions. These often move beyond advisory roles and make decisions that could have lasting impact. Such bodies related to sustainability include zoning and planning boards, development and redevelopment authorities, and commissions on shade tree planting, human relations, the homeless, and other social issues. Appointments are typically made by elected officials, but applications from interested residents are solicited and considered.

Recently, there have been numerous opportunities for individuals to comment on the Federal Clean Power Plan at the state level. A number of states are offering listening sessions and comment periods. On June 2, 2014, the US Environmental Protection Agency (EPA) announced the Clean Power Plan that included first-ever standards to reduce carbon pollution from existing power plants. The Clean Power Plan establishes pollution targets for each state, based on the state's particular fuel mix and emissions and reduction potential, with the goal to reduce national carbon emissions 30% by 2030. The EPA has empowered states to develop their own plans to comply with the federal standards, and many are soliciting input from residents and stakeholders to develop state-specific plans for compliance.

Individuals can get more directly involved with government by running for office or by supporting specific candidates or policies. The lobbying power of small groups of individuals at local and state levels should not be underestimated. However, as has been demonstrated with regard to energy and food policy, gun control, and many other issues related to sustainability, the power of individuals is often ignored, especially at the federal level, even when a clear public majority shares a particular stance. Because of the powerful influence of political action committees (PACs), politicians at the federal level often do not represent the long-term interests of their constituents but, rather, the interests of larger business or special interest entities. PACs are organizations that pool campaign contributions from members and donate those funds to campaigns for or against candidates, ballot initiatives, or legislation. PACs have strongly influenced farm and food-related legislation, and they have promoted legislation that either ignores climate change or actually expands the use of fossil fuels.

The climate change denier movement in the US Congress and an unwillingness to re-examine gun laws despite large public support are very much the result of the influence of certain PACs. Corporate money has resulted in policy that supports large agro-industry over small farmers or protects chemical and genetically modified organism companies from risk. It has resulted in the lack of a comprehensive sustainable energy plan, the removal of scientists from advisory capacity in the EPA, and inaction with regard to international conventions on climate change and other areas of sustainability. Because of PACs, there has been a lack of US federal legislation in these areas since the 1970s. The only changes we have seen have not been through acts of Congress but, rather, through executive orders or modifications within existing entities such as the EPA. Campaign finance reform that would limit the influence that larger organizations have on elected officials would return power to the individual, and it may be the only way to generate policy that would support our sustainable future.

Sustainable Investment

One way that individuals can make a difference with regard to sustainability is through their portfolio choices in their retirement plans or other investments. There are now many opportunities to invest in companies, organizations, and funds with the objective to both generate a measurable, beneficial social or environmental impact and provide a financial return for the investor. This type of investment is referred to as impact or socially responsible investing. It is also known as socially conscious, green, or ethical investing. This can include investment in assets such as stocks, exchange-traded funds, and mutual funds in which the underlying strategy seeks to consider both financial return and environmental and social sustainability objectives.

By directing retirement and other savings toward impact investments, individuals who wish to promote the goals of sustainability are able to do so by directly supporting corporate practices that promote environmental stewardship, consumer protection, human rights, and diversity through impact investing. This can include environmental or green investments in companies that are either developing solutions to environmental problems such as alternative energy technology or maintaining the highest environmental standards. Social-oriented investments may provide capital, credit, and training for economic development in low-income and other underserved communities, or they may invest in companies with a proven record of protecting and empowering workers.

Individual investors can use a variety of metrics to assess the positive impact of their investments. One major way is through the assessment of environment, social, and governance (ESG) criteria for specific companies or funds. The environmental criteria measure a company's performance in terms of environmental impact and stewardship, including accounting for externalities. Social criteria examine how a company manages relationships with its employees, suppliers, customers, and the communities in which it operates. Governance deals with a company's leadership, executive pay, audits and internal controls, and shareholder rights. Investors who want to purchase securities that have been screened for ESG criteria can do so through socially responsible mutual funds and exchange-traded funds. Most major investment firms provide independent ESG ratings. There has been some criticism of assessment criteria on ESG performance because reporting can be inconsistent. More detailed, uniform, and independently assessed criteria are needed.

There are other metrics for assessing the sustainability of a corporation or fund. Corporate Knights (http://www.corporateknights.com) and the Global Impact Investing Network (https://thegiin.org) provide resources that assist individuals who wish to maximize their impact through personal investment. For those interested in ensuring their investments are not put into high-carbon-emitting fossil fuel companies, the Carbon Underground 200 identifies the top 100 public coal companies and the top 100 public oil and gas companies and ranks them by the potential carbon emissions content of their proven reserves.[22]

Such impact investing is rapidly gaining in popularity. According to the US Forum for Sustainable and Responsible Investment, more than $1 out of every $5 under professional management in the United States—or $8.75 trillion—is invested in assets that promote social or environmental causes. In the United States, the overall number of mutual funds incorporating environmental and social benefits has increased 33% from 2015 to 2016.[23] Globally, interest in impact investing is even more impressive with annual investments well above $21.4 trillion and steadily rising. This has outpaced the growth of total professionally managed assets.[24] In many cases, socially and environmentally investments are either equally or outperforming those not in this class.

The changing composition of the workforce is beginning to drive patterns of investment even further toward those with positive impact. By 2020, millennials will comprise nearly half the working population and thus have the potential to lead the effort to impact corporate values through investment. According to a study by Morgan Stanley, in what is referred to as the "millennial effect," members of this demographic are twice as likely to invest in companies or funds that target sustainability-related outcomes compared to the total pool of investors with specific interest in investments that will directly address climate change or help reduce global poverty. Also, given their $2.5 trillion dollars of spending power, they are acting to promote corporate social and environmental responsibility by choosing brands that support these causes.[25]

Governments are also increasingly supporting impact investing. For example, in the United Kingdom, the government provides a 30% tax relief for social investments, which is anticipated to stimulate as much as GBP 500 million in additional investment over the next five years. The European Union created a regulation to formally recognize funds that invest 70% of investor capital into European Social Entrepreneurship Funds, enabling these managers to market and fundraise more effectively among impact investors. In addition, government-controlled pension funds are often very large players in the investment field, and they are being pressured by activists to adopt investment policies that encourage ethical corporate behavior, respect the rights of workers, consider environmental concerns, and avoid violations of human rights. One outstanding endorsement of such policies is the Government Pension Fund of Norway.

Another way that individuals can influence corporate behavior is through shareholder advocacy: proposing and representing resolutions that promote environmental, social, and governance change at annual stockholder meetings. As set forth by the US Securities and Exchange Commission, individuals have the opportunity to leverage their power of stock ownership to help further sustainability objectives. The organization As You Sow[26] promotes and supports shareholder advocacy in the areas of energy and water conservation, pollution prevention, waste reduction, environmentally and socially responsible sourcing, and community education and engagement.

Individual Action for Diverse Populations

One of the limitations of individual action in helping achieve sustainability in developed countries is that those sustainable choices or actions are often less accessible to the poor, disenfranchised, and less educated. Living simply and consuming less is often considered living more sustainably, and the simplistic living movement has grown within the context of sustainability. However, in many ways, this movement is confined to the privileged. People who choose to live more simply spend more money than the not-so-simple, usually because the simple and sustainable products tend to cost more than the mass-produced versions. Also, there is a major difference between choosing minimalism and having it forced upon one's daily life.

There are many examples of environmental injustice in which the poor suffer disproportionately from pollution, reduced resource quality and abundance, and proximity to municipal and toxic waste sites. The poor often feel disempowered, and the options for individuals to act or make choices as described previously are less available. Locally produced, farm-to-table, organic food, and fair trade and sweatshop-free products are unaffordable for many. Sustainable housing and home improvements also tend be very costly, and the return on tax incentives for items such as renewable energy does not benefit those with low incomes or who rent. The poor, who are often forced to work multiple low-wage jobs, are the largest consumers of unhealthy, unsustainably produced processed and fast food. Poverty can stifle the ability of individuals to act or volunteer, and it certainly leaves no opportunity for green investment when debt is more common than accumulation of savings. In addition, the poorest people often feel the least empowered.

Individuals from different ethnic groups tend to vary in the way they get involved with environmental issues. For example, American Latinos care very much about the environment but tend not to be engaged with environmental groups or with politicians over environmental issues. Recent surveys by the organization Latino Decisions[27] showed that 85% of surveyed Latinos in the United States believe that addressing smog and air pollution should be a priority for the President and Congress, 75% support establishment of national standards to prevent climate change, and even higher percentages support the development of clean energy and favor strengthening the Clean Water Act.

Despite their strong interest and support of environmental issues, most US Latinos do not join forces with mainstream environmental groups or movements. The surveys revealed that Latinos do not mobilize for a movement or cause but, rather, do so to improve their quality of life and the future of their children. Their environmentalism stems from poverty, religious beliefs, and interest in improving access to affordable and healthy food. Also, an overwhelming majority of survey respondents said they have never been contacted by an environmental organization and asked to join forces over shared environmental interests.

In order to strengthen the environmental and social movements, we need to think about how to unite diverse communities on common issues so that we can work together, share resources and expertise, and take advantage of the power of numbers to generate solutions. The best way for established organizations to engage diverse individuals is to partner with local organizations in their communities. This is starting to happen. For example, the Natural Resources Defense Council has begun partnering with local Latino community groups to achieve mutual goals, such as improving air quality and protecting children from pesticides. In addition, mainstream environmental organizations might be more successful at engaging diverse individuals if they reach out to existing organizations to learn about and collaborate with their existing initiatives rather than coming into the community as outsiders with their own agenda.

Providing the poorest people access to the tools, choices, and processes so that they can live their lives more sustainably must be a priority. This can occur through poverty reduction, which can be achieved through equitable economic development, especially in urban centers. Individuals and neighborhood groups need effective representation and government legislation to guarantee environmental justice. This should include increased transparency, accountability, accessibility, and community participation, as well as technical and financial assistance. Leadership training and community organization to support the participation of poor individuals in social movements and organizations are also essential.

Numerous organizations help support individuals and communities seeking environmental justice. One such group, the Center for Health, Environment & Justice (CHEJ), acts to build healthy communities nationwide, promoting social justice, economic well-being, and democratic governance by providing essential resources, strategic partnerships, and training to local leaders to achieve this mission. Through training and coalition-building, CHEJ mentors and empowers individuals so that they can have a say in the environmental policies and decisions that affect their health and well-being.

Increased access to sustainable housing and home improvement should be made available to the poor. The federal weatherization program and similar programs available through local power companies also help low-income individuals improve the sustainability of their households by helping them insulate their homes and improve the efficiency of their heating systems. A recent report from the US Department of Energy showed that energy savings from this program were substantially larger than the cost of weatherizing the homes and that there were significant improvements in health in those individuals who participated.[28]

Social entrepreneur Van Jones has supported government and for-profit and non-profit ventures that extend sustainability initiatives to those who typically have been unable to afford them. These initiatives have included weatherizing homes, installing solar panels, and increasing the availability of organic foods in low-income, urban neighborhoods. He has helped create programs to train inner-city residents for eco-friendly jobs in a program that he calls the "Green Corps," and he has promoted ambitious public spending programs on energy efficiency and renewable energy to stimulate the green economy and create good jobs for the poor and unemployed.[29]

Non-profit organizations with the mission of providing sustainable living opportunities for the poor are cropping up in many cities. One example in Buffalo, New York, is People United for Sustainable Housing (PUSH Buffalo). This is a local membership-based community organization fighting to make affordable, sustainable housing a reality in Buffalo's low-income neighborhoods. PUSH Buffalo was established to create strong neighborhoods with quality affordable housing by reclaiming empty houses from neglectful owners and redeveloping them for occupancy by low-income residents with an emphasis on green building. Much of its success is due to its practice of developing neighborhood leaders capable of gaining community control over the development process and planning for the future of the neighborhood. PUSH Buffalo and its affiliate, PUSH Green, also provide low-income residents with access to energy audits and low-cost financing for green improvements (Figure 9.6). In addition, PUSH works to empower the poor to lead direct action campaigns against corporations and government agencies whose practices contribute to the high poverty rates.

In Chapter 5, a number of ways to improve poorer people's access to more sustainable foods were presented. Examples included the development of inner-city community gardens, affordable farmers markets that add value to the Supplemental Nutrition Assistance Program benefits, and other ways to produce food in urban centers and other food deserts. The most successful of these programs are ones that include community residents as the primary stakeholders in developing and implementing the programs and offer them opportunities for income generation and poverty reduction.

Another challenge is that as we begin to make major economic shifts toward the goal of sustainability, the working class and poor may bear the brunt of that change. For example, as we shift from a fossil fuel–based economy to one based on renewable, lower carbon sources of energy, there is concern that ordinary workers, their families, and communities will bear the brunt of the transition to new ways of producing wealth as has happened with major economic shifts in the past. Shifting to a low-carbon economy will have major implications for individuals working in energy supply, industry, and transport and for everyone as consumers. To avoid this, a "Just Transition" framework that protects workers' livelihoods as we shift to a low carbon economy has been integrated into recent climate change policy.

In developing countries, a number of individuals have effected transformative change in the area of sustainability. We have already discussed the impact of Muhamud Yunnus, a Bangladeshi who was awarded the Nobel Peace Prize for pioneering the concept of microfinance. There are many other examples, such as Wangari Maathai, who led the fight to promote ecologically viable social, economic, and cultural development in Kenya and in Africa, and Guatemalan activist Rigoberta Menchú. Another, Chico Mendez, was a Brazilian rubber tapper, trade union leader, and environmentalist who fought to preserve the Amazon rainforest and advocated for the human rights of Brazilian peasants and indigenous peoples. He was eventually assassinated for his actions. According to the organization Global Witness, as with Mendez, the murder of environmental activists in developing countries (many of whom are indigenous people) is not uncommon, with a record number of 185 killed in 2015.[30]

Figure 9.6 Volunteers and staff from PUSH Buffalo/PUSH Green promote their free low-income weatherization program in Buffalo, NY. *PUSH Buffalo, Lonnie Barlow, Director of Communications.*

Individuals in developing countries have also come together in numerous social movements and organizations. The nature of such movements varies greatly according to their members' political, economic, and environmental histories and their approaches to social transformation. They include grass-roots organizations often initiated at the village level and organized networks of NGOs that work on a variety of causes, including indigenous peoples' rights, community forestry and conservation, river and watershed conservation, and other issues related to sustainability. There are clear differences between sustainability activists in developed and those in less developed countries. In developed countries, individuals who come together as activists tend to be middle class or wealthier, and it is their affluence that allows them the luxury to stand for environmental cases that go beyond meeting their basic needs. They also tend to rely on science, the legal system, lobbying, and the media. By contrast, in developing countries, such activists are fighting for environmental protection because these problems directly influence their livelihood and survival, and their activist approaches have mostly been as direct actions.[31]

In less developed countries, the power of individuals to contribute to sustainable change is limited, and their actions often have greater repercussions. Furthermore, with less economic opportunity, little or no access to energy or sanitation, and poor environmental conditions including indoor air pollution and depleted water and soils, the poorest people have much to endure and overcome. The technology and education to solve these problems are often deficient or inaccessible, as is the ability to finance sustainable improvements through credit. In many poorer nations, people, especially women, are often oppressed, and they often do not have the right to assemble, speak out, and protest; as described previously, they often put their lives at risk for doing so.

Any plan for sustainable development must increase access of members of the global poor to the solutions presented in this book and empower individuals to obtain the tools and participate in the processes. Social justice, human rights, and true democracy are fundamental to global sustainability, as is the opportunity to obtain low-interest credit through microfinancing. In every chapter, we have discussed how the poor are disproportionately impacted by the environmental effects of industrial society and the lack of resources and infrastructure. Individuals in less developed countries have a major role to play in sustainable development because public demand is the foundation on which transition to sustainable development can be accelerated. In other words, social demand can shape government policies. However, for this to occur, there must be opportunity for the exchange of knowledge, information, and the right for individuals to freely express this demand.

In addition to the basic rights of a free, democratic society, there must be mechanisms to share sustainable innovations not only from the industrial North to the South but also across the developing world. One successful example of how this is occurring is the Honey Bee Network founded by Anil Gupta of the Indian Institute of Management. This network collects traditional knowledge from isolated rural communities and spreads that innovation throughout India. It helps individuals share their ideas and put them to market.

The ability of individuals to share information is viewed as central to sustainable development. Thus, closing the **digital divide**, or difference in access to technology between developed and developing countries, has become an important component in the promotion of sustainable development. The goal of programs such as Communication Technologies for Development (ICT4D) is to close the global digital divide by providing technology such as smartphones and computers to further sustainable development (Figure 9.7). This technology allows for improved communication and information sharing, increased access to education, disaster response, support of disabled individuals, and the promotion of civic engagement. It can also enhance economic opportunities and entrepreneurship through online sales and purchasing, peer-to-peer lending, and microfinancing. It has been shown that increasing technology such as the availability of mobile phones in rural Africa can provide many of these benefits that have been directly linked to poverty reduction.[32]

Providing individuals in rural developing countries with access to technologies through the ICT4D initiative has become a priority in sustainable development and has become a funding priority for the World

Figure 9.7 Satellite Internet as seen here in the Ghana is a common form of connectivity in developing countries promoted ICT4D initiatives. *IICD/ Wikipedia.*

Bank and other public and private entities. There are a number of limitations and criticisms of ICT4D. Some argue that in the poorest nations, there are more pressing needs that electronic technology cannot solve. A lack of infrastructure, illiteracy, ongoing maintenance requirements, increased e-waste, and even the use of technology to exploit or violate the rights of others are viewed as major problems with this initiative.

Increasing the Power of Individuals

Central to our definition of sustainability is the empowerment of individuals to meet their own needs. We have seen that individuals can indeed make a difference in moving us toward a sustainable future. Education and exposure to global realities, and the alleviation of fear, guilt, and the feeling of powerlessness, can help individuals become agents of change, whether it is through small steps or in some cases radical transformation. Entrepreneurship, working with government, direct action, citizen science, and sustainable investment are powerful ways that individuals can make a difference. In poorer nations, the protection of basic individual rights, poverty reduction and opportunities for income generation, microfinancing potential, and information sharing are essential elements to empowering individuals to meet their own needs.

It is important to recognize that the objectives of sustainability cannot be met solely through individual action or behavioral change. When individuals with similar beliefs and attitudes organize and come together, their influence and impact can be amplified. Individuals can come together in social movements, organizations, and institutions. A **social movement** is a group of loosely organized people with shared values or ideologies who work together to achieve certain general goals.

Social movements usually result from spontaneously generated associations of people whose relationships are not defined by rules and procedures but, rather, by a common outlook on society. These groups are often able to sustain a campaign with the goal of carrying out, resisting, or undoing a social or political change. Examples are wide-ranging and include the civil rights, anti-apartheid, environmental, pro-labor, MoveOn, Black Lives Matter, and organic and slow food movements.

The success of social movements depends on a variety of factors. First, a group of engaged individuals organizes around a common issue, such as climate change. Social movements tend to be stronger when the group arises in solidarity in response to a moral imperative, a common sense of injustice, or an unpopular political direction, such as the climate denial movement or the recent US withdrawal from the 2016 Paris climate accord. Next, there needs to be organizational infrastructure including leadership, articulation of a specific mission, the ability to connect to the mainstream, and planned events such as the climate march on Washington, DC. Ultimately, for social movements to be successful, there must be political opportunity, such as the willingness of governments to commit to reducing carbon emissions because it makes economic sense and follows a moral imperative to slow or reverse climate change.

Social media platforms such as Twitter, Snapchat, and Facebook now play a major role in the development of social movements. Online groups help reduce obstacles to participation, such as time constraints, and encourage participation through the promotion of events, offering petitions to be signed, raising small amounts of money from a much larger pool of donors, and keeping members informed. Social media can be used to rapidly mobilize and organize groups for actions such as marches and letter-writing campaigns. The use of hashtags allows individuals to follow specific issues on a regular basis. For example, the hashtag #ActOnClimate has allowed for the rapid development of a resistance movement against climate denial in the United States.

Successful social movements may ultimately lead to the development of one or more organizations that support a specific aspect of the movement. An **organization** is a group of people who work together in a structured way for a shared purpose. They are typically well organized into subgroups such as committees, and they often have some sort of financial, physical, or virtual infrastructure through which they operate. They are generally led by a single person or a group of people either on a permanent or rotational basis who may be the organization's founders, drawn from its members, or brought in from the outside.

The purposes of organizations vary widely. There are political, social, financial, entertainment, education, health, development, environmental, and labor organizations, just to name a few. They can be governmental, NGOs, non-profit, and for-profit. Organizations sometimes join or support individuals in larger, less structured social movements. This book has presented numerous examples of organizations, many of which grew out of social movements, that are playing an important role in effectively addressing sustainability initiatives. In this chapter, we have discussed how individual action can contribute to progress in all areas of sustainability. As individuals come together in the grass-roots, bottom-up efforts of social movements, their impact can be even greater if not transformational. In addition to the work of individuals, social movements, and organizations, changes must also occur on institutional, municipal, regional, national, and global scales in order for us to sustain our planet and implement the solutions presented in this book. These types of change are the focus of the next three chapters, in which we first explore how institutions can and are becoming more sustainable. We then examine community and regional efforts to promote environmental protection and improve quality of life. Finally, we examine global sustainability, sustainable development, and the economic transformations that will make these possible.

Chapter Summary

- Although it may seem that individual action is insufficient to move us toward a sustainable future, there are many examples of how individuals have been effective agents of change through both incremental steps and radical transformation.
- A number of intrinsic and extrinsic factors cause individuals to act or change behavior. Intrinsic qualities such as openness, empathy, and compassion are instilled early in life and tend to occur in people who are committed to social and environmental causes. Extrinsic factors that cause people to act include financial incentives and the influence of other people or organizations. The major causes of inaction or denial of environmental or social problems range from fear-related to economic uncertainty and our culture of consumerism.
- Ambivalence or apathy toward environmental issues is caused by the overly pessimistic portrayal of environmental and social problems that leaves individuals with feelings of guilt and helplessness. Perceived constraints of daily life and structural issues that separate people from resources, environment, and the opportunity to witness injustice also result in apathy toward these issues.
- Knowledge of these positive and negative influences on individual action and behavioral change can guide us as we raise and educate a new generation of agents of change and creative problem solvers.
- Individual action often occurs within the context of the household, where decisions are often based on economics and return on investment. Top-down governmental financial incentives can make sustainability-related household changes more economically feasible.
- Social entrepreneurs or individuals who create businesses that have social and environmental missions can be very effective at furthering sustainability objectives. There are a number of models of social entrepreneurship, including for-profit and non-profit entities. Social enterprises that are developed in a consultative manner, focus on a specific need, or create a product or service that empowers individuals to provide for themselves are more likely to be successful at meeting their social and environmental missions.
- Individuals can increase their power and effectiveness by collaborating with others through direct action or protest, education, citizen science, and by working directly with government.
- Another way for individuals to further sustainability is by direct involvement in politics. Campaign finance reform that reduces power and influence from large corporations and special interests groups and restores it to the individual may be required to achieve policy change that will promote a sustainable future.
- Another way that individual action can make a difference is through environmental or socially responsible investment as well as through shareholder advocacy.
- A major limitation of the role of the individual is that sustainable choices or actions are often less accessible to the poor, disenfranchised, and less educated. The poor are disproportionately exposed to environmental injustice, and different ethnic groups vary in the way they become involved with social and environmental issues. Uniting diverse people who believe in similar causes and providing the poorest people access to the tools, choices, and processes so that they can

live their lives more sustainably are essential to achieving a sustainable future.

- Support of individuals in developing countries who are interested in effecting social and environmental change is also needed. In many cases, this will require improvements in basic human rights, poverty reduction, mechanisms for sharing sustainable innovations, and improved technology.
- When individuals come together in social movements and organizations, their power can be amplified. Social media is a powerful tool in helping achieve this.

Digging Deeper: Questions and Issues for Further Thought

1. Within the context of sustainability, what have been some barriers to personal change or individual action that you have experienced or observed?
2. What are the strengths of individual action? Can it be effective and lead to significant change? Under what conditions?
3. What are the limitations of individual action in achieving actual change?
4. Do the for-profit models of social entrepreneurship and social investment support or contradict our notion of sustainability? In what ways?

Reaping What You Sow: Opportunities for Action

1. Use one of the apps or programs discussed in Boxes 9.1 and 9.2 to measure your own environmental and/or social impact. Propose some personal commitments to reduce it.
2. Identify a governmental commission or advisory board related to sustainability in your community and attend a meeting. Identify its mission, recent successes, and ways that you might contribute.
3. Identify and research an example of social entrepreneurship that has resulted in a product or service that could contribute to sustainability like many of those presented throughout this book. Develop and give a pitch for it within the context of sustainability objectives.
4. Choose a specific sustainability objective and closely follow it on social media using a hashtag. Identify recent successes as well as barriers to success in achieving this objective. Post your views on the subject, and document the feedback that you receive.

References and Further Reading

1. Babcock, H. M. 2009. "Assuming Personal Responsibility for Improving the Environment: Moving Toward a New Environmental Norm." *Harvard Environmental Law Review* 33: 117–75.
2. Mooney, Chris. 2015. "The Surprising Psychology Behind Why Some People Become Environmentalists." *Washington Post*, March 19. https://www.washingtonpost.com/news/energy-environment/wp/2015/03/19/the-surprising-reason-why-some-people-become-environmentalists-and-others-dont
3. Mooney, "The Surprising Psychology."
4. Ajzen, Icek. 1991. "The Theory of Planned Behavior." *Organizational Behavior and Human Decision Processes* 50 (2): 179–211.
5. GfK Consulting. 2011. "The Environment: Public Attitudes and Individual Behavior—A Twenty-Year Evolution." GfK Consulting Green Gaugue US Survey, commissioned by SC Johnson.
6. Gladwell, Malcolm. 2002. *The Tipping Point: How Little Things Can Make a Big Difference*. New York: Back Bay Books.
7. Community Tool Box. http://ctb.ku.edu/en/table-of-contents/sustain/social-marketing/conduct-campaign/main
8. The NSMC. http://www.thensmc.com
9. Dietz, Thomas, Gerald T. Gardner, Jonathan Gilligan, Paul C. Stern, and Michael P. Vandenbergh. 2009. "Household Actions Can Provide a Behavioral Wedge to Rapidly Reduce US Carbon Emissions." *Proceedings of the National Academy of Sciences of the USA* 106 (44): 18452–56. doi:10.1073/pnas.0908738106
10. Chan, K., and C. C. Murray. 2010. "How Personal Actions Can Kick-Start a Sustainability Revolution." *Grist*. http://grist.org/article/2010-02-01-how-personal-actions-can-kick-start-a-sustainability-revolution
11. GfK Consulting, "The Environment."

12. Seth H. Werfel. 2017. "Household Behavior Crowds out Support for Climate Change Policy When Sufficient Progress Is Perceived." *Nature Climate Change* 7: 512–15.
13. Adam Briggle. 2015. *A Field Philosopher's Guide to Fracking: How One Texas Town Stood Up to Big Oil and Gas*. New York: Liveright.
14. Carlisle, Liz. 2015. *Lentil Underground: Renegade Farmers and the Future of Food in America*. New York: Gotham.
15. Steingraber, Sandra. 2010. *Living Downstream: An Ecologist's Personal Investigation of Cancer and the Environment*. Cambridge, MA: Da Capo Press.
16. Kirsch, V., J. Bildner, and J. Walker. 2016. "Why Social Ventures Need Systems Thinking." *Harvard Business Review*, July 25. Accessed May 19. https://hbr.org/2016/07/why-social-ventures-need-systems-thinking
17. Audubon. https://www.audubon.org/conservation/science/christmas-bird-count
18. scistarter. http://scistarter.com/project/776-Cicada%20Tracker
19. Piao, Shilong, Jianguang Tan, Anping Chen, et al. 2015. "Leaf Onset in the Northern Hemisphere Triggered by Daytime Temperature." *Nature Communications* 6 (April): 6911. doi:10.1038/ncomms7911
20. ISeeChange. https://www.iseechange.org
21. Environmental Advisory Council. http://eacnetwork.org/wp-content/uploads/sites/4/2014/05/PEC-EAC-Handbook-web.pdf
22. Fossil Free Indexes. n.d. "The Carbon Underground 200—2016 Edition." Accessed October 19. http://fossilfreeindexes.com/research/the-carbon-underground
23. US SIF. n.d. "The Forum for Sustainable and Responsible Investment: Community Investing." Accessed December 1. https://www.ussif.org
24. Global Sustainable Investment Alliance. 2014. "2014 Global Sustainable Investment Review." Accessed December 11. http://www.gsi-alliance.org
25. Morgan Stanley. 2017. "Millennials Drive Growth in Sustainable Investing." August 9. Accessed October 9. https://www.morganstanley.com/ideas/sustainable-socially-responsible-investing-millennials-drive-growth
26. As You Sow. http://www.asyousow.org
27. Ensia. 2015. "Latinos Care About the Environment. So Why Aren't Green Groups Engaging Them More?"December 9. Accessed December 11. http://ensia.com/articles/latinos-care-about-the-environment-so-why-arent-green-groups-engaging-them-more
28. Oak Ridge National Laboratory. "Weatherization and SEP Support Program." http://weatherization.ornl.gov
29. Jones, Van. 2008. *The Green Collar Economy: How One Solution Can Fix Our Two Biggest Problems*. New York: HarperOne.
30. Global Witness. 2015. "How Many More?" April 20. https://www.globalwitness.org/en/campaigns/environmental-activists/how-many-more
31. Gadgil, Madhav, and Ramachandra Guha. 1994. "Ecological Conflicts and the Environmental Movement in India." *Development and Change* 25 (1): 101–36. doi:10.1111/j.1467-7660.1994.tb00511.x
32. Masiero, S. 2013. "Innovation and Best Practice in Mobile Technologies for Development." January 1. Helpdesk Review for LSE Enterprise/UK Department for International Development (DfID). http://r4d.dfid.gov.uk/Output/194203

CHAPTER 10

ORGANIZATIONS, INSTITUTIONS, AND SUSTAINABILITY

In Chapter 9, we examined the important role that individuals can play in achieving the goals of sustainability. We identified the factors that motivate individuals to act, the many possibilities by which they can act, and discussed examples of the power of those actions. We also noted that one way to increase the power and effectiveness of individual action is to combine forces with others. This can be done by contributing one's effort to an entity or organization such as a governmental commission or a non-governmental organization (NGO) with a specific mission. Another way the power of individuals can be maximized is through the formation of social movements, in which individuals come together and collaborate for a common cause.

PLANTING A SEED: THINKING BEFORE YOU READ

1. What institutions are you affiliated with or do you interact with on a regular basis? Determine whether any of them have sustainability as part of their stated mission and what they are doing to meet sustainability objectives.
2. To what extent was sustainability a part of your kindergarten through 12th grade (K–12) educational experience? How would you change that experience to better incorporate sustainability priorities?
3. In what ways might teaching sustainability conflict with other curricular goals? In what ways would it complement those goals?
4. Name one large corporation that comes to mind with regard to commitment to sustainability. How is that commitment met in terms of practice?
5. What other types of institutions besides education and business have the potential or are already supporting sustainability objectives?

A social movement that only moves people is merely a revolt. A movement that changes both people and institutions is a revolution.

—Martin Luther King, Jr.

Beyond the level of the individuals and social movements are organizations and institutions. These larger entities can be much more powerful or even transformational in furthering sustainability objectives. This is because of the scale of their operations and their broad impact both on the environment and on large numbers of people. The words "organization" and "institution" are often used synonymously, which makes the semantics confusing. An **organization** is a group of people who work together in a structured way for a shared purpose. **Institution** is a broad term that includes many definitions. The term commonly applies to a custom or behavior pattern that is important to a society, such as marriage or law. Institutions can be long-standing traditions or practices that serve to maintain the status quo. Some of these practices, such as slavery or apartheid, are oppressive institutions. Other institutions are established to govern norms and behavior, to serve common problem-solving goals, and to foster the preservation and function of civil society. These include economic, governmental, educational, and religious institutions. Such social institutions are in essence a type of organization identified with a social purpose that transcends individuals and is established to maintain social order and govern the behavior of a set of individuals within a given community.

The sustainability of an organization or an institution can be assessed in terms of both its survivorship and its ability to further sustainability objectives. Regarding the former, individual organizations can come and go in a process that many liken to "natural selection," where only the "fittest" or most resilient organizations survive. This resilience is reflected in an organization's ability to adapt to the times through mission revision, rebranding, and maintaining financial solvency as audiences and the needs and interests of their constituency evolve in a competitive environment. Larger governmental, corporate, and educational institutions exhibit inertia, making the process of institutional change a challenge; however, because of their size and historical legacy, they often persist well after their peak in effectiveness.

In terms of furthering sustainability objectives, organizations and institutions can do so by designing or altering their infrastructure and operations in order to minimize their negative impact on the environmental, economic, and social spheres of sustainability. They can also provide a product or service that directly improves sustainability within these spheres. Institutions and organizations can also effect large-scale change by advocating, regulating, or educating for sustainability.

Throughout this book, we have discussed the major role that both governmental and non-governmental organizations play in the sustainability arena, and we continue to explore this in subsequent chapters. In this chapter, we focus on larger non-governmental institutions and their potential to contribute to a broader transformation toward sustainability. We primarily focus on two types of institutions—educational and business—and explore how each can contribute to sustainability objectives. Sustainability priorities, best practices, and tools, limitations, and resources for meeting them will also be offered. Today, there are very few institutions that do not incorporate some aspect of sustainability into their mission and/or operations. Here, we apply our criteria of sustainability and develop indicators as ways to more formally assess the sustainability of an institution and its contributions in order to illuminate areas for improvement.

Many other institutions, such as hospitals, religious institutions, non-profit organizations, and even the military, have similar opportunities in the area of sustainability—and the lessons learned in this chapter can be applied to them. In subsequent chapters, we examine governmental, economic, and international aid organizations and institutions in consideration of their role in supporting sustainability on community, regional, and global scales, including sustainable development.

Educational Institutions

Education and sustainability have been somewhat paradoxical. We think of learning as something that is intrinsically good. However, the environmental problems, social injustice and oppression, and economic inequities discussed in this book are not the cause of uneducated people but, rather, have resulted from the ideas and actions of our most educated. The purpose of education has been to offer a "better future" for our children through increased success as measured by the accumulation of wealth and material goods. The wealthiest nations that have the highest levels of education also have the highest per capita rates of consumption and the largest ecological footprints. Thus, as environmental educator David Orr has so eloquently argued, it is time to rethink the way we educate (Figure 10.1). We need to recognize that there is a responsibility to not only impart knowledge but also ensure that it is used well in the world and that knowledge is not wholly acquired until we understand the effects of it on other people and the planet.[1]

A sustainable future can only be realized if we directly address this paradox between education and sustainability. As suggested by Einstein, the significant problems that we face today cannot be solved with the same knowledge and way of thinking that we used to create them. In other words, our educational institutions need a paradigm shift that inspires students to think about the world, their relationship to it, and their ability to influence it in a completely different way. Students must be given the opportunity to obtain knowledge, skills, and values that are needed to understand and evaluate the environmental and social problems that we face, the role we play in creating them, and the ways in which we can solve them.

Figure 10.1 David Orr, the Paul Sears Distinguished Professor of Environmental Studies and Politics at Oberlin College, has been a leader in the sustainability in higher education movement. He helped launch the green campus movement starting in 1987, and proposed the goal of carbon neutrality for colleges and universities in 2000. Orr is pictured here at Oberlin's Center for Environmental Studies that was designed to function as a living machine. *DAVID MAXWELL/ The New York Times/Redux.*

There is also an education crisis in the developing world, in which currently 57 million children of primary school age do not have access to a school. Of these, 33 million are in sub-Saharan Africa and are mostly girls. Achieving universal primary education as a high priority was one of the United Nations' (UN) Millennium Development Goals for 2015, and it continues to be so for the current Sustainable Development Goals program. Obtaining a quality education is the foundation to poverty reduction, improving people's lives, and all the other goals of sustainable development; however, that education needs to be consistent with and directly address those goals rather than perpetuate unsustainable living and development.

The need to broaden our concept of education to include sustainability and sustainable development is explicitly stated in the UN's Agenda 21 generated during the 1992 Earth Summit in Rio[2] and by the 1996 President's Council on Sustainable Development.[3] Educational institutions, both K–12 schools and colleges and universities, are beginning to embrace educating for sustainability in terms of curriculum, infrastructure, and operations because of the recognition that a sustainable future depends on this. In the following sections, we examine how this is most effectively accomplished in primary and secondary education and at colleges and universities. A number of tools and resources for those committed to integrating sustainability and education are provided. Education is a vital part of the global sustainable development agenda. This is addressed in Chapter 12.

Sustainability in Primary and Secondary Education

Since the 1970s, the US Department of Education has become increasingly interested in sustainability, yet federal funding for sustainability education initiatives has been limited and inconsistent, especially for K–12 education. The National Environmental Education Act of 1970 recommended the integration of environmental and sustainability concepts into K–12 curricula, but because it was not well-funded and was poorly received by school administrators, this effort was discontinued in 1975. An extension of that act in 1990 treated K–12 sustainability education as a supplement

Figure 10.2 The No Child Left Inside movement emphasized student involvement in environmental education, courses, classwork, and field investigations to develop curiosity about and value for nature. *Phovoir/ Shutterstock.*

rather than as an integral part of the core curriculum. However, in response to recommendations from Agenda 21, there has been increasing funding and a recognition that there is a compelling responsibility to foster a transformation in education policy that embraces sustainability, particularly in primary and secondary schools. However, this commitment changes with political climate.

By the mid-1990s, some common principles on the most effective ways to educate for sustainability began to emerge. One of the most important of these is the recognition that all education relates to or has implications for the environment and sustainability. Thus, sustainability should be embedded within all courses of the core curriculum and also in the life of the school rather than as a stand-alone subject or unit. Doing this will enrich rather than detract from those core subjects because sustainability concepts provide context and opportunities for developing fundamental skills such as critical thinking and informed decision-making, including an understanding of how systems operate. Integrating sustainability throughout the curriculum also offers important multicultural and global perspectives, and it illustrates the interconnectedness of different subjects.

Teaching sustainability requires that students apply critical thinking skills and that they draw from their own experiences and world knowledge. Discussions about sustainability involve current events and often complex global connections. Thus, the concept can be used as a starting point for critical thinking exercises as students consider the many facets of an issue. In Chapter 9, it was shown that diverse populations vary in how they engage with sustainability issues. Therefore, it is important that sustainability education materials and programs include diverse cultural perspectives, reflect divergent cultural approaches to sustainability, and are made accessible to all interested communities.

Pedagogical approaches should include place- and project-based experiential learning that takes students out of the classroom and into nature. Students need to experience the natural world in order to develop an appreciation of it and the capability to understand, analyze, and solve environmental challenges. It is often said that you cannot save something you do not love, and you cannot love something you do not know, so to save nature, children must know it. This was the basis of the No Child Left Inside movement that encouraged regular field trips to local environmental areas, as well as the development of community gardens and nature areas on school properties (Figure 10.2). Project-based learning further helps students develop important skill sets, including collaboration and interdisciplinary thinking. When projects such as the development of a school recycling program or an environmental clean-up effort in their own community show positive results, students feel empowered and learn that they can improve the quality of their own environment and the well-being of their community.

GREEN SCHOOLS

Another component to educating for sustainability is the integration of the "greening" of school facilities and operations with curriculum across subjects and grades. Schools designed as sustainable or high performance provide safe, comfortable learning environments that help students and teachers perform at their highest level by providing a learning environment that is healthy. Studies have shown that sustainable lighting, improved indoor air quality, and other aspects of sustainable building improve student performance and reduce absenteeism.[4] Sustainable buildings that are more energy and water efficient and reduce waste are typically cost-neutral in upfront and ongoing operational costs for the life of the building. When integrated with the curriculum, green buildings become more than just a place of learning; they become an invaluable pedagogical tool.

Sustainable food service can also improve student health and learning, and it can serve as another important pedagogical tool. Not only is the nutritional quality of food improved but also greening school food service creates opportunities to teach students about how food, culture, health, and the environment are interconnected. Things such as locally sourced food, student gardens, and food waste reduction and composting projects offer opportunities for learning about sustainable food. One example is the California Food for California Kids program, in which participating school districts serve healthy, freshly prepared school meals made from locally sourced food. This program can and is being adapted in other communities and states.[5] One participant, the Sausalito Marin City School District, became the first in the nation to serve 100% organic lunches (Figure 10.3). In addition to serving nutritious lunches, unhealthy foods and beverages should not be available or not directly marketed in schools as frequently as they are today. The Campaign for a Commercial-Free Childhood and Corporate Accountability International are working to stop this type of marketing, and Food Mythbusters (http://foodmyths.org) provides resources to counter their unhealthy messages.

Thus, given the apparent value of integrated sustainability education, facilities, and food service, to what extent are they being adopted? What are the limitations, and what might help increase this effort? In a 2014 comprehensive survey of public schools in 12 states, 24% of school districts indicated that they included sustainability in their mission statement, and 37% had a set of policies to promote environmental and sustainability education. More than half had a green council or committee, energy savings and waste and recycling programs, a school garden, and a wellness policy.[6]

BARRIERS TO K–12 SUSTAINABILITY EDUCATION

The degree to which sustainability has made its way into curricular programs has been much less successful, and the shift from specific, discrete educational topics to a more integrated systems approach has not been easy. In the 2014 survey mentioned previously, less than 10% of schools had a sustainability literacy requirement, and only 35% of schools had integrated sustainability concepts across the curriculum. Another survey of more than 1,000 K–12 teachers who had already incorporated teaching for global sustainability into their classrooms showed that only 32% integrated global sustainability into core subjects (math, science, reading, writing, and social studies), whereas the majority taught it as an add-on or separate subject.[7]

Figure 10.3 Along with their 100% organic lunch offerings, Sausalito Marin City schools offer a garden and nutrition curriculum so that students learn where their food comes from. *Monkey Business Images/Shutterstock.*

Thus, although effective sustainability education is a growing movement in the United States, there is still much work to be done. The main challenges are teacher workload and time constraints, as well as funding limitations for teacher training resources and field trips.[8] Also, federal- and state-determined standards often do not include sustainability concepts. Much of the time constraints on teachers stem from an overwhelming emphasis on standardized testing and preparing for those tests. This has intensified as student performance on these tests is increasingly being tied to teacher evaluation. This emphasis on testing severely limits the ability of teachers to do the kind of innovative teaching described previously.[9] Also, teacher education, certification programs, and professional development are insufficient to transform our current mode of education into one that can realize all of the pedagogical and societal benefits of educating for sustainability. Finally, many schools and school districts lack strategic plans for green initiatives and creating a culture of sustainability.

To increase access to sustainability education, the previously mentioned limitations must be addressed. One way to increase access to professional development and resources for effectively teaching sustainability is by partnering schools with each other, colleges and universities, business and industry, NGOs, and community groups that share common sustainability-related objectives. An example of this is the Sustainable Design Project in Washington state. This is an interdisciplinary project bringing industry, business, community, and higher education partners together with K–12 teachers and students to design sustainable solutions to real-world challenges. Similarly, in Canada, the Sustainability and Education Policy Network (http://sepn.ca) is a national network of researchers and organizations that work collaboratively to advance sustainability in education policy and practice.

There are also many resources that can support teachers in their integration of sustainability throughout the K–12 curriculum. For example, Facing the Future (http://www.facingthefuture.org) provides curriculum resources on global issues, sustainable solutions, and problem- and inquiry-based learning. Its resources include student textbooks, lesson plans, assessment tools, simulation and problem-based learning exercises, and action- and service-learning information that will allow teachers to use global sustainability as a context for teaching core academic content and skills.

One of the challenges facing teachers is aligning sustainability education content and objectives with current academic standards. The Center for Ecological Literacy (http://www.ecoliteracy.org) provides resources for help in aligning sustainability education goals and approaches with standards such as Common Core State Standards; Next Generation Science Standards; College, Career, and Civic Life (C3) Framework for Social Studies State Standards; and National Health Education Standards.[10]

A sustainability strategic plan is essential for supporting environmental initiatives such as green building and renovation, sustainable lunch initiatives, and the creation of an overall culture of sustainability that can be integrated with curriculum. This is best achieved by creating either a school- or a district-wide sustainability committee charged with developing and overseeing the implementation of the strategic initiatives. Such a committee should reflect the diversity of stakeholders, including administrators, teachers, custodial and dining staff, members of the community, and of course, students.

The expansion of teacher certification programs that emphasize the integration of sustainability across the core in K–12 classrooms is also needed. This is starting to occur at the state level. This has been exemplified by Washington state's Professional Educator Standards Board, which recently adopted teacher preparation standards for colleges of education that seek to ensure that all students are prepared to be responsible citizens for an environmentally sustainable, globally interconnected, and diverse society. The expansion of these standards for colleges of education in other states can further the effort to create new generations of teachers equipped and motivated to integrate sustainability issues and the skills associated with that integration in their teaching. Leaders and faculty from college and university education programs can also play a role in infusing sustainability education into state and national standards.

In the United States, the effort to develop green curricula, facilities, lunch programs, and teacher education is expanding and being increasingly recognized at the federal level. In 2011, the Department of Education created the Green Ribbon Schools recognition award to honor those institutions that are demonstrating leadership in reducing environmental impact and

BOX 10.1

Policy Solutions: Green Ribbon Schools

When it is time to renovate existing schools or build new ones, doing so sustainably makes economic and pedagogical sense. It has also been shown to improve student and teacher performance. An example of a school district that has embraced this is the Gladstone School District in Oregon, which is located approximately 15 miles south of Portland. The district educates 22,000 students in kindergarten through 12th grade (K–12), more than half of whom are enrolled in the free and reduced cost lunch program based on economic need. The push for sustainability came from several enthusiastic teachers, a number of student-based initiatives, and administrator buy-ins. Because of this interest, as administrators began planning for facilities renovations in the district, they began to explore the advantages of taking a green approach, and the school board developed a sustainability initiative for the district.

The Gladstone sustainability initiative started with a $40,000 bond measure that was passed in 2006. The district then worked closely with the regional community throughout the implementation of the initiative. This resulted in the development of new partnerships with higher education institutions, statewide sustainability groups, and the renewable energy community. These sustainable building renovations then sparked innovation in teaching and learning. They also resulted in the development of a cross-curricular district committee to infuse sustainability themes across the entire K–12 curriculum that includes environmental, economic, and social equality issues and emphasized project-based learning. Many of the classes became involved with regional sustainability issues that brought students and teachers to environmental conferences.

The district became more committed to sustainability with its new construction projects for an Applied Science and Technology Center at Gladstone High School and the Gladstone Center for Children and Families. It was determined that these would be LEED-certified projects. LEED stands for Leadership in Energy and Environmental Design, and it has become the standard for green building certification. Developed by the non-profit US Green Building Council, LEED certification is based on a set of metrics for building design, construction, operation, and maintenance. Based on the number of points achieved, a project receives one of four LEED rating levels: Certified, Silver, Gold, and Platinum. Committed to LEED certification, the school district selected an architectural firm with experience in sustainable building design, and it assembled a large community advisory committee of experts to shepherd the building projects.

As part of its green building initiative, the Gladstone School District focused on reducing energy consumption in all of its buildings by making energy-efficient upgrades to lighting, HVAC, and other systems. It also installed solar photovoltaic electric systems on three of its buildings, including the new Center for Applied Science and Technology at the high school. These both served as pedagogical tools and provided energy cost savings that eventually paid for the improvements.

The Gladstone School District serves as a model for K–12 sustainability education. In recognition of this, it received a National Green Ribbon Award from the US Department of Education for emphasizing sustainability both in operations and in curriculum. Its approach has been even more instructive. Engaging stakeholders, forming partnerships, and working with local and regional energy experts have made this greening effort one that is truly sustainable.[1]

1 US Department of Education, Green Ribbon Schools. 2012. "2011–2012 Presentation of Nominee to the U.S. Department of Education." March 20. Accessed March 1, 2016. https://www2.ed.gov/programs/green-ribbon-schools/2012-schools/or-gladstone-high-school.pdf

costs, including waste, water, energy use, and alternative transportation; improving the health and wellness of students and staff, including environmental health, nutrition, and fitness; and providing effective sustainability education, including robust environmental education that engages all core subjects, develops civic skills, and prepares students for green careers (Box 10.1). In collaboration with the US Green Building Council, the Department of Education created the Green Strides website (http://www.greenstrides.org), which provides resources to help schools become Green Ribbon Schools.

Education for sustainability is extremely well-developed in other areas of the world. Japan was one of the first countries to integrate sustainability throughout the K–12 curriculum. Sustainability also plays a key role in curricula in Germany and most other European Union countries. Examples include England, where the NGO Sustainability and Environmental Education (SEEd) supports environmental and sustainability education focusing on eight themes: food and drink, energy and water, travel and traffic, purchasing and waste, buildings and grounds, inclusion and participation, local well-being, and global perspectives. SEEd's website (http://se-ed.co.uk/edu) provides numerous resources for teachers and administrators interested in sustainability and environmental education. We examine sustainability education in developing countries as part of sustainable development in Chapter 12.

Sustainability in Higher Education

In addition to the growth of sustainability education in K–12 schools, there has been greater endeavor to incorporate sustainability at institutions of higher education. The content and opportunities may be more sophisticated, but colleges and universities have similar curricular, pedagogical, and facilities and operations greening objectives for sustainability. In addition to infusing sustainability programming throughout entire curricula, as is the goal in K–12, there are focused programs on developing competent sustainability professionals. Colleges and universities have also developed sustainability research priorities, co-curricular education and outreach programs, and living/learning and study abroad programs, and they have made institutional commitments to sustainability.

The sustainability revolution in higher education is happening for a number of reasons. First, at most institutions, there already exists established expertise in areas related to sustainability, especially at land-grant universities that emphasize agriculture, science and technology, and engineering. Second, there has been financial and organizational support of research and teaching in these areas. Third, there has been an increased emphasis on interdisciplinary programs, which lends itself to sustainability education and research. Fourth, there is increased opportunity for making operations more sustainable, given that colleges and universities have large physical plants that include academic, dining, and residential buildings; transportation systems; and grounds. Fifth, although colleges and universities do face accreditation and are held accountable for outcomes, they are much less restricted in terms of specific standards and testing. This provides much more freedom and opportunity than in the K–12 environment. Finally, many institutions of higher education have as part of their mission the teaching of ethical and civic values, an emphasis on critical thinking and leadership, and the goal of having their graduates influence or even transform society in positive ways. They recognize the moral imperative to prepare students in ways that will allow them to contribute to the creation of a just and sustainable future through their professional contributions and through changes in their lifestyle.

Another reason for the growth of sustainability in higher education is that students have been demanding it, and colleges and universities are beginning to be compelled to respond to that demand. The increased competition for limited students and tuition dollars and the recognized importance of preparing responsible citizens and successful employees in the expanding green economy are at least in part driving this growth. In a 2008 survey of more than 10,000 prospective college and university students, almost two-thirds indicated that a college's commitment to sustainability would impact their decision to apply to or attend the school. Nearly one-fourth said this information would "strongly" or "very much" contribute to decisions about which schools to apply to or attend.[11] Other surveys show that incoming freshmen are increasingly more likely to choose their school based on sustainability concerns. A 2016 article in *USA Today* reported that student demand has forced colleges and universities to address climate change and to institutionalize sustainability.[12]

SUSTAINABILITY IN THE CURRICULUM

Sustainability enters the curriculum in two different ways. The first is through the development of sustainability-focused academic programs. The number of undergraduate majors and minors and graduate programs specifically focused on sustainability is on the rise. According to the Association for the Advancement of Sustainability in Higher Education (AASHE), in 2016 there were 1,530 of these programs at 564 campuses in 67 states and provinces in the United States and Canada. Among these programs, approximately half are sustainability-focused baccalaureate degree programs, and half are graduate programs. For those already in the workplace or with degrees, there are numerous certification programs offered by colleges, universities, and extension offices that focus on providing the knowledge and skills needed to connect an individual's current career to sustainability and the growing green economy.

Sustainability focused academic programs have been traditionally housed in disciplinary-based areas of study that emphasize the technical aspects of sustainability with an objective of preparing students for careers in these areas. These exist at both undergraduate and graduate levels in areas such as agriculture, resource management, law, business, architecture and design, public policy, and engineering and technology. Other more general undergraduate programs are more broadly interdisciplinary. They often represent the development of new programs or the revision or replacement of traditional environmental science/studies courses to better reflect all areas of sustainability rather than just an environmental focus. These programs stand alone or complement traditional disciplinary majors. Many of them offer both local and global perspectives and hands-on problem-solving, research, and design components. They also tend to emphasize community involvement. A number of study abroad programs offering students international, research-oriented programs in sustainability have also emerged (Box 10.2).

The second objective of educating for sustainability in higher education is the integration of sustainability throughout the curriculum in order to reach all students regardless of academic interest by broadly infusing sustainability throughout it. As with K–12 education, this has proven to be a major challenge. One way to accomplish this is to encourage faculty to integrate sustainability into existing courses throughout the curriculum that are part of general academic requirements or non-sustainability-related programs. This can be supported through in-house course development grants and faculty learning communities specific to sustainability, such as those at the University of Texas at Austin and many other institutions.

The successful integration of sustainability across the curriculum is made evident particularly by the rise of courses in the humanities with the goal to equip students with the skills and habits of mind necessary to move toward sustainability. Although rare, some colleges and universities mandate a sustainability course for all students as part of their general academic requirements, but this significantly lags behind requirements on diversity, global awareness, and foreign language. Dickinson College in Pennsylvania and San Francisco State University are examples of schools that have sustainability-related courses as part of their undergraduate general education requirements.

To further expose all students to sustainability, many schools offer non-academic sustainability programs to all students, typically through residential housing or offices of academic life. Sustainability has been integrated into new student orientation programming. Even more common are student peer-to-peer sustainability education programs. AASHE reports that nearly 100 colleges and universities currently offer such programs, which are often referred to as environmental resident, eco-rep, environmental ambassador, and residential sustainability coordinator programs.[13] These student-run programs work with campus residents in order to effect behavioral change related to things such as energy conservation, recycling and waste reduction, and composting. Many institutions also offer sustainability living/learning opportunities, community gardens, and energy conservation dorm competitions, and they have student groups focused on all or some aspects of sustainability as part of their co-curricular programs.

SUSTAINABILITY COMPETENCIES

Chapter 1 introduced some essential competencies in sustainable problem solving that were collaboratively developed by the sustainability education community.[14] They have been promoted as key components to and guides for the development of curricula for training the next generation of sustainability professionals that have been adopted by a number of sustainability programs. They include systems thinking, the capacity to anticipate how specific sustainability-related interventions or the lack of interventions may impact the future, and how to strategically create these interventions. It is also recognized that sustainable problem solving requires competencies in values or normative thinking for ethical decision-making based on integrity, justice, and in consideration of diversity and that strong interpersonal skills are needed for collaboration with stakeholders and other experts (Figure 10.4.).

Since the original development of these competencies, there has been much healthy and productive

BOX 10.2

Stakeholders and Collaborators: The School for Field Studies

A key component to educating for sustainability is offering students project-based, global experiences. Study abroad programs that focus on sustainability and offer research experiences can be invaluable in meeting this need. A leader in providing such experiences is the School for Field Studies (SFS). SFS offers semester-long and summer field study programs at its centers in Australia, Cambodia, Chile Costa Rica, Kenya, Tanzania, Bhutan, Peru, Panama, and the Turks and Caicos Islands. Its goal is to create transformative study abroad experiences through field-based learning and research at the human–environment interface.

Figure 10.2.1 SFS students participating in community-based research on biodiversity and resource management in Costa Rica. *Courtesy of School for Field Studies, https://fieldstudies.org/contact/*

One of the major strengths of SFS is its stakeholder approach to community-based research and the strong partnerships it has developed. Every five years, each SFS center develops a five-year research plan through a collaborative process among SFS faculty, local stakeholders and clients, and external research advisors. This research plan both defines the research agenda and provides a conceptual framework for the curriculum at each center. Students engage in high-quality, community-driven field research projects focused on social and environmental problems faced by each community in which the centers are located (Figure 10.2.1). Research results are shared with community members and other stakeholders, providing them with detailed and accurate information for decision-making and action. Many of these projects also result in scholarly presentations and papers. Through its programs, SFS provides unique opportunities for students to study abroad in ways that are consistent with our broad definition of sustainability.

Other sustainability-related study abroad programs include the Sea Education Association's Sea Semester, which offers international field research-based programs at sea aboard its oceanographic sailing vessels. Programs and research themes focus on climate change, cultural sustainability, ocean pollution, and environmental sustainability and policy. The School for International Training offers many sustainability-related undergraduate study abroad programs, master's degrees, and certificate programs. Also, most colleges and universities offer short-term study abroad experiences taught by their own faculty that involve an on-campus component and a shorter travel component. Many of them focus on issues related to sustainability. These programs are ideal for students who believe their program of study limits their ability to participate in semester-long programs abroad, and they offer students an international experience that they might not otherwise have had.

discussion about them. Some of the original competencies have been reworded and new ones have been added, but much of the original spirit remains. Certainly, if the sustainability in higher education movement is going to be sustainable itself, curricula must be readily adaptable as requirements for new competencies emerge. As the movement to educating for sustainability across entire curricula expands, new, more robust, and broadly applicable competencies are emerging. These include a focus on empathy, diversity, health and well-being, behavioral change, and affinity for all life.[15] The Sustainability Curriculum Consortium

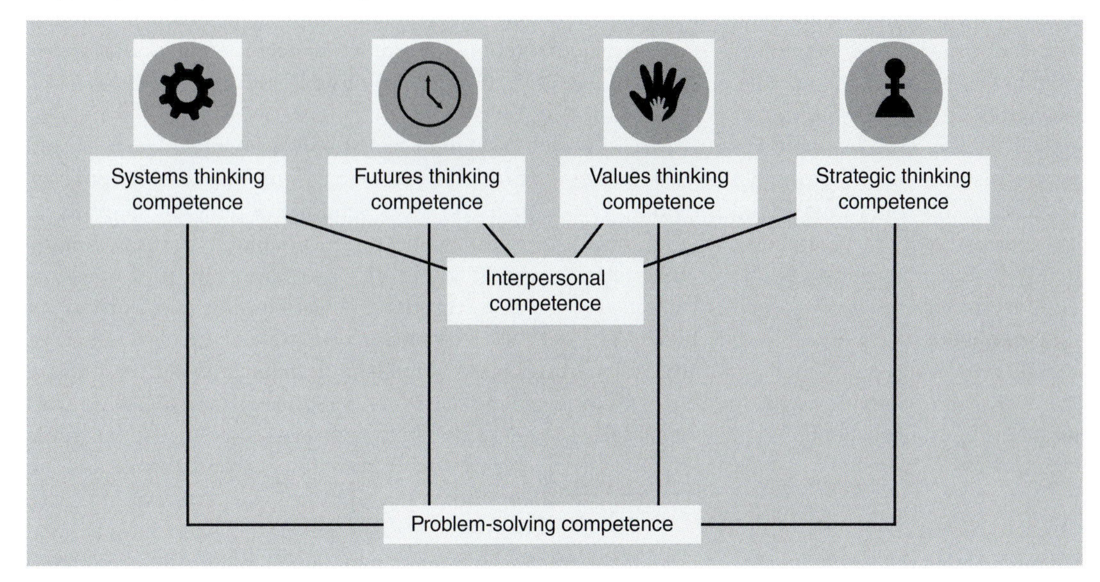

Figure 10.4 The key competencies for problem solving in sustainability as priorities for sustainability in higher education. (Provided by Arnim Wiek [copyright of icons is with Sustainability Science for Teachers, Arizona State University] – based on Wiek et al., 2011 and Wiek et al., 2015.)

(http://curriculumforsustainability.org) is an excellent resource for exploring these competencies and how to integrate them into both sustainability-focused programs and cross-curricular sustainability infusion initiatives in higher education.

RESEARCH ON SUSTAINABILITY

In addition to changes in the academic program, higher education has seen an increase in research in the area of sustainability in faculty undergraduate and graduate research programs. National, state, and local funding agencies as well as private–public partnerships support a variety of sustainability-related research priorities. Each year, the US Environmental Protection Agency (EPA) awards more than $4 billion in funding for grants and other assistance agreements for the advancement of human health and the environment, with water, air quality, and climate change as high priorities. The EPA also sponsors a national student design competition for sustainability focused on energy, built environments, materials and chemicals, and water. The National Science Foundation's Small Business Innovation Research/Small Business Technology Transfer program supports the application of basic scientific discoveries in ways that will offer societal and economic benefits by catalyzing private sector commercialization of technological innovations. These often are in the area of sustainability.

The rise in sustainability research is reflected by the increased number of journal articles on the subject—and this work is beginning to influence practical decision-making. One of the challenges, however, is the interdisciplinary nature of this work. Universities and funding agencies are beginning to prioritize and support interdisciplinary research in ways that break down disciplinary silos. This provides opportunities to draw on the strengths of researchers from specific disciplines, but also allows them to combine and integrate their knowledge around specific sustainability challenges. This can best be achieved when universities provide incentives and financial support for such collaboration. This can include prioritizing new hires in interdisciplinary positions and positive recognition of this type of work in the tenure and review process. The latter can be challenging when the process is housed within traditional disciplinary departments.

One way to facilitate interdisciplinary sustainability research is through the creation of umbrella institutes that bring researchers from different disciplines together for collaboration in both teaching and research. Such "sustainability institutes" are now extremely common. One example is the University of Oregon's Institute for a Sustainable Environment, which is a center for innovative, interdisciplinary research at the nexus of ecological, economic, and social sustainability. The center supports and provides technical support for collaborative research. It also helps researchers connect their work to public policy. The Sustainability Tracking, Assessment & Rating System (STARS) awards considerable credit for this type of institutional support of sustainability research by faculty and students.

SUSTAINABLE CAMPUSES

Because most higher education institutions are facilities- and operations-intensive, greening them is a huge opportunity to increase their contribution to sustainability efforts. This can offer institutions opportunities to save fixed and ongoing costs and to use facilities as a way to model and teach sustainability to students and the broader community, and it represents a significant opportunity to do the right thing by supporting all three areas of sustainability. Priorities include sustainable building operations and management, the design and construction of new buildings that reduce energy consumption and greenhouse gas (GHG) emissions, and building and grounds operations that reduce the use of toxic chemicals, divert materials from the waste stream, and increase sustainable landscaping, purchasing, and transportation. These efforts can be supported by the US Green Building Council's Leadership in Energy and Environmental Design (LEED) certification program. This certification is based on a set of metrics for building design, construction, operation, and maintenance. It offers one of four LEED certification rating levels: Certified, Silver, Gold, and Platinum.

Dining services can be made more sustainable through sustainable food purchasing policies that focus on local vendors, organic or naturally grown products, and producers that treat workers fairly and safely. There are also opportunities to reduce food waste through trayless dining and portion control. Food composting programs, the elimination of disposable utensils and tableware, and the reduction of single-use containers through reusable mug and to-go packaging programs are increasing trends in sustainable dining. A number of campuses promote vegetarian and vegan options and participate in programs such as "Meatless Monday." Others have banned bottled water and provide filtered water at reusable bottle filling stations (Figure 10.5 and Box 4.2). Contracted dining services should be examined for their commitment to sustainable practices, including the previously mentioned criteria, and the treatment of their workers. Most now demonstrate a commitment to sustainability, but one provider, Bon Appétit Management Company, is especially committed to the highest sustainability standards.

The "greening" of campuses is most often driven by campus sustainability committees that bring together representatives of all stakeholders, including students. Most campuses now employ sustainability coordinators who can work on projects such as energy audits, conducting carbon inventories, monitoring and educating about recycling programs, and training student sustainability educators. They can also be the conduit for sustainability-related suggestions from the

Figure 10.5 Muhlenberg College in Allentown, Pennsylvania, was one of the first colleges to limit bottled water use, provide reusable water bottles, and install water bottle filling stations as part of a program they call "Just Tap It." *Courtesy of Muhlenberg College.*

campus community. A seamless connection between the academic and operations sides of the institution will allow for greater opportunity to integrate teaching and research with campus greening efforts. It can also provide opportunities for unique collaborations among students, faculty, and plant operations staff. This may require a structural change in which the sustainability coordinator is located in an academic building or student life building rather than in a plant operations office.

SUSTAINABLE ADMINISTRATION

Commitment and leadership from top administrators and trustees are vital for achieving institutional sustainability. One way that leadership can be demonstrated is through formal pledges. The first opportunity to do this was the 1990 Talloires Declaration. This is a detailed action plan for incorporating sustainability and environmental literacy in teaching, research, and operations that as of 2016 had been signed by more than 500 university presidents and chancellors in more than 40 countries. Launched in 2006, the American College & University Presidents' Climate Commitment to address global warming is a pledge to neutralize institutional GHG emissions and accelerate the research and educational efforts of higher education on climate change. Currently, it has signatures from more than 700 universities and colleges in all 50 states, representing a student population of more than 5.6 million.

On November 19, 2015, the Obama White House launched the American Campuses Act on Climate Pledge in support of a strong international climate agreement in the UN COP 21 climate negotiations in Paris that was signed by more than 318 colleges and universities representing more than 4 million students. Recently, college and university presidents have endorsed a carbon pricing approach to reducing GHG emissions through the #PutAPriceOnIt campaign. In response to the recent withdrawal of the United States from the Paris agreement, college and university leaders have joined mayors, governors, businesses, and investors to declare continued support of climate action consistent with the accord. It is important, however, that these pledges are not used for institutional greenwashing or the mere appearance of taking steps toward increasing sustainability but, rather, serve as guides for action.

Institutional leadership can further the commitment to sustainability by incorporating its principles into mission statements and strategic and campus plans. Numerous institutions have created separate sustainability strategic plans that are frequently based on the criteria of the STARS reporting system. Others have created and adopted formal climate action plans to calculate, reduce, and mitigate the institutions' GHG emissions as per the Presidents' Climate Commitment. The social aspects of sustainability can be enhanced through actions that promote diversity and support faculty, staff, and student body. This can include prioritizing diversity initiatives in a diversity strategic plan and through just compensation and benefits, fair treatment of adjunct faculty, and support for underrepresented groups.

A more controversial approach for administrators and trustees to further sustainability objectives is through divesting institutional endowments from industries that detract from sustainability and reinvesting in socially responsible companies or funds. Endowments consist of donations that are held in perpetuity and are invested to create income for colleges and universities. The average and median endowment for all US and Canadian institutions are approximately $650,000 and $115,828, respectively, but nearly 100 institutions have endowments greater than $1 billion.[16] The earnings from these endowments can represent a large portion of operating funds and the global investment market.

The basic argument for divestment is that ethically driven institutions such as colleges and universities should not hold market positions that undermine the very values of those institutions. The main objectives are to send a signal to the market to slow cash flow into fossil fuel companies and to stigmatizing them. Such divestment is viewed as a moral imperative, and given the overall size of endowments, it could be quite effective in furthering sustainability. In the past, divestment proved to be successful when a large proportion of colleges and universities divested from South African companies in the late 1970s to end apartheid and also when they divested from Darfur in response to the genocide that occurred there in the 2000s.

Currently, there is a movement for educational institutions to divest their endowments from fossil fuel and related companies. This campus divestment movement is typically initiated by students. This has led to the emergence of more than 400 student-driven fossil fuel divestment campaigns on US campuses.

In 2011, Unity College became the first to divest. Currently, according to the organization 350.org, 40 US institutions of higher education among a global total of 124 have committed to either full or partial divestment from fossil fuels. The United Kingdom has the highest number of divestors.[17] Many of these divestment decisions are based on the Carbon Underground and other indices described in Chapter 9.

Although there have been some early movers on divestment of institutional endowment from fossil fuels, it remains controversial. A number of colleges and universities have been reluctant to shift their endowments away from fossil fuels to more sustainable investments despite student demand and the moral imperative to do so. The arguments against doing this have included the perception that divesting the endowment could diminish returns and that there would be substantial transaction costs given the size of these investments. Another problem is that individual investments in fossil fuel companies are typically embedded in multiple, more diverse funds, adding to the complexity of divestment. Also, as long as there are "unethical" investors in the market, those divested shares could be quickly purchased at a discount and sold for a profit. As a result, the market price would essentially stay the same, and the company would lose no money and perhaps not even notice a difference. Still others suggest that some fossil fuel companies are also heavily invested in the development of renewables so that divestment could negatively impact that transition.[18] Despite these concerns, for many institutions and their stakeholders, divestment is an important statement about the ethics of fossil fuel companies and their responsibility for climate change, and for this reason, the divestment movement continues to grow.

Another issue of concern is that limiting the divestment of institutional endowments to the fossil fuel industry neglects other businesses that violate sustainability principles. These businesses include those in industrial agricultural that are responsible for substantial GHG emissions, the private prison industry that promotes mass incarceration, or other businesses that might exploit workers or the environment. An alternative approach to divestment from a specific business or class of businesses that could address this issue would be to move endowments into social impact investment portfolios. One such class of investments includes businesses that are highly ranked in environmental, social, and governance (ESG) criteria. These were discussed with regard to individual investment in Chapter 9.

Many of the institutions that are not divesting from fossil fuels or moving funds into social impact investment portfolios make the case that their large financial commitments to sustainability through operations and the support of sustainability-related research and academic programs are more effective in furthering sustainability compared to divestment. They argue that if the amount potentially lost through transaction fees and perceived reduced returns were invested in actual sustainability initiatives, the positive impact toward carbon emissions reductions would be greater than that through divestment.

RATINGS AND RESOURCES

The growing interest in and demand for sustainability in higher education have led to the emergence of several campus sustainability rankings and a fairly extensive amount of resources to support implementation. The primary organization for both rankings and resources is AASHE, which is a non-profit membership organization with the mission to catalyze higher education to lead the global sustainability transformation. It is also responsible for the primary ranking and assessment system, STARS, which is a transparent, self-reporting framework for colleges and universities to measure their sustainability performance at four possible levels of achievement: platinum, gold, silver, and bronze. Other sustainability rankings for higher education include the Princeton Review's "green ratings," the Sustainable Endowment Institute's Green Report Card, the Sierra Club's "greenest college," and for MBA programs, the Aspen Institute's survey of those that offer programs that have social and environmental emphases.

These rating systems may push colleges and universities to consider or even prioritize sustainability efforts and could have marketing benefits, but there have been a number of criticisms, including the concern that the priorities they set are somewhat arbitrary and too general. Data collection for these ratings can be a substantial effort, which disfavors poorly resourced institutions that do not have the staff to contribute to this effort. Thus, it is problematic that the rating systems do not account for the wealth or endowment of the institutions, which may limit their ability to implement sustainability objectives. However, AASHE has been very receptive to feedback in generating future versions of its rating system, which has allowed the system to evolve throughout the years. Other rating programs, such as those of the Princeton Review and the Sierra Club, base their results on a subset of STARS data categories, which may favor some institutions that have made advancements in those particular areas and disfavor others.

AASHE also provides a major amount of support to further sustainability in higher education. Its annual conference is attended by thousands of higher education stakeholders from throughout the world who share innovations, strategies, and other initiatives that are changing the face of sustainability in higher education. AASHE also offers workshops, webinars, and other professional development opportunities and resources for all aspects of institutional sustainability. It also offers member institutions access to a recently designed research hub that provides extensive resources in a diversity of areas for campus sustainability.

Faculty development programs in sustainability have been shown to have lasting, institution-wide impacts. A long-term study of two such programs, Tufts University's Environmental Literacy Institute and the Piedmont Project at Emory University, showed that these programs have resulted in numerous new and redesigned courses and innovations in sustainability pedagogy. They also resulted in the development and funding of new interdisciplinary, sustainability-related research grants and publications, in addition to increased faculty involvement in local issues, problem solving, and practical actions. An ongoing program sponsored by AASHE at Emory University is focused on training campus sustainability leaders to develop such programs at their home institutions in order to widen these benefits and infuse sustainability throughout their curricula and operations (Figure 10.6). There are also a number of books that offer case studies and approaches that are very effective guides for those interested in greening all aspects of their campus.[19]

Informal Sustainability Education

There are a number of important organizations that serve to broaden the reach of sustainability education through programs that are not part of formal K–12 or academic degree programs. For example, the Cooperative Extension Program based at land grant universities has offices in most counties in each of the 50 US states. These offices provide to the public educational offerings in the areas of agriculture and food, home and family, the environment, and community economic development. The US Forest Service and the Natural Resource Conservation Service offer professional advice, training and continuing education programs, and small grants for sustainability-related projects. They also provide sustainability programing for children and resources for K–12 educators.

Assessment and Keys to Success

Once a commitment to sustainability at an educational institution is made and specific changes, programs, or projects are implemented to meet sustainability objectives, the next step is to assess their effectiveness.

Figure 10.6 In Sustainability Across the Curriculum faculty workshops promoted by AASHE, faculty from all disciplines are given the opportunity to explore how their courses connect to sustainability. *ProStockStudio/ Shutterstock.*

Sustainability education offers excellent opportunities for authentic assessment. Well-established methods of learning assessment can be applied to sustainability educational goals, and quantitative metrics are readily applied to operational aspects such as reduction in food waste, increased recycling, and reduction in energy use and climate emissions.

In K–12 schools, in which environmental and sustainability education is taught across all grade levels in an interdisciplinary manner as in the case of Washington state, the assessment of how well students meet sustainability standards can be integrated into core content assessments. The challenge is developing performance indicators for sustainability education. A good resource for this is the Cloud Institute for Sustainability, which has developed sustainability learning indicators that are aligned to specific standards, such as Common Core or Cultural Competency.[20] Institutional sustainability can also be indirectly measured through surveys of faculty, students, and parents. Sustainable Schools (http://sustainschools.org) provides sustainability questionnaires for both K–12 schools and colleges and universities to assess institutional sustainability in specific areas, such as curriculum administration, facilities, and outreach.

Most higher education institutions apply certain criteria areas from the STARS program or the Presidents' Climate Commitment to develop indicators to assess the extent to which they are meeting sustainability. These criteria areas include curricular and co-curricular education; operations; climate change mitigation and energy reduction, purchasing, waste, and transportation; dining services; energy; purchasing; and planning and administration. Many of these indicators are simple metrics based on specific goals, such as reducing GHG emissions by 10%, reducing food waste by 10%, or increasing the number of sustainability-related courses by 5%.

These indicators can then be used in a process referred to as **benchmarking**. This is a tool for comparing the environmental and sustainability performance of different organizations or units within an organization. For example, in terms of operations, energy consumption per square foot can be compared among buildings and among averages at similar-sized institutions, showing opportunity for improvement. In terms of education objectives, one could benchmark the percentage of graduates who have taken at least one sustainability-focused course or the number of majors in sustainability-related programs.

A more complete assessment of sustainability curricular and co-curricular education is often more sophisticated than benchmarking or the use of simple metrics. Penn State University is currently conducting a university-wide assessment of sustainability across all curricula. In addition to quantifying the number of sustainability-focused and sustainability-related courses, all faculty were surveyed about the extent to which they address environmental, social, and economic issues in their courses. Penn State has developed a formal rubric with specific criteria to first categorize the courses as sustainability-focused or -related and further describe the content, format, and assessment methods used in sustainability education courses. Among the criteria assessed are action skills development, environmental responsibility, and cultural diversity.[21]

In addition to surveying faculty about their courses, student learning can be assessed through either direct or indirect means. **Direct assessment** of student learning is accomplished by examining samples of student work that are either already part of a course or embedded in a course for the specific goal of assessment. Examples of the work to be assessed include targeted objectives reflected in exam questions and also student papers, presentations, or portfolios related to sustainability learning objectives. **Indirect assessment** is gathering information about student learning by examining indicators of learning other than student work output. For example, to assess the effectiveness of co-curricular sustainability education, all students could be surveyed regarding their attitudes, knowledge, and behavior in relation to sustainability upon arrival on campus and again in their final year in order to gauge improvement.

Increasing sustainability in educational institutions can have its barriers, but there are certain keys to success. A plan for sustainability must be an inclusive process that allows for participation of all stakeholders. At Wellesley College, the Sustainability Advisory Committee solicited ideas from the entire college community at a public "IdeaFest" and worked closely with key stakeholders across campus. Educating stakeholders that sustainability goes well beyond environmental issues and encompasses social justice, diversity, wellness, and workers' rights will broaden support for sustainability initiatives.

A strong case for the financial benefits of greening efforts is a valuable tool for pushing the agenda, as is showing the evidence that there is strong student and family demand and potentially novel donor interest for sustainability initiatives. Integrating sustainability into the curriculum and engaging all students, especially those not already sustainability-minded, are essential for achieving the substantial level of change required for a sustainable future. Finally, stakeholders and especially administrators need to be convinced that sustainability education is the right thing to do, it reflects the values of our intuitions, it enhances and does not detract from broader educational priorities, and it is essential for a sustainable future for all people and the planet. This is best achieved by "selling the sizzle," or selling your vision for sustainability in a positive manner rather than creating fear and painting pictures of doom.

Corporate and Business Institutions

Businesses and corporations are institutions as we have defined them. They are highly structured organizations with purposes that transcend individuals and pursue a particular endeavor, specifically the production and distribution of goods and services. They range from small, privately held owner-operated companies to publicly held multinational conglomerates.

We cannot ignore that businesses and corporations have been the primary drivers of unsustainable living and development. Corporate institutions have operated on models of exploitation of the environment, resources, and people in order to maximize profits. They have ignored externalities, and they have promoted mass consumption of resources and the production of waste. Corporations have had a long history of influencing governments in order to continue these actions without hindrance or repercussions, and they continue to do so. They have promoted conflict and war, and they have created a political climate change denier movement among politicians despite near unanimous global scientific evidence for it. Corporations have exerted influence over agricultural, energy, transportation, environmental, education, and international economic and development policy with the sole purpose of furthering their business interests and expanding profits.

This traditional model of exploitive, institutional capitalism has been anything but sustainable. However, we are beginning to see institutional transformation as businesses are becoming a major force in the promotion of sustainability. Throughout this book, we have considered the important role of innovation and entrepreneurship in sustainability and sustainable development. In most cases, these were small for-profit or non-profit businesses that developed sustainable solutions to specific problems. We have also considered the role of larger businesses in public–private partnerships and as stakeholders in sustainability-related programs and projects. This has included collaboration with educational institutions and their role in effecting policy change that results in sustainability outcomes and direct contributions to sustainable development. We have also seen businesses begin to consider sustainable industrial ideas, such as zero waste, vertical integration, and conversion from a linear to a circular economy.

Ray Anderson, the CEO and founder of Interface Carpets, has led the transition to sustainable business and is the author of *Confessions of a Radical Industrialist: Profits, People, Purpose—Doing Business by Respecting the Earth*. He and a growing number of corporate leaders believe that businesses will and must lead the sustainability revolution. Anderson argues that there is only one institution large enough, powerful enough, pervasive enough, influential enough, and wealthy enough to lead humankind out of the mess we have made and put us on the path toward sustainability. That institution, business and industry, happens to be the very one that got us into the mess we are in, but Anderson argues it is now poised to lead the sustainability revolution.[22]

In this section, we consider what it means for a business to be a sustainable institution, what might motivate a business to be so, and how this might best be achieved. By examining specific examples, we establish criteria for what makes a business sustainable within the context of our definition of sustainability, including corporate social responsibility. Tools, resources, and new approaches to business that increase sustainability are examined as well. Finally, we consider ways that current business paradigms can be transformed in favor of sustainability.

Sustainable Business

The term "sustainability" has been applied to business for a very long time. Historically, it referred to a business's economic viability or bottom line, which is the consistent ability to turn a profit. Relatively recently, the application of the term sustainability to business has come to incorporate economic, environmental, and social aspects. Now sustainable business refers to the triple bottom line (TBL), a phrase that was first coined in 1994 by John Elkington, the founder of a British consultancy called SustainAbility. The idea is that companies should be focused on three different bottom lines: (1) the traditional measure of corporate profit; (2) a measure of social responsibility, or how its operation treats or impacts people; and (3) a measure of the company's environmental impact or responsibility. TBL is thought to consist of three P's: profit, people, and planet. It aims to measure the financial, social, and environmental performance of a corporation over time and whether or not it is taking into account the full cost involved in doing business.

For a business to be sustainable, it must first recognize that its long-term interests are intellectually and financially consistent with resource efficiency, proactive health and safety practices, and responsible leadership. According to *Corporate Knights* (http://www.corporateknights.com), a business magazine that researches and rates corporate sustainability performance and promotes "clean capitalism," a sustainable business creates more wealth than it destroys. It does so by being grounded in human, financial, natural, and social capital. This can best be achieved by making supply chains, operations, transportation, and workplaces more sustainable and also through the design of sustainable products and services.

SUPPLY CHAINS

The network of organizations, people, activities, and information involved in extracting and transforming natural resources and raw materials into a finished product or service that is delivered from a supplier to the end consumer is known as a **supply chain**. An important way for a business to be sustainable is to maintain environmental and social standards throughout its entire supply chain. For example, as of 2015, Unilever, a large producer of household products, purchased all of its palm oil (more than 0.5 million metric tons per year) from certified, traceable sustainable sources. Non-sustainable production of palm oil is linked to major issues such as deforestation, habitat degradation, climate change, animal cruelty, and indigenous rights abuses, mostly in Indonesia and Malaysia. Now joined by Cargill, a larger trader of agricultural commodities, their combined practices have resulted in the conversion of more than 50% of all palm oil production to sustainable methods. Similarly, Mars Candy is developing genetic techniques to more than triple its cocoa production while cutting the land require for that production by 60%, significantly reducing the impact of that portion of its supply chain. Moreover, it is making this technique available in the public domain because it wants all companies to have access to more productive and more sustainable cocoa.

Improvements in sustainability by making sound choices about procurement along a supply chain may be furthered through vertical integration. **Vertical integration** is an arrangement in which the supply chain of a company is either partially or wholly owned by the company delivering the product or service (Figure 10.7). Vertical integration increases operating effectiveness and efficiency, and it reduces transaction costs by combining the acquisition of raw materials, manufacturing, and distribution capabilities into one unified organization. It can also allow a company to better monitor worker conditions and environmental impact at each stage of the process.

Vertical integration most effectively increases sustainability when the integrated processes occur in close geographical proximity to each other and when the waste produced during manufacturing can be cycled back into earlier parts of the chain. This reduces transportation and fuel use, but more important, it increases the possibility for consistent oversight of each stage of production in order to ensure sustainability. Although generally vertical integration is common in business, steps in the supply chain are less frequently in close proximity to each other. However, this is becoming more common in the food industry and with small-scale manufacturers. The location of parts of the supply chain in close proximity to each other even when not owned by the same corporation, referred to as **geographical agglomeration**, can further increase the sustainability of vertical integration.

Another approach of using supply chain integration to increase sustainability of manufacturing is referred to as **industrial ecology**. However, this takes integration one step further by designing industrial systems in order to allow for the exchange of materials between different industrial sectors, where the waste output of one industry becomes a raw material of another. The objective of industrial ecology is to change from more traditional linear supply chains

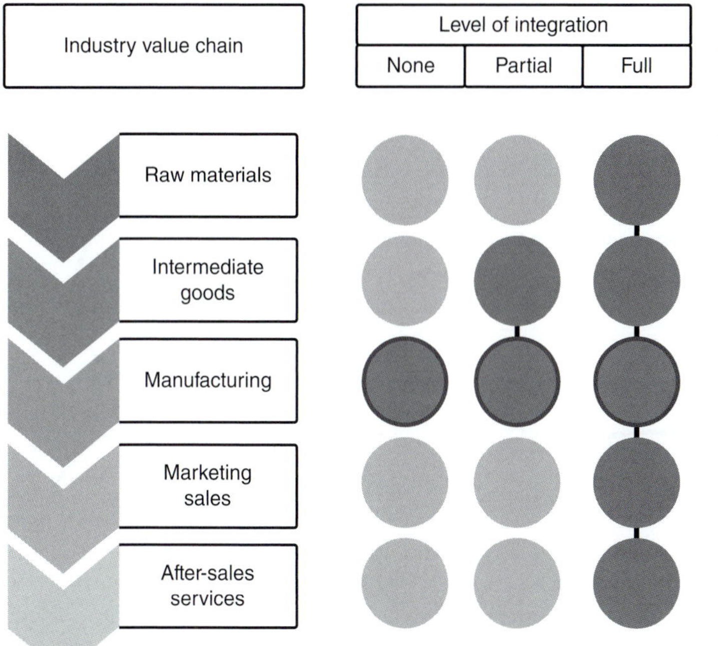

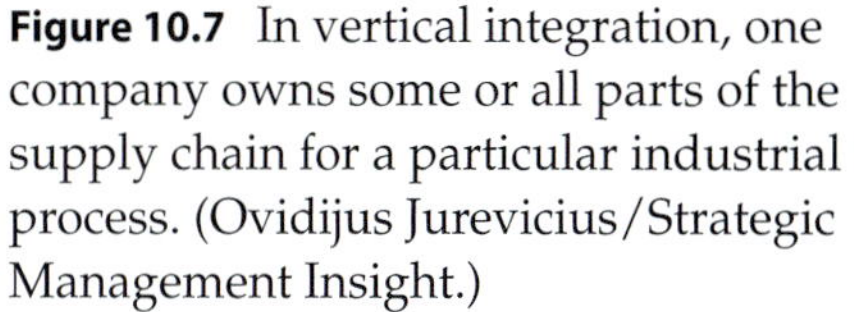

Figure 10.7 In vertical integration, one company owns some or all parts of the supply chain for a particular industrial process. (Ovidijus Jurevicius/Strategic Management Insight.)

to closed-loop systems that operate similarly to natural ecosystems. This was modeled on a small scale in Chapter 5 in the discussion of the Plant Chicago project (see Figure 5.8). A large-scale example of such an industrial ecosystem is in Kalundborg, Denmark; this ecosystem links multiple industrial processes in closed-loop systems (Figure 10.8).

FACILITIES AND OPERATIONS

A major part of building a sustainable supply chain is operational innovations that increase energy efficiency, reduce fossil fuel use and GHG emissions, and reduce waste. Much like educational institutions, increasing the sustainability of facilities can reduce upfront and ongoing costs. Green buildings and workplaces save money, provide healthier and more productive working environments, and offer easy opportunities to demonstrate corporate responsibility. Thus, many businesses have made sustainable management of facilities a priority, including the use of LEED certification. Building objectives include increased energy and water-use efficiency and reuse and the incorporation of facilities such as showers and storage units, which encourage bicycling to work. Sustainability improvements in the workplace, along with changes in culture and technology, have increased productivity and possibilities for collaboration and telecommuting even over long distances.

In addition to more sustainable facilities, improvements in operations have also increased sustainability. For example, FedEx was able to reduce its fuel consumption by approximately 40% while increasing its capacity more than 20%. This was achieved through the company's Fuel Sense Program, which involved converting its air fleet to more fuel-efficient aircraft and employing hybrid and smaller, more efficient vehicles on the ground. The company also developed software to help optimize schedules, routes, and fuel consumption, and it now powers a number of its distribution hubs with 1.5-megawatt solar-energy systems. FedEx was so effective at saving energy that it developed a stand-alone consulting business to help other companies increase their energy efficiency; this business generates additional profit for FedEx. There is also growth in water reuse and reclamation systems in industry. For example, Levi Strauss has developed a process for using 100% recycled water in parts of its jeans production, providing cost savings and reducing the impact on the world's water resources.

Waste reduction or elimination is another key component to making operations more sustainable because it can divert waste from the garbage stream and

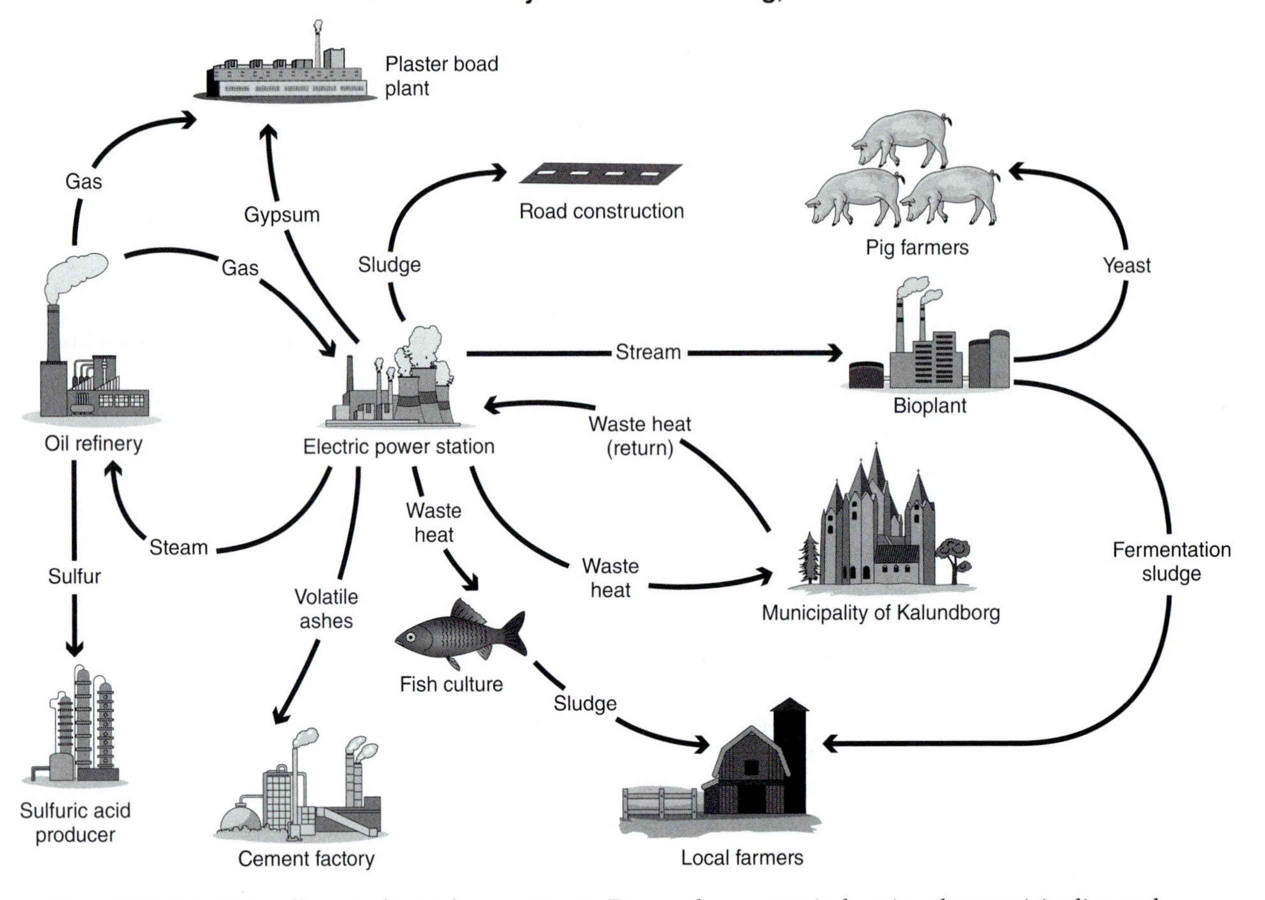

Figure 10.8 The Kalundborg industrial ecosystem in Denmark connects industries, the municipality, and local farmers in closed-loop systems in order to turn waste into resources for each other. (Nagilmer.)

decrease resource depletion with reuse and recycling. The practice of redesigning industrial and institutional systems and resource use so that all products or their components are reused, thereby resulting in no waste being sent to landfills or incinerators, is referred to as zero waste. As discussed in Chapter 8, a number of businesses are moving toward zero waste in order to eliminate landfill costs and allow for more efficient production. In early 2016, Unilever revealed that more than 600 of its sites across 70 countries have achieved zero waste to landfill status.

Reductions in product packaging and single-use, disposable products are also gaining traction in the business world. Reuse and repurposing returned products and exchange programs with upgrades are also an important aspect of waste reduction. In the United States, returns reduce corporate profitability by an average of approximately 4% per year, so there is much interest in reusing them either through sale as discounted, refurbished products or by recycling useable parts. Device trade-in programs associated with electronic upgrades are also on the rise. For instance, through Verizon Wireless' device trade-in program, more than 420 tons of e-waste was kept out of landfills in a single year through reuse of devices and/or harvestable parts. These types of efforts are consistent with the movement of our linear cradle-to-grave economy to one that is circular or cradle-to-cradle.

SUSTAINABLE PRODUCTS AND SERVICES

As the demand for eco-friendly and socially just products and services increases, businesses are responding. There has been a rapid rise in the number of sustainable product lines. To determine if a product is sustainable, one must examine both the product life cycle and the overall operations of a company. One example is the Green Works line produced by Clorox, which—unlike the company's traditional product line—is produced from renewable resources, is biodegradable, and uses no fossil fuel–based chemicals. This product line has contributed to the major expansion of the natural home cleaning market. One of the criticisms of so-called "natural" product lines is that no standard definitions or criteria have been established for them; this allows companies to assess the sustainability of their own products against their own standards. Also, these sustainable product lines may not reflect the overall operations and product manufacturing, which might be inconsistent with sustainability criteria.

In addition to sustainable products, there has been a rapid expansion of businesses that offer services that increase sustainability. This is particularly important because the service industry dominates much of the economy in developed countries. One example is the car share business, in which people rent cars for short periods of time. Car sharing is attractive to customers who make only occasional use of a vehicle. It has been shown that car sharing reduces car ownership and vehicle miles traveled, and it increases walking, biking, and the use of public transportation. Much like bike share programs, short-term car sharing is often point-to-point so that cars can be picked up or left at reserved, preferred parking spaces. It is estimated that for every car-sharing vehicle on the road, at least five private vehicles are removed.[23] Sustainability is further increased when those cars are electric. Autolib', an electric car sharing service in Paris, provides electric cars that are parked throughout the city at dedicated parking spots with charging stations. There has been slow but steady growth in car share business profits, with total US revenues of approximately $400 million in 2014. Although this is only a fraction of the traditional car rental market, many see huge market potential, especially in urban environments. Great market potential is seen in many other emerging sustainable service businesses, including peer-to-peer product sharing, and sustainable lawn care, dry cleaning, and food service industries.

NEW BUSINESS

In addition to greening existing product lines and services, and making current industrial operations more sustainable, there is a need for innovation and new technology that can move us toward sustainability. Entrepreneurship has led the way in developing new technologies and businesses that challenge conventional wisdom and create sustainable solutions to current problems. When this type of innovation creates a new market and eventually displaces established competitors, it is referred to as **disruptive innovation**. Many believe that disruptive innovation is the key to a sustainable economy, and examples include innovations in renewable energy, three-dimensional printing, and electrical vehicles and fuel cells. One new innovation that has the potential to be disruptive comes from a start-up company called Calera. It has developed technology to extract carbon dioxide from industrial emissions and bubble it through seawater to manufacture cement. This both reduces GHG emissions from power plants and other polluting enterprises and minimizes emissions during cement production. But in order to make a significant contribution to sustainability, this technology must be disruptive by displacing current means of cement production.

By definition, disruptive innovations start out small and eventually displace dominant competitors. The question is what role established industries can play in generating sustainable technologies. Entrepreneurial activity within an existing, large company is referred to as **intrapreneurship**. The problem is that large corporations are often managed in ways that do not encourage employees to take risks as intrapraneurs. Changes in management that are more likely to generate innovations that could contribute to sustainability include those that prioritize innovation, networking, and collaboration; create time for employees to pursue new ideas; provide funding for intrapeneurial activity; and eliminate management structures that inhibit risk-taking.

Successful examples of intrapraneurs include Luis Sota, who worked with the Mexican-based cement company Cemex to develop low-income housing solutions; Bob Annibale at Citigroup, who worked to develop microfinancing aspects of the company; and Andreas Eggenberg, who while working for Amanco, a leading pipe manufacturing company, led the development of affordable irrigation systems for low-income farmers.[24]

Another approach to stimulate intrapraneurial innovation is when large companies collaborate with universities or small start-up companies. For example, Shell Oil formed a collaboration with Saint-Gobain

Glass Company to develop photovoltaic (PV) solar panels using thin-film technologies. General Motors has partnered with the small company Coskata to develop new processes to cheaply make ethanol from agricultural and household waste.

PHILANTHROPY

Another way that corporations can contribute to sustainability efforts is through philanthropy. In 2013, Richard Branson, founder and CEO of Virgin Group, pledged that all profits from Virgin Transportation until 2021 will be devoted to the development of renewable energy. Media businessman Ted Turner has supported causes such as water quality, sustainable energy, biodiversity, and wildlife protection with more than $1.5 billion in donations. With approximately $30 billion in assets, the Bill and Melinda Gates Foundation is targeting US education, childhood deaths, malaria, polio, AIDS, and agriculture in poor countries. Increasingly, more modest donors are also contributing to sustainability efforts.

In many ways, this "green philanthropy" is used to build a sustainable legacy for these donors and their businesses, but some argue that although they are helping sustainability efforts in developing countries, these businesses may be neglecting or contradicting sustainability efforts in other areas. For example, there are businesses that generously support sustainability efforts in developing countries, whereas at home, the single bottom line leads to less worker flexibility, reduced benefits, and extreme income inequity between the highest and the lowest paid employees. Also, the Gates Foundation, which has generously devoted so much to sustainable development in the poorest countries, has no program to help prevent global climate change, which will disproportionately impact those same countries. Philanthropy alone does not make a business sustainable unless the same principles are woven through all of its actions.

What Motivates Sustainability in Business?

Sustainability is not necessarily considered counterproductive to profit objectives of corporations as it once was. Now, not only does sustainability offer potential to lower costs and increase revenues through new product lines and businesses, as described previously, but also the TBL approach is viewed as a major method of achieving competitive advantage and as an important driver of innovation. There are a number of reasons why this is the case and why nearly all corporations make the case that they are sustainable or actually commit to meeting sustainability objectives. Reasons for this are addressed here, and ways to differentiate between the two are presented later.

CONSUMER CHOICE

The number of consumers who seek out products and services that have better environmental, social, and health attributes has been steadily rising, and studies show that the majority of consumers feel personally accountable to address social and environmental issues as they purchase products. According to a 2013 Cone Communications Social Impact Survey,[25] 88% of Americans say that they would buy a product with a social or environmental cause or benefit, 89% are likely to switch brands to one associated with such a cause given comparable price and quality, and 54% did so in that year. In addition, 91% want even more products and services that are safer for health, environment, and that come from socially responsible companies; the respondents indicate that low availability of these products is the greatest barrier to purchasing more. This type of conscious consumerism has risen by 170% in the United States since 1993, and it is even higher in other areas of the world. This is driving many companies to produce more sustainable products, to operate more sustainably, and to market themselves as environmentally and socially conscious businesses.

INTANGIBLE ASSETS

Tangible assets are the sum total of the physical capital of a company, including buildings, equipment, and inventory. Intangibles or non-physical assets include people, relationships, knowledge, intellectual property, goodwill, brand recognition, and a company's reputation. Until the 1950s, approximately 80% of most business assets were tangible and 20% were intangible. Since then, however, the contribution of intangibles to the total value has steadily risen and in many cases is greater than the total of physical assets. For example, the success of a company such as Coca-Cola is highly dependent on its brand-name recognition and how the company is perceived by consumers. This rise in intangibles is also referred to as an increasing goodwill-to-assets ratio. The increasing importance and global influence of goodwill, branding, and reputation have motivated businesses to adopt more sustainable practices and product lines.

PRESSURE FROM INVESTORS

The chief executives of the world's largest investment firms and the financial sector as a whole have called for companies to build environmental, social, and governance management into their business models. This is because more investors are searching for firms with sustainable long-term strategies in place. This comes from the recognition that over the long term, environmental, social, and governance issues including climate change have real and quantifiable financial impacts. This has been a major force in driving the growth of sustainable business.

REGULATORY ENVIRONMENT

Environmental regulation may force companies into being sustainable. This has the potential to create opportunities for companies with sufficient foresight. Rather than adhering to the lowest environmental standards for as long as possible in order to reduce costs, it might be in a company's best interest to comply with more stringent standards if they are likely to be enforced in the future. This is especially true for product lines such as automobiles that take two or three years to develop and bring to market. For example, if an automobile company would have anticipated and adapted to emission standards in California when first proposed 2002, it would now have an extremely competitive advantage because the standards became law in 2016. In this way, environmental regulation can drive innovation, as businesses that focus on emerging norms have more time to develop technologies to meet them and come out ahead. This was the case when Hewlett-Packard realized that lead solders would eventually be banned due to toxicity, and it worked to develop alternatives that could go to market as soon as the ban occurred.[26]

Anticipated regulation with regard to reduced GHG emissions has the same potential to lead to a new green economy focused on renewables. Unfortunately, the largest companies are not leading this low-carbon revolution, perhaps because they have doubts about or are actually influencing (through political contributions) whether GHG emissions will ever be regulated, at least in the United States. Another reason may be that larger companies tend to be less nimble and are challenging to transform. Large firms often initiate sustainability programs only to disaffirm them later. For example, although Ford initiated a smart mobility fuel efficiency campaign, its least fuel-efficient model, the F-Series truck, continued to be its most heavily marketed and best-selling vehicle.

Innovation that will lead to the transition to a carbon-free, renewable energy–based economy will most likely come from smaller entrepreneurial companies such as SunPower, which produces high-efficiency crystalline silicon PV cells, roof tiles, and panels, making it one of the top US solar companies. Former president of Shell John Hofmeister believes that if and when the US government begins to lead the transition to a low-carbon economy and regulate carbon emissions, larger companies will begin to make the transition, as has occurred in other countries with carbon emission standards already in place.[27]

Assessing Sustainable Business

It is difficult to identify a business that does not make sustainability claims. Diverse corporations such as Halliburton, Ascent Solar Technologies, McDonald's, and Organic Valley Farms all make substantial sustainability claims on their websites, through other forms of marketing, and in their annual reports. So how do we best assess these claims as we make product and investment choices, and how do we offer suggestions to make these companies more sustainable or even hold them accountable? How do we differentiate companies employing greenwashing or deceptive sustainability marketing from companies that are truly committed to sustainability?

When assessing corporate sustainability, it is important to assess all criteria from our definition of sustainability over the range of all of a corporation's supply chains, facilities and operations, products and services, and corporate philanthropy. It is also important to determine whether sustainability objectives move beyond hopes, visions, and baseless claims toward real metrics in each of these areas. Assessment should not be limited to reviewing policies stated on the company's website or other marketing materials. Rather, the process of evaluating the sustainability of a company should consider third-party assessments, awards and recognitions for sustainable practice, affiliations or partnerships with established sustainability-related organizations, and sustainability rating systems such as those used for educational institutions. One should examine the extent to which sustainability is woven through the entire fabric of the institution and its business model as the ultimate achievement in sustainability.

THE TRIPLE BOTTOM LINE

As we assess a business for sustainability, we must consider all of the criteria from our definition. This includes not only the economic, social justice, and environmental bottom lines but also the principles of participatory processes, good governance, and measurable results. The aspects of the institution that should be assessed are its initiatives, actions, and outcomes in terms of its operations and products and services, not simply its message. It is essential to critically examine all aspects because a company may do quite well in one area that it markets as an overall success in sustainability, but it may have a less than stellar record in other areas. Here, we use Walmart as an example to demonstrate these sorts of disparities.

Walmart actively markets its comprehensive sustainability and global responsibility program that focuses on energy, waste, products, and sourcing. However, its own language employs words such as "envisioning," "aspirational goals," and "expectations" for each of its sustainability objectives more than real metrics of achievement, and a closer examination reveals that the company is doing better in some areas and not so well in others.

Walmart is probably the most efficient company in the history of American business. It has had success in converting to renewable energy, with 26% of its global electricity use coming from renewables. Although its stated goal was 100% and other companies are doing better percentage-wise, because of its size, Walmart has deployed more on-site solar capacity than any other US company. In so doing, it has made significant reductions in carbon emissions. Walmart has had even greater success in waste diversion from its stores, and it proposes to ultimately have its stores approach zero waste; however, this is contradicted by the sale of billions of dollars of cheap, throwaway goods, often with excessive packaging. Walmart has the stated goal to sell greener products, but this is difficult to assess because its approach to product sustainability is not transparent, especially because its supply chain is mostly in China and other less developed countries. However, Walmart is a major supplier of organic and local foods, energy-efficient lighting, and other environmental products. Making these and other products more affordable has both environmental and social benefits.

Walmart's environmental initiatives, even when they have lower rates of achievement than those of other businesses, do have a significant beneficial impact because of their scale. However, in other realms, the company does not do as well. Its efficiency and low prices, which are considered to be on the positive side of the sustainability equation, do come at a cost. Low product prices result from global sourcing, and because of Walmart's purchase volume, the company has been able to squeeze suppliers to produce goods very cheaply, making it increasingly challenging for them to be profitable. Thus, suppliers have been forced into reducing already low wages to stay in business.

Walmart's global sourcing of products has also come at the cost of US manufacturing jobs. A report shows that three local jobs are lost for every two Walmart creates, and the company reduces retail employment by an average of 2.7% in every county it enters. In the United States, Walmart is well known for paying its own employees extremely low wages with minimal or no benefits such as health care. Depending on the state, there is a 26–30% wage gap between Walmart and other large retailers. According to a recent Human Rights Watch report, Walmart has a record for aggressive anti-union tactics, some of which are illegal, including spying on employees and creating a climate of fear among them. Also, because of its low pricing and aggressive positioning of its retail stores, Walmart is viewed as a detriment to communities in which it outcompetes and causes the closure of small independent businesses, essentially destroying local economies.

As can be seen from the Walmart case, assessing business sustainability is tricky. Because of its tremendous scale, its improvements in environmental performance have really made a difference, and Walmart should be acknowledged for that. Assessing global supply chains is often challenging. Although Walmart does provide products—some of which are green—at very low prices, it seems to be doing more social damage than good. It is easy to pick on Walmart, and in some cases, criticisms of the company are considered classist because its market is the lowest income bracket. However, this business model based on efficiency and global sourcing from lean supply chains has come to dominate all sectors of the economy. Many higher end stores, internet shopping, and service and food franchises use the same model and have similar negative sustainability consequences, all of which thrive on consumer choice.

There is an alternative model that is more sustainable across all criteria in which the business can generate a profit while providing higher quality, greener products and embracing a commitment to environmental stewardship, superior wages, and worker treatment. These businesses have integrated sustainability into every aspect of their operation, service, and product offerings, and they are still able to generate a profit. Sustainability is essentially in their DNA. In some cases, these are companies that have existed for a long time and made the decision to change course toward sustainability. An example is Interface Company, whose founder, Ray Anderson, decided in 1994 to completely embrace sustainability and make a momentous shift in the way the company operates its business and views the world. The result is a carpet system composed of replaceable tiles that are 100% recycled, incorporate biomimicry in their design pattern, and are manufactured using a climate-neutral process while maintaining high human rights and labor standards.

Other companies have been built based on the concept that sustainability and profitability go hand-in-hand. Their founders had the vision to create new businesses that tied together profitable growth with environmental and social responsibility. An example of a large multinational company is IKEA, which takes the group approach to sustainability by designating dedicated sustainability leaders for each business area (products, customer engagement, supply chain, workers, and community) who are supported by a central group of sustainability specialists. A number of smaller businesses started out with sustainability at the core of what they do while yielding double-digit rates of annual growth. Companies such as Green Mountain Coffee, Stoney Brook Farms Yogurt, Common Threads Garments, Patagonia, and New Belgium Brewing Company (Box 10.3) have shown that sustainability can be an opportunity for success.

Many of these businesses share best practices in an effort to both show the world that the triple bottom line is achievable and transform the consumer goods and service industry into one in which sustainability is central. They view this transformation as "the next industrial revolution"—one that is essential for our future. There are some organizations that can help provide support both in getting started and in assessing sustainable businesses. One is the Sustainability Consortium (http://www.sustainabilityconsortium.org). Members of the Consortium include manufacturers, retailers, suppliers, service providers, NGOs, civil society organizations, governmental agencies, and academic institutions. Each brings valuable perspectives and expertise to work collaboratively in building and making available science-based decision tools that address sustainability issues throughout a product's supply chain and life cycle. They provide specific product category metrics for assessing sustainability and have developed standards for sustainable products.

THIRD-PARTY ASSESSMENTS AND RATING SYSTEMS

Dozens of different rating and ranking systems have emerged that have their own unique methods to assess corporate and product sustainability. Many argue that this makes things complicated and confusing, and what are needed are simpler, more transparent systems. It is agreed that rating systems do play a role in improving the performance of individual companies, but much like sustainability in education rating systems, providing information can be time-consuming and costly. However, the number of rating systems is much higher in the business world, and there is often significant overlap, a lack of specificity, and unclear distinction among them.

A "rate the raters" research initiative by the consulting organization Sustainability identified a number of qualities of good rating systems. Rating systems that are most highly valued tend to be those that are the most transparent in their methodology and simple to understand. Those that not only focus on current and past performance but also assess a company's position to deliver sustainable value in the future are also favored. Some of the most preferred ratings are the Carbon Disclosure Project and Climate Counts, which aim to motivate companies to disclose their environmental impacts and offer a record of their efforts to address climate change. The Dow Jones Sustainability World Index and Ethical Investment Research Services (EIRIS) offer corporate sustainability reports and rankings mostly from an investing perspective that track stock performance of the world's leading companies in relation to economic, environmental, and social criteria. A number of sources offer ratings of companies and investment funds based on ESG metrics. There are also a number of easy-to-use consumer sustainability guides and tools, which were discussed in Chapter 9.

BENEFIT CORPORATIONS

A corporation that has a positive impact on society, workers, community, and the environment can

BOX 10.3

Innovators and Entrepreneurs: New Belgium Brewing Company

New Belgium Brewing Company, a craft brewery located in Fort Collins, Colorado, has infused sustainability into its business model from the very beginning. Founded in 1991 by Jeff Lebesch and Kim Jordan, New Belgium has successfully produced world-class craft beers while proving that business can be profitable and promote the core values of sustainability, both of which have been central to their strategic plan since they started the company. The result has been a business that has experienced annual double-digit growth, making it the third largest craft brewery and the eighth largest overall brewer in the United States, thus proving that a business can be very successful while being a positive force in society.

Figure 10.3.1 New Belgium Brewery a certified B Corp. *Courtesy of New Belgium Brewery, http://www.newbelgium.com/Sustainability/Stories/sustainability-stories/2014/09/15/We-like-beer-But-do-you-know-what-we-love-more*

Now a Certified B Corporation (Figure 10.3.1), New Belgium is a model and recognized for its efforts as a sustainable business. Included among these efforts are its 100% employee ownership, 99% waste diversion rate, water conservation and reclamation, local sourcing of ingredients, energy conservation, renewable energy use, and reductions in greenhouse gas emissions—all of which are carefully measured and documented. New Belgium's Fort Collins brewery was built on a reclaimed urban brownfield, as is its newer brewery in Asheville, North Carolina, which is LEED certified as well. New Belgium also has an impressive record of philanthropy, environmental advocacy, commitment to community, and promoting cycling as healthy recreation and transportation.

Always seeking to improve, New Belgium is conducting extensive waste audits in an effort to identify barriers in its pursuit of zero waste. It produces at least 13% of its electricity on-site and has upgraded lighting and cooling to improve energy efficiency by 50%. The company produces its own energy using a 96-kilowatt solar photovoltaic array with nearly 1,200 solar panels on top of its buildings. It also powers its brewery with biogas that is produced as the company treats all its waste water using innovative technology that produces methane gas as a by-product, which is reclaimed to generate electricity. New Belgium even charges itself an internal energy tax based on the amount of energy it purchases to fund future energy efficiency and renewable projects. In 1999, New Belgium became the country's first brewery to purchase 100% of its electricity from wind power when its employee-owners voted to use part of their bonus pool to subscribe to the city of Fort Collins' wind program at a premium of 2.5 cents more per kilowatt-hour than the cost of electricity generated by fossil fuels.

This commitment to making sustainability a strategic core value by redefining success to include responsibility for people and the planet is not unique to New Belgium. Rather, it is part of a global movement of benefit corporations that envisions sustainability as a driving force for innovation and as a way for businesses to add value to and expand the market for their products and services while creating a more equitable and lasting prosperity for all.

now be legally designated as a benefit corporation in 30 US states and the District of Columbia (Box 10.3). This designation provides a higher level of legal protection, accountability, and transparency than existing for for-profit entities. Benefit corporations are required to prepare an annual benefit report detailing their activities and an assessment of their overall social and environmental performance against a third-party standard adopted by the corporation's directors or shareholders. Although there is no tax advantage from this designation, it sends a clear signal to consumers and legally protects corporations from shareholder suits against them for valuing the sustainability as much as profit. Similar legal designations have been or are under development in other countries. Examples include Italy's Società Benefit and Britain's community interest company designation.

B Corps are benefit corporations that are certified based on a complete assessment review by the nonprofit organization B Lab. They must meet rigorous standards of social and environmental performance, accountability, and transparency. There are more than 1,400 certified B Corps from more than 42 countries representing 120 industries, and this number is steadily increasing. The B Corp certification further substantiates the social and environmental mission of benefit corporations and distinguishes them from greenwashed companies. It also insulates them from investor pressure to pursue profit over their benefit mission. The B Labs website (https://www.bcorporation.net) provides a list of certified companies, a B Corp jobs board, and other resources for companies interested in certification.

The Role of Other Institutions

This chapter has focused on the very positive role that two institutions, education and business, can play in creating a sustainable future. There are certainly opportunities for other institutions to make significant contributions as well. Health care, government, and religious institutions typically have large infrastructure that can be made more sustainable, as in the case of schools and businesses. The fight for reform of health care policy that will make it universally accessible and affordable is gaining traction throughout the world.

Governmental institutions will always be essential for achieving sustainability objectives and will continue to play a central role in sustainable development. We have seen the importance of government regulation in protecting and improving the environment and that such regulation actually drives innovation and stimulates economic growth. However, putting this type of regulation in place is becoming more challenging as some governments become less functional in being able to pass regulatory legislation. In fact, in the United States, we have entered, at least for the short term, an era of deregulation. But this can be changed. One way is through campaign finance reform that will limit corporate or special interest control over government. Another is by getting young people who have been raised with an understanding of environment and justice to engage in politics and run for office. This is starting to happen at the local level, and ultimately, those elected to local offices are able to gain influence over the values and platforms of their political parties and work their way up to higher office.

With more than 80% of people in the world identifying with religious groups that have sustainability-related values in their doctrines, religious institutions have a major potential role in furthering sustainability objectives. This, however, will require them to translate their doctrines into actions that will improve the human and environmental condition and that will also foster a decoupling of consumption from success and happiness, as is the philosophy of many religions. Charitable work that supports impoverished communities and serves development must be done in a participatory manner in which the beneficiaries of this service have a say in how they will be served. It should also be done in a way that is sensitive to local values and beliefs in an effort to avoid the imposition of religious or cultural change.

Religions have a responsibility to engage in what is referred to as **public theology**. This is an approach to religious leadership that both cultivates a sustainability ethic in individuals and leverages public policy to support that ethic. Recently, this has been quite visible in the Catholic Church, with Pope Francis condemning climate change deniers and policies that violate social justice, such as the repeal of Deferred Action for Childhood Arrivals (DACA). The Catholic Church and other religious institutions are also joining the movement of divesting their endowments from the fossil fuel industry as seen in higher education. This type of leadership is spreading across all religions, giving hope that there will be opportunity for significant change in support of sustainability.

New Institutional Paradigms

In educational, business, and other types of institutions, a real paradigm shift has begun to emerge. We are seeing new models such as the development of circular economies, measures of success that are broader than the accumulation of wealth, and a new ecological approach to institutions that views them as interacting entities in a broader system. These transformational changes are being driven not just by cost savings and consumer choice but also by an understanding by institutional leaders that sustainability is the best model for success, not only for their institutions but also for the planet as a whole.

In their book *The Solution Revolution*,[28] William Eggers and Paul Macmillan argue that government alone cannot tackle our ever-growing environmental and social problems. They recommend a new economic system that erases public–private boundaries. This new economy is based on the convergence of stakeholders from business, government, philanthropy, and social enterprise to solve major problems and create public value. This solution economy is unlocking trillions of dollars in social benefit and commercial value. Also, tension is beginning to rise between the efficiency business models based on low cost and global sourcing, mastered by companies such as Walmart, and an innovation model based on higher wages, new products, and internalizing external costs.

Others have called for an economic revolution that recognizes that our fiscal, environmental, and social crises are the product of our broken institutions. Our current economy, created in and for the industrial age, embraces the efficiency business model, which places an emphasis on having more, having it cheaper, and getting it quickly. This opulent economy that our institutions promote is not sustainable because it creates resource addictions, encourages unhealthy lifestyles, ignores happiness, and offers a false sense of prosperity. It is dependent on the exploitation of resources and workers, destroys communities, and does not take into account external costs. Thus, economists such as Umair Haque of the *Harvard Business Review* and Havas Media Lab suggest that sustainability will only be achieved after there is a eudaimonic revolution. This revolution will redesign our institutions and economy in ways that will foster human accomplishment within the context of *eudaimonia*, or the sense of living life in a full and deeply satisfying way.[29] Eudaimonic prosperity is about mastering a new set of habits that ignite the art of living meaningfully and is not based on what one has but, rather, on what one is capable of doing.

This redefinition of prosperity can be part of a vibrant economy, as seen in successful businesses that have already embraced eudaimonia by creating high-quality products that allow people to live, work, and play better (see Box 10.3). A sustainable economy cannot be based on consumption and having more but, rather, should build communities and measure growth in ways that move beyond simple economic indicators such as gross domestic product and the Human Development Index. In the following chapters, we explore how community, regional, and global changes that embraced these ideals are fostering sustainability and sustainable development.

Chapter Summary

- An institution is an organization identified with a social purpose that transcends individuals and is established to maintain social order and govern the behavior of a set of individuals within a given community. Institutions can be governmental, economic, educational, and religious, among others. When considering the sustainability of an institution, one must assess the extent to which its infrastructure, operations, actions, and products meet sustainability objectives in the environmental, economic, and social spheres and whether it specifically promotes, supports, educates, advocates for, and practices sustainability within these spheres.
- Education and sustainability have been somewhat paradoxical because our current environmental and social problems have resulted from the ideas and actions of our most educated people. A sustainable future can be realized only if we directly address this paradox. In both K–12 and higher education, educating for sustainability in terms of curriculum and infrastructure and operations is becoming a priority because of the recognition that a sustainable future depends on this.
- Sustainability education in primary and secondary schools not only involves changes in pedagogy and curriculum but also requires that those changes be integrated into the core curriculum. Sustainable buildings, operations, and dining are other components of educating for sustainability.
- K–12 sustainability education can be increased by reducing the emphasis on standardized testing, aligning sustainability education content and objectives with current standards, through teacher professional development, and through the

expansion of teacher certification programs that emphasize sustainability.

- There has been a substantial increase in the commitment to incorporating sustainability at institutions of higher education. There are well-developed rating systems for assessing, and resources for implementing, sustainability education at colleges and universities.
- Sustainability in higher education is being integrated into curricula, research, campus facilities and operations, dining, co-curricular offerings, and commitment and leadership from top administrators and trustees.
- Business and industry can be sustainable institutions when they focus on the triple bottom line composed of the traditional measures of corporate profit and also measures of social responsibility and environmental impact or responsibility.
- Sustainable businesses are grounded in human, financial, natural, and social capital. This is best achieved by making supply chains, operations, transportation, and workplaces more sustainable; through the design of sustainable products and services; and through corporate philanthropy.
- Sustainability in business is motivated by consumer choice, the increase in the proportional value of intangible assets, pressure from investors, and government regulation.
- Assessment of sustainability in business requires the use of sustainability metrics and third-party rating and certification programs. Businesses that have embraced sustainability principles, especially those that start with sustainability as a core value, have achieved significant and measureable sustainability outcomes while maintaining profit and growth.
- A corporation that has a positive impact on society, workers, community, and the environment can now be legally designated as a benefit corporation. B Corps are benefit corporations that are certified based on a complete assessment review by the non-profit organization B Lab to meet rigorous standards of social and environmental performance, accountability, and transparency.
- Other institutions, such as government, religion, and health care, have a major role to play in working toward a sustainable future through operations, policy, and activism.
- New institutional paradigms are being developed and offer hope for a sustainable future. These changes are being driven by cost savings, consumer choice, and the understanding by institutional leaders that sustainability is the best model for success not only for their institutions but also for the planet as a whole.

Digging Deeper: Questions and Issues for Further Thought

1. How are social movements, organizations, and institutions different from each other? What are the strengths and weaknesses of each as a change-maker for sustainability?
2. Religion is an institution. Research your religion's views on different aspects of sustainability. How has your religion effected change with regard to these aspects? If you belong to a religious congregation, what sustainability efforts has it made?
3. Does your current or prior academic institution have sustainability objectives incorporated into its mission? How are they stated? In what ways is it specifically addressing those objectives?
4. Educational, religious, and other institutions that have endowments are being urged to consider, are considering, or are in the process of divesting their endowments from fossil fuel companies. How does this connect to sustainability? What are the limitations of divestment from fossil fuels? Are there other industries that should be considered for divestment for the same reasons?
5. What motivates a large corporation to be more sustainable?

Reaping What You Sow: Opportunities for Action

1. Visit the webpage of one corporation with which you are familiar. Using the information it provides, assess that corporation in all three major areas of sustainability. Identify areas in which information is missing or might contradict sustainability objectives. Write to the corporation and have it address these lapses.
2. A first step toward reducing institutional waste is to conduct a waste audit. There are numerous online resources on how to do this. Conduct a

waste audit at your institution (school, business, government building, or place of worship). Once you have your results, develop an institutional waste reduction plan and campaign.

3. There are many non-profit NGOs with missions related to sustainability. Research some of those that are the most influential. Identify one that seems the most effective to you in terms of both mission and action, and identify ways that you can get involved with it in order to support its efforts.
4. Developing a sustainability strategic plan is essential for institutional sustainability and should be a transparent process that involves all stakeholders, of which you are one. Research the status of the strategic planning process at your institution and find ways to contribute to it. If none is in place, explore ways to initiate the process.

References and Further Reading

1. Orr, David. 1994. *Earth in Mind: On Education, Environment and the Human Prospect*. Washington, DC: Island Press.
2. United Nations GAOR, 46th Sess., Agenda Item 21, UN Doc A/Conf.151/26 (1992).
3. President's Council on Sustainable Development. 1996. *Education for Sustainability: An Agenda for Action*. Washington, DC: US Government Printing Office.
4. Olson, S. L., and S. Kellum. 2003. "The Impact of Sustainable Buildings on Educational Achievements in K–12 schools." November 25. Madison, WI: Leonardo Academy.
5. Center for Ecoliteracy. https://www.ecoliteracy.org/article/california-food-california-kids
6. Chapman, Paul. 2014. *Environmental Education and Sustainability in US Public Schools*. Berkeley, CA: Inverness Associates.
7. Church, W., and L. Skelton. 2010. "Sustainability Education in K–12 Classrooms." *Journal of Sustainability Education* 1: 1–13.
8. Church and Skelton, "Sustainability Education in K–12 Classrooms."
9. Council of Great City Schools. 2015. "Student Testing in America's Great City Schools: An Inventory and Preliminary Analysis." Washington, DC: Council of Great City Schools.
10. Center for Ecoliteracy in Partnership with National Geographic. 2014. "Big Ideas: Linking Food, Culture, Health, and the Environment: A New Alignment with Academic Standards." Berkeley, CA: Center for Ecoliteracy.
11. The Princeton Review. 2008. "The Princeton Review 2008 College Hopes and Worries Survey Report." New York: The Princeton Review.
12. Eller, Donnelle. 2016. "At Students' Behest, Colleges Add Efforts to Address Climate Change." *USA Today*, January 8. Accessed February 6, 2016. http://www.usatoday.com/story/money/business/2016/01/08/students-behest-colleges-add-efforts-address-climate-change/78431048
13. Association for the Advancement of Sustainability in Higher Education. n.d. "Student Peer-to-Peer Sustainability Education Programs." Accessed February 8, 2016. http://www.aashe.org/resources/peer-peer-sustainability-outreach-campaigns
14. Wiek, A., L. Withycombe, and C. L. Redman. 2011. "Key Competencies in Sustainability: A Reference Framework for Academic Program Development." *Sustainability Science* 6 (2): 203–18. https://doi.org/10.1007/s11625-011-0132-6
15. Glasser, H., and J. Hirsh. 2016. "Toward the Development of Robust Learning for Sustainability Core Competencies." *Sustainability* 9 (3): 121–34.
16. NACUBO: Public NCSE Tables. n.d. Accessed October 3, 2017. https://www.nacubo.org/Research/2009/Public-NCSE-Tables
17. Fossil Free. n.d. "Commitments." Accessed October 18, 2017. https://gofossilfree.org/commitments
18. National Association of Scholars. 2015. "Inside Divestment: The Illiberal Movement to Turn a Generation Against Fossil Fuels." Retrieved October 3, 2017. https://www.nas.org/projects/divestment_report
19. Barlett, Peggy F., and Geoffrey W. Chase, eds. 2013. *Sustainability in Higher Education: Stories and Strategies for Transformation*. Cambridge, MA: MIT Press; Barlett, Peggy F., and Geoffrey W. Chase, eds. 2012. *Higher Education for Sustainability: Cases, Challenges, and Opportunities from Across the Curriculum*. New York: Routledge; Barlett, Peggy F., and Geoffrey W. Chase, eds. 2015. *Sustainability in Higher Education*. Waltham, MA: Chandos; Evans, Tina. 2012. *Occupy Education: Living and Learning*

Sustainability. New York: Lang; Barlett, Peggy F., and Geoffrey W. Chase, eds. 2004. *Sustainability on Campus: Stories and Strategies for Change*. Cambridge, MA: MIT Press.

20. The Cloud Institute for Sustainability Education. http://cloudinstitute.org/cloud-efs-standards
21. The Pennsylvania State University, Sustainability Institute. n.d. "Sustainability Course Assessment." Accessed February 12, 2016. http://sustainability.psu.edu/learn/faculty-staff/sustainability-course-assessment
22. Anderson, Ray C., and Robin White. 2009. *Confessions of a Radical Industrialist: Profits, People, Purpose—Doing Business by Respecting the Earth*. New York: Macmillan.
23. Transportation Research Board. 2005. "Car-Sharing: Where and How It Succeeds." http://www.nap.edu/catalog/13559/car-sharing-where-and-how-it-succeeds
24. Skoll Foundation, Allianz, and IDEO. 2008. *The Social Intrapreneur: A Field Guide for the Social Intrapreneur*. Accessed July 15, 2018. https://www.allianz.com/v_1339502342000/media/current/en/press/news/studies/downloads/thesocialintrapreneur_2008.pdf
25. Cone Communications. 2015. "2015 Cone Communications/Ebiquity Global CSR Study." Accessed February 23, 2016. http://www.conecomm.com/2015-global-csr-study
26. Nidumolu, Ram, C. K. Prahalad, and M. R. Rangaswami. 2009. "Why Sustainability Is Now the Key Driver of Innovation." *Harvard Business Review*, September. Accessed November 24, 2015. https://hbr.org/2009/09/why-sustainability-is-now-the-key-driver-of-innovation
27. Gunther, M. 2016. "Can Large Companies Lead the Low-Carbon Revolution?" *Environment 360*, Yale University. Posted February 9, 2016. https://e360.yale.edu/features/can_large_companies_lead_the_low-carbon_revolution
28. Eggers, William D., and Paul Macmillan. 2013. *The Solution Revolution: How Business, Government, and Social Enterprises Are Teaming Up to Solve Society's Toughest Problems*. Boston, MA: Harvard Business Review Press.
29. Haque, Umair. 2011. "Is a Well-Lived Life Worth Anything?" *Harvard Business Review*, May 12. https://hbr.org/2011/05/is-a-well-lived-live-worth-anything

CHAPTER 11

SUSTAINABLE COMMUNITIES, CITIES, AND REGIONS

In the previous two chapters, we examined the potential for individuals, organizations, and institutions to act or change in order to create a sustainable future. These entities interact with each other in communities, cities, and across regions that are in many ways the fundamental units of sustainability. In this chapter, we consider how these units can and are becoming more sustainable for the benefit of people and the planet. We highlight successes, best practices, and limitations by examining a number of specific cases at each of these levels of organization. Our focus here is on developed countries. In Chapter 12, we explore how sustainability at these levels is being addressed in developing countries in relation to sustainable development.

PLANTING A SEED: THINKING BEFORE YOU READ

1. What is the definition of community, and what does it take to be a member of a community?
2. Given our definition of sustainability, what would be the characteristics of a sustainable community?
3. In what ways is the community in which you live sustainable or not sustainable? How could it be made more sustainable?
4. What is the relationship between community sustainability and climate change?
5. Why has suburban sprawl been a common pattern of development in the United States? Are there more sustainable alternatives?
6. What does urban development have to do with environment and sustainability?

Suburbia is where the developer bulldozes out the trees, and then names the streets after them.

—Bill Vaughn, American columnist

Community Sustainability

A **community** is broadly defined as a group of people, often with diverse characteristics, who live in the same geographical area or setting. To varying degrees, we consider that group of people a community when they are linked by social ties, share common perspectives, and engage in joint action. Other less commonly perceived characteristics of community include **pluralism** or coexistence of more than one distinct cultural tradition and also shared concern for policy decisions and social problems such as crime, poverty, and environmental damage. It is also recognized that even within the context of the previously mentioned shared connections, there can exist contrasting elements of agreement and divisiveness within a community.[1] Communities include small villages or towns, neighborhoods within cities, cities themselves, and larger metropolitan areas.

A sustainable community is one that can reconcile differences and unite its members in order to meet the three main criteria in our definition. The Institute of Sustainable Communities defines a sustainable community as one that is economically, environmentally, and socially healthy and resilient. It is also one that manages its human, natural, and financial resources to meet current needs while ensuring that adequate resources are equitably available for future generations.[2] In addition, a sustainable community minimizes external impacts of its own actions on areas outside of its boundaries. The development of a sustainable community also depends on transparent governance and decision-making. In the following sections, we examine some of the common objectives of sustainable communities. We also discuss resources and recipes for success, including ways to assess community sustainability. Later, we examine community sustainability within the context of broader regions, discuss barriers to community and regional sustainability, and examine specific cases in which these barriers have been overcome.

The Goals of Sustainable Communities

There are a variety of goals and objectives pursued by communities that make them more sustainable. An overarching one is for them to become more livable through economic development, job creation, and improved quality of life. It is also recognized that a community's actions are both influenced and impacted by large-scale issues as described in previous chapters. Thus, sustainable communities should be built and designed in ways that limit resource use, depletion, and destruction; reduce waste and pollution; and mitigate and adapt to climate change in order to provide a healthier, more livable environment.

COMMUNITY ECONOMIC DEVELOPMENT

National and international economic policies typically do not directly assist economies on a local level. Existing large, free market economies and policies that have tended to maintain low wages and weaken labor have often resulted in economic dead zones within communities and urban neighborhoods. A remedy to the resultant community decline is community economic development (CED). The objective of CED is to improve the quality of life in struggling communities by making them more livable through enhancing economic opportunities, empowering individuals, and improving social conditions. This is achieved by providing more access to capital in order to create local businesses and by expanding the local entrepreneurial base. This in turn creates local employment opportunities and wages, and it increases local access to consumer goods and services, improving both the economic and the social condition of neighborhoods.

The services needed to accomplish the previously mentioned objectives are provided by local governments, non-profit economic development corporations, and employment and training centers that prepare community members for these new opportunities. Capital can be provided by the local or state government, community development microfinancers, credit unions, banks, and venture capital funds that specifically support local business development. For example, the largest US residential solar power provider, SolarCity, received $750 million from New York state to construct its solar panel factory in Buffalo on the brownfield where the now extinct Republic Steel once stood. The company will spend more than $5 billion during the next decade to build and operate the plant, creating 5,000 jobs with above-average wages in the economically distressed upstate New York economy.

Another commonly used approach to CED is to establish empowerment or enterprise zones that provide grants, tax relief, low-interest loans, and other benefits in order to attract business to specific, often distressed areas. Larger investments by state and local governments, such as the development of sports arenas or other entertainment venues and attractions, coupled with tax relief for businesses associated with or located near those projects have been used to stimulate large-scale community or urban development. An example is the Neighborhood Improvement Zone created in Allentown, Pennsylvania, to encourage development and revitalization, the centerpiece of which was the construction of an 8,500-seat arena by the city. Through state law, taxes generated in the zone are used to pay debt service on any financed development by private businesses and to pay the debt on the $224 million in municipal bonds sold to build the arena.

Business incubators run by local economic development organizations often provide support for smaller, new start-up companies. They typically provide low-rent or free space, shared resources, and mentoring, training, and advisory services. Entrepreneurs seeking entrance into an incubation center apply by providing a business plan that outlines how the business will grow within the center, and they eventually graduate and move into their own location within the community. After a specific time period, when the new company is ready for its next stage of growth, the center helps the company transition into that new location. Incubation centers also provide "maker spaces" or "hives" where entrepreneurs can work in a collaborative environment to develop new ideas, products, and technologies.

EQUITABLE COMMUNITY ECONOMIC DEVELOPMENT

A major criticism of the types of previously described CED is that they often do not directly benefit or serve the needs of low-income individuals and families. For example, the use of public subsidies to stimulate private businesses to invest in and develop specific zones within a city, such as those in Buffalo and Allentown, primarily benefit those businesses and employees directly in the zone, but not the community as whole. Often, existing businesses relocate with their employees to the zone because the subsidies allow the developer to offer reduced rental rates to them. As a result, few new high-paying jobs are created. This can create enriched areas where businesses prosper while poverty in surrounding communities increases.

One way to remedy this inequity is through community benefits agreements (CBAs) or community benefits ordinances (CBOs). A CBA is a contract between community groups and real estate developers which requires the developer to provide specific improvements to the existing community or neighborhood in or near that development. Site-specific CBAs ensure that particular projects create opportunities for local workers and communities. Typical examples include requirements for local hiring, wage increases, the development of affordable housing, local purchasing and contracting, and support to schools, parks, and community organizations.

CBOs are very similar to CBAs, but are written as government ordinances rather than as agreements. Both CBAs and CBOs can ensure that specific development projects create opportunities for local workers and communities, and more effectively result in poverty reduction. The Partnerships for Working Families (http://www.forworkingfamilies.org) offers resources for those interested in developing these tools for a more just form of development. Another way that CED can meet the economic needs of low-income individuals and families is through the establishment of workforce training and employment centers that equip and place individuals in newly created businesses within the community. These centers and programs are usually operated through private–public partnerships that include local community and technical colleges, chambers of commerce, and development corporations. An example related to the SolarCity project in Buffalo, New York, is the creation of a one-semester certificate program in semiconductor manufacturing technology at a local community college.

Federal governments help improve the equity of CED by offering grants that are administered at the community level. Community Development Block Grants from the US Department of Housing and Urban Development (HUD) and Community Service Block Grants (CSBGs) offered by the US Department of Health and Human Services provide resources and empower communities to address community development needs. CSBGs also provide support services and activities for low-income individuals that alleviate the causes and conditions of poverty. The Community and Economic Development Network in Canada functions in a similar way.

COMMUNITY WEALTH BUILDING

There is increasing interest in a model of economic development that does not rely on public subsidies to induce corporations to bring mostly low-wage jobs to distressed urban centers. Instead of this trickle-down approach, community wealth building focuses on economic inclusion and building a local economy from the bottom up. It accomplishes this by spreading the benefits of business ownership through the formation of worker-owned cooperatives. Through these cooperatives, worker-owners have both share and voice in the company. Such democratic ownership keeps jobs and profits local. In addition to local, worker-owned job creation, these cooperatives are often committed to being sustainable, green, and democratic workplaces and also committed to CED.

The model works by identifying anchor institutions such as hospitals and universities that are already located in the distressed city and are not likely to move. These institutions spend billions of dollars on the goods and services they use on a daily basis, but much of that tends to be spent outside of the urban center. Examples of such expenditures include fresh produce, clean power, cleaning supplies and services, and linens—usually through long-term contracts. Worker-owner businesses are established near these anchor institutions to meet these needs locally. This is attractive to the institutions because many of them have integrated sustainability into their mission, as discussed in Chapter 10, and working with these businesses helps them accomplish this mission. Each cooperative is linked together by a larger non-profit that provides technical assistance, job training, strategic guidance, and affordable and creative financing to support home ownership and entrepreneurship in the community (Figure 11.1).

The first major success employing the community wealth-building approach is the Mondragon Corporation, which is a federation of worker cooperatives based in the Basque region of Spain. It was founded in the town of Mondragón in 1956 by graduates of a local technical college. Today, it is composed of more than 100 cooperatives that employ nearly 84,000 workers.

A more recent use of the community wealth-building approach to economic development is the Evergreen Cooperative in Cleveland, Ohio. It is now referred to as the "Cleveland model" because it is serving as a template for urban development in many cities. It is anticipated that within the next few years, up to 10 new for-profit, worker-owned cooperatives will be created to serve the anchor institutions in the Greater University Circle area of Cleveland, an otherwise impoverished area. These cooperatives will employ low-income residents. Financial projections indicate that after approximately eight years, a typical Evergreen worker-owner could possess an equity stake in their company of approximately $65,000. This approach is already beginning to stabilize neighborhoods by increasing home ownership and catalyzing small business development.

The Cleveland Model

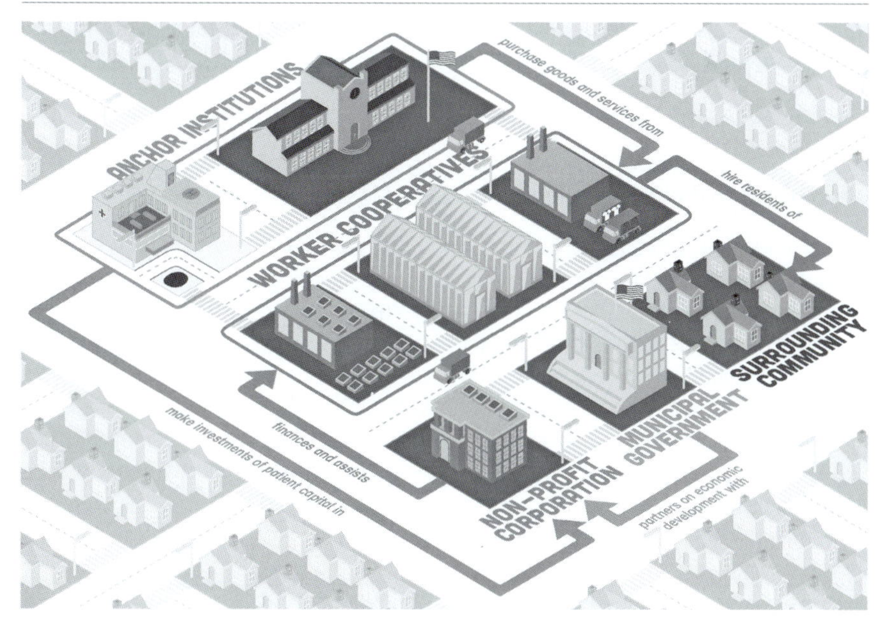

Figure 11.1 The Cleveland Model builds community wealth by linking anchor institutions to worker cooperatives, and by offering financial and other kinds of assistance from governmental agencies and non-profit corporations. *Courtesy of The Democracy Collaborative and Community.*

There is growing evidence that the community wealth-building approach such as that being implemented in Cleveland is a sustainable solution. This is because it is improving the ability of communities and their residents to own assets, create jobs, expand public services, and ensure local economic stability. Some argue it may be the next American Revolution; others argue that it does not go far enough in reducing poverty and broadening home ownership.[3]

HOUSING AND COMMUNITY STABILITY

One of the problems associated with CED is that as the incomes of residents in distressed communities increase with economic development, these people often flee the distressed neighborhoods, further degrading the quality of life by creating a more transient community. This flight can be prevented by creating attractions such as libraries, parks, community centers, retail spaces, restaurants, and other gathering places that represent diverse neighborhood interests and preferences. Also, broader efforts in community beautification such as mural projects often improve communities, making them more attractive to stay. These approaches are most effective when community members are viewed as primary stakeholders, giving them a sense of pride and ownership about such improvements.

Another problem associated with economic revitalization is that it can make the community a more attractive place to live for those with greater income. This then causes sharp increases in rents and home values that displace low- and moderate-income residents, contradicting the social and equity objectives of sustainability. This process of **gentrification** can limit the availability of housing that is affordable to those with lower incomes, often forcing them out of their own community as it develops.

Gentrification can be prevented by ensuring that a substantial part of community development includes the creation of affordable housing. This can be done by mandating mixed-income housing development construction by local governments through their zoning and planning commissions. Such **inclusionary zoning** encourages or requires developers to include affordable housing units in new developments. These requirements usually range from 10% to 25% of new housing development in a particular area. This type of zoning has been put in place in more than 500 US cities and counties with much success at creating affordable housing while promoting racial integration. However, critics believe that it does not serve those with the lowest incomes and may restrain overall rates of development because of additional costs.[4] Another way to ensure the availability of affordable housing in improving neighborhoods is to maintain low real estate taxes and to provide rent control for long-time residents or young families seeking their first home.

Affordable and equitable access to housing is a cornerstone to community sustainability. It has been a priority of the US government since the early 1960s with the creation of HUD and the evolution of housing policy since then (Box 11.1). As originally conceived, housing was made affordable through the

BOX 11.1

Policy Solutions: The Evolution of a Sustainable Housing Policy

United States housing policy has had a slightly checkered past but has been evolving as an effective tool to promote sustainable communities and housing. Starting with the US Housing Act of 1937, the federal government established national goals for decent living environments and began, through local public housing agencies, to improve living conditions for low-income families. It also funded the removal of "slums" and established the first national public housing programs and urban renewal projects. In 1965, housing policy was expanded with the Housing and Urban Development Act, which instituted several major expansions in federal housing programs, including the creation of the US Department of Housing and Urban Development (HUD) as a Cabinet-level agency. In 1968, the Fair Housing Act banned discrimination in housing. This established the Government National Mortgage Association (Ginnie Mae) to expand availability of mortgage funds for moderate-income families using government-guaranteed mortgage-backed securities.

Despite the good intention of making housing available and affordable to people of all incomes and races, many of the initiatives of early housing policy had limited long-term success and were heavily criticized. As originally conceived of in the 1950s and 1960s, public housing led to the concentration of poverty in ghettos referred to as "the projects" that led to crime, drug use, and the failure of schools. Denounced by Vice President Al Gore in 1996 as "crime-infested monuments to a failed policy that killed the neighborhoods around them," HUD began to rethink its public housing strategy as part of its HOPE VI and other programs in order to create strong, sustainable, inclusive communities and quality affordable homes for all.

Recently, there has been a radical shift in housing policy that integrates affordable housing with the concept of community sustainability that promotes designs such as garden cities and scattered-site housing, in which publicly funded, affordable homes are scattered throughout diverse, middle-class neighborhoods. There has been a movement away from public housing and toward providing subsidies for private developers to construct affordable homes, offering reduced interest mortgages to low-income buyers, and programs to transition families out of public housing. Also, it is becoming more common for affordable housing projects to incorporate sustainable features that reduce the cost of construction and maintenance and help empower residents and connect them to outside resources and the community.

HUD's newer holistic approach to expanding economic opportunity includes the Promise Zone and Choice Neighborhood initiatives that establish partnerships between the federal government, local communities, and businesses to create jobs, increase economic security, expand educational opportunities, increase access to quality affordable housing, and improve public safety. This is done through a competitive granting process such as Community Development Block Grants.

Reflecting this community approach, HUD is now collaborating with the Department of Transportation and the Environmental Protection Agency to provide citizens with access to affordable housing, more transportation options, and lower transportation costs while protecting the environment in communities nationwide. This has allowed for better integration of transportation, urban, and land-use planning. Through this collaboration, funding opportunities are available to help communities realize their own visions for building more livable, walkable, and environmentally sustainable regions.

Through this evolution of housing policy, the US homeownership rate reached a record high, surpassing 6.9% in 2005. However, the 2008 US financial crisis caused by deregulation of the financial industry led to a sharp decline in homeownership that has yet to rebound. This decline was caused by an increase in mortgage default rates due to a decline in home values, job loss, and wage reduction. The subsequent economic recovery was not inclusive, experienced mostly by the highest wage earners, and has thus led to an ever-growing inequity in wealth. This inequity, coupled with decreased funding to support low-income housing, is limiting the possibility of that rebound.

Despite the ongoing decline in homeownership, there is still much that housing policy can do to improve community sustainability. Proposed initiatives include funding for a new Clean Communities competitive grants program that would provide support for transit-oriented development, reconnected downtowns, brownfield clean-up, and the implementation of Complete Streets policies. By 2016, economic opportunity had been expanded through the creation of 20 new Promise Zones and increased support for Promise Zone tax incentives to stimulate growth and investment in targeted communities, such as tax credits for hiring workers and incentives for capital investment within the zones.

There is also increased emphasis on supporting private sector partnerships and investments that play a key role in strengthening communities, including a New Markets Tax Credit, which promotes investments in low-income communities. Other initiatives are being developed to improve climate and disaster resilience, including competitive grants to spur investments that bolster resilience to climate impacts. Cutting-edge projects would incorporate resilience strategies, such as adaptive materials, risk-sensitive design, and next-generation transportation and logistics technology. Through these innovative programs and continued multiagency partnerships such as Partnership for Sustainable Communities, more than 1,800 communities nationwide are implementing integrated place-based initiatives that improve community sustainability. This is occurring due to the recognition that breaking through agency silos and closely coordinating with local businesses and non-governmental organizations will yield the greatest success.

development of low-income or public housing administered by federal, state, and local agencies. Early on, these public housing projects consisted of complexes of low-rise and high-rise apartment buildings that led to numerous negative social consequences. One of these was the concentration of poverty, primarily for people of color. This occurred because of the resistance to low-income public housing in wealthier, commercially vibrant urban neighborhoods and the exodus of residents from those neighborhoods in which it was being developed. This led to what amounted to racially and economically segregated "ghetto neighborhoods," often referred to as "the projects," with high crime rates, poor health conditions, a lack of quality education for residents, and increasingly poorer public perception and negative stereotypes associated with public housing. This concentration of poverty was further enhanced as international trade policy led larger businesses to switch from local to global supply chains. This led to the massive decline of well-paying manufacturing jobs in US communities.

The negative consequences of public housing began to be remedied through policy changes and shifts in development approaches (see Box 11.1). New designs emerged, such as self-contained communities surrounded by green areas, referred to as garden cities. Another approach was scattered-site housing, in which publicly funded, affordable homes were scattered throughout diverse, middle-class neighborhoods. Eventually, there was a shift away from public housing. New approaches then included providing subsidies for private developers to construct affordable homes, offering reduced-interest mortgages to low-income buyers, and programs to transition families out of public housing (see Box 11.1).

Recently, affordable housing has been integrated with the concept of community sustainability. The American Institute of Architects and HUD offer awards that encourage and recognize affordable housing for energy efficiency, integration of recreation, common and green spaces, and aesthetics. This has resulted in a radical shift in affordable housing models to incorporate sustainable features that reduce the cost of construction, maintenance, and help empower residents and connect them to outside resources and the community, a trend that started in Europe.

There has also been a focus on retrofitting existing housing to in order to improve both affordability and energy efficiency. A recent study showed that these improvements can result in cost-effective energy savings of 15–30%, saving building owners and residents $3.4 billion per year[5]. For example in Ann Arbor, Michigan, the local housing commission is completing renovations in the Baker Commons public housing facility in order to reduce energy use by 20%. The Pittsburgh non-profit ACTION-Housing, Inc., retrofits homes that are capable of generating their own renewable energy for young adults phasing out of foster care and low-income workers.

The challenge with retrofitting homes so that they are both affordable and energy efficient is financing upfront costs. HUD, state governments, and non-governmental organizations (NGOs), are beginning to finance affordable housing upgrades. In California, Sacramento's Housing Agency Municipal District obtained $1 million from their publicly owned, non-profit electric utility Sacramento Municipal Utility District to reduce energy consumption in multifamily housing by 25–30%. In Boston, combining financing from federal, state, and local programs with private sources and utility incentives enabled the Boston Housing Authority to capture 31.5% total energy savings.

There are other organizations that provide resources to communities that create opportunities for people with economically and racially diverse backgrounds to live in sustainable, safe, healthy, and affordable housing. In 2015, NeighborWorks America supported a network of 240 non-profit organizations to assist 355,900 families in obtaining affordable housing and created more than 21,700 new homeowners.[6] Other organizations like PUSH Buffalo (Chapter 9) and the bcWORKSHOP in Dallas are examples of community-based organizations that reclaim abandoned, neglected homes and develop them as affordable, sustainable housing. Habitat for Humanity is well-known for partnering with people in need to build simple, decent, and affordable housing.

In spite of these efforts, affordable housing remains a problem as wage increases often do not keep pace with home rental and mortgage prices. HUD defines the baseline cost for affordable housing as 30% of income. In many cities throughout the industrialized world, costs often exceed 40% of family income. For example, in London, England, a two-bedroom rental costs 40–60% of the average income of people in their 20s and 30s, making housing there unaffordable for most young families.[7] The problem is even worse in larger US cities, where the average hourly wage needed to rent a typical two-bedroom unit is $19.35. That is more than two and a half times the federal minimum wage, and $4 over the national average wage of $15.16 earned by renters.[8] Thus, a sustainable community needs to integrate economic development that results in wage increases with approaches to making housing affordable.

EDUCATION

Much was said about sustainable education in Chapter 10. But it is important to mention here that declining housing and communities, and the concentration of poverty, impose a tremendous limit on equitable access to education. For example, a comparison of two Chicago communities located less than an hour's drive apart shows that one spends more than three times per student on education than the other. This is because the bulk of a school's funding comes from the local real estate base, so available funding is much lower in poorer areas, which in turn tend to have declining and poor-performing schools with significantly reduced student outcomes and achievement. This further leads to flight from these areas by those with children who can afford to do so, further eroding the tax base.

There is no protection in the US Constitution for reasonably equal access to education. Shifting funding from local real estate taxes to regional, state, and federal sources in conjunction with better housing and economic development could reverse this trend by decoupling school funding from community affluence. Also, fair funding formulas that reallocate taxes to reduce the gap between wealthy and poor districts are being proposed in states such Pennsylvania, where that gap is one of the largest. Although politically challenging, this would improve schools in impoverished neighborhoods and in turn make these neighborhoods more attractive places for families to live. The US Department of Education's Promise Neighborhoods program supports equitable education through the collaborative development of integrative cradle-to-career solutions of both educational and family and community support programs with excellent schools at the center. The goal of the integrative approach is to break down agency "silos" so that solutions are implemented effectively and efficiently across them.

FOOD SYSTEMS

As discussed in Chapter 5, our current food system is global in nature, dominated by large agro-industry, is fossil fuel and chemically intensive, depletes and degrades water and soil, is a major contributor to climate change, deprioritizes healthy foods, and generates substantial food waste. It has drastically decreased the economic viability of small and mid-sized farms and farm-related businesses, as well as the number of young people interested in becoming farmers. This global, industrial food system does provide consumers with inexpensive foods, but it favors those that are processed, calorie-rich, and nutritionally deficient. This is fostered by a federal farm policy that subsidizes commodity crops that favor agro-industry and processed and fast-food companies while disincentivizing the production of healthier crops such as fruits and vegetables. This is in direct contradiction to government and medical nutritional recommendations (see Box 5.1, Figure 5.1.1). In addition, there is an inequitable distribution of food retail sources, which are disproportionately absent from inner-city, low-income communities, many of which are now designated as food deserts. This is because of the perception that poorer neighborhoods lack purchasing power. This type of food system is anything but sustainable.

The sustainability of a community is highly dependent on its food system. A food system is the sum of all activities required to make food available to consumers. A sustainable food system integrates all of these elements to enhance economic, environmental, social, and nutritional health for all. A sustainable community food system protects rather than exploits workers, is profitable to all who work in it, and supports the health of consumers by providing equitable access to nutritious foods. One way that these advantages can be achieved is through community-based agriculture that links food production with economic and community development. This can offer financial and physical access to affordable, nutritious food that was produced and transported in ways that minimize environmental damage and maximize health for all members of culturally and economically diverse communities. A sustainable food system must also be resilient to unpredictable climate, increased pest resistance, and reductions in water availability (Figure 11.2).

The previously mentioned conditions of a sustainable community-based food system are achievable. It should be recognized, however, that the lack of purchasing demand for nutritional food in poorer communities is a misperception. For example, the potential market in inner-city Detroit is nearly $1 billion

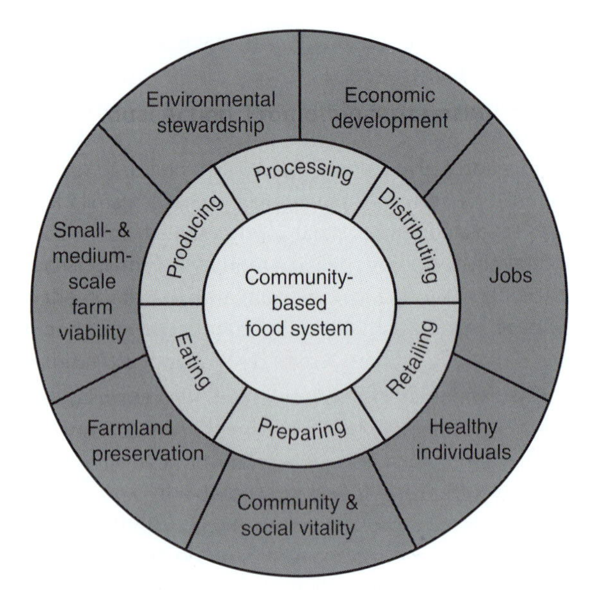

Figure 11.2 The characteristics of a community-based food system. (Michigan Good Food Initiative.)

per year, but because of the lack of food retail in this food desert, that money is spent outside of the community.[9] Community food systems can be made more sustainable through the development of food production and retail businesses that can be equitably accessed, eliminate food deserts, provide the healthiest and most environmentally friendly food at affordable prices, reduce or eliminate food waste, and encourage food spending within the local community. In Chapter 5, many of these solutions were presented, including community-supported agriculture programs, urban agriculture, community gardens, community kitchens, and the sale of local and organic food. Other examples are food retail development and food buying cooperatives. Many communities have also developed farmer training and small farm incubation programs in order to increase the number of new farmers and the local food economy.

HEALTH CARE

In Chapter 10, it was demonstrated how educational institutions can be sustainable. As in education, health care institutions must not just improve sustainability in their facilities and operations in order to increase efficiency and cost savings; they should also integrate sustainability, mission, and management in order to meet the goals of a sustainable health care system. These goals for which there are measureable indicators include access, efficiency, equity, quality, and overall indicators of health in the population. Improvement in these areas is not simply dependent on local providers and hospitals. They will need to cooperate more broadly with government, NGOs, medical and public health professionals, corporations, and private and public insurers in the delivery of sustainable health care.

Access to health care is typically driven by national policies. Ideally, all citizens of a particular country would have access to health care and protection from financial risk due to health care needs. Such universal health care is typically offered through federal legislation and taxation, with some costs paid by the patient. Among all 33 developed nations, the United States is the only one that lacks universal health care.

In the United States, the 2010 Affordable Care Act mandated that all citizens purchase insurance if it is not provided by their employer or through other social programs to make health care and health insurance more affordable and more available. It has reduced the number of those without insurance from 15.5% to 11.7% but has still left 32 million uninsured. Medicare, a federal program, provides health coverage for the elderly and those with severe disabilities; Medicaid, a state and federal program, provides health coverage for those with very low income. There are also state and local programs such as those in Milwaukee, Wisconsin, and Pennsylvania that provide health insurance for children in families below a minimum income. Despite these programs, there are communities in which individuals still assume the financial risks of not having health insurance and lack access to adequate care.

Other measures of a sustainable health care system include the efficiency of delivery of effective health care relative to costs and also equity in health care delivery across income levels. A lack of equity is made evident when those with below-average income report not visiting a physician when sick; not getting a recommended test, treatment, or follow-up care; or not filling a prescription or skipping doses when needed because of cost. Finally, quality and overall health outcomes, including effective, safe, coordinated, and patient-centered care, as well as reduced rates of mortality and illness and increased life expectancy, are other important indicators of the sustainability of health care systems.

Most of the outcomes described previously are the result of federal and, in some cases, state policy.

However, there are a number of community-level approaches to improving health care, especially among the poor and uninsured. These can also decrease costs. One way is through the establishment of free community health clinics. One example is the free clinic in Ithaca, New York, which serves uninsured working people who might have otherwise either not sought treatment or would have gone to the emergency room—a much more costly prospect.

Community-based health care that provides health care assistance at home can be effective at preventing illness and reducing costs. For example, community health workers who provide prenatal health care in the home reduce infant mortality and complications related to pregnancy. Other community initiatives can also improve the health of citizens, including programs that address safe housing, improved access to nutritious foods, expanded opportunities for safe recreation and exercise, and increased health education. Finally, communities and their hospitals must be prepared to increase their resilience to climate change, which has numerous potential health consequences such as newly emerging diseases, and impacts of increased heat, drought, and weather-related disasters.

Given the current US political climate, the future of US health care policy is uncertain. There is an active movement led by President Trump to dismantle many of the improvements described previously either by undoing the policies or by defunding them. There is also a growing political movement pushing for universal health care through a single-payer system such as that in many other developed nations. The implications of these agendas have yet to be seen, but many are not optimistic about the future of inclusive access to affordable and quality health care.

OPEN SPACE

A sustainable, livable community provides equitable access to open or natural spaces such as parks, greenways, and gardens. These spaces can contribute to environmental, economic, and social components of sustainability. In terms of environment, open spaces offer many benefits. They can provide ecosystem functions such as reduced flooding through the absorption of large amounts of storm water on pervious surfaces, and they can also protect or improve water quality as water enters natural aquatic systems. In addition, they have the potential to remove pollutants from the air, take up atmospheric carbon, and reduce urban heat island effects and noise while protecting natural habitats, watersheds, and biodiversity. However, these environmental functions can be constrained by the size, the amount of area left green or forested, and the location of these spaces within the context of the surrounding developed area. But in parks that have large forested areas, trees can remove as much as 15% of the ozone, 14% of the sulfur dioxide, 8% of the nitrogen oxide, 0.05% of the carbon dioxide (CO_2), and significant amounts of particulates from the air.[10] Because people who live near parks have greater opportunities to connect with nature, they often become advocates not only for their park but also for the environment in general.

The environmental benefits of parks are often difficult to quantify, as are the mechanisms behind them. There is evidence that parks can have negligible and sometimes negative environmental impacts. For example, a study of a park in a heavily urbanized area of Hong Kong showed that it did little to improve air quality, suggesting that the environmental functions of parks and open spaces in dense cities might be limited.[11] Even when a park and open space offer direct environmental benefits, they can actually cause some unintended negative effects. In some cases, the overuse of parks may end up harming some of the environmental elements they were intended to protect, especially when use is more active rather than passive. Camping and picnic areas with car access as well as dirt paths can result in soil compaction, making them impervious, and playing fields and golf courses often rely on intense fertilization. Thus, when these amenities are located near streams or wetlands, there is potential for negative impact.

Even in cases in which the environmental benefit of open space is not fully realized, there are many other benefits that improve the sustainability of communities and make them more livable. Parks often provide substantial economic benefits to an area. Many larger parks attract tourists simply as a destination or by offering concerts, races, and sporting and charitable events. This in turn can generate revenue for businesses in the surrounding area. Parks bring economic vitality to surrounding areas by attracting new businesses and housing development, and they raise property values in neighborhoods of all economic levels. For example, Chicago's Millennium Park has been attributed to a $1.4 billion boost to local residential development and millions more in tourist dollars. This includes the attraction of new retail, hotels, and restaurants that use the park as a marketing device.[12]

In Salem, Oregon, land adjacent to green space was estimated to be worth approximately $1,200 an acre more than land only 1,000 feet away.

Parks can also provide social benefits and assets to quality of life in areas where they are well maintained and equitably distributed. Parks provide green space for neighborhood socializing and foster community building among nearby residents. One study revealed that individuals living adjacent to green common spaces were more social, knew more of their neighbors, felt more strongly connected to their neighborhood, and were more likely to develop interracial and ethnic relationships, especially in parks with cultural and recreational amenities.[13]

Recreational facilities such as basketball courts provide youth with safe venues to socialize and have been shown to deter crime, reduce juvenile arrests, and increase academic performance in low-income neighborhoods. There are clear public health benefits, especially for children and teenagers. One study showed that higher concentrations of community public parks with various amenities can cause a 25% increase in the number of people who are physically active at least three times per week. This in turn causes measurable reductions in obesity and morbidity.[14]

In order for the social benefits of open space and parks to be fully realized, there must be equitable access to them. One negative consequence of the economic benefits that these spaces provide is that as property values near parks rise, gentrification may occur and decrease access for middle- and low-income families. This is yet another reason for addressing this issue in the same manner as described in the section on housing. Access is also influenced by park safety and maintenance. Residents will avoid parks that do not look or feel safe, have high crime rates, or are poorly maintained. Such parks actually can negatively affect property values and sense of community.

Because health and quality of life disparities often follow the same pattern as green access disparities, maintaining equal access to parks is an important component of community sustainability. Park maintenance and security need to be justly applied. With regard to parks and open spaces, the environmental, economic, and social benefits can be in conflict with each other. Smaller well-distributed parks might improve access and more widely distribute benefits, but the extent of ecosystem services becomes constrained. One way that these conflicts can be resolved is through the creation of networks of small parks that are linked together through greenways or corridors. These green connections often become attractions themselves; such as the Brooklyn Strand, a proposed 21-acre network of open spaces that is currently being developed (Figure 11.3). The Comox–Helmcken Greenway currently under construction in Vancouver will connect parks, schools, community centers, neighborhoods, and retail with greenways that promote

Figure 11.3 The Brooklyn Strand, a proposed 21-acre network of open spaces with large central green spaces connected by greenways. *Based on Brooklyn Strand Urban Design Plan https://thislandisparkland.com/2015/06/05/parkifying-the-city-through-making-connections/*

walking, cycling, and social interaction. A number of cities are converting blighted property or brownfields into parks. The new Atlanta Beltline is an example of upcycling abandoned urban space that will include 1,300 acres (520 hectares) of new parkland.

Park systems that are designed and managed in consultation with the communities that they serve tend to be more successful at meeting sustainability objectives. In addition to locating parks in a way that provides for equitable access, the management, maintenance, and security of these places need to be centralized and uniformly applied. This, coupled with empowering neighborhoods to support their parks through "friends of"–type organizations, is also important and can maximize the environmental, economic, and social benefits of open space.

THE ARTS

The creative and performing arts are another important catalyst in sustainable community development. The arts often bring diverse members of communities together in meaningful ways. Artists beautify and help create a sense of identity and pride in communities through public art displays. They also attract people into communities, providing additional contributions to economic development. Creative expression is often used to directly promote sustainability, including social justice and environmental protection. Artists can also ensure their own practices are sustainable with the support of organizations such as the Broadway Green Alliance and the Center for Sustainable Practice in the Arts.

A number of organizations support the role of art and culture in contributing to community sustainability. The Pratt Center for Community Development is an organization that produces innovative culture, arts, media, and organizing strategies that seek to engage neighborhood residents and artists to promote sustainability activism. The Philadelphia-based theater group Just Act promotes participatory, socially just community planning by helping individuals learn about themselves and their assets through storytelling and other modes of performance. For example, it is working with Chester, one of the oldest and poorest communities in Philadelphia, to develop a "cultural asset map" through story-based performance. This will guide the redevelopment of the community, with the main cultural corridor as the centerpiece. There are also a number of academic degree programs focused on arts and sustainability, such as those at Arizona State University and Hofstra University, and many other institutions are including the arts as they weave sustainability throughout their entire curriculum.

TRANSPORTATION

Sustainable transportation is a key component to a sustainable community because it links housing, employment, recreation, open space, food, health care, education, and other elements of a sustainable community. As the number of people who work and live in a community increases with economic development, transportation systems typically have difficulty meeting increased need. Thus, transportation assessment and planning must be an integral part of community development. Without a sustainable transportation system that links community elements together, efforts to improve those other elements will be limited.

A sustainable community transportation system should be **multimodal**, offering travelers a diversity of options that are all resource efficient, affordable, and healthy. It should be inclusive by meeting mobility needs of non-drivers, especially low-wage earners, and workers with disabilities or who are older. The various components of community transport might include roads and parking for drivers, car and ride sharing opportunities, buses and light rail, and opportunities that encourage safe walking and bicycling. Each of these components should be part of an integrated system that connects with regional transportation. For example, most bus and rail systems now accommodate bicycles. Other priorities for sustainable transportation planning include efficient use and preservation of natural resources, including land, air, and water, which contributes to climate stability.

There are economic benefits to a good public transit system. Households that take advantage of public transit and reduce their number of cars from two to one save more than $6,000 per year compared to households with two cars that do not use public transportation. This savings is comparable to annual household expenditures on food or housing. In communities in which a significant number of residents have just one car or no car at all, the cumulative cost savings is large. For example, in Portland, Oregon, where there is an extensive and affordable integrated transit system, residents drive an average of four miles per day less than the average for other cities. For a city the size of Portland, this results in approximately

2.9 billion miles of reduced vehicle travel per year, offering an accumulated direct cost savings of $1.1 billion and time cost savings of $1.5 billion. Much of this savings stays within the local economy, making the development and operation of an efficient transportation system a sound investment.[15] Walkable and bikeable communities also encourage support of local businesses and increase local economic development, especially in the area of retail.

A sustainable transportation system should also be assessed in terms of its environmental benefits. Transportation is the second largest contributor to greenhouse gas (GHG) emissions in the United States, accounting for approximately 28% in 2016. More than half of transportation-related GHG emissions are from private vehicles such as cars, light-duty trucks, sport utility vehicles, trucks, and minivans. Public transit systems can drastically reduce the number of vehicles used within a community and hence reduce GHG emissions. Many transit systems amplify this effect by running their vehicles on renewable and cleaner forms of energy. Communities that promote public transit, the use of green vehicles, cycling, and improved walkability are also successful at reducing pollution and road congestion.

There are many social benefits to a sustainable community transportation system. There are health benefits through reduced pollution, increased walking and cycling, and reduced stress from not having to drive. Injuries and deaths from road traffic accidents—the number one global cause of mortality for those between 15 and 29 years of age—are also reduced through increased use of public transit.[16] All of this translates into economic benefits through reduced health care costs. There is also increased social interaction within the community, resulting in much improved personal relationships and community health.[17] Effective, sustainable community transport systems also promote social equity by both providing access and reducing environmental health impacts to those with lower incomes, who tend to be disproportionately exposed to them.

Europe and Asia have been more reliant on mass transit primarily because of the lower availability of resources or land required for automobile travel and the lack of an automotive culture that predominates in the United States. This car culture has been perpetuated though mass marketing, the preferential development of roads and highway systems over transit systems, as well as gasoline subsidies. As a result, transport emissions of US cities are approximately 4 times higher than those of Europe and approximately 24 times higher than those of Asia. However, well-developed transport systems such as that in Portland coupled with incentives such as economic benefits, convenience, and speed are resulting in increased use and expansion of mass transit in some US cities.

In addition to the development of effective multimodal public transportation systems, there are a number of other transportation improvements that increase community sustainability. Creating an environment that is more conducive to walking is fundamental for reasons described previously and because public transit users are pedestrians at either end of their transit trip. A number of design features can contribute to a community's walkability, including mixed-use development that places housing within walking distance to locations of employment, retail, and other services. The following also encourage walking: designing streets with pedestrian safety in mind that discourages high-speed driving, creating continuous connectivity among walkways, increasing lighting and security, and making pedestrian spaces more appealing with landscaping, benches, and locations for public gathering. There should also be appropriate accommodations for people with disabilities. The city of Rochester, New York, achieved these objectives by filling a sunken section of expressway and converting it into a city street that includes walking and bike paths that allow for greater connectivity among neighborhoods and the city center and also promote the development of a more livable mixed-use community. This project was completed in 2017.

Communities can be made more bikeable by removing obstacles to bicycling and creating bicycling infrastructure that makes it easier and safer for cyclists. These changes include making safer bike routes by creating bike lanes, separate bike paths, or reducing automobile speed limits on bike routes. Creating safer routes and encouraging biking and walking to school increase physical activity and improve the health and academic performance of children. The National Center for Safe Routes to School provides resources to help develop safe routes and encourage cycling to school.[18] Providing places to lock or store bikes and, in some cases, shower facilities for those who commute to work by bicycle also promotes cycling. Buildings that incorporate these features receive credit in the Green Building Council's Leadership in Energy and Environmental Design (LEED) certification process.

A number of cities have also launched affordable lock-and-dock bike share programs that allow riders to borrow a bike from one location and return it to another. Many bike share systems offer subscriptions that make the first 30–60 minutes of use either free or very inexpensive. This encourages the use of bikes as transportation rather than casual riding for the day, which might be better served by bike rental. This also allows each bike to serve several users per day. For many systems, smartphone mapping apps show nearby stations with available bikes and open docks. Many private bike share programs partner with public transit systems, adding to the diversity of sustainable transport options.

Like bike share programs, car share programs provide cars by the hour, often including gas and insurance with membership. This makes cars available to those who have only limited need, allowing them to save on car ownership. It also reduces the total number of cars on the street, relieving congestion and making community transport more sustainable by being less car intensive; each short-term rental car replaces approximately 15 owned cars. Also, car sharing members drive less and spend approximately 10% of what car owners spend per year on car travel. A number of car share programs offer dedicated parking spots; use only fuel-efficient, electric, or hybrid cars; and are integrated with other modes of transportation by having pick-up and drop-off locations at bus and rail stations and at airports. In 2014, the global number of car sharing members was approximately 5 million; as it grows, it is anticipated to further reduce total vehicle travel and associated carbon emissions.

Similar to car sharing, ride sharing includes shared carpooling and for-pay rider share companies such as traditional taxis, in addition to newer ride share programs. Car- or vanpooling is an important part of transportation sustainability in rural areas. Washington state has one of the largest vanpool programs, supporting nearly 20,000 commuters. With the support of state grants, hundreds of companies in Washington have created vanpooling programs because vanpooling reduces their parking costs, helps them retain employees, can offer tax benefits, and meets total commute trip reduction sustainability goals. Participants also save on transportation costs. High-occupancy vehicle lanes that are limited to vehicles with at least two or three passengers at peak travel, convenient park and ride facilities, and preferred parking and other employer incentives promote carpooling.

Also, part of the sharing economy are companies such as Uber and Lyft that are similar to taxis but offer a greater level of convenience by using smartphone apps to connect riders with private drivers and process cashless payments. There is some question as to whether this type of car service is sustainable. Some argue that these services are detracting riders from public transport. They may or may not reduce the number of car trips taken in major cities or car ownership on the whole. Also, their employment model classifies workers as independent contractors, not employees, so that the companies are not responsible for expenses, payroll taxes, fuel, car insurance, and sick pay. The California Labor Commissioner recently ruled that these drivers are employees and that the companies are required to cover these expenses; however, this was overturned in federal court. The Natural Resources Defense Council in partnership with the University of California at Berkeley's Transportation Sustainability Research Center plan to analyze the contribution or detriment to community sustainability of this kind of transportation service.

ENERGY AND CLIMATE CHANGE

A sustainable community reduces its use of fossil fuels, mitigates climate change by limiting GHG emissions, makes energy more affordable to all its members, and must be resilient to the effects of climate change. In so doing, it also reduces its external impacts on areas outside of the community and global ecology. In addition to promoting energy-efficient transportation, this can occur by making government buildings and operations more energy efficient and when local energy utilities commit to transitioning to renewables. Local governments everywhere are taking action to reduce energy use and on climate. The government sector is at the forefront of the green building revolution as it designs, builds, and operates its facilities using green building principles both to improve environmental health and to reduce government spending, with 30% of all LEED projects owned or occupied by government entities.[19] These visible public projects also model and encourage the private sector to build green. Other efforts to green government operations include transitioning to light-emitting diode (LED) bulbs in all street, traffic, and other public lighting. Many communities are also converting their transportation fleets to more energy-efficient vehicles.

The US Environmental Protection Agency (EPA) has developed the Local Climate Action Framework as a guide to plan, implement, and evaluate community climate, energy, and sustainability projects, programs, and specific policy proposals. This program has invested $20 million to support 50 showcase communities. The expected savings from that investment were expected to surpass that amount by 2015 while significantly reducing GHG and other air pollutant emissions and making those communities more livable. By demonstrating these benefits and through networking and information sharing, communities outside of the showcase program are beginning to adopt similar programs.

One example of a showcase project is the Duluth, Minnesota, energy-efficiency program that focuses on outreach and support to reduce energy use and costs for single-family and multifamily housing, including low-income advanced energy retrofits. Its Building Performance Institute trains contractors and do-it-yourself individuals interested in making their own energy savings improvements. This $500,000 project will result in $680,000 in energy savings, reduce CO_2 emission by 7,012 million tons, and create at least 10 new green jobs. More information about the Duluth project can be found at http://www.duluthenergy.org, and all of the showcase projects are featured on the EPA website.[20]

The transition to renewables and the mitigation of climate change can occur at the community level when local energy utilities commit to transitioning away from fossil fuels to renewables. This typically only occurs when the utility is publicly owned by the municipality. Until the 1980s, most utilities were public. With the deregulation and privatization movement in the 1980s and 1990s, almost all utilities were sold off to large, private corporations whose infrastructure is dependent on fossil fuels. Driven by profit, these energy conglomerates are reluctant to adapt to renewables because of the cost of changing to a renewable infrastructure and the relatively low cost of government-subsidized fossil fuels. Some municipalities have attempted to negotiate with energy providers to offer renewables, but they have had little success given the power and reluctance of these companies.

Some communities have overcome this barrier to renewable energy by converting their energy sector into a non-profit public entity. This municipalization of the utilities allows the community to have autonomy in choosing the source of energy and to reduce costs by eliminating the expense of marketing, executive pay, and the shareholder drive for profit. The state of Colorado has led the country in this municipalization, with 30 municipalities that now have publically controlled utilities. This has also been a common trend in Europe. For example, in Germany, 170 municipalities and 650 worker cooperatives are providing a major proportion of the country's energy.

The municipalization of energy utilities has led to a proliferation of renewable energy, especially solar and wind. This is because communities that want to transition to renewable sources now have the agency to do so, and many have the goal of achieving 100% renewable energy in the coming decades. Many of these municipal utilities also invest in reducing energy consumption in their communities (e.g., in Sacramento, California). Municipalization has not been an easy process, as power companies spend millions of dollars on campaigns to prevent the resultant loss of market share. The ability of communities such as Boulder, Colorado; Sacramento, California; and Hamburg, Germany to take back control over their energy represents a success story of local democracy and environmental sustainability.

In addition to the municipalization of private energy corporations, communities can shift to sustainable energy through the process of **community choice aggregation (CCA)**. CCA has been adopted into law in a handful of US states, allowing individual customers to bundle their purchase of electricity in order to secure alternative energy supply contracts on a community-wide basis. CCAs have been successful at transitioning to renewable energy while reducing consumer cost. An example is the City of Evanston, Illinois, which is partnering with Homefield Energy for its CCA to provide participating residents and businesses with 100% renewable energy. Some CCAs use the policy as a way to transition to local renewable energy sources rather than using the more typical approach of purchasing renewable energy credits to put more renewables on the national grid. This shift to local production has been adopted by a number of communities in Sonoma County, California.

Both private and public utilities often help those with lower income meet their energy needs through a variety of residential energy-efficiency programs. One way is through offering rebates for efficient appliances and heating systems. Another is to provide home energy audits and assistance with home improvements that reduce energy use. Federal and state tax incentives for increased efficiency or home conversion to renewables also increase community sustainability in the area of energy.

A sustainable community is one that mitigates climate change through the reduction of GHG emissions by transitioning to renewables and increasing energy efficiency through improvements in housing, industry, agriculture, and transportation. In addition, a sustainable community must also prepare for the anticipated effects of climate change. These include flooding as sea levels rise, increases in extreme whether events, increased temperature and drought, threats to water and food security, and the emergence of new diseases. These will seriously stress communities and cities and thus must be incorporated into today's decision-making processes. In response to this, New York City has launched a $19.5 billion climate resiliency plan that includes constructing adaptable floodwalls, storm-surge barriers, and dune systems and revising building codes throughout the city. **Geodesign** is a tool that can guide planning for resiliency by integrating geographical information system (GIS) mapping and zoning and development decision-making. The Rockefeller Foundation has made catalyzing attention, funding, and action to promote resilience to climate change worldwide a major priority.

Nature-based solutions (NbS) that are inspired and supported by nature are an important component to building community resilience with regard to climate change. NbS can offer multiple benefits, including flood management and securing improved health outcomes. Examples include parks and green roofs that reduce heat stress, lagoons and canals that store water, and the use of permeable surfaces such as rain gardens that allow for the infiltration of water.

To realize the full potential of NbS for sustainable urban development and community resilience in a changing climate, the NATURVATION project started in 2016 with 7.8 million euros from the European Commission. It takes a cross-disciplinary, international comparative approach to capture the multiple impacts and values of these solutions. It does so by supporting collaboration among university researchers, city governments, NGOs, and businesses throughout the world. The goal of NATURVATION is to develop nature-based innovations and help implement them on the ground to best address the urban challenges associated with climate change in a way that contributes to economic activities and social well-being.[21]

CIVIC ENGAGEMENT

Communication in the form of civic engagement is fundamental to community sustainability. Civic engagement is the way in which citizens participate in the life of a community in order to effect change and help shape its future. It is the participatory process that is necessary to build social capital, or the networks of relationships among people, which in turn are required for the effective pursuit of sustainability. Communities with the greatest and most diverse citizen participation are often the best able to achieve community sustainability objectives and are the most resilient. Engaged citizens are more informed, feel more a part of the community, and are the most capable to either directly generate significant change or sustain it through buy-in and ownership.[22]

Civic engagement can occur by recruiting diverse community stakeholders to participate in planning processes and activities through inclusive workshops referred to as *charrettes*. Civic engagement can also be more proactive through the activity of grass-roots movements that create opportunity for community change through the democratic process. This can occur through a **referendum**, which is the legal mechanism for voters to repeal or accept a law passed by the state legislature. Referenda enable citizens to bypass their local legislature by putting new policies or laws on the ballot for direct vote by the people. The municipalization of the energy utility and the subsequent transition to renewables such as that in Boulder, Colorado, is an example of a grass-roots movement that took control of its sustainable future through referendum. Another example is Referendum 52, which authorizes the state of Washington to sell $500 million in bonds to fund comprehensive energy efficiency retrofits in schools.

There are a number of ways to improve opportunity for civic engagement in communities and cities. One of the most effective ways is through the creation and participation of neighborhood associations and cultural societies. Civic engagement is increasingly enhanced through social media, rebranding efforts that help reinvigorate the associations and educate and involve citizens at minimal cost, and partnerships with local education organizations. A successful example that employs many of these approaches is a collaboration between the University of Oregon and the city of Salem to work to leverage resources and expertise, increase civic engagement, and bring community sustainability initiatives to fruition. An excellent resource for those wishing to encourage civil discourse and involvement in the process of democratic governance in order to make their communities more sustainable is Sustainable Communities Online.[23]

Smart Growth, New Urbanism, and Regional Sustainability

In many cases, it makes sense to think about and act on sustainability issues beyond the level of the community and apply these principles to broader metropolitan areas, networks of cities, and regions. A regional approach could make more sense when it comes to planning, broadening transportation efforts, equitable education funding, shared public resources, and the generation and distribution of a suite of renewable energies. A number of movements have grown out of this recognition, including smart growth, new urbanism, and regional approaches to advancing sustainability that bring together stakeholders from broader geographical regions. Success at this broader geographical level can be more challenging because it requires connecting people, communities, and governments to have them think and act beyond their immediate locale.

Smart Growth

Smart growth is a theory of land development that recognizes that growth and development will continue to occur and therefore seeks a comprehensive approach to directing that growth in sustainable ways. The movement grew out of the alarming, uncontrolled rate of sprawl or the expansion of development and the movement of human populations from centralized cities and towns into the countryside (Table 11.1). Between 1982 and 1997, the United States converted approximately 25 million acres of rural land, forests, rangeland, pastures, cropland, and wetlands into subdivisions, roads, and highways into what we think of today as suburbia or suburban sprawl (Figure 11.4). This development has tended to be unplanned, without regard to consequences for community, livability, the environment, and other aspects of sustainability. It is a dispersed form of development that is typically low-density, single-use, and auto-dependent, and it reduces open space, scenic areas, and farmable land. It has also resulted in the economic decline of inner-city communities.

Smart growth principles are directed at developing sustainable communities, cities, and regions. These principles include mixed-use, clustered land development that locates a range of affordable housing opportunities near businesses and places of work. By clustering development, open space can be preserved and shared by nearby residents as opposed to each home having its own large lot. In so doing, open space, parks, farmland, and critical environmental areas can be both preserved and made accessible

TABLE 11.1 Total Population of US Metropolitan Areas, 1970–2010

YEAR	SUBURBAN	URBAN	NATIONAL SPRAWL INDEX
1970	49,101,068	84,283,519	36.81
1980	69,967,436	84,680,392	45.24
1990	85,239,692	92,431,065	47.98
2000	101,295,542	102,952,391	49.59
2010	114,357,186	111,554,393	50.60

Total Population of US metropolitan areas, 1970-2010, National Sprawl Index is based on the percentage of the county population living at low suburban densities compared to medium to high urban densities. (Data from Smartgrowth America.)

Figure 11.4 Suburban Sprawl. *trekandshoot/Shutterstock.*

to communities (Figure 11.5). Smart growth provides for a variety of transportation options, including local and regional public transportation, and it enhances opportunities to walk and bike. It encourages redevelopment in existing communities or developed land rather than the development of farmland or natural lands. Smart growth development must also engage citizens and stakeholders in the decision-making process.

Smart growth is ideally growth that enhances the environment, the economy, and the well-being of the community equally. In terms of environment, smart growth reduces travel time. The proximity of people to their jobs and the town centers, improved opportunities for biking and walking, and effective public transit result in less driving and reduced emissions, including GHGs. Smart growth also protects water quality by preserving natural spaces and reducing paving. This enhances the reabsorption of storm water directly back into the Earth and reduces runoff and residual pollutants. By building on land that has previously been developed, such as brownfields and farmland, habitats and biodiversity can better be preserved.

Smart growth also offers economic benefits. It streamlines municipal budgets by reducing the per capita cost of infrastructure such as roadways and water systems, reduces maintenance costs, and reduces the costs associated with urban decline, including the reuse of vacant properties and brownfields.

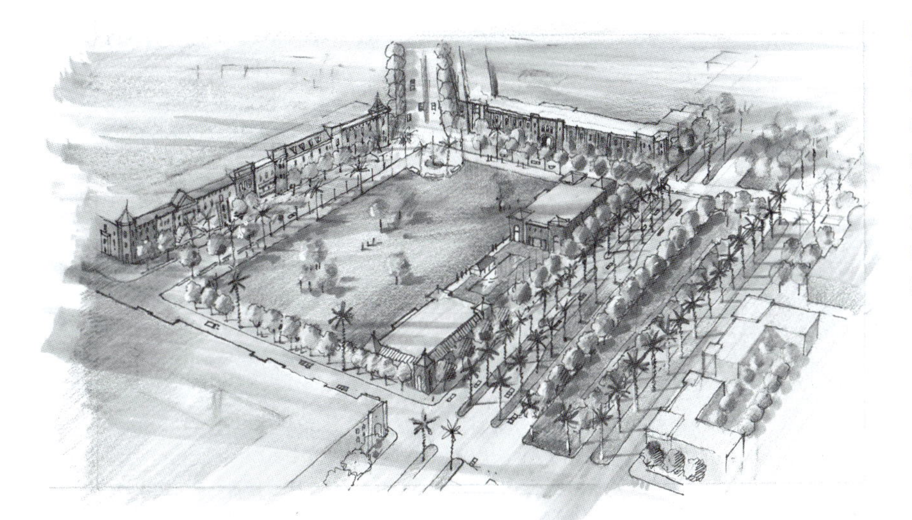

Figure 11.5 In smart growth, clustered mixed-use development allows for more open space as shown for the newly designed town center for Southside Savannah. *Courtesy of David M. Schwarz Architects, Inc.*

Because development is more concentrated, it reduces the costs of public services such as fire, police, and waste management. These municipal savings are not insignificant. A 25% shift from low-density sprawl-like growth to more compact development would result in a national savings of $4.2 billion in maintenance during a 25-year period.[24] Other economic benefits include fewer missed workdays by employees because they can more easily get to work. Accommodating walkways and mixed use allow for more foot traffic past storefronts and other businesses, which stimulates and sustains local and regional economies. Compared to sprawl-type development, housing tends to be more affordable because of reduced land ownership, cost of transportation, and shared resources and infrastructure.

Sprawl economically isolates low-income urban residents from areas of economic growth. Improved transportation, urban redevelopment, and the creation of mixed-income housing with smart growth help reduce that economic isolation by offering greater access to jobs and reducing the expense for transportation to and from work. A range of different housing types makes it possible for senior citizens to stay in their neighborhoods as they age, young people to afford their first home, and families at all stages in between to find a safe, attractive home they can afford.

There are social and health benefits of smart growth. One of the goals of smart growth is the development of vibrant, healthy communities. Because neighbors are less isolated from each other, it allows for increased community cohesion and creates social capital that in turn can stimulate civic engagement. It has been shown to reduce crime. Smart growth provides obvious health benefits because the promotion of walking and cycling is an integral part of the design. Less driving reduces risk from car accidents—a major source of injury and mortality—and there is less exposure to pollution. There is a clear relationship between sprawl and reduced physical activity. People living in areas of sprawl-type development are less likely to walk and will suffer greater levels of obesity and mortality compared to people who live in less sprawling counties. People in walkable neighborhoods on average do approximately 35–45 more minutes of moderate-intensity physical activity per week and are substantially less likely to be overweight or obese compared to similar people living in low-walkable neighborhoods.[25]

Despite its many potential advantages, smart growth is not without its critics. Some have contested the idea that smart growth reduces crime, and there is likely variability among regions. Others have argued that smart growth negatively affects lower income and minority families because it results in increased housing costs that decrease the availability of affordable housing and therefore displace or at least disproportionately disadvantage a large segment of the population. However, many smart growth strategists elect to include provisions for affordable housing in their planning. In addition, when considering proximity to work and transportation costs, overall expenses are often less despite higher housing prices. Some simply object to what they think of as excessive land-use regulation and go as far to blame the severity of the international finance and mortgage crisis on smart growth. These criticisms should be considered in any development process so that a robust economic, environmental, and healthful region with a variety of social benefits can be developed in a way that is more inclusive and thus more sustainable.

New Urbanism

New urbanism applies the smart growth principles to urban development and redevelopment. New urbanism is based on the principles of how cities and towns had been built in the past. It is another example of how sustainable innovation often incorporates historical approaches, a term we coin as *retrovation*. New urbanism focuses on human-scaled urban development with walkable blocks and streets, housing, and shopping in close proximity designed for people. Place-making, or the creation of public spaces such as plazas, squares, sidewalks, cafes, and porches to host daily interaction and public life, is a high priority.

The redevelopment of underutilized and neglected places is also a focus of new urbanism. One way this is accomplished is through HUD's HOPE VI and Choice Neighborhoods programs aimed at revitalizing the worst public housing projects in the United States into mixed-income communities with access to vital community services such as transportation, health care, healthy foods, green spaces, and quality schools. The latter is being achieved by linking different agencies through the Department of Education's Promise Neighborhoods Program.

AGRARIAN URBANISM

Agrarian urbanism is a new concept that integrates numerous food-related activities, including small farms, shared gardens, agricultural processing, and markets, into new urbanism design. It is different than urban agriculture in that it moves beyond the practice of farming to allow food to guide the structuring of the places in which we live. It places numerous food-related activities, such as small farms, shared gardens, agricultural processing, and a central market, into the urban center. It integrates agriculture at social and cultural levels. The goal is to promote sustainability by maximizing the economic, environmental, and social benefits of agrarian life.[26] A unique example of agrarian urbanism is the Southlands project in Tsawwassen, British Columbia, which is being designed to integrate urbanism with all scales of agricultural production (Figure 11.6).

Regional Sustainability

A regional approach to sustainability makes sense for resources or services that occur over a wider geographical area. Examples include energy, transportation land use, and agriculture. For example, with energy, the temporal variation in supply and demand for different sources of renewable energy can best be managed by integrating a diversity of renewable sources of energy across a region and managing it with smart grid technology. Regional transportation systems that connect cities with outlying and other metropolitan areas will enhance mobility while offering economic and environmental benefits. Land use decisions should also be considered at a regional scale to avoid redundancy and allow for a broader integration of land use types. For example, in US states that are commonwealths, there are hundreds of municipalities that each individually control their own zoning, and often each municipality is required to offer all zoning designations. A regional approach would allow for the concentration of specific land uses such as housing, urban development, industry, agriculture, and conservation areas over a broader geographical range.

Regionalism is particularly relevant to food systems. Local food and small producers are viewed as the cornerstone of a sustainable food system. However, over a broader geographical region, each farm might vary in volume of supply, food needs, variety, supply chains, and markets. A network of small producers can pool their buying power for inputs, plan in a way to diversify product offerings, and increase access to larger markets, resulting in a more sustainable, robust food system. This is referred to as **a food value chain**. One important component of a food value chain is the food hub. Smaller producers—one of the fastest growing segments of the local food market—tend to lack the capacity to access retail, institutional, and commercial food service markets on their own. By offering a combination of aggregation, distribution, and marketing services at an affordable price, food hubs make it possible for many producers to gain entry into new, larger volume markets. This increases income potential for small farmers, and it makes locally produced food available to a broader market.

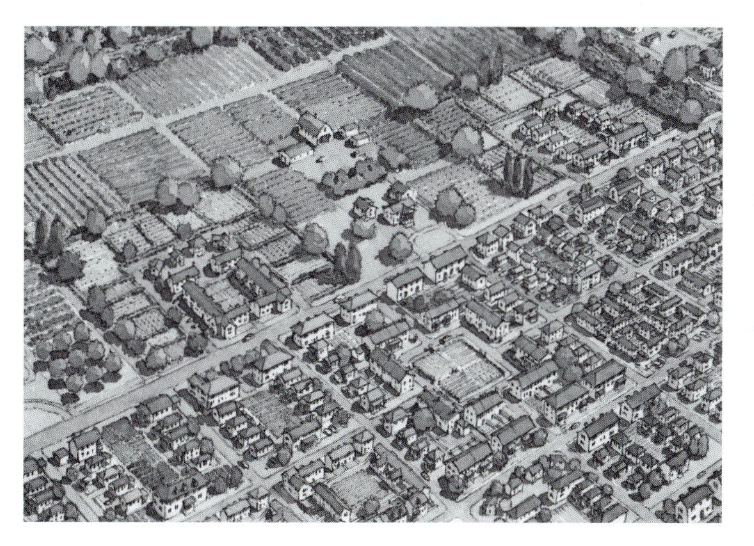

Figure 11.6 An artist's take on the aerial view of the proposed Southlands Development in Tsawwassen. This development is based on the concept of agrarian urbanism that that integrates numerous food-related activities including small farms, shared gardens, agricultural processing, and markets into New Urbanism design. © *Courtesy of DPZ CoDesign.*

Barriers to Community and Regional Sustainability

Now that we have an idea of what a sustainable community, city, and region each look like and all of the benefits they offer (Figure 11.7), it is difficult to imagine why they tend to be so rare. The problem is that there are a number of barriers to achieving sustainability at the community and regional levels, including subsidies, taxes, and policies and hidden costs that favor sprawl over smart growth. Other common barriers are the corruption and misuse of funds allocated to support community development projects and also a growing contempt of any government or social program even when it is effective. Declining infrastructure and lack of funding to invest in improvements, corporate interests that are in opposition to community sustainability efforts, and cultural barriers also limit progress at these levels. Overcoming these barriers will require strong, visionary leadership and public and private commitment. In this section, we discuss some of these barriers and then provide examples of how they have been overcome in creating sustainable communities, cities, and regions worldwide.

Federal Financing and Sprawl

Federal finance policies favor single-use suburban development over mixed-use, multifamily urban development and redevelopment, and they concentrate poverty in one area. Regulations on federally backed loans limit financing on larger buildings that include both residential and commercial use despite the evidence that loan performance in walkable, mixed-use neighborhoods is less risky than that in single-use family neighborhoods. The result is that 81% of federal loans and loan guarantees support single-family homeownership. These finance rules also reinforce the concentration of poverty by limiting investment in poorer areas, leading to deteriorating housing stock, fewer jobs, higher crime, and declining schools. The lack of financing for mixed-income housing and rehabilitation of deteriorating housing stock, coupled with the process of gentrification, further concentrates poverty.

Although there is frequent strong criticism of the use of tax dollars to stimulate economic and other types of development in urban centers, particularly from those who live in outlying suburban areas, the truth is that more public dollars are spent on subsidizing suburban development than urban

Figure 11.7 An idealized sustainable community that can be developed only by eliminating existing barriers. © *Coalition for a Livable Sudbury 2014.*

redevelopment. This is primarily in the form of one of the largest federal tax expenditures of $70 billion per year for the mortgage interest deduction. This expense does little to achieve the goal of expanding homeownership to mid- and low-income families because nearly 80% of the benefit goes to higher income households that are eligible to itemize deductions. This in turn encourages them to purchase more expensive, large homes in the suburbs. This subsidy discourages those wishing to purchase older, less-expensive properties in cities with the intent to rehabilitate them. For example, instead of buying a home for $50,000 and investing $100,000 to restore it, purchasing a new suburban home for $150,000 is more economically favorable because of the greater tax benefit.

Sprawl development is further subsidized when federal and state funds are used to construct and maintain roads and highways, regional transportation systems, water and sewer systems, and other infrastructure needed for that development. Also, electric rates tend to be the same for urban and suburban homes and businesses, but the costs of delivering that electricity to the suburbs is much greater, so in essence urban energy consumers are subsidizing those in the suburbs. People who live in the suburbs drive more often and consume more energy in their homes. A substantial part of that cost is paid for by taxpayers through the more than $40 billion dollars in fossil fuel subsidies.

A number of policy changes could help remove this barrier. One is the conversion of the mortgage interest deduction to a tax credit and lowering the maximum amount of interest it covers. This would benefit homeowners whether they itemize or claim the standard deduction, and instead of being based on the household's marginal tax rate, the tax benefit would be a fixed percentage of the household's mortgage interest, thus making the mortgage subsidy more equitable. Other policy changes that would help reduce the disincentive to developing sustainable communities rather than sprawl include raising non-residential caps on loans to mixed-use projects and minimizing risk by allowing shorter loan periods or requiring larger down payments.

Investment in Infrastructure

It is no secret that basic infrastructure required to maintain community function and connection between US communities is experiencing massive deterioration. This includes outdated water and waste management systems; crumbling roads, bridges, and highways; and increasing numbers of blight properties and industrial sites or brownfields. Relative spending on infrastructure has declined or not kept pace with inflation, and it can no longer meet current need. The average public investment in infrastructure during the past 20 years has been at 1.9% of total US economic output per year adjusted for inflation. Even to maintain this low level of investment during the next decade, federal, state, and local governments would need to boost spending by a total of $863 billion above current levels.

Currently, there is strong opposition to meeting this funding demand, but what most people do not realize is that the costs of not doing so will be even greater. Each year, US drivers spend 5.5 billion hours in traffic, resulting in costs of $120 billion in fuel and lost time. Poor transportation infrastructure costs US businesses $27 billion in additional freight costs. Due to the electric grid's low resilience, weather-related outages cost the United States up to $33 billion each year. Also each year, there are approximately 240,000 water main breaks. These result in service interruptions, property damage, and the need for expensive repairs that often are temporary fixes. Despite these high costs, investment in maintenance and improvements in infrastructure made by all levels of government have declined sharply in recent decades. For example, total public spending on capital improvements on US drinking water and waste water utilities peaked in 2010 and decreased 8% between 2010 and 2014.[27] This is the most significant four-year decline in real expenditures since 1956. Moreover, the burden for this investment has shifted drastically from the federal to state and local governments (Figure 11.8).

Increased funding and innovative financing will be required to meet these needs.[28] One example of this is the Build America Investment Initiative, which is designed to expand private investment and collaboration in major infrastructure. In public–private partnerships, private equity investors fund infrastructure projects in exchange for a return while also taking on some of the projects' risks and responsibilities. They also allow governments to introduce private sector capital, management, and technical expertise into the project. An additional benefit to increasing investment in essential infrastructure is that it results in substantial job creation.

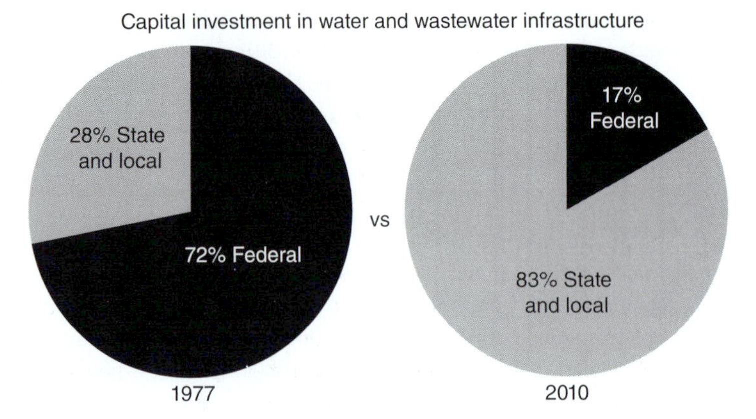

Figure 11.8 Federal capital spending for drinking water and sewer infrastructure has declined significantly since the 1970s, according to the Congressional Budget Office. State and local agencies now foot most of the bill for new treatment plants. (Kaye LaFond/Circle of Blue. http://www.circleofblue.org/2014/world/u-s-water-systems-deteriorated-slow-change-need-new-strategy-money/)

Corporate Barriers

In many instances, corporations have fought to halt progress for community sustainability through extensive lobbying and expensive public campaigns that use fear to dissuade individuals from supporting these kinds of efforts. Corporations have spent millions of dollars to retain or obtain control of fossil fuel–based utility systems, to privatize public water authorities, and on food policy that has had negative consequences for local and regional food systems. With regard to food, large corporate entities have begun to co-opt and capitalize on the "organic" and "local" labels. The result of this has been an over-bureaucratization and increase in costs of the certification processes that restrict the ability of small farmers to achieve those labels. Large agribusiness has also made it more difficult for small farmers to compete with producers that have large-scale supply chain marketing and distribution apparatuses. The creation of value chains and food hubs, as described previously, is one solution to this barrier to sustainable food systems.

Transparency and Corruption

When public funds are spent on development projects, there is potential for corruption or the perception of political favor. Often referred to as "crony capitalism," businesses can benefit from close relationships with government officials, resulting in favoritism in the distribution of contracts, permits, special tax breaks, or other forms of government interventionism in relation to publicly funded projects.

There are examples of community economic development projects in which only a few stood to gain the most financially from the projects. Sometimes there has been outright theft of government program funds. This is in direct conflict with our notion of sustainability, even if a project has successful outcomes. For example, a 2½-year investigation of HUD's Community Development Block Grants led to 159 indictments for misuse or misappropriation of funds. This theft and corruption diverts funds from community benefit and tarnishes what might otherwise be effective tools to achieve community sustainability. It also leads to the misallocation of resources, negatively affects public and private sector development, and distorts public policy. The primary mechanism for abolishing corruption and removing it as a barrier to community sustainability is cooperation by government, citizens, and the private sector to ensure transparency.

Cultural Barriers

Probably the most difficult barriers to overcome in achieving sustainability in communities and across regions are cultural. Rampant individualism and a lack of knowledge of how to work within an institution involving all stakeholders can limit effectiveness even among those who are most committed to the cause. Another cultural barrier to community sustainability is the distrust and growing contempt of government or social programs even when they are effective. Racism, greed, consumerism, and obsession with the car still dominate US culture and are barriers to sustainability. There are also many misperceptions and inconsistencies in what many consider a livable, sustainable lifestyle. For example, a recent poll of US baby boomers revealed that 78% prefer a cul-de-sac rather than a connected street, two-car garages, and houses with a large footprint and lawn; however, they overwhelmingly want to be near retail space without realizing that the pattern of development that they favor makes that impossible.[29]

Chapters 9 and 10 discussed many ways that individuals and institutions can change, including those that are cultural. Improved sustainability education especially in the early years, limiting the mass marketing of consumption, the sustainable business movement, and better demonstration of models of good programs and sustainable initiatives will contribute to the necessary cultural shifts. So will growing interest in the desire to increase quality of life by reducing consumption and having more livable lives in sustainable communities. However, as previously discussed, political and economic structures often limit this possibility. So the removal of cultural barriers will require both individual behavioral change and changes in institutional structures that better facilitate the transition to sustainable living. Examples of communities on a variety of scales that have overcome these barriers to sustainability are described in the next section with the goal of illuminating the ways in which they have done so.

Removing the Barriers to Community and Regional Sustainability

There are numerous success stories in which communities or regions have overcome the barriers to sustainability described in the previous section. In some cases, these were due to grass-roots, citizen-based efforts. In others, the accomplishments were the result of strong and visionary leadership and government action. The most successful accomplishments have occurred when individuals and the public and private sectors have come together to achieve sustainability objectives that benefit all stakeholders. Many of them required innovative technologies and approaches. All of them required leaders with strong interpersonal competence to effectively engage and bring together individuals and groups, often with very different values. This section features examples of how these barriers were overcome in order to achieve specific sustainability objectives.

Curitiba, Brazil: Transportation

Located in Parana state of Brazil, the city of Curitiba has implemented several innovative approaches to creating a sustainable city as it transitioned from an agricultural center with less than 200,000 people until 1950 to a manufacturing and economic center that is now home to more than 2 million people and supports a regional metropolitan population of 3.5 million. The key to Curitiba's sustainable growth was a detailed sustainability master plan adopted in 1968 that implemented several innovative systems to create jobs, improve public transportation accessibility, promote housing development, and improve waste management.

Curitiba's branching transportation system consists of five main arterial traffic roads into and out of the city, each with a dedicated central two-directional bus lane. The system uses triple-articulated, large-capacity buses that are color-coded based on function and route. There is only one fare, and interchanges are located throughout the city so people can change directions and busses without additional cost. Central to the design are unique elevated glass boarding tubes for shelter and efficient ticket purchase. The bus doors are wider and open directly into the tube, maximizing access for all users including the disabled. This allows for faster loading and unloading on the busses, resulting in less idling and reduced travel times. During peak hours, buses arrive every 60 seconds and are always full. The bus companies are paid per kilometer driven, not per passenger, so they run services on less popular routes and off-peak hours without a loss.

This system has been a tremendous success. Upon its implementation, an initial 25,000 passengers grew to more than 2 million passengers per day. Eighty percent of the daily trips made in Curitiba are through the bus system, resulting in a 30% fuel reduction and 25% lower carbon emissions per capita than in Brazil's other major cities. Per capita economic loss due to time spent in severe congestion in Curitiba is reduced by 10%. The system is economically sustainable as well. It is privately operated and completely user-funded without government subsidy.

Integrating this transportation system with the overall urban plan has led to numerous other benefits, including economic development that has resulted in a per capita income that is 66% higher than the Brazilian average. There are 50 m^2 of parkland per person offering open space, recreation, and cultural and environmental services. This open space is complemented by compact affordable housing. There is also an established public housing program that provides housing assistance and more than 50,000 housing opportunities to low-income families. The city also has a program that supports its poorest people, keeps the city clean, and diverts waste by providing bus tokens, food, and cash in exchange for collected trash and recyclables. This is part of a larger waste management program that has resulted in a 70% recycling rate (Figure 11.9).

Curitiba is regarded as a model sustainable city. In a recent survey, 99% of Curitiba's residents said they were happy with their city. In 2010, the Global Sustainable City Award was given to Curitiba. The award was introduced to recognize those cities that excel in sustainable urban development. So how did Curitiba overcome the barriers to community sustainability? There are many reasons, but two are very obvious. First, there was strong visionary leadership of then-mayor Jamie Learner, who believed that if you want to make life better for people, cities must become better through improved mobility, sustainability, and identity. This then led to a detailed master plan that integrated all of the elements of community sustainability and was the key to Curitiba's sustainable growth and success.

Amsterdam, the Netherlands: Waste

In the 1980s, the people of Amsterdam and the Netherlands as a whole began to express concern about their sustainable future. The Netherlands was becoming one of the most densely populated and highly industrialized countries in the world. The production and consumption of consumer goods were rising, and with that, their increased waste production was filling diminishing landfill space. Outdated incinerators and overused landfills were increasing toxic levels in soil, water, and air and beginning to threaten quality of life.

This unsustainable pattern of growth, consumption, and waste production led to what is known as the Netherlands "social response." This was an effort to unify the people, businesses, and government to develop more sustainable communities. This was achieved through goal setting and also legislation, regulation, implementation, and enforcement to ensure those goals were met. Central to this was a national waste policy that was based on the waste disposal hierarchy (see Chapter 8), waste disposal and treatment standards, and a focus on limiting waste streams through prevention, reuse, and recycling. The Netherlands is the first country to require producers and importers to be responsible for the collection and disposal of used goods that they originally produced.

Amsterdam, the most populous city and capital of the Netherlands, has embraced the national social response and national waste policy, having one of the highest rates of waste avoidance, recovery, and recycling in the world. Its waste management plan views waste as a raw material as the first link in a value chain. Its waste-to-energy system (see Chapter 8) creates nearly 100% clean energy that powers all government buildings and public lighting. It also uses sewage sludge to generate biogas that can also be used to generate heat and electricity. The city produces

Figure 11.9 Curitiba, Brazil, designed as a sustainable city with public transportation routes that radiate from the city center while maintaining green space. *FRED.*

high-quality materials such as concrete, calcium silicate bricks, de-icing salt, and asphalt from the materials recovered from waste, including the bottom ash or incombustible remains from incineration and the fly ash from their stack scrubbers. It also focuses on reuse and recycling of products and durable goods. At the center of this is an ingenious system of underground garbage collection that addresses a severe lack of space for street-side placement of garbage. Rather, residents bring their waste to the closest green or gray box, open the door, and drop it down into a container underground, where it can be stored and then emptied once or twice a week. Underground storage and transport of waste are rapidly becoming recognized as the future of waste management and being developed in a number of communities and cities (Box 11.2).

Amsterdam's waste management program results in a 99% diversion rate, making it essentially a zero waste community. The barriers to this path to sustainability were overcome by involving all stakeholders, public and private, in building a national agenda to what was perceived as an impending crisis. An eye to the future, along with collaboration, leadership, private–public partnerships, and enforcement of regulations and standards, is the cornerstone to the success of this program.

Samsø Island, Denmark: Energy

Samsø is a 112 km^2 (43 mi^2) Danish island with a permanent population of approximately 4,000. From 1997 to 2005, Samsø successfully transitioned to becoming 100% self-sufficient with renewable energy through a combination of wind and solar for electricity and geothermal and plant-based energy for heating. Samsø generates more power from renewable sources than it consumes, and it sells the excess electricity to the national utility. Samsø's CO_2 footprint is an amazing net negative 12 tons per inhabitant, which includes the 10 offshore turbines that were built to compensate for carbon emissions from transportation. The average CO_2 footprint in Denmark is 10 tons per inhabitant. If the offshore turbines were not included, the Samsø footprint would be 4.5 tons per inhabitant.

How was this accomplished, and can it serve as a model for other communities? A key to the success of this project was ownership of the wind turbines. Five of the offshore wind turbines were purchased by the municipality, three by local farmers, and the rest by an investment company selling shares to stakeholders. In other words, the transition to 100% renewable energy generates income for the municipality, which lowers taxes, and it provides income to the hundreds of residents who have full or partial ownership of the turbines. But the real story is how the cultural shift was made in order to get locals to embrace this project. That was achieved through the visionary leadership of local Søren Hermansen.

Hermansen worked with the municipality to develop a winning proposal for a competition sponsored by the Danish Ministry of Environment to find a community that could change from fossil fuel to 100% renewable energy. The next step was to convince the diverse members of this island community to invest in the project. Hermansen knew Samsø islanders were tight-knit and conservative. He viewed that as an advantage because he knew that once he convinced enough potential first movers to act, the rest would follow. Hermansen accomplished this by showing up at every community or club meeting to give his pitch for transitioning to renewables, emphasizing both the environmental and the economic potential benefits. He often brought free beer. By engaging stakeholders in a personal way, he was able to garner trust and then widespread support.

Thanks to Hermansen, Samsø has become a global example of how to create a sustainable community through local ownership and community engagement. Samsø has attracted global attention for its accomplishments. Hermansen recognized that the greatest challenge to accomplishing change is the engagement of a diversity of people. His approach has been to search for collaborative partners and to ask the question, "What can *we* do?" He linked commons and community into what he calls "commonity." Once the diverse people of Samsø recognized that their commonity can sustainably share the economic and environmental benefits of common resources, the cultural barrier to sustainability transformed into a cultural revolution that made Samsø the model for community transition to renewables.

Noto, Japan: Agriculture

Although only 20% of Japan's land is suitable for farming, agriculture has been a major part of the economy in terms of the number people who engage

BOX 11.2

Innovators and Entrepreneurs: Underground Waste Collection

In Amsterdam, garbage is placed in underground storage systems throughout the city, where it is collected typically on a weekly basis. This preserves limited sidewalk and street space, and it increases aesthetics and improves sanitation of the city. The reduced rate of pick-up saves on fossil fuels and GHG emissions. This type of underground disposal and storage of garbage is leading the way to even more environmentally friendly approaches to garbage disposal and transport.

A new innovation that eliminates the use of trucks to pick up garbage is the underground pneumatic waste collection system. In this type of system, users deposit their garbage into ideally situated waste inlets that are accessible 24 hours a day. Each collection point serves 100–150 residences. Each type of refuse (mixed waste, organic waste, and paper waste) goes into separate inlets that are clearly labeled. By applying a massive amount of negative pressure, the deposited garbage is sucked through the pipes, where it may travel 8 or 9 km at up to 50 km/hour to transfer stations; it is then stored in large containers at the stations (Figure 11.2.1). The system is remotely monitored and controlled, and no personnel are needed in the collection and transportation to the transfer stations. Then depending on the type of waste, containers are shipped to specific locations for further processing, typically using existing public, underground railway networks. Ultimately, organic wastes are converted into biogas, recyclables are processed, and other waste is cleanly incinerated to produce electricity or heat.

Although the initial investment was high, there are a number of economic benefits that will offer a payback period within 10–12 years. Benefits include reduced personnel costs and reduced waste vehicle and fuel costs. Environmental benefits are decreased fuel consumption and CO_2 emissions, reduced chance of garbage spilling into the streets, as well as less traffic and disturbance from vehicular garbage removal. There is also evidence that the waste collection points of the pneumatic system increase rates of recycling.

There are approximately 1,000 pneumatic systems currently in operation throughout the world, including China, Southeast Asia, Korea, the Middle East, Europe, and to a lesser extent, North America. Major cities in which the system is operating include Copenhagen, Barcelona, London, and Stockholm. In the United States, this type of system has been installed in several locations, including Disney World and Roosevelt Island, New York. Systems are being planned as part of urban development plans in Carmel, Indiana; Montreal, Canada; and Mecca, Saudi Arabia.

Figure 11.2.1 An underground, pneumatic waste system eliminates the need for garbage pickup, keeps streets clean, increases recycling, saves energy, and reduces GHG emissions. *Courtesy of Envac.*

in it. However, the agricultural economy accounts for less than 1.3% of Japan's gross national product, and Japan's economic boom that began in the 1950s left farmers economically far behind the rest of the nation. However, until recently, government policy that provided for large agricultural subsidies and restrictions on imports made farming somewhat viable and encouraged farmers to increase the output of all crops, thereby increasing Japan's food security. These policies are now being eroded somewhat because of US pressure and international trade policy that requires high levels of liberalization or relaxation of government involvement in economic activity. As a result, the future of Japanese farming is in peril.

The Japanese agricultural model is becoming increasingly unsustainable. The average income of farmers is less than one-fifth the average household income in Japan. The number of farmers is rapidly declining, and farmers are aging, with an average age of 65 years and more than one-third older than age 75 years. Domestic farmers now only meet 44% of total consumption, with imports making up the deficit. A small number of farms have become large producers, claiming a major share of total production.[30] All of these factors directly contradict criteria for sustainable food systems.

The issue of food security and sustainability is being addressed in the Ishikawa Prefecture, or district, of Japan by taking a regional approach to economic development that includes agriculture. The Ishikawa Prefecture is on the northern part of the main island of Japan and consists of three main regions: Kanazawa, Kaga, and the Noto Peninsula. Noto is characterized by a mosaic of socioecological production managed systems that employ traditional integrated agriculture. As such, it has been designated as a Globally Important Agricultural Heritage System (GIAHS) by the Food and Agriculture Organization (FAO). The aim of the GIAHS program is to identify threatened agricultural systems and conserve them much in the same way that designated historic monuments and natural areas are conserved. This is done by integrating the preservation of agricultural systems and knowledge, conserving biodiversity and unique habitats, and protecting and promoting sustainable livelihoods on a regional basis.

In Noto, many of the customs and cultures are tied together with the life of agricultural communities. The approach in Noto is to maintain the link to these 4,000-year-old cultural traditions to produce a variety of affordable, organic, healthy foods as opposed to rice monocultures, which have come to dominate Japanese agriculture. Farmers follow what they refer to as the "mottainai spirit," or the ethic that things are too precious to waste. Because farms are small, the focus has been on increasing yield for a diversity of regionally traditional vegetable and rice varieties using permaculture techniques that were born in this region of the world. They also promote farming methods that help conserve biodiversity, such as maintaining water on rice paddies during winter to provide feeding grounds for the crested ibis and other rare birds and amphibians.

The designation of Noto as a GIAHS is linked to regional sustainable development. This is achieved by the regional promotion of sales and production of crops by branding agricultural products as organic and traditional. The traditional agricultural sites are promoted as part of a broader regional tourism program including green tourism and experiential learning in order to increase regional income and generate demand for the agricultural products. The Ishikawa Prefecture offers a diversity of tourism opportunities that have created demand for the local agricultural products, creating more jobs and higher wages in the agricultural sector.

The GIAHS program in Noto helped make more sustainable a community whose food system was threatened by national and global policy. Its focus on connecting traditional, diverse, small farming approaches to a regional effort to both conserve biodiversity and increase economic development through tourism has enhanced agriculture and food security for local communities throughout the Ishikawa Prefecture. FAO's partnership with national and international agencies as well as Japanese academic institutions provided necessary information and financial resources to achieve these objectives.

Although the Noto case is focused on community and regional sustainability, it must be recognized that the threat to local sustainability that prompted this effort was caused by broader national and international policy. These include government subsidies, international trade agreements, and development approaches that emphasize privatization, fiscal austerity, deregulation, and free trade. Such global policies will directly impact local sustainability, and they must be considered within this context. In Chapter 12, we explore global sustainability from the perspective of sustainable development, and we discuss how different types of approaches can either foster or hinder sustainability objectives.

Chapter Summary

- A community is a group of people, often with diverse characteristics, who live in the same geographical area or setting linked by social ties, common perspectives, joint action, and other aspects of life. Sustainable communities are made more livable through economic development, job creation, and improved quality of life. They also minimize their impact on the environment both within and outside of their immediate location.
- Community economic development improves the quality of life in struggling communities by making them more livable through enhancing economic opportunities, empowering individuals, and improving social conditions. Affordable and equitable access to housing is a cornerstone to community sustainability and has had mixed success as housing policy has evolved. The use of local taxes to fund education has created large inequities among communities in the quality of schools.
- The sustainability of a community is highly dependent on its food system. A food system is the sum of all activities required to make food available to consumers. A sustainable food system integrates all of these elements to enhance economic, environmental, social, and nutritional health for all. A sustainable community relies on access to health care systems that integrates sustainability, mission, and management, including equitable access.
- A sustainable, livable community provides equitable access to open or natural spaces such as parks, greenways, and gardens. These spaces can contribute to environmental, economic, and social components of sustainability. The creative arts also have a major role to play in sustainable community development. Sustainable transportation is a key component to a sustainable community because it links housing, employment, recreation and open space, food, health care, education, and other elements of a sustainable community.
- A sustainable community reduces its use of fossil fuels, mitigates climate change by limiting GHG emissions, makes energy more affordable to all its members, and must be resilient to the effects of climate change. In so doing, it also reduces its external impacts on areas outside of the community and global ecology.
- Communication in the form of civic engagement is fundamental to community sustainability. Communities with the greatest and most diverse citizen participation are often the best able to achieve community sustainability objectives and are the most resilient.
- A regional approach to sustainability could make more sense with regard to planning, broadening transportation efforts, equitable education funding, shared public resources, and the generation and distribution of a suite of renewable energies. A number of movements have grown out of this recognition, including smart growth, new urbanism, and regional approaches to advancing sustainability that bring together stakeholders from a broader geographical region.
- There remain a number of potential barriers to achieving sustainability at the community and regional levels. These include subsidies, taxes and policies, and hidden costs that favor sprawl over smart growth; corruption and misuse of funds allocated to support community development projects; a growing contempt of any government or social program, even when it is effective; a declining infrastructure and lack of funding to invest in infrastructure that would be more sustainable; corporate opposition to community sustainability efforts; and cultural barriers.
- There are a number of cases in which the barriers to sustainability have been overcome through visionary leadership and planning, participatory processes, and programs that promote sustainability.

Digging Deeper: Questions and Issues for Further Thought

1. Online communities moderated through social media have become very common. What aspects of community do online communities provide? What aspects of community are missing?
2. What are the elements of a sustainable community, and what are the barriers associated with sustainable community development?
3. What are the causes of sprawl? In what ways is it inconsistent with sustainability objectives? Are

there realistic and attractive alternatives to living in large-scale suburban developments?

4. What are the impacts of various modes of development on the sustainability of our food systems? How can development proceed in a way that preserves such systems?
5. What are the connections between economic inequity and barriers to community and regional sustainability? What are the ways in which such barriers can be overcome?

Reaping What You Sow: Opportunities for Action

1. Identify a specific community of interest. Conduct research on potential problems or opportunities for improvement within that community. Identify relevant community stakeholders. Contact as many of them as possible to gain perspective on any problems and opportunities for solving them. Propose a solution to the problem, including plans for implementation and assessment.
2. Identify both local and regional planning commissions in your area. Attend a meeting, and observe what kinds of issues they are working on. Try to ascertain if their work is consistent with community and regional sustainability objectives by asking questions at the meeting.
3. Write and send a letter to one or more of your local politicians and specifically ask them what their vision is for the sustainability of your community and/or region.
4. Construct a map of bikeable and walkable areas of your community. Identify areas that limit this type of transportation, and propose solutions to improve this aspect of sustainable living.

References and Further Reading

1. MacQueen, Kathleen M., Eleanor McLellan, David S. Metzger, et al. 2001. "What Is Community? An Evidence-Based Definition for Participatory Public Health." *American Journal of Public Health* 91 (12): 1929–38.
2. Institute for Sustainable Communities. http://www.iscvt.org
3. Iuviene, N., A. Stitely, and L. Hoyt. 2010. "Sustainable Economic Democracy: Worker Cooperatives for the 21st Century." A publication series produced by the MIT Community Innovators Lab with support from the Barr Foundation.
4. Mukhija, V., L. Regus, S. Slovin, and A. Das. 2010. "Can Inclusionary Zoning Be an Effective and Efficient Housing Policy? Evidence from Los Angeles and Orange Counties." *Journal of Urban Affairs* 32 (2): 229–52.
5. "Engaging as Partners in Energy Efficiency: Multifamily Housing and Utilities | ACEEE." Accessed March 22, 2016. http://aceee.org/research-report/a122
6. http://www.neighborworks.org
7. Barr, Caelainn, and Shiv Malik. "Young Families Priced out of Rental Markets in Two-Thirds of the UK." *The Guardian,* March 14, 2016, sec. World news. http://www.theguardian.com/world/2016/mar/14/young-families-priced-out-rental-markets-in-two-thirds-uk
8. "This Is The Hourly Wage It Takes to Afford a 2-Bedroom Apartment." *The Huffington Post.* http://www.huffingtonpost.com/2015/06/01/minimum-wage-apartment-hourly-salary-housing-costs_n_7472472.html
9. University of Michigan Urban & Regional Planning Capstone Project. 2009. "Building a Community-Based Sustainable Food System: Case Studies and Recommendations." Ann Arbor: University of Michigan Urban & Regional Planning Capstone project.
10. Sherer, P. M. 2006. "The Benefits of Parks." San Francisco: The Trust for Public Land.
11. Lam, Kin-Che, Sai-Leung Ng, Wing-Chi Hui, and Pak-Kin Chan. 2005. "Environmental Quality of Urban Parks and Open Spaces in Hong Kong." *Environmental Monitoring and Assessment* 111 (1–3): 55–73. doi:10.1007/s10661-005-8039-2
12. Goodman Williams Group. 2005. "Millennium Park Economic Impact Study." Prepared for the City of Chicago, Department of Planning and Development.
13. Kuo, France E., William C. Sullivan, Rebekah L. Coley, and Liesette Brunson. 1998. "Fertile Ground for Community: Inner-City Neighborhood Common Spaces." *American Journal of Community Psychology* 26 (1): 823–51.

14. Ewing, R., T. Schmid, R. Killingsworth, A. Zlot, and S. Raudenbush. 2003. "Relationship Between Urban Sprawl and Physical Activity, Obesity, and Morbidity." *American Journal of Health Promotion* 18: 47–57.
15. Cortright, Joe. 2007. "Portland's Green Dividend: A Whitepaper for CEOs for Cities." Chicago: CEOs for Cities.
16. World Health Organization. 2015. "Global Status Report on Road Safety 2015." Accessed March 21, 2016. http://www.who.int/violence_injury_prevention/road_safety_status/2015/en
17. Giles-Corti, B., S. Foster, T. Shilton, and R. Falconer. 2010. "The Co-Benefits for Health of Investing in Active Transportation." *New South Wales Public Health Bulletin* 21 (6): 122–27.
18. National Center for Safe Routes to School. "Safe Routes." http://www.saferoutesinfo.org
19. US Green Building Council. 2011. "Roadmap to Green Government Buildings: Guide for Government Professionals Implementing Green Building Programs and Initiatives." Washington, DC: US Green Building Council.
20. US Environmental Protection Agency. "Energy Resources for State, Local, and Tribal Governments." https://www.epa.gov/statelocalclimate/climate-showcase-communities-program
21. Centre for Environmental and Climate Research. "Naturvation." http://www.cec.lu.se/research/naturvation
22. Portney, Kent. 2006. "Civic Engagement and Sustainable Cities in the United States." *Public Administration Review* 65: 575–91.
23. Sustainable Communities Online. "Civil Engagement." http://www.sustainable.org/creating-community/civic-engagement
24. Smart Growth America. http://old.smartgrowthamerica.org/issues/economic-prosperity/municipal-budgets/
25. McCann, B. A., and R. Ewing. 2003. "Measuring the Health Effects of Sprawl: A National Analysis of Physical Activity, Obesity and Chronic Disease." Washington, DC: Smart Growth America.
26. Duany, Andres, and Duany Plater-Zyberk & Company. 2011. *Garden Cities: Theory & Practice of Agrarian Urbanism*. London: The Prince's Foundation for the Built Environment.
27. Congressional Budget Office. 2015. "Supplemental Data for the Public Spending on Transportation and Water Infrastructure, 1956 to 2014 Report." CBO Publication No. 49910. March.
28. US Department of the Treasury Office and Economic Policy. 2014. "Expanding Our Nation's Infrastructure Through Innovative Financing." https://www.treasury.gov/press-center/press-releases/Documents/Expanding%20our%20Nation%27s%20Infrastructure%20through%20Innovative%20Financing.pdf
29. "New Study Confirms That Boomers Are Clueless." MNN—Mother Nature Network. Accessed April 13, 2016. http://www.mnn.com/your-home/remodeling-design/blogs/new-study-confirms-boomers-are-clueless
30. Tokyo Foundation for Policy Research. "Japan's Agriculture and the TPP." Accessed April 20, 2016. http://www.tokyofoundation.org/en/articles/2013/japan-agriculture-and-tpp

CHAPTER 12

SUSTAINABLE DEVELOPMENT AND GLOBAL SUSTAINABILITY

In this chapter, we examine how sustainability can be achieved at the global level. We do this first by applying our definition of sustainability to development and then consider indicators that are used to measure it. We then explore the different approaches to sustainable development by examining the variety of players and programs. We also take another look at global climate change in relation to global sustainability by examining challenges and progress in dealing with it. We then examine barriers that must be overcome to achieve global sustainability, and some recommendations on how to overcome them are presented. Finally, we explore some common themes that have emerged throughout this book that will bring to light some of the essential elements of effective sustainable solutions and sustainable development.

PLANTING A SEED: THINKING BEFORE YOU READ

1. What are the various ways that we measure development? To what extent do these measures account for our criteria for sustainability?
2. What qualities come to mind in considering whether a country is developed or developing?
3. What is meant by sustainable development? How is it related to your working definition of sustainability?
4. In what ways is individual consumption related to sustainable development?
5. What percentage of the US federal budget is allocated to international aid and development?

The principle of common but differentiated responsibilities is the bedrock of our enterprise for a sustainable world.

—Narendra Modi,
Prime Minister of India

Sustainable Development

In Chapter 2, we walked through the process of developing a working definition of sustainability. Each time this exercise is conducted with a different group of people, a slightly different definition emerges. The definition that you arrived at might also be different. That is okay because the process asks us to incorporate common elements from all the ways that sustainability has been framed as the concept developed. By incorporating these elements through a participatory process with diverse stakeholders, we end up with a working definition from which we can develop, apply, and assess sustainability objectives in any given area.

When we apply our definition of sustainability to development, some obvious objectives emerge. The first is to create ways for people who are not sufficiently having their basic needs met to do so. This must be done without compromising the ability of future generations to meet their needs. As we develop ways to accomplish this, it is important to use a participatory, inclusive process within the context of science. Sustainable development must maximize renewal, encourage reuse, address consumption, minimize waste, protect the environment, and in particular, address climate change. Central to sustainable development is the creation of equitable economic opportunity that offers all people the capacity to meet their own needs. Sustainable development must result in an elevated standard of human well-being, including, but not limited to, improved health, improved nutrition, access to clean air and water, and universal access to basic human rights. Finally, we have to develop ways to assess any effort to achieve sustainable development so that we can adapt and improve our approach based on real measures of success and failure.

How We Measure Development

Approaches to sustainable development must be measurable so that we can determine where needs are the greatest, identify barriers, and gauge the successes and failures of development efforts. In this section, we explore traditional measures of development, examine their strengths and limitations, and then discuss some alternatives that are more consistent with sustainability objectives, including measures of human well-being.

ECONOMIC INDICATORS

A commonly used indicator of a country's economic health, standard of living, and level of development is **gross domestic product (GDP)**. GDP is the monetary value of all final goods and services produced within a nation in a given year. It is the sum of the value of all consumer spending or consumption, all government spending, all investment within a nation, and the value of a nation's net exports (i.e., the difference between total exports and imports). It is typically calculated on an annual basis. The 20 highest ranking countries in 2016 had GDPs greater than $500 billion. The top 3 countries were the United States with $18.5 trillion, China with $11.2 trillion, and Japan with $4.9 trillion. The average of the 100 lowest GDPs was approximately $10 billion, or three orders of magnitude less than those with the highest GDPs, reflecting the wide global variation in national wealth and development (Figure 12.1).

The use of GDP as an economic development indicator has many advantages. It is easy to calculate, and because it is done so uniformly from country to country, it can be used as a basis of comparison. It can also be adjusted for the rate of inflation in each country as well as for exchange rates among currencies. GDP is highly correlated with other measures of development, such as life expectancy, education, and literacy rates. However, it is not a direct measure of the overall standard of living or well-being of a country because it does not take into account factors that reveal the distribution of that wealth among all its citizens.

There are a number of other drawbacks of the use of GDP as an indicator to compare economies and development among countries. First, it does not consider income earned by businesses or individuals outside of the country or income earned within the country by foreign interests, which is becoming a growing portion of most national economies. **Gross national product (GNP)** takes this into account, and it is often used as an economic indicator in place of GDP. Neither of them, however, accounts for a nation's debt, which can be a significant drag on an economy.

Although GDP and GNP measure monetary flow for a given country, when compared among countries they are limited indicators of standard of living or development. For a valid comparison of these indices among countries, they should be adjusted for differences of individual purchasing power. The latter adjustment is referred to as **purchasing power parity (PPP)**. PPP is how much of a local good a person can buy in their country and thus reveals more information about the standard of

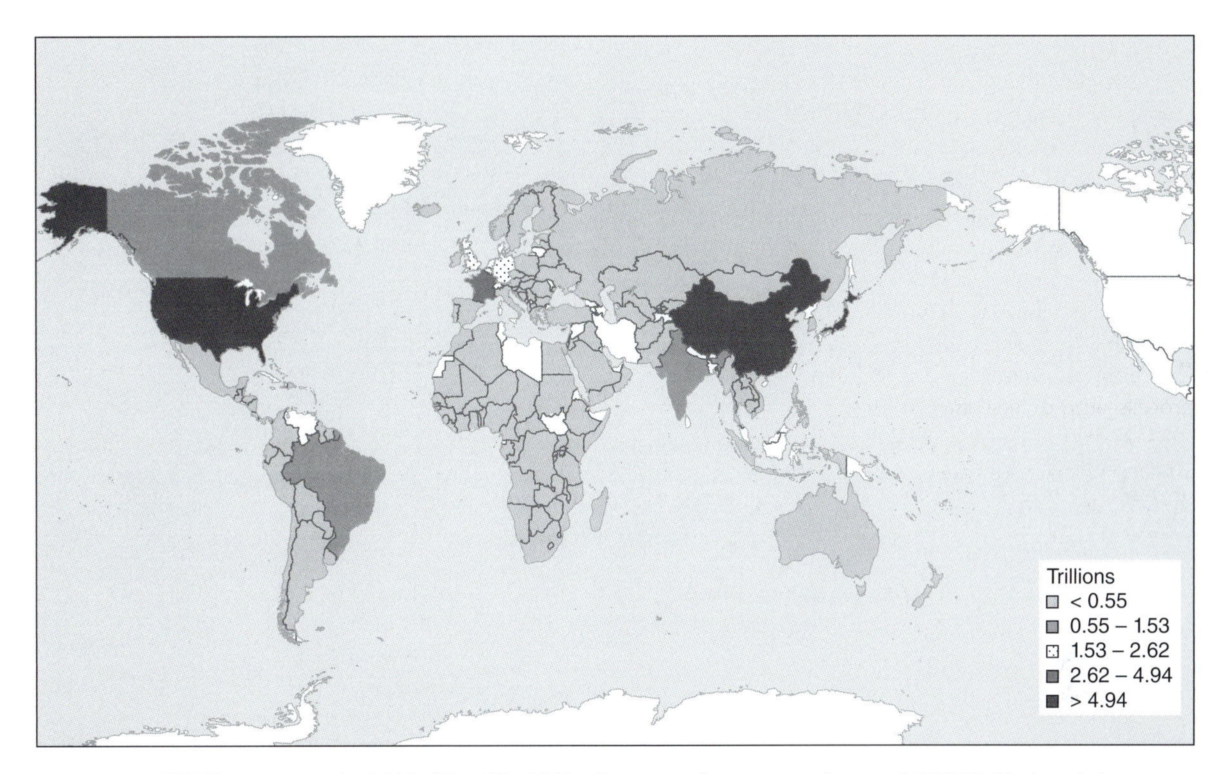

Figure 12.1 GDP by country in 2016. (The World Bank national accounts data and OECD National Accounts data files, https://data.worldbank.org/indicator/NY.GDP.MKTP.CD?view=map)

living of individuals compared to GDP alone. When a country has a high GDP but low GDP per capita PPP, it means that the majority of its citizens are sharing a smaller portion of that value and have less purchasing power. Countries with low per capita GDP PPP tend to have lower standards of living and higher poverty rates.

Although comparing GDP or GNP in relation to per capita PPP among countries reveals differences in the average wealth and purchasing potential of each person, it does not reveal anything about the distribution of income and economic gain among individuals, which can be highly inequitable. For example, two countries with similar GDP or per capita GDP PPP may be similar, but there may be larger inequities in one country in which only a small percentage of the population experiences increases in wealth (Figure 12.2). Thus, it is not fair to say that countries are equally well-off based on GDP per capita data alone. Increases in GDP over time may be associated with increased inequity. For example, open trade policy has resulted in increases in GDP, but only a small, wealthy minority is receiving the largest share of that increase in wealth, resulting in greater economic inequity.

Even when used on a per capita basis and accounting for individual purchasing power, GDP and GNP are limited as measures of progress or development, especially sustainable development. They measure progress based solely on consumption and on things that cost money. They do not consider externalities, a central premise to sustainability. They only measure environmental damage if we invest in restoration or clean-up. They do not consider natural resources as assets or whether growth is achieved through sustainable use or depletion of those resources. They do not value unpaid work such as child care services provided by a parent; the gardens we grow; cooking, repairs, and cleaning we do for ourselves; or charity, leisure time, or leave from work for family needs. Also not considered are the effects of economic growth on aspects of health such as obesity and stress-related disease, two factors that are positively correlated with GDP. To measure sustainable development, we need indicators that take these factors into account.

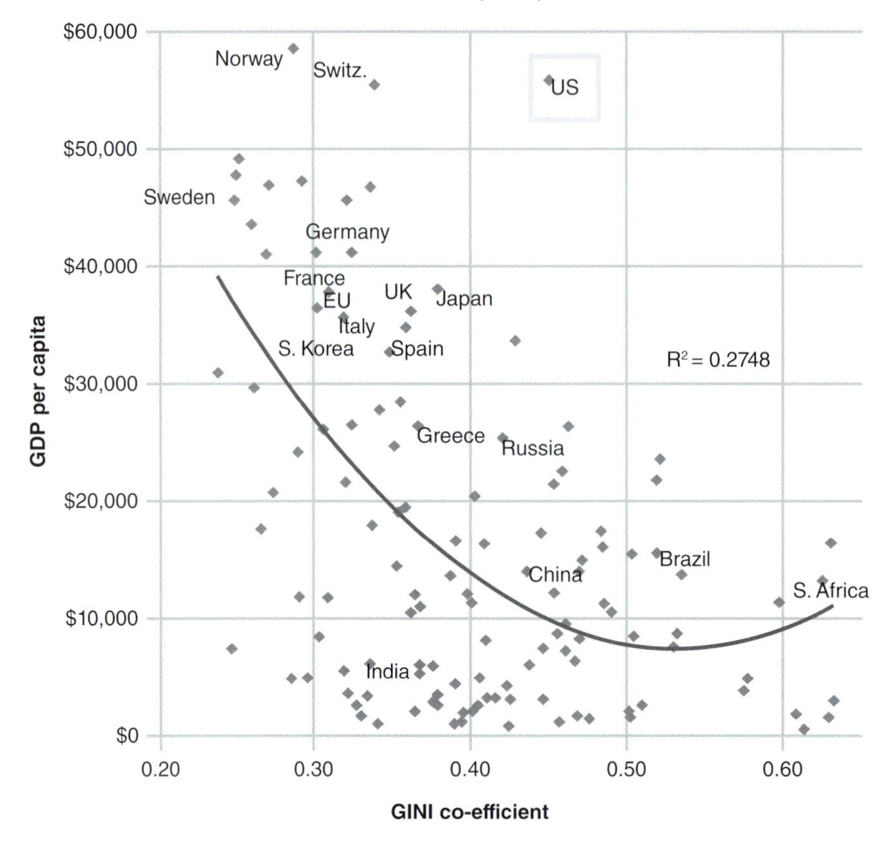

Figure 12.2 GDP per capita vs. income inequity as measured using the Gini Index (income equality = 0). The figure shows that GDP per capita and economic equity are at best loosely related to each other and the United States is a big outlier. For example, the United States, Switzerland, and Norway all have high GDP per capita, but the United States has much greater economic inequity. India, which has a very low GDP per capita, has very low inequity (greater equity) on par with Switzerland. (Data from CIA World Fact Book 2014, based on chart by Sumana Manohar and Hugo Scott-Gall of Goldman Sachs, http://www.businessinsider.com/goldman-sachs-chart-shows-us-stands-out-income-inequality-2016-8)

HUMAN DEVELOPMENT INDEX

In an attempt to better address sustainability objectives not accounted for in traditional national income approaches to measuring development, a number of new indicators have emerged. For example, the United Nations (UN) computes the **Human Development Index (HDI)**, which ranks countries based on not only per capita indicators of economic growth but also average life expectancy and average education levels and literacy rates. The HDI is an overall index of economic development that offers a rough ability to make comparisons that better reflect the economic condition and welfare of individuals compared to using just GDP and related composite statistics. Thus, it has become the most widely accepted measurement of development.

However, the HDI also has a number of limitations. Critics argue that as with any composite index, it is extremely difficult to construct objectively determined criteria and methods for weighting them. Other limitations are that it does not account for wide variation within a country. For example, economic and social welfare depend on several other factors—such as war, human rights, gender inequity, and levels of pollution—not accounted for in this index. Also, because a large component of the HDI is determined by traditional per capita indices of economic growth, many of the problems associated with them, such as not accounting for high levels of economic inequity, are still not reflected in the HDI. In fact, GDP and GNP tend to drive the HDI, as demonstrated by the 0.95 positive correlation between the two, especially for low-income countries.

THE "HUMAN DEVELOPMENT REPORT"

Each year since 1990, the UN has published the "Human Development Report" based on the HDI. This report is viewed as an independent, analytically and empirically grounded discussion of issues, policies, and trends in development. The 2016 report emphasized the synergy between human development and work, broadly defined as paid employment, unpaid care work, and voluntary and creative work. It also differentiated between positive and negative work. Positive or sustainable work not only provides material wealth but also fosters community and knowledge, strengthens dignity and inclusion, and is central to human progress. It provides positive outcomes such as empowerment of women, innovation, and improved health and opportunity. Examples of negative work are forced and child labor and hazardous work, and it may involve human trafficking, violate human rights, threaten freedom, and shatter dignity. This annual report sheds light on trends on development within this context of work.[1]

OTHER INDICATORS

Less commonly used, but perhaps more meaningful, indicators include the Genuine Progress Indicator, which incorporates income distribution, housework, volunteering, higher education, leisure time, crime rates, resource depletion, environmental damage, defense expenditures, and more. The **Gross National Happiness Index (GNH)** surveys the public on the extent to which they are happy. GNH was first introduced in Bhutan, where more than 90% of citizens rank themselves as happy despite a low GDP. GNH grew out of Bhutan's commitment to building an economy that would serve its culture based on Buddhist spiritual values instead of Western material development gauged by GDP. GNH has inspired a modern political happiness movement, which has been adopted by other countries and has resulted in the placement of happiness on the global development agenda. The Better Life Index, which is used by member countries of the Organization for Economic Co-operation and Development (OECD), measures 11 categories of development: housing, income, jobs, community, education, environment, civic engagement, health, life satisfaction, safety, and work–life balance.

THE SOCIAL PROGRESS IMPERATIVE

Given the interest to move beyond GDP and to include social performance as a measure of development, the **Social Progress Index (SPI)**, which embraces social and environmental development priorities, was created through extensive discussions with stakeholders throughout the world and across sectors by the Social Progress Imperative. The SPI measures national progress in three broad dimensions—basic human needs, foundations of well-being, and opportunity—each with specific indicators (Figure 12.3). Produced annually, it is a useful tool to track changes in social progress in relation to economic growth, and it has been shown to catalyze efforts to improve human well-being.

There is a strong positive relationship between GDP per capita and overall social progress; however, economic growth does not tell the whole story. The non-linear relationship between GDP per capita and social progress (Figure 12.4) shows that as countries reach higher levels of income, the rate of change in social progress slows. However, among poorer countries, small differences in GDP per capita are associated with large advances in social progress. The latter offers hope within the context of sustainable development.

When we examine SPI and per capita GDP together, we see that some countries have very high social progress despite low GDP (overperformers), and others have a high GDP yet comparatively low levels of social progress (underperformers). For example, the Netherlands and Saudi Arabia have similarly high levels of GDP per capita, but the Netherlands achieves a significantly higher level of social progress (see Figure 12.4). Similarly, at the low end of global GDP per capita, Bolivia has a far higher SPI score than Angola even though its GDP per capita is slightly lower (see Figure 12.4).

The reasons for over- or underperformance vary greatly among income classes and among countries within an income class. A closer examination of the

Social progress index indicator-level framework

Basic human needs	Foundations of Well-Being	Opportunity
Nutrition and basic medical care	**Access to basic knowledge**	**Personal rights**
Undernourishment	Adult literacy rate	Political rights
Depth of food deficit	Primary school enrollment	Freedom of expression
Maternal mortality rate	Secondary school enrollment	Freedom of assembly
Child mortality rate	Gender parity in secondary enrollment	Private property rights
Deaths from infectious diseases	**Access to information and communications**	**Personal freedom and choice**
Water and sanitation	Mobile telephone subscriptions	Freedom over life choices
Access to piped water	Internet users	Freedom of religion
Rural access to improved water source	Press freedom index	Early marriage
Access to improved sanitation facilities	**Health and wellness**	Satisfied demand for contraception
Shelter	Life expectancy at 60	Corruption
Availability of affordable housing	Premature deaths from non-communicable diseases	**Tolerance and inclusion**
Access to electricity	Suicide rate	Tolerance for immigrants
Quality of electricity supply	**Environmental quality**	Tolerance for homosexuals
Household air pollution attributable deaths	Outdoor air pollution attributable deaths	Discrimination and violence against minorities
Personal safety	Wastewater treatment	Religious tolerance
Homicide rate	Biodiversity and habitat	Community safety net
Level of violent crime	Greenhouse gas emissions	**Access to advanced education**
Perceived criminality		Years of tertiary schooling
Political terror		Women's average years in school
Traffic deaths		Inequality in the attainment of education
		Globally ranked universities
		Percentage of tertiary students enrolled in globally ranked universities

Figure 12.3 The Social Progress Index is based on indicators related to basic human needs, foundations of well-being, and opportunity. (Social Progress Imperative's Social Progress Index 2017, http://www.socialprogressimperative.org)

SPI scores for each underperformer can help explain underperformance and in so doing shed light on areas of reform that might help advance social progress. This could be extremely helpful among middle- to low-income countries. However, some underperformers at lower income levels have dictatorial or pseudo-democratic forms of government, are engaged in war or revolution, or are places where there are significant human rights violations, making such reform challenging. Full annual reports on this global index can be accessed at http://www.socialprogressimperative.org.

Sustainable Development Initiatives

Given the sustainability objectives such as those built into the SPI, how is sustainable development achieved? What organizations and programs have allowed for increased equitable economic opportunity, improved environmental protection, growth in well-being, and other elements of sustainability? In this section, we examine the different types of organizations that play a role in sustainable development. We then consider some specific programs and tools that have been implemented. We also identify barriers to sustainable development, and ways to overcome them are suggested.

Sustainable Development: The Players

With an understanding of sustainable development, the question of who or what is implementing, promoting, and supporting it arises. The answer ranges from domestic initiatives to non-governmental organizations (NGOs), foreign aid, international finance institutions, and the UN. Here, we discuss institutional and government support of sustainable development from global to very local efforts.

THE UNITED NATIONS

With the recognition of the direct linkage between environmental protection and human development

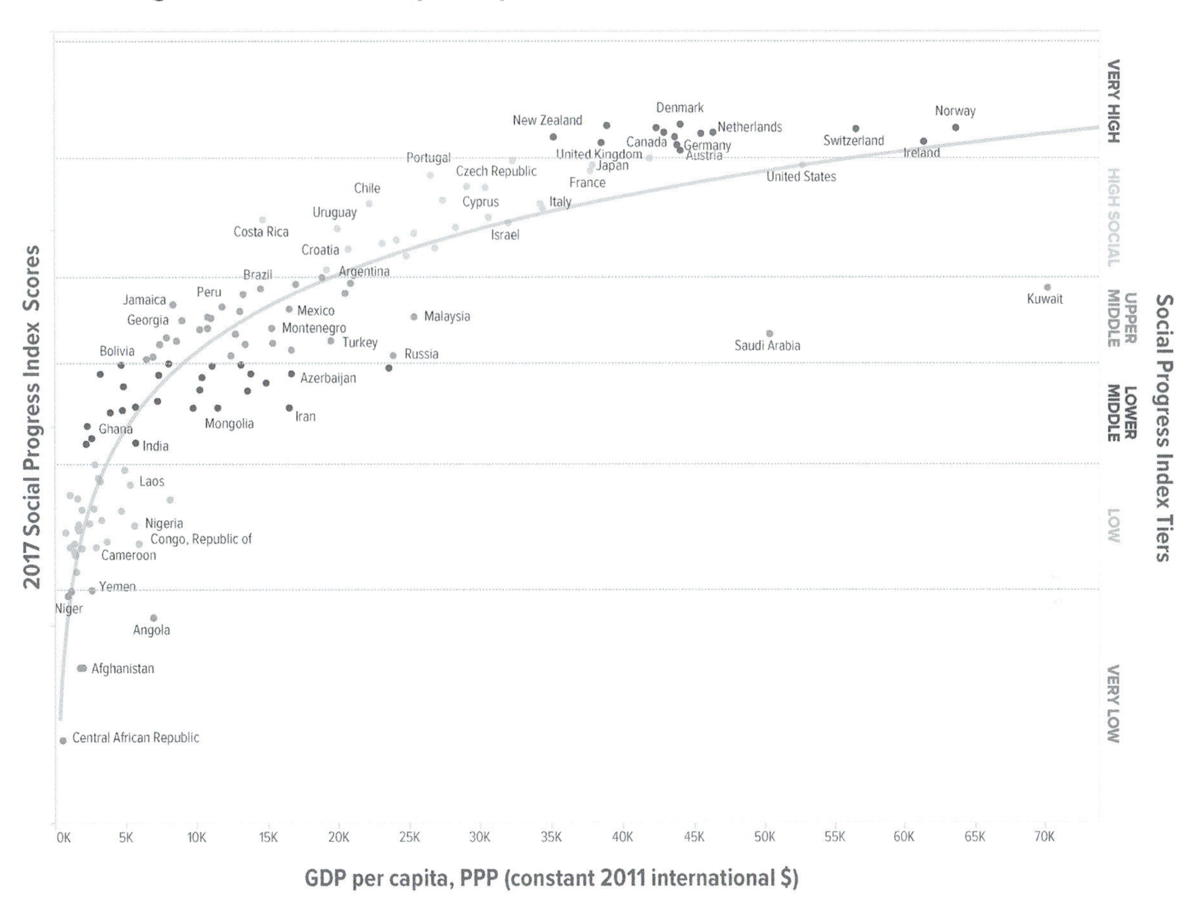

Figure 12.4 Social Progress Index vs. per capita GDP (PPP) for 2017. The graph shows that at similar GDPs there is variation in the degree of social progress. (Social Progress Imperative's Social Progress Index 2017, http://www.socialprogressimperative.org).

at the 1972 UN Conference on the Human Environment in Stockholm and the emergence of the concept of sustainable development with the 1987 Brundtland Report (see Chapter 2), the UN began its major role in fostering sustainable development. The 1992 Earth Summit in Rio led to Agenda 21, which became the UN's program of action for global sustainable development. The idea was to create a powerful, permanent body of the UN dedicated to sustainable development on a par with the UN Security Council. This new council would defuse threats to sustainable development the same way the Security Council tries to do so with threats to peace. However, there was limited support for that level of ambition, and various components of sustainable development and Agenda 21 fell under a variety of programs under the UN Economic and Social Council.

The UN did have a number of successes through the implementation of Agenda 21. It brought the concept of sustainable development to the forefront of thinking about development in general and created broad awareness about it, essentially making it a household phrase. It also resulted in the inclusion of sustainable development as a guiding principle in subsequent UN agreements and documents as well as other conventions and laws on biodiversity, climate change, gender equity, agriculture, and forestry.

The action framework for the UN's Agenda 21 and the Rio Summit relied heavily on the goodwill and action of nation states. The three-pronged approach included (1) developed countries taking the lead in changing production and consumption patterns, (2) developing countries adapting their development goals to include sustainable development paths and methods, and (3) the commitment of developed countries to support the sustainable development of developing countries through finance, technology, transfer, and reforms to global economic financial structures and practices. Despite the well-meaning intent of this plan, little progress has been made toward implementation, especially with regard to consumption patterns and funding. Until very recently, subsequent intergovernmental agreements on climate change have failed as well. Also, trade issues and agreements have not favored developing countries. As a result, developing countries have been following an unsustainable model of development while developed countries increase their pressure on the planet and are limiting improvements on the global human condition with their increasing rates of consumption, need for market expansion, and by exploiting the people and the environment in developing countries.

In an effort to deal with the lack of improvement in social and economic conditions, in 2000 the UN established the Millennium Development Goals (MDGs), which are detailed later in this chapter. The MDGs were established to encourage development by improving the social and economic conditions for the global poor. Although there was demonstrated success with many of the goals, the MDG approach again has focused the sustainability effort on the developing countries and ignored the larger sustainable development agenda. Recently, the UN has led the effort to achieve climate and sustainability commitments through COP 21 in Paris and with the establishment of the Sustainable Development Goals (SDGs) as a follow-up to the MDGs.

THE INTERNATIONAL INSTITUTE FOR SUSTAINABLE DEVELOPMENT

The International Institute for Sustainable Development (IISD) is an international non-profit organization that was established in 1990 to promote human development and environmental sustainability. It does this through innovative research, communication, and partnerships that integrate environmental and social priorities with economic development. IISD applies scientific and policy approaches to solving problems related to sustainable and equitable development, universal access to clean water and sanitation, and carbon-free energy while maintaining ecosystem integrity and improving community resilience. It also provides data and information that support sustainable change. IISD funds its program with its multimillion-dollar budget through the support of a variety of public and private sources.

INTERNATIONAL FINANCE INSTITUTIONS

The UN views international finance institutions (IFIs) as critical partners in sustainable development. From 2010 to 2015, IFIs contributed $550 million to UN development programs either as direct grants or as loans to national governments. The UN then offers extensive country presence and expertise and accountability in project implementation. The World Bank is an IFI that consists of the International Bank for Reconstruction and Development and the International Development Association. They provide loans to developing countries for development programs that focus on poverty reduction. The World Bank is a component of the World Bank Group, which is part of the UN system but operates separately.

Another IFI is the International Monetary Fund (IMF). The IMF is an organization of 189 countries that fosters global monetary cooperation in order to secure financial stability, facilitate international trade, and promote sustainable economic growth. Criticisms of the World Bank and the IMF include their emphasis on finance and development over environment and human rights and their promotion of international trade and privatization. These are discussed further in the section on barriers to sustainability.

Other IFIs that work with the UN and the World Bank are multilateral development banks, financial institutions, subregional banks, and aid coordination groups, all of which coordinate aid policies and practices at the global, regional, and national levels. An example of a multilateral development bank is the African Development Bank, which focuses on promoting sustainable economic development and social progress in African nations. There are other development banks and aid institutions with the same mission that represent other regions of the world.[2]

THE OECD AND FOREIGN AID

The OECD is a collective of 34 democratic countries with market economies that work with each other to promote economic growth, prosperity, and sustainable development. OECD member countries account for 63% of world GDP, 75% of world trade, and more than 50% of the world's energy consumption, yet they comprise only 18% of the world's population. They are the major global economic powerhouses.

The mission of the OECD is to achieve high levels of sustainable economic growth and employment and a rising standard of living in member countries. It does this in order to maintain global financial stability to facilitate the development of the world economy through sound economic expansion that will improve the economic and social well-being of people among the global poor. The OECD accomplishes this by promoting the market economy and domestic and international policies that promote development. Member countries also provide 95% of world official development assistance. In 2016, development aid from OECD countries reached a new peak of $142.6 billion. This was an increase of 8.9% from 2015 after adjusting for exchange rates and inflation. A rise in aid spent on refugees in donor countries boosted the total, but even when accounting for this, aid still increased 7.1%. Despite this progress, the 2016 data show that aid to the least developed countries declined by 3.9% in real terms from 2015, and aid to Africa declined 0.5% because some members did not meet their commitment to reverse past declines in flows to the poorest countries.[3]

Complementing the support through the OECD, individual governments often send aid directly to support development in specific countries. In the United States, the US Agency for International Development (USAID) provides support for health-related projects, economic development, humanitarian relief, or to directly support government budgets. Providing direct governmental aid can empower those governments to develop their own means to help their citizens and promote development, and it can also strengthen international relationships. Some drawbacks to this are transparency and manipulation. Because these funds go directly into the general budgets of recipient governments, it is challenging to determine the direct result of this aid. In 2017, USAID contributed approximately $22 billion in non-military foreign aid. Although many people object to this level of foreign aid, most do not realize that it accounts for less than 1% of the US federal budget.

Direct aid to foreign governments from the United States and other developed nations can be very useful for funding sustainable development. Many view it as the best way to generate large sums of money for the transition. It also can provide real agency for governments to self-determine how to proceed with sustainable development. However, direct foreign aid opens the door for corruption from governments that receive this money. It can also be used paternalistically to control countries through the threat of revoking aid. Some question whether direct aid to federal governments actually reaches and benefits local communities. Also, the provision of foreign aid often comes with restrictions that benefit the donor country. For example, aid may stipulate that all expenditures and contracts be made through corporations from the donor country, making it an indirect corporate subsidy. However, such direct aid can be very effective in promoting sustainable development if it is managed transparently, filters down to local communities, and is provided though non-manipulative processes (Box 12.1).

MULTINATIONAL ENTERPRISES

Multinational enterprises (MNEs) are large corporations that have their facilities, markets, and other assets in more than one country. Many MNEs have budgets that are larger than those of smaller countries. Because MNEs account for a large portion of global trade and investment and are large employers, they can significantly influence patterns of economic development, environment, and social well-being in the places in which they operate. Although this varies among companies, MNEs have the potential to contribute to sustainable development by bringing jobs, capital, and technology to the countries in which they operate.

Foreign investment by MNEs may not always be beneficial for developing countries, particularly in the long term. The benefits of this investment can be tarnished by poor labor and environmental practice. This is especially the case when there is pressure for MNEs to make their supply chains leaner in order to reduce costs in a competitive market. MNEs are also criticized for having excessive political influence over governments, their exploitation of developing countries, and their contribution to job losses in their home countries when they move operations elsewhere for economic reasons. In addition, MNEs tend to invest in low- to middle-income countries. As a result, there is very little investment in the poorest nations. Also, as development progresses, MNEs sometimes leave

BOX 12.1

Stakeholders and Collaborators: Partnerships, Participation, and Transparency in Sustainable Development in Bangladesh

Direct aid from international agencies is most effective in promoting sustainable development if it is managed transparently and filters down to local communities where individual families are empowered. One example of this is the collaboration among the UK Agency for International Development, the Australian Government, and the Bangladesh Government's Ministry of Local Government, Rural Development and Co-operatives, which work directly with families who have settled on the char river islands in northwestern Bangladesh.

Chars are river islands that are formed by the sand, mud, and silt that are deposited by the large river systems of Bangladesh. These newly formed islands are inhabited by 50,000–70,000 poor, landless people seeking refuge from poverty in hopes of a sustainable livelihood. Many of these so-called **char dwellers** are climate refugees and migrants. As a result, they lack state administrative structures to provide basic public goods, and they have little or no agency to demand rights from the state. These are vulnerable people in desperate need of climate-related development interventions.

This collaboration among the foreign aid agencies, the Bangladesh Government, and the local people of the chars has led to the development of the Chars Livelihoods Program (CLP).[1] The CLP's first phase ran from 2004 to 2010, and its second phase ended in 2016. The project targets the most vulnerable communities in the chars of northwestern Bangladesh, a region suffering from intense poverty and the negative impacts of climate change.

Figure 12.1.1 Families who live on chars (river islands formed from sedimentation) are extremely vulnerable to natural disasters and increased flooding associated with climate change. This family wades through floodwaters left behind after heavy rains caused major rivers to overflow their banks in northern Bangladesh. *http://www.ipsnews.net/2014/10/bangladeshi-char-dwellers-in-search-of-higher-ground/*

1 All information on the CLP is derived from http://clp-bangladesh.org. Accessed May 7, 2016.

The goal of the CLP is to reduce extreme poverty among program beneficiaries through a comprehensive package of support that includes livelihood and income generation and also health, education, and legal services. In unison, these interventions are meant not only to lift the char dwellers out of extreme poverty but also to enhance their resiliency against climate change and an unpredictable economy.

Working with the local community, the CLP carefully selects eligible households to participate in 18-month program cycles. Families are chosen based on need and their potential for success. The CLP provides the households with an income-generating asset worth approximately $200. Typically, these are agricultural assets such as equipment, seed, or livestock. The goal is that such assets will provide the basis of a self-sustaining livelihood that will eventually lead to further economic diversification. This diversification of income sources in turn increases resilience to withstand environmental and economic shocks. The CLP also provides participants with a modest monthly stipend to ensure that the assets are not sold during economic downturns. Additional support in areas of water and sanitation, health and nutrition, social norms, and flood protection is also provided to empower program participants to break the shackles of multiple dimensions of poverty.

According to the CLP, the first phase of its program covered 55,000 of the indigent households and benefitted more than 900,000 people. The second phase of the program was anticipated to lift another 78,000 households out of poverty.[2] The CLP carefully assesses the value of household productive assets, household income, household expenditure, and household cash savings to measure success of the program. More than 85% of the program participants have graduated successfully by developing independent, sustainable livelihoods that have lifted them out of poverty.

The CLP's activities have targeted some of the most climate and economically vulnerable of Bangladesh's population and have empowered them through livelihood development. The CLP has also enhanced the Bangladeshi state's capacity to serve its ultra-poor citizens through developing partnerships with donor agencies and fostering public–private collaborations. The CLP's success has helped Bangladeshi ultra-poor citizens increase their resilience to cope with distresses associated with climate and economic downturns. The program's success has in large part been determined by its careful consideration of local conditions, knowledge, and culture and by its insistence on placing people at the center of development.

—Contributed by S. Mohsin Hashim, PhD
Muhlenberg College, Allentown, Pennsylvania

2 Source: http://clp-bangladesh.org/wp-content/uploads/2014/03/CLP-leaflet-01.pdf. Accessed May 7. 2016.

those countries for cheaper labor and lower overhead. This was seen with the 2015 departure of the manufacturing side of the US company Intel from Costa Rica. Intel, the country's largest foreign manufacturer, was located in Costa Rica for 15 years and was viewed as its most important stimulus of development. However, as the standard of living improved in Costa Rica, Intel moved its manufacturing to Asia, presumably in search of cheaper labor and other cost reductions.

The economic development of MNEs can be made more sustainable through compliance to a set of human rights and environmental guidelines established by the OECD that may be adopted by countries in which these companies operate. This comprehensive set of recommendations for good corporate behavior is not legally binding, but MNEs with headquarters in adhering countries are bound to comply—and adhering governments are required to deal with alleged violations. Compliant MNEs have been shown to pay wages 40% higher than average wages of local firms in Asia and Latin America. More than 42 countries, including all OECD members that together represent 85% of global foreign investment, are adhering to the guidelines. In addition, a number of companies with global supply chains, such as the sustainable clothing company Threads-4-Thought, market themselves as sustainable companies, ensuring sound environmental and labor practice by working with factory partners that are committed to ethical manufacturing. International certification programs such as Fair Trade and Sweatshop-Free encourage sustainable practice by MNEs operating in developing countries.

NON-GOVERNMENTAL ORGANIZATIONS

As early as the 1991 Earth Summit in Rio, the role that NGOs could and should play in sustainable development was well established. Agenda 21 made it clear that NGOs play a vital role in the shaping of participatory democracy and sustainable development through a variety of approaches (Table 12.1). NGOs have well-established and diverse expertise and capacity. They are often more effective because of their independence from national and foreign governments and businesses that are often not trusted by other stakeholders. Even in cases of extreme poverty and government disarray, NGOs have been successful at pushing for sustainable development with on-the-ground programs, influencing policy changes, and driving intergovernmental negotiations that have included diverse significant results, such as the regulation of hazardous wastes, global bans on land mines, and the elimination of slavery.

NGOs have also focused their attention on powerful corporations that can often have greater resources and influence than the governments in the nations in which they operate. They have increased awareness about the environmental and social externalities of business activity in developing countries. Because of corporate concern about brand reputation, NGOs have been able to successfully challenge and influence the labor, environmental, and human rights records of MNEs and effect real change by influencing customers and shareholders.

TABLE 12.1 Influential Sustainability NGOs

ORGANIZATION	ACTIVITIES
Ceres	Ceres promotes sustainable business practices and solutions by working with more than 80 companies, from auto companies to financial services providers (one-third are *Fortune* 500 firms), as well as 130 member organizations. In 2003, Ceres launched the Investor Network on Climate Risk, which has grown to include 100 leading investors collectively managing more than $11 trillion in assets.
Conservation International (CI)	CI works with scientists, local communities, and practitioners in the field to protect nature, global biodiversity, and human communities. It strives to protect natural wealth, promote sustainable business, and foster effective governance. CI has supported the creation, expansion, and improved management of nearly 50 million acres of marine and terrestrial protected areas, and its data collection has led to the discovery of more than 1,400 species new to science.
Doctors Without Borders (Médecins Sans Frontières)	Doctors Without Borders provides emergency medical aid to people affected by conflict, epidemics, disasters, or exclusion from health care. Since 1971, the organization has treated tens of millions of people in more than 80 countries. In 1999, it received the Nobel Peace Prize.
Food and Water Watch	Food and Water Watch works to make food, fish, and water safe, accessible, and sustainable. It has raised consumer awareness of the environmental and economic costs of bottled water and has also helped dozens of communities—from Stockton, California, to Trenton, New Jersey—fight the privatization of public water supplies.
Heifer International	Heifer International has provided more than 20.7 million families—105.1 million men, women, and children—with animals and training in sustainable agriculture so that they can feed and care for themselves. Founded more than 70 years ago by a US farmer, the organization focuses on ending hunger and poverty.
Natural Resources Defense Council (NRDC)	NRDC's more than 350 lawyers, scientists, and other professionals work with businesses, elected officials, and community groups in the United States and internationally on issues including curbing global warming, clean energy, reviving the world's oceans, defending endangered wildlife and wild places, pollution prevention, ensuring safe and sufficient water, and fostering sustainable communities.

ORGANIZATION	ACTIVITIES
The Nature Conservancy (TNC)	Focused on conserving land and species throughout the world, TNC has protected more than 119 million acres of land and thousands of miles of rivers worldwide. It also operates more than 100 marine conservation projects worldwide.
Ocean Conservancy	Since 1972, the Ocean Conservancy has worked to protect the health and vitality of the world's oceans, including the species that call it home and the humans whose livelihoods depend on them. Through its International Coastal Cleanup program, the organization has removed 144,606,491 pounds of trash from the world's beaches during the past 25 years.
Oxfam	An international confederation of 17 organizations, Oxfam fights poverty and injustice in more than 90 countries. It works on interconnected issues such as human rights, emergency response, and sustainable development.
Sierra Club	Founded in 1892 by conservationist John Muir, the Sierra Club is one of the oldest and largest environmental organizations in the United States. It has protected millions of acres of wilderness and has helped pass key environmental legislation, including the Clean Air Act, the Clean Water Act, and the Endangered Species Act. It is also leading efforts to move away from the use of fossil fuels.
Slow Food International	As its name implies, Slow Food stands for the opposite of fast food: clean, fair, and healthy food for all; regional traditions; gastronomic pleasure; and a slow pace of life. Begun in Italy in the 1980s, Slow Food has members in 160 countries and promotes the principles of its Slow Food Manifesto through local and international events, its University of Gastronomic Sciences, and more.
World Resources Institute (WRI)	WRI works with leaders to turn information into action, with a focus on issues such as climate change, energy, food, forests, water, cities and transportation, governance, business, and finance. WRI has more than 450 experts and staff working throughout the world.
World Wildlife Fund (WWF)	The WWF works in 100 countries to conserve nature and protect biodiversity. Founded in 1961, it is now supported by nearly 5 million members worldwide.

Source: Courtesy of Sustainability Degrees, Degree Prospects, LLC.

The effectiveness of NGOs in supporting sustainable development is not without challenge. Many of them are constrained by the availability of funds and must place considerable effort on raising them. Their effectiveness is also complicated because as external agents of change, NGOs risk not having complete knowledge about context. There are limits to what they can do, and they are often contested by powerful local actors. International NGOs that partner with and empower local stakeholders are thus often the most successful.

LOCAL GOVERNMENTS

Although sustainable development is a global process fostered and supported by international organizations, the work to address and achieve it is often most effective at the local level. This is because many of the problems and solutions related to sustainable development have their roots in local activities. Thus, the developing countries themselves can often be the most effective agents of sustainable development, and their local and regional governments are essential for promoting inclusive sustainable development within their territories. Local governments, which are largely responsible for economic, social, and environmental infrastructure, are an essential part of planning processes, establish local policies and regulations, and are central to the implementation of national policies. As the level of governance closest to the people, they play an essential role in educating, engaging, and empowering local stakeholders in the implementation of projects and programs designed to promote sustainable development.

Although we have seen considerable external support for sustainable development projects, domestic financial support within developing countries is actually far larger than external support. In addition to providing financing, there are a number of ways that IFIs such as the World Bank support developing countries as the primary agent of their own sustainable development. They can help poorer countries more effectively mobilize their domestic private financial resources. One way that this can be done is by helping them improve their tax collection. Because of weak tax collection, developing countries collect much lower proportions of their GDP in tax revenue than do developed countries.

When IFIs and national governments invest in domestic small and medium-sized enterprises, the results are more sustainable. This is because such enterprises are a fundamental part of the economy of developing countries. Local governments play a crucial role in furthering innovation and prosperity. When credit through domestic microfinance networks is available, these enterprises can provide loans that can empower the poor, especially women, who typically have less access to conventional credit.

One successful example of a government-backed small enterprise is the Infrastructure Development Company Limited in Bangladesh. This company has installed solar panels in more than 3.95 million off-grid homes with the help of microfinancing offered in partnership with numerous grass-roots institutions, such as the Grameen Bank founded by Mahumud Yunnus (see Chapter 9). By powering lighting, cell phones, water pumps, health clinics, and small businesses, measureable improvements in the lives of the poor have been made.

Sustainable Development: Programs

Sustainable development has been fostered by the work and financial resources of a variety of organizations over a wide range of scales from the global to the local. In this section, we explore examples of specific projects aimed at promoting aspects of sustainable development. Through these examples, recipes for success and conditions for failure will begin to emerge. This will allow us to begin to identify barriers to sustainable development and seek solutions that will help eliminate them.

THE MILLENNIUM DEVELOPMENT GOALS

In 2000, the UN established a declaration that detailed eight Millennium Development Goals (MDGs)—ranging from eradicating extreme poverty and hunger to ensuring health, education, sustainability, and the development of a global partnership for development—to be achieved by a target date of 2015 (Table 12.2). The MDGs served as a blueprint agreed upon by the international community that represented an unprecedented effort to meet the needs of the world's poorest people.

There were a variety of ways in which the MDG efforts were supported and funded. The World Bank, the

TABLE 12.2 Millennium Development Goals, Target Areas, and Achievements (From the UN MDG)

GOAL	TARGET	WHAT HAS BEEN ACHIEVED?
MDG 1: Eradicate extreme hunger and poverty	Halve the proportion of people living below national poverty line Halve the proportion of people who suffer from hunger	Number of people living on less than $1.25/day has been reduced from 47% to 14%. Proportion of people suffering from hunger has been reduced from 23.3% to 12.9%.
MDG 2: Achieve universal primary education	Ensure that children everywhere, boys and girls alike, will be able to complete a full course of primary schooling	Enrolment rate has increased from 83% to 91%.
MDG 3: Promote gender equality and empower women	Eliminate gender disparity in primary and secondary education, preferably by 2005, and in all levels of education no later than 2015	Two-thirds of developing countries show gender equality in primary education.
MDG 4: Reduce child mortality	Reduce by two-thirds the mortality rate for children younger than age 5 years	Child mortality was reduced from 90 to 43 deaths per 1,000 live births.
MDG 5: Improve maternal health	Reduce the maternal mortality ratio by three-fourths	Maternal mortality rate has decreased from 380 to 210 deaths per 100,000 live births.

MDG 6: Combat HIV/AIDS, malaria, and other diseases	Have halted and begun to reverse the spread of HIV/AIDS	Number of new HIV infections declined by 40%.
	Have halted and begun to reverse the incidence of malaria and other major diseases	Malaria cases increased after 1990, peaked in 2003, and have since slightly decreased.
MDG 7: Ensure environmental sustainability	Integrate the principles of sustainable development into country policies and programs and reverse the loss of environmental resources	Vague language; difficult to quantify.
	Halve number of people without access to drinking water	2.6 billion more people have access to improved drinking water (achieved as of 2010).
	Have achieved by 2020 a significant improvement in the lives of at least 100 million slum dwellers	Vague language; difficult to quantify.
MDG 8: Develop a global partnership for development	Develop further an open, rule-based, predictable, nondiscriminatory trading and financial system	Vague language; difficult to quantify.
	Address the special needs of the least developed countries	Vague language; difficult to quantify.
	Address the special needs of landlocked developing countries and small island developing states	Vague language; difficult to quantify.
	Deal comprehensively with the debt problem of developing countries through national and international measures in order to make debt sustainable in the long term	Vague language; difficult to quantify.
	In cooperation with developing countries, develop and implement strategies for decent and productive work for youth	Vague language; difficult to quantify.
	In cooperation with pharmaceutical companies, provide access to affordable essential drugs in developing countries	Vague language; difficult to quantify.
	In cooperation with the private sector, make available the benefits of new technologies, especially information and communications technologies	Vague language; difficult to quantify.

IMF, and the African Development Bank canceled $40 billion to $55 billion in debt owed by heavily indebted poor countries in Africa to allow them to redirect resources to programs for improving health and education and for alleviating poverty. The IMF also provided policy advice, technical assistance, direct financial support, and advocated for increased foreign aid and the opening of markets to exports from developing countries. An MDG multipartner trust fund was established that generated more than $750 million from public and private donors. Developed UN member countries committed to a contribution of 0.7% of their GDP, although not all completely met that commitment.

MDG implementation and assessment was led by the UN Development Programme (UNDP) and achieved through the collaboration of relevant international organizations and target countries' national and local agencies, as well as NGOs. Many remarkable gains were achieved for each MDG by the target year 2015 (see Table 12.2). The greatest successes were in the areas of hunger and poverty reduction and increasing access to drinking water. For example, the number of people who live on less than $1.25 per day and rates of child mortality were decreased globally by more than 50%. One of many successful cases for improved access to drinking water occurred in Lahore, Pakistan. Here, drinking water, sanitation and hygiene conditions, and related high rates of waterborne diseases were worsening due to high levels of poverty, environmental and groundwater pollution, and lack of political will/interest in improving water and hygiene. Through the MDG program, an innovative bottom-up and participatory approach was applied, engaging young women, educational institutions, religious leaders, and children. This significantly increased access to clean drinking water by nearly 50% and reduced rates of waterborne diseases.

Although there were many other success stories, results for each goal varied widely and were uneven among geographical regions. For example, in most African nations, there was much less progress across all MDGs. It is clear that despite the measured gains, there is much more work to do in order to improve the condition of the poor and eliminate global inequities. Critics of the MDGs argued that there was a lack of justification behind the chosen objectives and that there was uneven progress among them. Others questioned the approach of introducing local change through external innovations and financing. They argued that these goals would have been better achieved through community initiative building from bottom-up stakeholder decision-making and implementation within existing cultural and government structures. Some believed that global goals were unrealistic and focused attention away from what could be accomplished through sound domestic policies, aid, and other forms of cooperation. However, by allowing the MDGs to serve as symbols of the possible outcomes, the development community became energized and focused on both achievement and accountability.

SUSTAINABLE DEVELOPMENT GOALS

Another problematic aspect of the MDGs was that they separated environmental sustainability as a single goal when in reality the program objective was the pursuit of sustainable development in a way that addressed and integrated social progress, economic growth, and environmental protection. Thus, when the 2015 MDG deadline was reached, a new post-2015 agenda was framed as the Sustainable Development Goals (SDGs).

The 17 SDGs set for 2030 follow and expand on MDGs in a broader sustainability agenda that strives to reflect the lessons learned from them, build on successes, and put all countries firmly on track toward a sustainable future (Figure 12.5). They go much further than the MDGs in addressing the root causes of poverty and the universal need for development that

Figure 12.5 The 17 UN Sustainable Development Goal initiated in 2015 for 2030. *MariaGershuni.*

works for all people. They recognize that sustainable development can occur, but only within the planetary limitations and within the context of climate change. They also address population growth, responsible consumption, infrastructure needs, rapid urbanization, and global partnerships for sustainable development.

Under the SDGs, the UNDP will support developing countries in three different ways, through the MAPS approach: mainstreaming, acceleration, and policy support. Mainstreaming is the integration of the new global agenda in national development plans and policies. This work is already underway in many countries at national request. Acceleration is the support of countries to accelerate progress on SDG targets based on the experience of the MDG implementation. The UNDP will make its policy expertise on sustainable development and governance available to governments at all stages of implementation. NGOs will also play a major role. For example, the Rainforest Alliance is promoting SDGs, especially Goal 15 (see Figure 12.5), through its forest and agricultural product certification programs. These programs, together with their support of community-based forest programs such as those in the Maya Biosphere Reserve, help maintain the highest forest management standards, including zero deforestation.[4] New technology will improve communication among partners, thus strengthening implementation, data acquisition, and monitoring and assessment. With these new, more broadly based goals and approaches, it is hoped that the SDGs can go well beyond the successes and address the limitations of the MDGs.

SLOWING POPULATION GROWTH

The world population has surpassed the 7 billion mark and is projected to grow to more than 9 billion by 2043. Population growth rates vary among countries and regions, where many of the poorer countries continue to experience rapid population growth. In addition, complex international migration patterns place additional pressures on countries with already high population growth rates, especially in urban environments. Population growth, mass migration, and urbanization negatively impact all three areas of sustainability, making it challenging to meet basic human needs. Through increased consumption and production, expanding populations degrade environments and deplete already limited resources. Urbanization puts increased pressure on insufficient infrastructure. Economically, increasing populations increase poverty and further widen the inequitable distribution of wealth. They also limit the ability to provide social protections, including access to health care, education, housing, sanitation, water, food, and energy.

Can sustainable development occur within the context of rapidly expanding populations? Population dynamics influence all aspects of development at the local, national, regional, and global levels. So there is no question that slowing population growth must be part of the sustainable development agenda. The question is how to do so. Population dynamics are the result of individual choices and opportunities. However, restricting individual rights through forced population control contradicts fundamental sustainability objectives. Thus, the UN Population Fund supports solutions to the problem of rapid population growth that expand rather than restrict individual rights and choices. It accomplishes this by adopting policies that are human rights based, gender responsive, and that empower individuals to have more control over their own destiny and those of their families and communities.

Examples of policies that are effective at slowing population growth are those that promote universal access to sexual and reproductive health and rights. This includes providing family planning, access to birth control, and sexuality education. This can be challenging because these are often viewed as immoral by certain groups that then fight to eliminate support for organizations that provide these services. However, these measures significantly improve the lives of people and offer many societal benefits, including reductions in unwanted and teenage pregnancies and unsafe abortions. They curb infant and maternal mortality, and they reduce gender-based violence. They also help combat HIV/AIDS and other sexually transmitted diseases that lead to high mortality rates.

Policies that improve education, increase access to jobs and better working conditions, and increase access to essential services such as child care can also slow population growth rates and can influence personal decisions about family size. Universal access to free education, particularly for girls and women, has been shown to slow population growth rates. Evidence for this can be seen in India, where in the south women who have an increased literacy rate are making decisions to postpone marriage and child birth, and they are reducing the number children they have. This has significantly reduced the average fertility rate compared to that of the northern area of the country, where access to education and the literacy rate for girls and women are much lower.[5]

This same human rights approach is also crucial in constructing migration policies and urbanization and settlement patterns. Migration is stimulated by conflict, violence, human rights violations, and now climate change. It allows people to escape from these deplorable conditions and is fundamental to global justice. There are immigration policies that allow for migration to generate large economic and social gains for the destination country while ensuring quality living and working conditions for migrants. Sustainable immigration policy realizes the developmental benefits of migration for both sending and receiving countries, and it should ensure the rights and safety of migrants and limit discrimination against them. The problem of rapid urbanization must be addressed by creating cities with sustainable communities that can accommodate increased demands for livelihoods and services and by strengthening the linkages between rural and urban areas. Implementation of sustainable development initiatives must be based on sound population data and projections.

SUSTAINABLE CONSUMPTION

The issue of population growth cannot be addressed in the absence of considering rates of consumption. The greatest challenge of sustainable development is to solve the problem of meeting the increasing needs and expectations of a growing population while addressing consumption. Past and current patterns of consumption are not sustainable and are destroying the environment, depleting natural resources in an inequitable manner, contributing to social problems such as poverty, and hampering sustainable development efforts. Transitioning to global sustainable consumption and production (SCP) is an essential prerequisite to sustainable development. However, the burden of SCP cannot be placed solely on developing countries. This is because of the massive consumption gap, where the richest one-fifth of the world accounts for 86% of consumption while the poorest one-fifth accounts for less than 1% of consumption and in the most developed countries per person consumption rates are as high as 30 times greater than those in the poorer countries.

Fundamental changes in the ways societies produce and consume are indispensable for achieving global sustainable development. All countries need to take action—with developed countries leading the way—to close the consumption gap through decreased consumption and the achievement of global development within the framework of SCP. Unfortunately, recent trends contradict this. The gap is narrowing, but not because of a reduction in consumption in developed countries. Rather, rates of consumption are increasing in many developing countries. From 1971 to 2011, global per capita GDP increased by a factor of 11. With this, individual incomes and rates of consumption have risen in emerging countries such as Brazil, China, India, and Russia. These increased rates of consumption added to the high rates in developed countries have caused net global consumption to increase.

SCP implies that a sustainable future will require significant change both in the way we consume and in the way products are produced. This will require a substantial cultural shift in developed countries, in which the bulk of products are consumed and in which these products are made affordable due to supply chains that exploit the environment and labor. It will also require a departure from an economic model in which growth is based on market expansion that depends on increases in global consumption of unsustainable products. This can only be achieved through the global application of sustainable solutions that we have discussed throughout this book, including transitioning to renewable resources, reducing the excessive amounts of food waste and moving toward zero waste in general, and eliminating subsidies and tax incentives that encourage wasteful consumption by creating artificially low prices.

The barriers to SCP on the consumption or demand side can be overcome by educating consumers about responsible consumption. As discussed in Chapter 10, this can be achieved through education programs that decouple measures of success from the accumulation of material goods and that provide an understanding of the effects of our actions on other people and the planet.

Policies that require companies to pay for the true environmental and social costs associated with the goods and services that they produce will place these at the same price point as more sustainable products, making sustainable products more attractive. This will also promote innovation and the expansion of alternative sustainable options. As the market for such products increases, prices will decline.

Another approach is to incentivize responsible consumption. One way that this has been accomplished is with green loyalty purchase card programs such as the one piloted in the Netherlands. This program awards points to consumers for sustainable behavior such as recycling and for purchasing sustainable goods and services. Earned points can be redeemed for other sustainable goods and services. Such cards are now emerging throughout the developed world and have been shown to be effective at promoting sustainable purchases, but they are limited in their ability to reduce overall consumption. The expansion of technology-based sharing economies has and will continue to be effective in reducing consumption. In developed countries, consumption of animal products and processed foods that have serious environmental consequences has stabilized or declined, whereas demand for local and organic products has increased. This has been motivated by both health and environmental concerns.

On the supply side, the trend of corporations changing to sustainable models has been on the rise. New efforts to redefine capitalism in order to better serve society's needs are emerging, and many companies are working to embed and activate greater purpose in their brands. According to a recent survey of CEOs of global companies, 24% of companies have recently integrated sustainability into their organizational purpose in response to stakeholder expectations, and a further 12% are considering doing so in the near future.[6] Business practices and policies that address externalities and incentivize sustainable practice will help move developed nations toward SCP. Policies that encourage increases in efficiency and support small, local businesses and the development of sound sustainability rating programs will support this goal. Remaining challenges will require the collaborative efforts of consumers, businesses, researchers, educational institutions, NGOs, and politicians to work toward the common goal of sustainable consumption.

In developing countries, the challenges are very different. Eradication of extreme poverty is an essential first step, as reflected by its placement as the first of the 17 SDGs for 2030. This can only happen when developed nations stop exploiting these poorer nations and assist them in sustainable economic development, including sound resource management and improvement of social conditions. One strategy is to improve the sustainability of agricultural systems. For example, in Ethiopia, the Agricultural Transformation Agency, with support from the UN, is working with 12.8 million smallholder farmers to improve their yield and marketing for their products. Efforts to improve public health and family planning increase the quality of life and working potential of families. This in turn results in the reduction of poverty and population growth. The UN and other organizations are bringing together diverse stakeholders to create green economies, often through the use of microfinancing to promote sustainable consumption and production. In addition to policies and organizations focused on poverty reduction, sustainable agriculture, public health, and green economic development, SCP must be integrated into expanding equitable education programs in developing countries.

GENDER AND SUSTAINABLE DEVELOPMENT

As stated in the MDGs and SDGs, gender equality is a fundamental human right that is a necessary foundation for a sustainable world. There is considerable evidence indicating that gender inequality has a high economic cost by denying women opportunities to work and generate income. It also leads to greater environmental degradation throughout the world. Providing women and girls with the following will slow population growth rates, stimulate sustainable economies, and benefit humanity as a whole: equal access to health care and education; freedom from oppression; a secure, safe, and violence-free lifestyle; decent opportunities to earn a living wage; and representation in decision-making processes. Sustainable development cannot happen without gender equality and the full and equal participation of women in decision-making and the process of development.

A section of the Institute of Development Studies in the United Kingdom known as Bridge is one of many organizations working to advance gender equality, women's rights, dignity, and empowerment in development. Its approach is to work as part of a global network to increase access to relevant, quality knowledge on gender and development; create well-developed, influential networks for change; and furnish actors with the evidence and tools they need in order to advocate for policy and decisions that will result in greater gender equity. These tools include cutting-edge programs that deal with gender issues in relation to food security, access to education and health care, and microfinance for economic development. Bridge provides support for women and girls in mass migrations, mechanisms for reporting conflict-related violence, and involvement of women in the demilitarization and peace processes. Recognizing that gender equality will not be achieved without the involvement of men and boys, Bridge also offer opportunities for engaging them in work for gender equality. It has developed indicators to measure gender inequality and has connected global social movements for gender equity through provided technology and social media.

COMMUNITY SUSTAINABILITY

As in developed countries, much of the work of global sustainable development occurs at the community level with many of the same priorities, including community and local economic development; improved housing and infrastructure; sustainable food systems; universal access to education, health care, and renewable energy; and dealing with climate change. Sprawl is less of an issue; the trend in developing countries is rapid urbanization, which makes urban planning and infrastructure a priority. Open space is a priority but more for the purposes of preserving environmental services, biodiversity, and cultural heritage.

Communities are a priority in sustainable development. In addition to one of the SDGs directly focused on sustainable communities, the remaining goals need to be implemented at community and regional levels. The World Bank has taken the lead in assisting local governments in poorer nations to prepare local economic development strategies. It also finances infrastructure development and provides financial support to community-level local economic development agencies. Community economic development has also been an objective of the US Peace Corps, whose volunteers work with development banks, NGOs, and municipalities to strengthen infrastructure and encourage economic opportunities in communities.

One example of efforts to improve community sustainability in developing countries is the provision of affordable housing. However, this has proven to be even more of a challenge than that in US cities. Globally, more than 1 billion people currently live in inadequate housing, with 835 million of them in urban areas. This is further complicated by the lack of land ownership rights in many countries. Access to affordable, secure places to live is a basic human necessity that is a key aspect to advancing sustainable development. The non-profit organization Ashoka is developing new business models called "hybrid value chains" that connect for-profit businesses and non-profit organizations to solve these complicated housing issues in the poorest areas of the world (Box 12.2). In addition, international development finance institutions such as the Overseas Private Investment Corporation are playing a role in developing low-income housing by providing affordable mortgages to poor families.

BOX 12.2

Innovators and Entrepreneurs: Hybrid Value Chains and Affordable Housing

Currently, 1.1 billion people live in urban slums and lack adequate housing. By 2030, the number of people who lack adequate housing will increase to 3 billion. In response to the current and future housing crises, the organization Ashoka has launched Housing for All initiatives in India, Columbia, Brazil, and Egypt. The goal of the initiatives is not to provide free housing as a handout but, rather, to enable people to be owners of their homes. This turns them into full economic citizens who are then able to participate in and sustain the local economies.

The goal of Housing for All is to deliver affordable, higher quality homes to more than 1 billion people through the use of hybrid value chains. A typical value chain is the set of activities a business performs to produce a product or service for profit. Each step in the chain produces more total value than the sum of the value of each individual activity. In contrast, the hybrid value chain is a series of collaborative activities of both private business sectors and non-profit organizations that are attempting to solve critical social problems. The hybrid value chain allows each to gain more economic and social value than either actor could have generated alone. It provides businesses with access to untapped trillion-dollar markets of low-income populations to which they deliver needed goods and services in a more cost-effective way. Hybrid value chains have a greater potential for sustained social and economic gain for all parties compared to traditional business models. In addition, low-income populations improve their livelihoods as their basic human needs are met and new economic opportunities arise.

In the case of the Housing for All initiatives, the hybrid value chain links businesses to the potential $6.4 trillion low-income global housing market while delivering improved lives to the underserved. This is accomplished by linking businesses with social organizations that have housing as their mission. Through an inclusive process, businesses and organizations provide microfinance, design, development, and supporting services to those in need of housing. This collaboration results in the construction of affordable housing that meets the goals and objectives of these potential home owners. They often incorporate small businesses into the design of the home to create new economic opportunities. Through the use and development of hybrid value chains, Ashoka's initiative has thus far led to more than $11.7 million in sales, creating approximately 28,000 affordable housing solutions in Columbia, with similar results in India, Brazil, and Egypt.

The hybrid value chain is viewed as a sustainable way to simultaneously solve social problems and expand markets that will reach everyone, including the 4 billion people who are not yet part of the world's formal economy. This business model has been providing jobs more than three times faster than traditional businesses. In addition to providing affordable housing, it is providing affordable, intraocular lenses to restore sight to cataract patients in India, a chain of computer schools serving hundreds of slums across Latin America and Asia, drip irrigation systems to millions of farmers in Mexico, and many other win–win solutions.[1]

1 Drayton, Bill, and Valeria Budinich. 2010. "A New Alliance for Global Change." *Harvard Business Review*. September 1, 2010. https://hbr.org/2010/09/a-new-alliance-for-global-change

EDUCATION FOR SUSTAINABLE DEVELOPMENT

In Chapter 10, we discussed why and how sustainability is being integrated into kindergarten through 12th grade and higher education curricula in the developed world. Globally, sustainable development including SCP cannot be achieved by technological solutions, policy change, or development aid alone but, rather, must also include changes in individual thinking, action, and values. This requires quality education and learning for sustainable development at all levels and in all social, cultural, and economic contexts. Thus, effective education for sustainable development (ESD) must become part of the global sustainability agenda of the UN Educational, Scientific and Cultural Organization (UNESCO) and all UN member states.

Much like in developed countries, ESD in developing countries requires both a transformation of the learning environment and integration of sustainability principles across the curriculum through whole-school approaches. The content and pedagogical strategies are essentially the same as in the developed world, focusing on critical thinking skills, project-based learning, and systems thinking in order to empower individuals to transform themselves and the society in which they live through the transition to green economies, more sustainable lifestyles, and the creation of global citizens who are capable of understanding and resolving local and global challenges. The goal is to educate and empower citizens to be active contributors to creating a more sustainable world while offering futures with equitable economic opportunities.

Although there is a global recognition of the need for ESD, only limited progress has been made. This has been due to a lack of vision or awareness at local levels and a lack of policy or funding to support infrastructure and teacher training. ESD in the poorest countries is often not even a consideration, given insufficient access to education in general, even at primary levels and especially for girls and women.

ESD is best achieved through government leadership, international support from organizations such as UNESCO, and community understanding and acceptance of sustainability. The latter is vital and occurs through public participation processes in which all stakeholders examine the needs and desires of a community and what they value as essential elements of basic and secondary education. For example, in some communities in which tourism is viewed as a major component of their development plan, educating for all aspects of that industry is considered an essential component of ESD. The Barefoot College in India is another example of an educational program based on community need. It trains primarily women in areas of sustainable technology that they can then bring back to their own rural communities with an emphasis on empowering and creating income opportunities for women in some of the poorest communities.

Financing and implementing ESD and universal access to education have been objectives of the MDGs and now the SDGs. It will require international and domestic financial resources, government leadership, support from NGOs, partnerships with universities, and international cooperation. An example of a regional, international cooperative approach to training teachers for ESD is the Learning for a Sustainable Environment: Innovations in Teacher Education Project in Southeast Asia (Box 12.3).

TECHNOLOGY AND SUSTAINABLE DEVELOPMENT

In many ways, the relationship between sustainability and technology is paradoxical. This is because of the role that modern technology has played in leading us down this unsustainable path of exploitation and excessive consumption. However, it is important to recognize that sustainable development can be greatly enhanced through the use of technology. Throughout this book, we have discussed many examples of technical solutions that improved sustainability across all sectors, including ways to generate clean water and energy, reduce waste, and improve food production. Moreover, implementing and assessing sustainability are driven by data, and technology is required to collect, access, and analyze those data.

The ability of individuals to share information is central to sustainable development. For example, increasing access to education relies on technology such as the ESD learning modules described in Box 12.3, which are distributed on the internet. Communication technology is also needed for disaster response, energy, health, agriculture, and the promotion of civic engagement. Recently, the Global Center for Food Systems Innovation created a smartphone app that allows farmers in Kenya to learn about fertilizer applications through simulation. Smartphones and computer and internet access also enhance economic opportunities and entrepreneurship through online sales and purchasing, peer-to-peer lending, and microfinancing. The World Bank and other organizations, such as the Great Transition Initiative, have made closing the digital divide between developed and developing countries a priority because of technology's proven role in poverty reduction and furthering sustainable development even in the most isolated, rural areas of developing countries.

MICROFINANCE AND SUSTAINABLE DEVELOPMENT

Microfinance is increasingly being considered as one of the most effective tools for reducing poverty and promoting sustainable development. It does so by bridging the gap between large, formal financial institutions and the rural poor. Microfinancing offers small loans directly to villagers, microentrepreneurs, impoverished women, and poor families. This allows them to create small, local businesses that generate income and offer services and products that improve the quality of life for the entire community. The rates of repayment on these loans average approximately 95%, much better than those in traditional credit markets.

The goal of microfinancing is to reduce poverty by providing credit at a grass-roots level to create microentrepreneurial opportunities. A study of the effectiveness of microfinance in sub-Saharan Africa revealed mixed results depending on gender and number of dependents of the borrower, the type of business, and the extent to which microfinance institutions penetrate a country and offer equitable access. But in many cases, microfinance has been shown to be an essential tool in creating the kinds of

BOX 12.3

Stakeholders and Collaborators: Learning for a Sustainable Environment: Innovations in Teacher Education Project

An excellent model of international cooperation for training and supporting teachers is the Learning for a Sustainable Environment: Innovations in Teacher Education Project. This program brings together educators from 29 Asian countries to work at the UNESCO Asia–Pacific Centre of Educational Innovation for Development in Bangkok, Thailand, and at Griffith University of Brisbane, Australia.

The project has two goals. The first is to develop a model for providing effective professional development in the knowledge, skills, and values of ESD for educators throughout the Asia–Pacific region. The second is to develop, pilot, and provide culturally sensitive workshop materials that can be used as the basis for professional and curriculum development activities by teacher educators in their home countries. These materials are organized as modules and employ a case study approach. They are currently used in both pre-service and in-service teacher education programs throughout the Asia–Pacific region, are available on the internet, and have served as the framework for national training workshops for teacher educators in other regions of the world.

microenterprises that help lift people out of poverty, especially in regions with few formal employment opportunities. In sub-Saharan Africa, microfinance has increased ownership of consumer durables and business assets but surprisingly without significant increases in consumption and waste production—a key indicator of sustainable development.[7]

BIODIVERSITY AND SUSTAINABLE DEVELOPMENT

The development of rural, tropical regions can be a potential threat to biodiversity. But in reality, because of its value, the preservation of biodiversity has offered opportunity for sustainable economic development through multiple-use conservation programs such as the Man and the Biosphere and carbon sequestration projects (see Chapter 7) and also through the development of ecotourism projects. Ecotourism adds value to the existence and preservation of habitats, biodiversity, and cultural heritage. Sound ecotourism provides economic opportunities for the local people living in communities in or near these habitats and offers them an alternative to unsustainable extraction of resources that provides a higher, more sustainable standard of living. It also educates tourists about sustainability and creates in them a sense of environmental stewardship that lasts well beyond their experience.

A number of international organizations, including the UN and Conservation International, support ecotourism as a component of their sustainable development and environmental conservation strategies. The International Ecotourism Society provides guidelines and standards, training, technical assistance, and educational resources to ensure sound practices so that ecotourism provides positive economic and social outcomes for the local community while preserving biodiversity and ecological sensitive areas. For ecotourism to contribute to sustainable development, it must involve community members in the decision-making processes and provide opportunity for local business ownership. It must also provide rising standards of living through fair wages paid to workers ranging from those with service jobs to trained guides and management. In addition, it must minimize biological impact. Ecotourism projects owned and operated by foreign interests that do not pay fair wages, and where the majority of the profit is removed from the community, deter sustainable development.

Tortuguero National Park in Costa Rica (Figure 12.6) is home to a major sea turtle nesting site. The local economy was transitioned from one based on harvest, direct consumption, and sale of turtle meat and eggs to what is now among the oldest turtle ecotourism sites in the world. This conversion involved training hunters to become guides and conservationists, and in so doing it improved the standard of living for the entire community. It has become a major tourist destination. In 2004, changes were made in order to reduce potential negative impacts of rapidly increasing tourist activities on nesting turtles.

In the past, tourists and their guides walked the beach at night with muted flashlights searching for nesting turtles. In 2004, the national park changed the system so that tourists are prohibited from entering the beach until one of a handful of park-certified turtle spotters communicates to a guide when and where turtles are ready to be viewed; this system thus results in minimal impact on nesting behavior. The Tortuguero story illustrates both the challenges and successes of ecotourism. Its rapid growth as a tourist destination was beginning to threaten the main attraction, but with adjustments these threats have been mitigated. Tortuguero remains a leader in turtle conservation through creation of an economy based on income generation from tourist dollars rather than through consumptive uses of turtles as in the past.

CLIMATE CHANGE AND SUSTAINABLE DEVELOPMENT

The issue of climate change must be central to any discussion of global sustainability and sustainable development. It is a global problem with both consequences and solutions that fall within the economic, environmental, and social contexts of sustainability. It is also an equity issue. The global poor who have emitted the least amount of greenhouse gases (GHGs) will be disproportionately impacted by climate change, and they lack the resources to deal with its negative impacts. Industrialized nations have emitted the bulk of climate changing GHGs, but they are better able to adapt to the impacts.

Much like the problem of consumption, the gap in emissions between the wealthy and the poor must be addressed. This must include both reductions in emissions in the industrialized areas of the world and ensuring that development in the poorer areas of the world is done in a way that does not result in a net increase in emissions. The industrialized nations that profited through their disproportionate release of GHGs should also address the inequity of climate change by

Figure 12.6 Tortuguero National Park in southeastern Costa Rica, and example of sustainability through ecotourism. (Meletis, Z. A., & Harrison, E. C. 2010. Tourists and turtles: Searching for a balance in Tortuguero, Costa Rica. *Conservation and Society*, 8(1), 26.)

supporting poorer countries in their efforts to both mitigate and adapt to an already changing climate.

Climate-related policy and financing must be a core priority and integrated with broader sustainable development strategies. This is because the impact of climate change and the ability to respond to it through mitigation, adaptation, and disaster relief affect the ability of countries to achieve SDGs and other development goals. For example, climate change increases the costs of development in the poorest countries by 25–30%—costs that are not currently accounted for in the cost estimates of SDG implementation. Conversely, development goals could either negatively or positively affect the success of climate policy depending on how those goals are pursued.

It is clear that the sustainability of the entire planet will require immediate and drastic action on climate. Today, 85% of all energy consumed is still from fossil fuels, which are subsidized at a rate more than 40 times larger than renewables. Carbon emissions have risen exponentially since the Industrial Revolution, and we currently emit more than 110 million tons of heat-trapping GHGs every 24 hours. This has directly resulted in significant increases in global temperatures and has begun to negatively impact agriculture, forests, oceans, and other ecosystems. It has already resulted in the greater frequency of intense and often disastrous weather events, and sea levels are on the rise, impacting coastal and island communities. Climate-related changes are having geopolitical consequences, so much so that the US Department of Defense has warned that the consequences of climate change, including refugees, food and water shortages, and the global spread of disease, are creating global instability and increasing the potential for conflict. In addition to becoming a global security issue, the climate crisis is also now regarded as the greatest risk to the global economy if nothing is done.

The reality is that something is being done, and there is great optimism that we will be able to slow and eventually mitigate this climate crisis. Renewable energy is gaining ground. We are witnessing a global energy transformation that is reflected by record-breaking additions in new renewable energy capacity since 2000. According to the International Renewable Energy Agency's 2017 global renewable energy report,[8] renewable power capacity increased 102% from 2000 to 2016, with 2016 as the strongest year ever for new capacity additions (8.7%). Solar energy had the greatest gains, followed by wind. Both are growing exponentially.

We have reached global **grid parity**, or the point at which producing electricity through renewable energy has become cheaper than doing so from fossil fuels. As of 2016, biomass geothermal, hydropower, solar photovoltaic, and on- and offshore wind were all within the range of fossil fuel cost for utility scale power. The cost of solar has been decreasing at a rate of 10% per year, energy storage technology has greatly improved, and record levels of investment in renewables have well exceeded that in fossil fuels. As a result, more renewable energy is being added to the grid each year than all that from fossil fuels and nuclear combined.

The renewable energy revolution that is underway has stimulated the economy through job creation. In 2015, 9.8 million people were working in renewable energy, up from 7.1 million in 2012. This number is expected to double in the next 10 years. The energy transition will boost global GDP by 0.8% by 2050, adding a cumulative $19 trillion to the global economy. Approximately 300 million people now receive electricity through off-grid renewables, and continued investment in this area will provide millions more access to energy for the first time.

The provisions of the 2016 COP 21 Paris agreement provide an excellent framework for future change. They set the goal of limiting global temperature increase to 2°C above the temperature during preindustrial times and urge efforts to limit the increase to 1.5°C. The agreement established binding commitments by all countries to pursue domestic measures aimed at reducing emissions and to commit to regular reporting on progress made in achieving them. The objective is to limit the amount of GHGs emitted by human activity to the same levels that trees, soil, and oceans can naturally absorb starting in 2050 and continuing through 2100.

COP 21 also reaffirms the binding obligations of developed countries and encourages voluntary contributions from emerging developing countries to support the least developed countries in their efforts to mitigate and adapt to climate change. The current goal for this support is $100 billion per year from 2020 through 2025, with a new, higher goal to be set for the period after 2025.

The commitment to addressing climate change is there. Even before the ratification of COP 21 and despite the fact that some provisions are not binding, many countries were already moving forward with change. China is already adopting a nationwide cap-and-trade system that will probably connect with existing programs in the European Union (EU). Countries such as Germany are already generating more than 80–100% of their electricity from renewable resources, and many others are generating more than half from renewables. Numerous countries, cities, regions, states, utilities, educational and public institutions, and businesses either have shifted or are committed to shifting to 100% renewable energy within the next few decades. As of 2017, even Google was operating on 100% renewable energy.

In the United States, future plans to construct coal plants have been canceled, and numerous existing coal plants have been retired. In 2015, nearly 75% of the investment in new electricity generation in the United States was in renewable energy, mostly wind and solar. This is all happening regardless of President Trump's recent withdrawal of the United States from COP 21 and his stated commitment to "jump starting" the US coal industry. The global rise in investment in renewable energy and the precipitous drop in coal prices and consumption before COP 21 reveal that economics are driving the transition to renewables. They also demonstrate that this recent, most likely temporary, policy shift is misplaced.

Many of the solutions proposed in this book are helping us solve the climate crisis. Technological advances, new policies, and the strong will of the people are responsible for current and future success. Educating for sustainability and sustainable development is working. We must also learn how to effectively communicate with climate change deniers to put an end to this irrational movement (Box 12.4). Climate issues are

BOX 12.4

Individual Action: Promoting Sustainability by Changing the Way We Talk About Problems

Most of us find ourselves in debates over sustainability issues on a regular basis. This is particularly true in the area of climate change. Have you ever found yourself in a conversation with a "climate denier" in which you tried to convince the person about the realities of climate change? What sorts of evidence did you use? Were you able to change the person's mind? The answer to the latter is most likely no, and both sides were probably left feeling that the divide over climate change can never be closed. But opposing evidence is emerging. We can close the divide and come together on important issues, but to do so, we need to change the way we frame the conversations about these issues.

According to Texas Tech Climate Scientist Katherine Hayhoe, our first mistake is that we often approach the conversation in a condescending, confrontational manner. We label the person with whom we are speaking a "climate denier"—as an adversary who needs to be converted. This immediately labels and dismisses that person, and it typically shuts down the dialog right from the start. Hayhoe recommends a less confrontational exchange using words that are less accusatory and that carry less baggage. This in turn can open up the discussion in a way that allows all parties to equally express their values and concerns, thereby opening the possibility of a more productive dialog.

As we engage in the debate over climate change, we typically make our case by using scientific evidence. We present graphs and maps that illustrate rising levels of GHGs and our contribution to those increases. We show the increase of global temperatures over time, how melting ice caps are shrinking, and how sea level is rising. A recent study revealed that this is ineffective, but we probably already know that from experience. The study revealed that the likelihood of "conversion" using scientific evidence is limited because these attitudes increasingly reflect ideological positions rooted in deeply held values. The authors found it more productive to identify negative impacts of climate change and ways to mitigate them that deniers find important.[1] Hayhoe agrees and recommends that we approach the conversation by identifying shared values and talk about solutions that are consistent with those values. As Stanford psychologist Robert Willer notes, the basis of the polarization that divides us on issues such as climate change is a wider moral divide. He adds that the most productive way to come together on these issues is to first understand and then frame the conversation within the context of the other's ideologies.

Our scientific approach to the debate often tries to convince people about climate change by showing them the impending doom that confronts us if we do not act. It incites fear and guilt, which according to Hayhoe tends to cause people to disassociate from the issue. Another problem is that our presentation of melting ice caps and stranded polar bears distances the problems from us. A better way to approach the discussion is to talk about how climate change will affect us and the things *we* value. By sharing positive solutions to these problems that already exist, we can instill a sense of purpose and hope. This will be much more effective at convincing people that climate change threatens things that they value and that with their support, solutions that already exist can be applied to address those threats.

One of the competencies emphasized in sustainability education is interpersonal communication. It may be the most important, and the one that we least effectively teach.[2] We need to develop our abilities to more productively talk about sustainability in order to bring about positive, real-world change. We have seen throughout this book that sustainable solutions require effective communication with collaborators and diverse stakeholders. Our success as sustainability professionals and advocates will require us to leave our academic comfort zones such as science or policy and to communicate with those with whom we do not necessarily agree. We must learn to do this with mindfulness and compassion in order to generate real sustainable solutions.

1 Bain, P. G., M. J. Hornsey, R. Bongiorno, and C. Jeffries. 2012. "Promoting Pro-environmental Action in Climate Change Deniers." *Nature Climate Change* 2 (8): 600–03. https://doi.org/10.1038/nclimate1532

2 Brundiers, K. and A. Wiek. 2017. "Beyond Interpersonal Competence: Teaching and Learning Professional Skills in Sustainability." *Education Sciences* 7 (1): 39.

taking center stage in elections throughout the world and gaining recognition from the actions of committed individuals such as those who have participated in the climate marches held in cities throughout the world since 2012. There is hope for global sustainability—and by overcoming existing barriers, that hope can be further realized.

Barriers to Global Sustainability

Throughout this book, technical, policy, market-based, and business solutions have been presented that are moving us in the direction of sustainability. Much progress is being made, and optimism is justified. However, a number of barriers to sustainable progress exist and must be overcome in order to achieve sustainability objectives. This section highlights these barriers to sustainability at the global level in order to reveal how we might overcome them.

The Vocabulary of Development

Words matter, and the vocabulary that we use to label countries as developed or developing may actually hinder sustainable development. The UN uses the term "developing country" more than any other organization, yet it does not have a definition for it. The World Bank historically used the bottom two-thirds of GDP as a strict cut-off between developing and developed countries. However, it is becoming more difficult to differentiate among these two categories because there is so much heterogeneity among countries that we label as developing. For example, China, Russia, India, and Haiti are often grouped together as developing countries, but they have drastically different levels of industrialization, GDPs, and access to basic needs and rights.

One problem is that as many developing countries continue to make economic gains, they often have reduced access to development assistance. For example, many low- to middle-income countries still have large numbers of people living in extreme poverty despite overall economic gains. As these countries become labeled "more developed," it is important to maintain international financial assistance to deal with the extreme poverty and improve economic equity.

Another related problem associated with the vocabulary of development is its focus on national economic growth and lack of consideration of overall improvements in quality of life. Typically, developed countries are thought to have universal health care, innovative educational systems, high rates of literacy, and improving infrastructure. They are also thought to have decreasing economic inequity, increasing earning potential, and a rising middle class. Ironically, these qualities are seen as much or more in some countries that are not typically considered as developed, and they are beginning to diminish in some that are classified as the most developed, including the United States. It is important to integrate all of these elements in our development language by employing indicators such as those from the Social Progress Initiative.

When we consider goals and indicators for development, it is important to recognize that they may be applicable to all countries, not just those that meet the traditional criteria of "developing." For example, the MDGs were meant to be for developing countries. The developed countries were viewed as the providers and the developing countries as the ones that needed help. Recognizing the problems of this dichotomy expressed previously, the SDGs view every country as needing some form of development. Aspects of sustainable development apply to every country, and a sustainable future requires solutions to these problems regardless of global measures of wealth.

Foreign Debt

Foreign or external debt is an outstanding loan that one country owes to another country, institutions within that country, or IFIs. Developing countries throughout the world have debt obligations that are often major barriers to any form of sustainable development. Many countries in their immediate post-colonial years took out huge loans in order to develop and provide social services in years of deficit. Many of these loans were from their original colonizing countries that once again stood to benefit through their exploitation. For example, South Africa accumulated $137 billion in outstanding debt during its post-colonial/apartheid history.

Unfortunately, most borrowing countries have not been able to repay these external loans because of a lack of economic growth for which the loans were intended. Large foreign or external debt is a barrier to sustainable development. This is because the enormous interest payments on these debts divert funds away from sustainable development objectives. Debt often results in worsening economic, social, and environmental conditions. In addition, austerity measures or required limits to government spending and restrictions on imports are often conditions on loans from IFIs such as the World Bank. These measures result in wage cuts and reductions in funding for social programs that disproportionately impact the poor.

Debt cancellation, restructuring of debt, or partial relief would free this money to be used for sustainable development. Providing a social safety net for the poor to reduce the impacts of austerity measures often associated with loans is also important. In order for debt relief to effectively contribute to sustainability, there should be transparent management of the application of these funds in partnerships with local communities and organizations. The strengthening of local economic institutions is also required for debt relief to have a significant impact. In the absence of transparency and strong local financial institutions, debt relief may help immediate needs in the poorest countries, but it would likely be less successful in supporting lasting improvements in quality of life, in stimulating investment, and in long-term economic growth and improvement in other areas.[9] However, when all of these factors are taken into consideration, debt relief has been shown to promote growth in countries with stagnated economies. It has improved the quality of life and environmental conditions for billions of people.

One example of a successful use of debt relief as a financial tool in sustainable development occurred in 1989 through what is referred to as the "Brady Plan." At this time, US Treasury Secretary Nicholas Brady negotiated agreements between private creditors and 16 low- to middle-income nations for debt relief and economic reform as proposed by each individual country. By linking debt relief with participatory economic reform in countries with relatively stronger financial institutions, the Brady Plan was very effective at restoring investment and growth in these countries. Also, it was discussed in Chapter 7 how one approach, debt-for-nature swaps, has the potential to provide economic opportunity through the preservation of biodiversity. It was also previously mentioned that the World Bank, the IMF, and the African Development Bank canceled $40 billion to $55 billion in debt owed by heavily indebted African countries to allow them to redirect resources to programs that supported MDGs. However, the long-term benefits of this may be limited by the weakness of their financial institutions and by the conditions placed on that debt relief by these IFIs.

Trade Policy

The latest trends in trade policy have in many cases served as a barrier to sustainable development. Neoliberalism has been the underlying principle of most trade policies since the early 1980s. The goal of neoliberalism is to create a global economy by making international trade easier. This has been accomplished by opening or deregulating trade in order to increase the movement of goods, resources, and enterprises across borders. This also allows businesses to find cheaper resources to maximize profits and efficiency.

On the surface, it appears that neoliberal trade policies such as the North American Free Trade Agreement (NAFTA) resulted in economic development or growth. GDP among countries that participate in such agreements has increased. However, as discussed previously, GDP hardly accounts for our sustainability criteria, and such policies have had negative effects. Neoliberal trade policies have drastically increased economic inequity, reduced worker wages, and increased unfair labor practices and environmental degradation in poor countries. They are also responsible for the loss of well-paying jobs from the manufacturing sector in what was once the most industrialized area of the world.

This book has presented numerous examples of how neoliberal economic policy has served as a barrier to sustainability, including the privatization of water, the export of toxic waste to poorer nations, and the decline of inner cities and communities in the United States that occurred with the loss of manufacturing jobs. A recent example of how such trade policy impedes sustainability efforts, including the ability to meet carbon reduction goals as prescribed by agreements made at COP 21, involves India's solar policy. The Indian government has developed an ambitious program to create its own solar energy. In just five years, India's federal solar program allowed the country to transition from basically having no solar capacity to having one of the world's fastest growing solar industries. To allow the solar program to also contribute to the the country's overall objective of sustainable development, the government has incentivized the installation of locally produced solar cells and modules. This in turn has boosted the solar industry, created jobs, and alleviated poverty.

Despite the success of India's solar program, in terms of both implementing climate change recommendations and sustainable economic development, it has been viewed as a direct violation of neoliberal trade policy. As such, the United States raised a case with the World Trade Organization (WTO) against India's program claiming that India was unfairly restricting access to American suppliers even though imported solar technology currently has a dominant share of the market in India. New trade policy must move beyond antiquated objectives of raising GDP and focus on opportunity for sustainable development. It should also support developing countries in their efforts to meet climate commitments rather than penalize them.

Funding and Aid Policy

Depending on how it is applied, international funding and aid policy can also serve as barriers to sustainable development. Previously in this chapter, we examined a variety of ways in which sustainable development, including carbon emissions reduction, can be funded. This included loans through IFIs, foreign aid from wealthier nations, investments by multinational and other private enterprises, and funding from NGOs and local governments. Non-military foreign aid from wealthier countries has been a major part of this. The target for direct aid from OECD countries has been 0.7% of GDP. This has most often been applied to humanitarian support and immediate crisis relief rather than long-term sustainable development.

The economic realities of the past decade—and the fact that non-military aid is often linked to security interests rather than sustainable development—have resulted in stagnant or decreasing levels of support to countries that desperately need it for sustainable development initiatives. Also, private investment in the poorest countries is often viewed as too risky and therefore less common. Because of these factors and the increasing needs of poorer nations, current rates of funding are insufficient. Economic experts have suggested a number of possibilities for increasing such funding, including global environmental taxes such as a carbon-use tax, international trade taxes, the development of global bonds, and fostering ways for immigrants to send money back to their country of origin. Also, public finance should be used to leverage private capital and fill gaps in the markets where private finance is not reaching, especially in the poorest countries.

The funding barrier is not simply about quantity. Conditions associated with that funding, the strength of financial institutions and their ability to apply those funds where they are most needed, and broader economic policy impede rather than support sustainable development. It has been shown throughout this book that restrictions on the ways that funds can be used and a lack of transparency in their dispersal reduce the effectiveness of that funding. Extreme austerity measures required with external funding disproportionately impact the poor. Also, the imposition of neoliberal economic policy as described in the previous section is often a condition of loans or aid from the WTO, IMF, and other international aid organizations. This is evident in much of sub-Saharan Africa, where neoliberal policy has been a condition of external funding of development. This has had negative consequences because it required the elimination of government social programs and reductions of government subsidies on food, medicine, and education. It also resulted in increased unemployment, greater income inequality, decreased availability of social services for the most needy, and social turmoil.

In contrast, the nation of Eritrea in northeast Africa was able to maintain public–private partnerships as part of its development strategy. As a result of their government's involvement in economic growth—including support of small-scale private businesses and state support for domestic industries, communities, and local businesses—improvement of infrastructure and social services has been significant. This has reduced their dependency on foreign aid and increased economic equity and social justice. Decoupling foreign aid from trade and other types of policy that benefit the supporting countries and their corporations more than the aid recipients will be vital for achieving sustainable development.[10]

Failing States

A failed state is one in which the political or economic system is so weak that the sovereign government is no longer functioning properly. It is a country that is essentially in the process of disintegration. Characteristics of failed states include loss of control of territory, inability to interact with the international community, and failure of the government to make collective decisions and provide public services and security.

The Fund for Peace, a non-partisan, non-profit research and educational organization, has developed

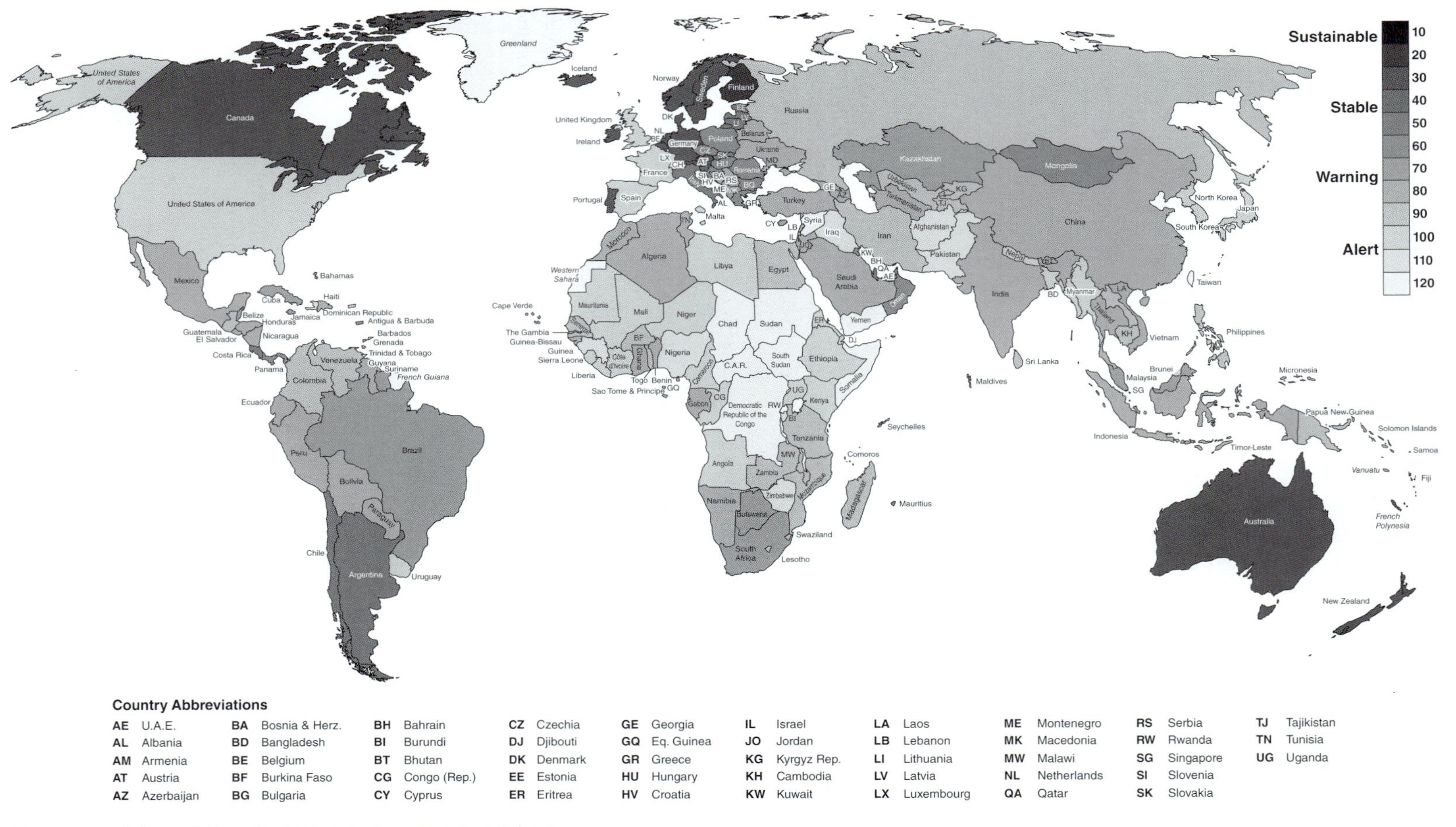

Figure 12.7 Map of Fragile States Index. *The Fund for Peace.*

an index to quantify the extent of state fragility for UN member states. The Fragile States Index is based on a variety of social, political, and economic factors that show dire conditions in 16 countries and numerous others at risk for failure (Figure 12.7). These countries often are or have been confronted with war, violence, and genocide; human rights abuses; mass migrations and refugee crises; and extreme hunger and other deplorable conditions.

The large number of failing and fragile states is an obvious threat to global security and a barrier to global sustainability, and it is perhaps one of the most challenging barriers to overcome. In January 2016, the World Bank's Fragility, Conflict, and Violence Group recognized that a lack of recognition of issues related to fragile states was a major underlying reason why the MDGs were not transformative. This led to a specific SDG on peace and security and also the development of specific indicators for monitoring success toward this goal. The challenge will be putting this rhetoric into reality. One key factor will be linking sustainability objectives and sustainable development with peace-making and state-rebuilding activities.

An example of where this has not happened is in the US-led post-war economic restructuring in Iraq. This did not happen in the post-war economic restructuring in Iraq where farmers had been practicing seed saving for thousands of years. United States law required that new seed could be supplied only by private companies such as Monsanto on which Iraqi farmers were made permanently dependent because saving seed from these crops is illegal.[11] This has destroyed an agricultural tradition and replaced it with a much less sustainable approach. In contrast, a more positive result was achieved by the group Green Iraq. Composed of Iraqi and international ecologists, the group was able to garner international funding to support the restoration of marshes destroyed during the war. The UN Environmental Programme concluded that roughly 58% of the marsh area present in the mid-1970s had been restored and that there had been a remarkable rate of re-establishment of native plants and animals in these re-flooded marshes.[12]

Political Inertia

Significant change or transformation is required for global sustainability. The ongoing lack of will for such change is a major barrier. This can be due to a variety of factors that range from the individual to collective action and institutional design. It is believed that individuals are innately or biologically wired for routine and that there is a fundamental fear of change. The way large institutions such as political systems are designed also emphasizes stasis over change. As a result, they often persist even when their flaws are obvious.[13] Political inertia is a challenge to overcome, essentially paralyzing governments from making change. Corporations or others with special interests that benefit from the status quo also make every effort to maintain current conditions.

Political inertia has been a major barrier to working toward sustainability in the United States, particularly with the lack of legislation regarding the use of fossil fuels and climate change. We have discussed many other examples, such as policies that favor unsustainable agriculture and promote consumption. Change is also difficult to achieve when there are spatial and time lags between known problems and what people experience. Problems such as climate change are subject to such time lags, making it easier to pass the problem on to future generations. Being disconnected from sources of energy and food, and low prices due to subsidies and not accounting for externalities, also makes it challenging to for people to recognize problems and effect change.

Sometimes political inertia can be overcome, and the change that ensues can be quite negative within the context of sustainability. The recently successful right-wing, populist movement that emerged in the US election of 2016 and the Brexit vote in which British voters decided to leave the EU are examples of such change. Although the political rhetoric suggests a swing in the opposite direction of sustainability, there is still considerable legislative inertia and bringing about such change has proven difficult. The power of the US president to repeal and defund sustainability-related policy through the use of executive orders such as backing out of the COP 21 Paris agreement is a real threat to a sustainable future and a threat to democracy as well.

One positive outcome from the recent political climate has been the commitment of individual states and institutions to adhere to the Paris agreement regardless of the federal stance. It has become abundantly clear that economics now favor such action and that the majority of Americans favor it. President Trump's decision to retreat from such policy represents a missed opportunity to lead the economic transition to renewable energy by doing the right thing. Another positive result is that the current political climate has caused people to become more engaged as activists and as participants in politics in what is referred to as the "resistance movement," including running for local office. History has shown that oppressive politics will eventually cause the political pendulum to swing back in the other direction.

There is evidence that political change is achievable. Recent multinational commitments to climate accords with near-universal agreement, that originally included the United States as a signatory, are a sign that change is possible. There is a greater global concern for environmental issues and social justice. This is happening through education, the work of NGOs, the use of social media, and because of demographic shifts—particularly in the United States, where younger generations are becoming the largest voting bloc. However, the political change required for sustainability will also depend on campaign finance law that will limit the power of corporations over government, create term limits for politicians, and instill pressure to eliminate corruption at all levels of government. Such reform will allow policy to meet the needs of the greater good and promote longer term thinking, a necessary step for overcoming this barrier to sustainability. It will also require increased political engagement among those of us who favor a sustainability agenda.

Sustainability Science

A lack of interdisciplinary scientists working on broad solutions-based research has been a barrier to sustainability, but this is changing. The field of sustainability science is rapidly expanding, and some have argued that this is an indicator of society's progress toward the goal of sustainable development. A recent analysis found that the rate of research publications in sustainability science is growing almost twice as fast as the average across all research disciplines—approximately 7.6% annually.[14] Increasingly, all key areas of sustainability are being addressed throughout the world. Given the magnitude, complexity, and global nature of problems that we face, an interdisciplinary, internationally collaborative, systems approach to sustainability science is vital.

The urgency and high stakes of sustainability science are changing the traditional role of these scientists as discipline-based producers of knowledge who are typically reluctant to apply or advocate for the implications of their research. The pressing need for scientists to deliver solutions to major societal problems and make high-cost decisions has forced them to move out of their disciplinary silos and into modes of scholarship that have not traditionally been valued by the broader academic community.

Sustainability scientists often find themselves in untraditional partnerships with varied stakeholders, such as those with businesses. For example, the Environmental Defense Fund has connected scientists with large corporations to help them achieve sustainability goals such as the use of safer chemicals, reducing GHG emissions, and improving the sustainability of global supply chains. Scientists are also increasingly interacting with policy communities and linking their knowledge with action to promote sustainability.

Given the urgent and ethical need to inform the decision-making process, sustainability scientists are apt to apply the precautionary principle. Thus, when the stakes are very high, they may be willing to accept slightly more uncertainty when drawing conclusions or making recommendations than they normally would in what is sometimes referred to as **post-normal science**. In this way, post-normal science can serve as a bridge between complex systems and environmental policy. Given the urgent need for their conclusions, sustainability scientists also need to communicate their results in more rapidly published venues that offer open access.

A barrier to effective sustainability science is that these untraditional modes of scholarly work and dissemination have often been undervalued by scientific and academic institutions; however, this is beginning to change. Interdisciplinary teaching and research have become prioritized. Colleges and universities are starting to modify traditional "publish or perish" academic models to include other kinds of recognition. These newly recognized outcomes include leadership in policy interventions, the creation of technological platforms that are implemented with decision-makers, and participation in and development of collaborative planning workshops. There is also growing acceptance of less traditional but more rapid means of reviewing and disseminating quality scientific results, including open-access and electronic formats.

One organization that is supporting the role of sustainability science in delivering solutions is Future Earth. It does this by organizing scientific communities and ensuring that the knowledge they generate is done in partnership with the users of science, such as policymakers, the UN, and climate conventions. Future Earth has created global hubs and regional centers that promote networking and catalyze new research in sustainability. New programs in higher education also cater to the need for training solutions-oriented sustainability scientists. For example, Harvard University's new program in sustainability science links research and policy communities, and it builds capacity to link science with action to promote sustainability.

Another barrier to applying science to policy and change is an outright rejection of it, particularly in the United States. There is a tendency for people to reject the validity of a scientific argument because its conclusion contradicts their deeply held views. There is also a broad misunderstanding of science as a belief system rather than an objective process that employs deductive inference to test hypotheses. Those who lack knowledge of science make statements such as "It is only a theory" without recognizing that a theory is an explanation that is already supported by a large body of evidence generated through repeated scientific testing that guides future research. Also, the tendency is for scientists to be skeptical of their own data—a good thing—but those with strong non-scientific views use this self-imposed skepticism to infer the weakness of a particular scientific conclusion. Scientists need to develop more effective ways to communicate to policymakers and the broader public (see Box 12.4). Finally, industry and special interest groups, such as the Heartland Institute with its anti-climate change agenda, cherry-pick specific data that help them further their agenda and in so doing mislead policymakers and the public.

The solution to this rejection of science is additional and improved science education that goes beyond just teaching factual information to actually engaging students in the process of science. As students engage in science and learn what the standards are for scientific evidence, they will learn how to differentiate between sound science and pseudo-science. They then will be empowered to make scientifically based decisions. A recent study showed that once high school students understand science, they are able to interpret scientific information relatively independently of ideological constraints. The study showed that with an understanding of climate science, students were able to deduce anthropogenic causes of climate change even when their teachers overtly rejected that idea.[15]

Big Data

Another barrier to sustainable development has been the availability of and access to data. The systems approach to sustainability science requires large data sets. Referred to as "big data," these data occur over large time and spatial scales. The large data sets include both environmental and socioeconomic data, and they often integrate these data. They are viewed as essential for sustainable development and community resilience. The human appropriation of net primary production (HANPP) is an example of what may comprise one of these large data sets. HANPP is an integrated socioecological indicator that quantifies human-induced changes in productivity through the consumption of biomass such as food, paper, and wood and fiber. This aggregated indicator is used to quantify human effects on the composition of the atmosphere, levels of biodiversity, energy flows within food webs, and the provision of important ecosystem services on a global scale.

The future of agriculture will also be very data-dependent. However, farmers are currently overwhelmed with large amounts of incompatible data from a multitude of sources. To address this issue, the Open Ag Data Alliance has created ways for farmers, academics, and service providers to share and have full access to compatible, open-source data. For example, a farmer can access real-time weather and soil moisture data to develop planting and harvesting schedules in order to optimize production.

One of the challenges associated with the need for and use of big data in sustainable development is the risk of marginalizing the world's poorest populations that typically do not have access to such data or the training needed to collect or use them. This can be addressed by sharing technology and data and also through training programs. The organization Vital Signs is helping African countries collect their own ecosystem and socioeconomic data to help inform development decisions. It has trained teams of researchers in Tanzania, Ghana, Kenya, Rwanda, and Uganda to collect data in a variety of ways, including household surveys on farming practice, nutrition, and where they get their water and fuel wood. Others are collecting and analyzing soil samples or measuring trees and canopy cover to help calculate the amount of aboveground carbon a landscape holds. A simple solution to establish credit history for microfinance institutions has been the use of mobile purchase data.

Another problem related to data and poverty that has been a major barrier to sustainable development is that there are no public records for approximately 2.4 billion people who lack official identification. Children whose births have never been registered and many women represent the largest proportion of this group. Only 19 of 198 economies provide a unique ID at birth and use this consistently in civil identification and public services. Legal identity is basic to many of the rights set out in the Declaration of Human Rights and the Convention on the Rights of the Child. National identification programs are vital for providing basic social and health services and offering equal rights to economic resources, including property and finance. ID programs also improve the control over property, land, and financial assets including bank accounts.

The role of identification programs in providing social and economic opportunities in sustainable development is so important that it is part of SDG 16, urging states to ensure that all have free or low-cost access to widely accepted, robust identity credentials. The World Bank, through its Identification for Development (ID4D) program, is providing technical support to help meet this goal. One of the largest efforts has been in India, where the country's unique identification authority called Aadhaar has taken the lead with this effort. It has implemented a nationwide program to register and assign a 12-digit identification number linked to a biometric screen such as a fingerprint or iris pattern to all citizens. This has already resulted in improved delivery of government services and individual access to bank accounts. Despite the benefits of ID4D programs such as Aadhaar, many have raised concerns and questions regarding privacy issues and potential for abuse; however, protections for such abuse are being strengthened.

Greed and the Growth Paradigm

The dominant world view among developed nations is that their economic futures depend on unbridled economic growth that is rooted in continued escalating rates of global consumption. In Chapter 10, we recognized that an educational system with the purpose of offering a "better future" for our children through increased success as measured by the accumulation of wealth encourages this view, as does the constant barrage of marketing disposable, upgradeable products that represent wants more than needs. This way of thinking perpetuates greed, inequity, and consumption, and it may be our greatest barrier to sustainability and an improved future for the many as opposed to the few.

The question of whether this capitalistic model of growth, consumption, greed, and inequity can be changed cannot be ignored as we contemplate our sustainable future. We have a choice: Either we create an ecologically and socially sustainable version of capitalism or we wait for the prevailing conditions to precipitate its collapse. We are beginning to make the right choices. There are inspired individuals whose actions are making a real difference. Our education system is beginning to embrace the idea that students require knowledge to solve problems rather than perpetuate them. Many business leaders who once worked at the heart of the finance world are now proposing specific ways to fix a failing system. We are seeing the creation of sustainable communities and regions, and we have well-defined sustainable development and consumption goals toward which progress is being made.

Greed, excessive consumption, and increasing economic inequity are indeed barriers to sustainability, but research shows that individuals are not inherently greedy. There are a number of forces that are inciting small but significant change. Our education and religious institutions are increasingly emphasizing compassion and empathy, and they are helping us to better understand the real meaning of well-being. We can break the work–spend cycle that is marketed to us on a daily basis that creates these false views of happiness. This can be done by directly limiting that marketing and eliminating the existing tax deduction on those marketing expenses.

Businesses leaders are recognizing that we need a fixed version of capitalism—a new economic model. And there is some truth to Ray Anderson's argument (see Chapter 10) that there is only one institution big enough, powerful enough, pervasive enough, influential enough, and wealthy enough to accomplish this and lead humankind out of the mess we have made and put us on the path toward sustainability. That institution, the one that made that mess to begin with, is business itself. Sustainability has not only become a priority but also is beginning to be woven into the fabric of many of businesses.

Government will have a major role to play in this corporate transformation. It will need to rebuild public infrastructure, contribute to economic planning, and regulate international trade in ways that protect the environment, equity, and social well-being. This is in stark contrast to the neoliberal approach that scorns government regulation and intervention. NGOs such as the New Economics Foundation are already helping shape a new economy. So are grass-roots movements consisting of students, members of trade unions, and members of environmental, social justice, and faith groups. Many of these groups have often had disparate agendas, but they are now coming together in support of common ideals.

So what will a new economy that supports sustainable development look like? The jury is still out, but some common themes are emerging. It will likely require a shift from an emphasis on unlimited economic growth to one that favors the growth of human potential through improved conditions and increased quality of life. Some argue for a steady-state economy with mildly fluctuating levels in population and consumption of energy and materials. A sustainable economy will have to account for externalities and incorporate the environmental and social costs into the prices of products and services. It may benefit from a bioregional approach in which economies are embedded within natural social units determined by ecology rather than by economics. This makes them more self-sufficient in terms of basic resources, products, and services.

One organization, the Next System Project, is working with a broad group of researchers, theorists, and activists to help develop comprehensive alternative political–economic system models. One of a number of models they promote is called economic or stakeholder democracy. This system continues to favor the role of markets but shifts decision-making power from corporate managers and shareholders to a larger group of public stakeholders that includes workers, customers, suppliers, neighbors, and the broader public. Another model incorporates the concept of **common property regimes**. These are institutional arrangements for the cooperative or collective use, management, and sometimes ownership of natural resources such as forests that are managed for long-term benefits.

None of this change will come about without further development of a solutions-based economy that is fueled by disruptive innovation through the work of creative individuals such as those featured throughout this book. Sustainable development will depend on solutions generated by social entrepreneurs or change-makers. Any new system must accelerate, not stifle, such creativity and innovation. This is already occurring through the support of organizations such as Ashoka and B Lab that serve the global movement of using business as a way to achieve sustainability objectives. Sustainable development will also require a continued emphasis on green and social investment so that our retirement plans, savings, and institutional endowments do not continue to grow at the expense of other people or the environment but, rather, in support of them.

Finally, this new economy will require each of us to change. We will need to become comfortable with new models of success that favor generosity over greed, understanding over ignorance, compassion over indifference, and humility over arrogance. The actions of individuals such as those featured in this book will continue to make changes that transform, build community, and influence businesses through our own choices. We will need to come together to create a strong new power base to effect change through the democratic process. To overcome the barriers to sustainable development, we will need a transformation that engages not just top-down or bottom-up approaches but instead comes from all directions.

Our Sustainable Future

As we embark on this path of change toward a sustainable future, there are some important themes to keep in mind. These ideas consistently emerged as we examined ways to generate sustainable solutions and overcome barriers and challenges to them. They are essential elements of any sustainable solution, whether it is a program, institution, policy, or the way in which we use resources. These are a few essential items that should be in the "toolbox" of the current and next generation of problem solvers:

1. Environment, economics, and social justice are directly linked, and the failure of one will bring all of them down. This principle should guide our decisions and our actions, and it should be central to the way we think about our future.
2. Collaboration, participation, and ownership are essential for success. When stakeholders with different views are involved in the decision-making process, we move closer toward sustainability. We have seen that when potential beneficiaries from programs play a role in developing them, they work better. Opportunity for ownership creates buy-in and individual responsibility that sustain projects. Cooperation is more effective than antagonism.
3. There must be transparency in our decision-making process so that there is knowledge and trust and also so that greed, corruption, and injustice are exposed and ultimately prevented. The good news is that with the ubiquitous

presence of cameras and instantaneous communication, there is ever-increasing global transparency. Transparency is such an important component of sustainability and sustainable development that it is now being tracked with indices developed by organizations such as Publish What You Fund and Transparency International.

4. We need to continue to rethink education at all levels and in all places. We must directly address the paradox between education and sustainability by recognizing that the problems that we face today cannot be solved using the same way of thinking that was used to create them. There needs to be a paradigm shift that inspires students to think about their relationship with the world and the others in it and how they can influence those relationships in positive ways.
5. Innovation, creativity, and entrepreneurship are essential elements of problem solving. We have discussed many examples throughout this book, and there are many more. Any change in our economic, political, or social organization and the process of development should ensure that there are mechanisms that will encourage, inspire, and support these important elements.
6. Systems thinking is the necessary approach for both understanding our complex problems and solving them. In every chapter, we found that there is no single solution to a given problem and that the issues we face are all interconnected. We must analyze each potential solution using the tools of life cycle and full-cost analyses and systems thinking to determine how they may either directly or indirectly influence other aspects of sustainability. By doing this, we can maximize the benefits and minimize the harm by combining the best elements of them in a multifaceted approach to achieving sustainability outcomes.
7. We need to measure what we do. Every effort must have clear objectives, as well as indicators that can be used to measure both our successes and our failures. These objectives should be based on backcasting principles to ensure they represent what we must do, not simply what we are willing to do. It is not enough to simply assess what we do. We also have to be willing to adapt and change course based on our successes and failures.
8. We cannot do this without government. We have discussed many examples of effective government policies that have improved sustainability. Such policies, even when they have been regulatory, have improved environmental and social conditions while stimulating the economy. Public–private partnerships have been one of the most effective ways to deliver lasting sustainable solutions. We have also seen that deregulatory, neoliberal policy that restricts the role of government can create greater economic inequity and can have negative social and environmental consequences. Government is best able to contribute to sustainability efforts when it is transparent and participatory and develops reforms that reduce the influence of corporations and opportunities for corruption so that it can justly serve all people.
9. The power of individual action is not to be underestimated. This can be through our daily actions, our consumer choices, redefining our own values, educating each other, the democratic process, or by taking part in or leading larger social movements. We must continue to develop our own interpersonal competence and that in others so that we can better make the case and convince others of the need for meaningful and just change (see Box 12.4). In these ways, individuals have and will continue to transform the world. Our sustainable future will rely heavily on the metamorphosis of individuals from states of apathy, indifference, or helplessness to ones of inspiration, empowerment, engagement, and leadership.
10. The major challenges we face, such as poverty, climate change, gender inequality, and conflict and the failure of states, are global problems. These issues will only be solvable through a commitment to global citizenship. Global citizens are people who think of themselves first and foremost not as members of a state, nation, or tribe but, rather, as members of the human race. As global citizens, we need to reflect on the global implications of every decision we make and take action to ensure economic, environmental, and social justice for all people.

The previous ideas represent common themes that have emerged as we defined sustainability, identified problems related to it, and explored solutions to those problems. The list is not complete, and the burden is ours to explore other important elements that will facilitate the generation of sustainable solutions. It is important to remember that sustainability is not an end point. It is a set of objectives—a movement toward continuous improvement in all aspects of sustainability. That movement is well underway, and in some areas progress is even exceeding expectations. It will be up to current and future generations of sustainability problem solvers to continue to identify the barriers to sustainability and how to overcome them by using creative solutions to keep us moving in that direction.

Chapter Summary

- Sustainable development provides ways for people who are not sufficiently having their basic needs met to do so without compromising the ability of future generations to meet their needs. It is development that maximizes renewal, encourages reuse, minimizes consumption and waste, protects the environment, and addresses climate change. Sustainable development occurs through the creation of equitable economic opportunity that offers all people the capacity to meet their own needs, resulting in an elevated standard of human well-being through participatory, inclusive processes within the context of science.
- Development is measured in a variety of ways, including GDP, and by more inclusive indices that reflect human development, economic equity, happiness, and social progress.
- The major players in sustainable development include the UN, the International Institute for Sustainable Development, international finance institutions such as the World Bank and the IMF, the OECD nations, multinational enterprises, local governments, and NGOs.
- Major programs in sustainable development include the UN's MDG and SDG programs, efforts to slow population growth and consumption, and the promotion of gender equity, community sustainability, and education for sustainable development.
- Other tools in sustainable development include increasing technology to reduce the digital divide and the provision of low-interest microfinancing for entrepreneurial activity.
- The preservation of biodiversity, the use of nature-based solutions, and the mitigation of and adaptation to climate change are central to sustainable development.
- Barriers to sustainable development include issues with vocabulary, neoliberal trade policy, current funding and aid policies, foreign debt, the proliferation of failing states, and both political inertia and rash political change. Others include the young state of sustainability science and outright rejections of science by policymakers and the public, in addition to availability and access to large data sets. The ultimate limitations to sustainable development are greed and the current growth paradigm.
- A sustainable future is not impossible, and a number of encouraging trends—even in the face of negative political climates—offer hope. A number of important ideas have emerged that will guide us on this path to a sustainable future.

Digging Deeper: Questions and Issues for Further Thought

1. What is meant by the opening quote of this chapter, "The principle of common but differentiated responsibilities is the bedrock of our enterprise for a sustainable world"?
2. What are the limitations of using standard indicators of economic growth such as GDP in assessing development?
3. What are some issues related to the language of "developed" and "developing" countries?
4. Explore the social progress data for the past four years at www.socialprogressimperative.org. What trends do you see? What factors are driving these trends?
5. How are the SDGs different from the MDGs?
6. What is neoliberal trade policy, and how does it promote and/or limit sustainable development?
7. In what ways have the recent "right-wing, populist" political changes affected sustainable development objectives, and why is this movement having difficulty advancing its agenda against many sustainability principles?

8. What are hybrid value chains, and how do they help support sustainable development?
9. What are some common recipes for success in overcoming barriers to sustainability and sustainable development that have emerged throughout this book?

Reaping What You Sow: Opportunities for Action

1. Organize and promote a pop-up film festival on critical issues related to the sustainability of our global food system in conjunction with Real Food Media. Its website, http://realfoodfilms.org, offers support and toolkits both for local screenings and for organizing communities to take action on food security issues.
2. Using the resources available from The Next System project at http://thenextsystem.org/next-system-teach-ins, organize a teach-in at your school or in your community that explores what is required to deal with systematic challenges regarding our sustainable future.

References and Further Reading

1. United Nations Development Programme. 2015. "2015 Human Development Report." http://report.hdr.undp.org
2. World Bank. 2013. "About Us." http://web.worldbank.org/WBSITE/EXTERNAL/EXTABOUTUS/0,,contentMDK:20040612~menuPK:8336267~pagePK:51123644~piPK:329829~theSitePK:29708,00.html
3. OECD. 2016. "Development Aid Rises Again in 2016 but Flows to Poorest Countries." http://www.oecd.org/dac/development-aid-rises-again-in-2016-but-flows-to-poorest-countries-dip.htm
4. "Breathing Life into Global Sustainability Goals." n.d. Rainforest Alliance. Accessed August 6, 2018. https://www.rainforest-alliance.org/articles/breathing-life-into-global-sustainability-goals
5. Mukherjee, Saswati. 2013. "South India Lags Behind National Fertility Rate, Slows Population Boom." *The Times of India*, May 28. Accessed May 17, 2016. http://timesofindia.indiatimes.com/india/South-India-lags-behind-national-fertility-rate-slows-population-boom/articleshow/19249154.cms
6. Coalition for Inclusive Capitalism. "About Us." http://www.inc-cap.com/about-us
7. Banerjee, Abhijit Vinayak. 2013. "Micro-credit Under the Microscope: What Have We Learned in the Past Two Decades, and What Do We Need to Know?" *Annual Review of Economics* 5 (1): 487–519. doi:10.1146/annurev-economics-082912-110220
8. International Renewable Energy Agency. 2017. "REthinking Energy 2017: Accelerating the Global Energy Transformation." Abu Dhabi, United Arab Emirates: International Renewable Energy Agency.
9. Arslanalp, Serkan, and Peter Blair Henry. 2006. "Policy Watch: Debt Relief." *Journal of Economic Perspectives* 20 (1): 207–20. doi:10.1257/089533006776526166
10. Amahazion, Fikrejesus. 2016. "Neoliberalism and African Development." *Madote*. Accessed October 13, 2016. http://www.madote.com/2015/09/neoliberalism-and-african-development.html
11. Order 81, paragraph 66 - [B], issued by L. Paul Bremer, The American Administrator of the Iraqi Coalition Provisional Authority of the Iraqi Government. 2002.
12. Richardson, C., and N. Hussain. 2006. "Restoring the Garden of Eden: An Ecological Assessment of the Marshes of Iraq." *BioScience* 56 (6): 477–89.
13. Pierson, Paul. 2004. *Politics in Time: History, Institutions, and Social Analysis*. Princeton, NJ: Princeton University Press.
14. "Sustainability Science in a Global Landscape." 2015. A report conducted by Elsevier in collaboration with SciDev.Net.
15. Stevenson, K. T., M. N. Peterson, and A. Bradshaw. 2016. "How Climate Change Beliefs Among US Teachers Do and Do Not Translate to Students." *PLoS One* 11 (9): e0161462.

Appendix: List of Abbreviations

AASHE: Association for the Advancement of Sustainability in Higher Education
AC: alternating current
BOD: biological oxygen demand
BRT: bus rapid transit
Bt: *Bacillus thuringiensis*
C3: College, Career, and Civic Life
CAA: Clean Air Act
CAFE: Corporate Average Fuel Economy standards
CAC: Conservation Advisory Commission
CBA: community benefits agreement
CBO: community benefits ordinance
CCA: community choice aggregation
CCBA: Climate, Community & Biodiversity Alliance
CED: community economic development
CERCLA: Comprehensive Environmental Response, Compensation, and Liability Act
CFC: chlorofluorocarbon
CFCs: chlorofluorocarbons
CHEJ: Center for Health, Justice & Environment
CI: Conservation International
CLA: Clean Water Act
CLP: Chars Livelihood Program
COP 21: Conference of the Parties 21 or 2015 United Nations Climate Change Conference in Paris
CSA: community-supported agriculture
CSBGs: Community Service Block Grants
CWA: Clean Water Act
DACA: Deferred Action for Childhood Arrivals
DC: direct current
DDT: dichlorodiphenyltrichloroethane
EAC: Environmental Advisory Council
EITI: Extractive Industries Transparency Initiative
EPA: US Environmental Protection Agency
ESD: education for sustainable development
ESG: environment, social, and governance
ETS: emissions trading schemes
EU: European Union
FAO: UN Food and Agriculture Organization
FDA: Food and Drug Administration
FFV: flexible fuel vehicle
GDP: gross domestic product
GHG: greenhouse gas
GIAHS: Globally Important Agricultural Heritage System
GIS: geographical information systems
GMO: genetically modified organism
GNH: Gross National Happiness Index
GNP: gross national product
GPWM: Global Partnership in Waste Management
GWP: global warming potential
HANPP: Human Appropriation of Net Primary Productivity
HAPs: hazardous air pollutants or air toxics
HCFCs: hydrochlorofluorocarbons
HDI: Human Development Index
HOA: homeowner association
HUD: US Department of Housing and Urban Development
IAQ: indoor air quality
ICT4D: Communication Technologies for Development
ID4D: Identification for Development
IFC: International Finance Corporation
IFI: international finance institution
IISD: International Institute for Sustainable Development
IMF: International Monetary Fund
INBio: Cosa Rican National Biodiversity Institute
IPCC: Intergovernmental Panel on Climate Change
IPM: integrated pest management
JMP: Joint Monitoring Program for Water Supply and Sanitation
KPCA: Kimberley Process Certification Scheme
LAGI: Land Art Generator Initiative
LCA: life cycle assessment
LED: light-emitting diode

LEED: Leadership in Energy and Environmental Design
MAB: Man and the Biosphere Programme
MAPS: mainstreaming, acceleration, and policy support
MBR: Mayan Biosphere Reserve
MDGs: Millennium Development Goals
MNE: multinational enterprise
MSW: municipal solid waste
NAFTA: North American Free Trade Agreement
NbS: nature-based solutions
NEON: National Ecological Observatory Network
NFT: nutrient film technique
NGO: non-governmental organization
NRDC: Natural Resources Defense Council
NTFP: non-timber forest product
ODS: ozone-depleting substances
OECD: Organization for Economic Co-operation and Development
OPEC: Organization of Arab Petroleum Exporting Countries
PAC: political action committee
PAYT: pay as you throw
PCB: polychlorinated biphenyl
PET: polyethylene terephthalate
PPP: purchasing power parity
PM: particulate matter
PUSH: People United for Sustainable Housing
PV: photovoltaic
RAS: recirculating aquaculture systems
REC: renewable energy certificate
REDD+: Reducing Emissions from Deforestation and Forest Degradation
RERED: Rural Electrification and Renewable Energy Development Project
ROI: return on investment
RPR: reserve-to-production ration
SCP: sustainable consumption and production
SCR: selective catalytic reduction
SEEd: Sustainability and Environmental Education
SFS: School for Field Studies
SDGs: Sustainable Development Goals
SPI: Social Progress Index
SNAP: Supplemental Nutrition Assistance Program
STARS: Sustainability Tracking, Assessment, & Rating System
TBL: triple bottom line
TNC: The Nature Conservancy
UN: United Nations
UNCED: United Nations Conference on Environment and Development (Rio Earth Summit)
UNDP: United Nations Development Programme
UNEP: United Nations Environmental Programme
UNESCO: United Nations Educational, Scientific and Cultural Organisation
UNICEF: United Nations Children's Emergency Fund
USAID: United States Agency for International Development
USDA: United States Department of Agriculture
UV: ultraviolet
VOCs: volatile organic compounds
VCS: Verified Carbon Standard
WASH: water, sanitation, and hygiene
WIC: Women, Infants, and Children
WRI: World Resources Institute
WF: water footprint
WHO: World Health Organization
WTE: waste-to-energy
WTO: World Trade Organization
WWF: World Wildlife Fund
ZEECs: zero emissions tax credits

Index

Page numbers in bold indicate figures, boxes, or tables.